THE SCIENTIFIC REVOLUTION

H. FLORIS COHEN

THE SCIENTIFIC REVOLUTION

 A

HISTORIOGRAPHICAL INQUIRY

THE UNIVERSITY OF CHICAGO PRESS
Chicago & London

H. Floris Cohen (born 1946 in Haarlem, the Netherlands) studied history at Leyden University. He served as a curator at the Museum Boerhaave from 1975 to 1982. Since 1982 he has been professor of the history of science and technology at the University of Twente.

The University of Chicago Press, Chicago 60637
The University of Chicago Press, Ltd., London
© 1994 by The University of Chicago
All rights reserved. Published 1994
Printed in the United States of America
03 02 01 00 99 98 97 96 95 94   1 2 3 4 5

ISBN 0-226-11279-9 (cloth)
ISBN 0-226-11280-2 (paper)

Library of Congress Cataloging-in-Publication Data

Cohen, H. F.
    The scientific revolution : a historiographical inquiry / H.
Floris Cohen.
        p.   cm.
    Includes bibliographical references and index.
    1. Science—History.   2. Science—Europe—History.
    3. Science—Historiography.   I. Title.
    Q125.C538   1994
    509.4′09032—dc20   93-41784
                                                            CIP

⊛ The paper used in this publication meets the minimum requirements
of the American National Standard for Information Sciences—
Permanence of Paper for Printed Library Materials,
ANSI Z39.48-1984

It is not by renouncing the apparently impossible
and unattainable goal of knowing the real, but,
on the contrary, by boldly pursuing it that science
progresses on its endless path towards truth.

Alexandre Koyré

It is better for a historian to be
wrong than to be timid.

Lynn White, Jr.

For Marja

For David Omar
born 21 January 1994

To the memory of R. Hooykaas
1 September 1906–4 January 1994

❧ CONTENTS ❧

PART TWO
THE SEARCH FOR CAUSES OF THE SCIENTIFIC REVOLUTION

Part Three
Summary and Conclusions: 'The Banquet of Truth'

❧ ACKNOWLEDGMENTS ❧

S INCE THE AIMS AND SCOPE OF THIS BOOK are set forth in an introductory chapter, I confine myself here to the pleasant task of thanking all those whose cooperation and help have been so valuable to me.

Some formal acknowledgments first. While working on this book, I used a few portions of an early version for congress appearances, and I thank Kluwer Academic Publishers and the Director of the Japan-Netherlands Institute for permission to recycle here "Beats and the Origins of Early Modern Science" (in V. Coelho [ed.], *Music and Science in the Age of Galileo* [Dordrecht: Kluwer Academic Publishers, 1992], pp. 17–34) and "The Emergence of Early Modern Science in Europe; with Remarks on Needham's 'Grand Question', Including the Issue of the Cross-cultural Transfer of Scientific Ideas" (*Journal of the Japan-Netherlands Institute* 3, 1991, pp. 9–31), respectively. I further thank Cambridge University Press for permission to reproduce in chapter 6 a diagram from Joseph Needham's *Clerks and Craftsmen in China and the West.*

I am grateful to helpful librarians at my home base, the University of Twente, but also at the Instituut voor geschiedenis der natuurwetenschappen, Utrecht (with a special thank-you to Lian Hielkema), at the Museum Boer-haave, Leiden (idem to Peter de Clerq and Harry Leechburch), and at the Library of Congress.

Thanks are due to all those who helped me secure a sabbatical leave that enabled me to devote the year 1987 to writing a first draft of Part I. Four months of that year were spent as a resident fellow at the Woodrow Wilson International Center for Scholars, Washington, D.C. I thank the staff of the West European Program and also my research assistant, Melissa Kasnitz, for their support, and the entire Center for its hospitable and stimulating atmosphere (to which co-fellows William Christian, Carole Fink, Peirce Lewis, and Marx Wartofsky contributed in no small measure). I gratefully recall a stimulating exchange with Mordechai Feingold at that time.

The Department of History at the University of Twente deserves ample thanks, too: Petra Bruulsema for excellent secretarial help, and Christiaan Boudri, Marius Engelbrecht, Casper Hakfoort, Paul Lauxtermann, Hans Sparnaay, Tomas Vanheste, and Péter Várdy for their ongoing readiness to exchange ideas on a pleasantly informal basis as well as for their perseverance in holding on against ever more increasing odds.

I thank the 1989/90 junior history of science class at Princeton University supervised by my former colleague at Twente Nancy J. Nersessian for trying out a much more primitive version of large portions of this book; also their teacher for her suggestions and invariably constructive criticisms.

I equally thank for similar contributions all those other friends and colleagues who read the book or portions thereof in advance: H. Achterhuis, K. van Berkel, I.B. Cohen, H.J. Cook, W.Th.M. Frijhoff, A. Van Helden, J.C. Kassler, M.E.H.N. Mout, M.J. Osler, L. Pyenson, B. Theunissen, R. Wentholt, R.S. Westfall, and J.G. Yoder. Many others, too, offered helpful suggestions during conversations too numerous to record. Rob's and Sam's written and spoken comments, in particular, exuded welcome encouragement in that time and again they appeared to grasp much better than the author what he was aiming at. So did R. Hooykaas, who to my deep regret did not live to see the book in print. I shall never forget the scholarly as well as personal interest he took in its somewhat uneven progress over time.

With respect to the University of Chicago Press I am happy to acknowledge eye-opening criticisms from two readers, the dedication of the staff, and, in particular, the tenacious devotion and generous enthusiasm of Susan E. Abrams and the unbelievable accuracy and loyal cooperation of Pamela J. Bruton.

My last and warmest and greatest thanks are to my daughter, Esther, and to my wife, Marja, for their presence and for much else.

Utrecht, February 1994

⚜ LIST OF PRINCIPAL AUTHORS ⚘

This list has been prepared for handy, not for exhaustive, reference. Those authors are listed here whose ideas on the birth of early modern science have been treated at some length in the book. The list has been drawn from those sections which are wholly or partly devoted to their ideas (not, as a rule, from introductory, concluding, or summing-up chapters and sections). Thus, the list is merely an extract from the Index at the end of the book, where everyone whose name occurs in the book is listed with every single page where the name occurs.

⁂ ONE ⁂

'ALMOST A NEW NATURE'

Is it not evident, in these last hundred years (when the Study of Philosophy has been the business of all the *Virtuosi* in *Christendome*) that almost a new Nature has been reveal'd to us? that more errours of the School have been detected, more useful Experiments in Philosophy have been made, more Noble Secrets in Opticks, Medicine, Anatomy, Astronomy, discover'd, than in all those credulous and doting Ages from *Aristotle* to us? so true it is that nothing spreads more fast than Science, when rightly and generally cultivated.[1]

THESE WORDS WERE WRITTEN IN 1668, not by a scientist, but by the poet John Dryden. From the vantage point of 1668 Dryden expresses succinctly the view, widely held at the time, that the 17th century has been witnessing an unprecedented upheaval in science. What is more: now that the right way to do science has been found, so he says, science is bound to expand rapidly.

Posterity has concurred in both judgments. Yes, ever since Dryden wrote these sentences science has spread over ever more domains of human thought and activity, to the point of transforming our daily lives in ways no one could have foreseen in 1668. And yes, the 17th century has continued to be taken for that specific period in history when modern science was born. Not our present-day science, to be sure, but a mode of doing science still quite recognizably akin to present-day conceptions. The awareness that something unprecedentedly new happened in 17th-century science has been with us for as many centuries as have since passed.

The character of that awareness, though, has undergone radical changes. In their own time, the second half of the 17th century, notions like Dryden's emerged as polemical points in the famous 'Quarrel of the Ancients and the Moderns', the protagonists of the latter point of view adducing science as one key example of how far the moderns had been able to surpass the achievements of the ancients. In Enlightenment days, individual men of science (more often than not identified with Descartes and Bacon) were celebrated as the pioneers of a new age in which Enlightenment thinkers felt themselves to be fully participating—a new age of light and reason now that the forces of superstition and darkness had been vanquished. It was in this particular mental frame that aspects of 17th-century science were being investigated by a number of pioneers.

Not until the 19th century, however, did the birth of modern science begin to be regarded as a thing of the past about which such questions could be asked as how it took place and what made it possible. For the time being, such historical questions turned up inside the wider framework of philosophical viewpoints which tended to yield answers already determined by those very viewpoints. As distinct from such an approach, which made the history of 17th-century science subservient to overall conceptions of what science is about, the desire to understand the emergence of modern science as a historical issue in its own right goes back to the first decades of our own century. Pierre Duhem's celebrated thesis that modern science has really been born in the Middle Ages marked the turning point. Efforts to arrive at a historical understanding of the birth of modern science speeded up considerably between 1924 and c. 1950, as a result of two closely related events. One was the rise and subsequent rapid spread of the concept of the 'Scientific Revolution' as an analytical tool expressly forged for grasping the essence of the emergence of modern science. The other was the history of science turning into a professional academic discipline—a process that, coincident with fresh academic opportunities to be sure, crystallized largely out of the new concept of the Scientific Revolution. From then on, the professional study of the Scientific Revolution has been an acknowledged area of specialized scholarly concern in its own right.

Thus, with a prehistory not to be neglected, historical interpretations of the birth of modern science have now been with us for almost a century. Solidly grounded in a rapidly increasing number of specialized inquiries into the thought and the mental and social worlds of individual 17th-century scientists, such studies of the Scientific Revolution as a historical phenomenon have together yielded an immense expansion in our understanding of how modern science came about and what key characteristics distinguished it from earlier efforts at coming to grips with the secrets of nature. These studies, however, have never been mapped. No overview is available of the vast and flourishing literature devoted as such to the two root questions of what the Scientific Revolution was all about and what made it possible. There exists no comprehensive, critical analysis of the body of that literature; no systematic tracing of the lines that connect over time the many and variegated ideas of historians on the Scientific Revolution with one another; no systematic attempt, in short, at taking stock of where we are now and how we got where we are. This, then, is what the present book sets out to accomplish.

Such a definition of the aim I have set myself in this book is rather formal. Ten quite different books, all falling under this broad definition, might still be written by ten individual authors. I should like to prepare the reader further for what he or she may expect the present study to accomplish through a brief sketch of how it has grown out of certain intellectual preoccupations of mine over the past twenty years or so. In this way I hope to clarify the leading ideas that inform the book, that set it apart from a mere chronicle or literature over-view, and that keep its component parts together.

Some leading ideas of this book and their genesis. It all started with a highly unconventional political thinker, whose name has remained virtually unknown outside the Netherlands: Jacques de Kadt (1897–1988).[2] When I first met him in 1967, his political career had just come to an end. The first stage of that career had been spent in the communist movement of the early twenties. Leav-ing the movement for good in 1927 was for him a decisive act of liberation, and he rapidly turned into one of the most knowledgeable and penetrating early critics of Leninism and of Stalin's emerging policies. In the thirties this attitude widened into a staunch antitotalitarianism, which, coupled with an acute sense of power politics, made him a very early warner against the threat of National Socialism, too. In a marvelous book written in 1938 he predicted with uncanny precision the coming of the Second World War, its ultimately victorious out-come, and the emergence of the United States and the Soviet Union as the two dominant world powers.[3] The analysis of the fascist threat undertaken in that book was not confined to political issues, though—the problem was con-fronted in the larger framework of an investigation into the vitality of Western civilization. De Kadt saw the West as the unique bearer of such values as plu-ralism, open-minded research, and the critical investigation of whatever sub-ject appeals to one's curiosity. For him, the West was the unique creator of a civilization marked by 'affirmative scepticism'.[4] Modern science and technol-ogy made possible for the first time what so far had eluded every expansively minded society: the spread of Western civilization all across the globe. But this process of ongoing Westernization was a dual-edged affair. The key issue faced by de Kadt already in the thirties was: Are the products of science and technol-ogy going to be absorbed outside the West together with the pluralist values they embody and that brought them forth, issuing in an open, dynamic world of lively intercultural competition? Or is Western civilization in the process of exporting the very means which enable the fiercest opponents of these values, in the totalitarian spirits exemplified by Nazism and Leninism/Stalinism, to overrun all that the West stands for?

Surely de Kadt has not been alone in addressing such key questions in the grand manner, although the down-to-earth, political-realist fashion in which he attacked them has always remained very rare. What his thought did for me, a twenty-one-year old undergraduate student majoring in political and social his-tory, was to focus my attention on the key role science and technology play in the ongoing process of worldwide Westernization. I came to share de Kadt's insight into the double-edged nature of that process and of the unflinching defense against totalitarianism necessary to ensure the outcome of an open world animated by 'affirmative scepticism'. But I cared more for history than for an active pursuit of politics, and my acutest curiosity came to be concentrated for better or worse on the historical origins of the process of Westernization that has been sweeping the globe ever since the transformative powers of science were unleashed. 'What are the sources of the modern world?' I wanted to know.

Textbooks gave ready answers to that question. They sought the origins of the modern world in the 16th century, when European expansion overseas led to early conquests; when the unity of the medieval Church fell apart; when an economy based partly on commercial entrepreneurship began to replace feudal relations; when the nation-state arose, warfare was modernized, and modern science began. Yes, modern science was mentioned in some of those textbooks of modern history—in brief, isolated sections to be sure. To make a long story short, I soon began a personal quarrel with virtually all literature on the origins of the modern world that has yet to cease. I came to think then, as I continue to think now, that in what has set the West apart from other civilizations, modern science is little less than the key element. This is not at all to deny that other historical events and processes have equally helped bring about "the differentiation of the West from the Rest" (to borrow David Landes' apt phrase).[5] Max Weber's work, in particular, is replete with in-depth studies of such events and processes (to be taken up at various places in the present book). But my point here is that in the process of differentiation modern science was the key factor that has made the process virtually irreversible. After all, modern science gave rise to a science-based technology, and this is what caused in large measure Europe's ascendancy among the world's civilizations (which was not originally built upon such a technology) to be cemented for centuries to come. Modern science is not just a thought-construction among others—it entails both an intellectual and an operative mastery of nature. Whereas empirical technology is a feature of every major civilization, the systematic application of scientific insights to change our natural environment ('to conquer Nature by obeying her', as Francis Bacon phrased it) is a creation of Europe alone. When exactly the 17th-century promise of a technology based upon science began to turn into reality—more specifically, to what extent portions of modern science went into the making of the Industrial Revolution—is a matter of ongoing historical debate (discussed in the present book in section 3.4.3).[6] What is certain is that the industrial and the postindustrial worlds are inconceivable without the prior emergence of modern science. But for this particular source of supremacy for the West, the perennial ebb and flow of cultural ascendancy would have followed its natural course through world history unabated as it already had with the rise and fall of so many civilizations of the ancient and the medieval worlds.

Thinking along such lines—by then much more vaguely drawn, to be sure—it began to dawn upon me that a large portion of the riddle of "the differentiation of the West from the Rest" that had come to intrigue me so lay hidden in this unique process—the emergence of modern science. Hence my sense of joyous recognition when I came across a resounding articulation of my student thoughts in a line which I found later to have become quite famous:

> [The scientific revolution] outshines everything since the rise of Christianity and reduces the Renaissance and Reformation to the rank of mere episodes, mere internal displacements, within the system of medieval Christendom.[7]

That is Herbert Butterfield, the historian, in his book *The Origins of Modern Science* (1949; discussed in the present book in sections 2.4.3 and 3.6.3). Butterfield elaborated the idea in his chapter 10, "The Place of the Scientific Revolution in the History of Western Civilisation," which is still one of the very few specific, historically informed treatments of the topic. Among historians of 17th-century science the line just quoted enjoys quite a reputation in that it expresses an acute sense of their being at work, somehow, at one of the key junctures of world history. But that awareness is rarely articulated beyond the level of sentiment, let alone elaborated into a full-fledged study of what modern science did for Western civilization and of how Western civilization could give rise to modern science. Such a study still remains to be written. No doubt the task is a daunting one, and the present book has no wider ambition than to provide some—in my opinion indispensable—spadework for its further preparation. The need for the fulfillment of the full task has never, I believe, been more succinctly set forth than by the historian of the Scientific Revolution Richard S. Westfall in the opening passage of a brief teaching guide:

> The Scientific Revolution was the most important 'event' in Western history, and a historical discipline that ignores it must have taken an unhappy step in the direction of antiquarianism. For good and for ill, science stands at the center of every dimension of modern life. It has shaped most of the categories in terms of which we think, and in the process has frequently subverted humanistic concepts that furnished the sinews of our civilization. Through its influence on technology, it has helped to lift the burden of poverty from much of the Western world, but in doing so has accelerated our exploitation of the world's finite resources until already, not so long after the birth of modern science, we fear with good cause their exhaustion. Through its transformation of medicine, science has removed the constant presence of illness and pain, but it has also produced toxic materials that poison the environment and weapons that threaten us with extinction. . . . I am convinced that the list describes a large part of the reality of the late twentieth century and that nothing on it is thinkable without the Scientific Revolution of the sixteenth and seventeenth centuries. . . . I have yet to see the work that presents, in one integrated argument, the full position I just sketched so briefly, the position that offers the ultimate justification for the inclusion of the history of science prominently in any academic course that presumes to explain the origins of the world in which we live.[8]

Written in 1986, these words articulate far better than I ever could have at the time what it was that made me decide, after completing a doctoral thesis on the social-democratic movement in the Netherlands during the twenties, to shift my professional attention toward what I had now come to regard as the core of the discipline of modern history: the history of science.

All this is not to say that my interest in the history of science was solely determined 'externally', in the sense of an exclusive interest in the causes and the effects of modern science at the detriment of the *content* of science—its shifting ideas and concepts and theories and tools; its creative dead ends and its even more creative triumphs. On the contrary, I admire science greatly as one of the outstanding adventures of the human mind. There is a long-standing controversy between 'internalists' and 'externalists' in the history of science. I discuss the controversy at length in section 3.5 and at many other places in the present book, seeking in particular to help rescue it from sterile either/or positions. For me personally the dual aspect of science has never involved a contradiction. Not doubting for one moment that science is a social phenomenon, I have also never doubted that it is at the same time much more than that. I have become ever more aware of modern science as one, or even the, key motor of social transformation over the past two centuries—its 'motor of social change' aspect. I am equally impressed with the power of science to reveal, at least to some extent, the secrets of nature—its 'truth' aspect.

It is from this dual position that I set out to acquaint myself better with the Scientific Revolution. Three lines of inquiry presented themselves. The first followed naturally from my new job at the National Museum for the History of Science at Leyden (now called the Museum Boerhaave). Although I have never ceased to like scientific ideas even better than scientific instruments, serving as a museum curator for seven years surely helped to sharpen my awareness that modern science is a matter of the hand almost as much as of the mind. Second, an early opportunity arose to make myself familiar in some depth and on a firsthand basis with the various modes of thought of the pioneers of the Scientific Revolution. This opportunity first took shape as an examination of Christiaan Huygens' musical theories, and terminated in a coherent presentation, in my book *Quantifying Music,* of a vital episode in that frequently ignored scientific discipline of the 17th century among others: the science of music.[9]

Meanwhile, the third and most obvious line of approach to the Scientific Revolution was to read what its historians had to say about it. That proved an exhilarating voyage of discovery in its own right, and a most revealing one—I suppose most, if not all, of my fellow historians of science can testify to that. I found out how relatively recent the Scientific Revolution as a conceptual tool of historical analysis really is; out of what depths of philosophical and historical understanding the concept had been forged; how quick and radical have been the changes it has undergone since; how immensely rich an event the

Scientific Revolution was all by itself; how wide the net is that has been cast by those courageous enough to set out to explain the Scientific Revolution one way or another; and, finally, how little the diverse participants to this ongoing debate over the nature and causes of the Scientific Revolution appear to be aware of the full scope and meaning the debate has acquired over more than a century of historiographical pursuit.

The moment when I finally became aware of this last point was the moment when I decided to write this book.[10] I reasoned that it made little sense even to consider adding one more overall interpretation of the Scientific Revolution to all that has already been said about the event without first taking apart and reassembling in ordered fashion much of what *has* already been said about it. With a tool of historical analysis like the concept of the Renaissance, or with such a world-historical event as the French Revolution, comprehensive efforts in historiographical mapping and analysis have not at all been uncommon. The history of historiography may on occasion be a rather tedious genre if undertaken in too much of a bookkeeper's spirit—this does not obliterate the vital contribution the genre may make to further historical analysis. Yet for the Scientific Revolution no such effort has ever been undertaken on anything like the scale I envisaged.

Related work.   What has been said so far is surely not to ignore the existence of some related work on the historiography of the Scientific Revolution, much of which I have gratefully made use of in my own researches. There is I.B. Cohen's ongoing research, most of which focuses on the history of the idea of scientific revolutions in a general sense, but with some helpful hints about the origins of the notion of the Scientific Revolution, too.[11] (Much confusion has arisen between these two distinct terms, *scientific revolutions* and *the Scientific Revolution,* and their conceptual and historical disentanglement will form one subtheme in the present book.) Further, there is the chapter "The Problem of Cause" in A.R. Hall's *The Revolution in Science.* There are some broad lines in the history of the historiography of science drawn by Henry Guerlac, Thomas S. Kuhn, Helge Kragh, John Christie, and Arnold Thackray, among others.[12] There is the chapter by Roy Porter, "The Scientific Revolution: A Spoke in the Wheel?" in which the question is examined of whether the Scientific Revolution fits in with the overall picture of *Revolution in History* (the title of the book in which the chapter finds its place). And recently a chapter by David Lindberg, "Conceptions of the Scientific Revolution from Bacon to Butterfield: A Preliminary Sketch," has appeared.[13]

Another, related historiographical cluster consists of studies devoted to individual authors of historical theories on the Scientific Revolution and to their pertinent books. This literature is considerably larger than the paragraphs and sections and chapters indicated just above and there is no need to be exhaustive in my listing here. There are, for example, commemorative articles on just about every historian of the Scientific Revolution who is no longer among us, and *laudationes* in honor of many of them. There are, in the scholarly journals, essay reviews and state-of-the-art surveys. There are reviews of virtually all the books on the Scientific Revolution that come up for scrutiny in the present study (I am far from claiming that I have examined all such reviews, or even attempted to). There are books and essays on Pierre Duhem, on Alexandre Koyré, and on Joseph Needham, among others—naturally, the contribution each of these men made to our understanding of the Scientific Revolution forms one topic among many in such studies. There are, finally, ongoing debates over certain specific interpretations of the Scientific Revolution that have attained the 'thesis' state: the Merton thesis, the Hessen thesis, the Yates thesis spring to mind among a few others. One remarkable thing about such debates is that they tend to be held in virtual isolation from one another or, *a fortiori,* from the larger framework of the debate on the Scientific Revolution to which they at least partially belong. Or, phrased differently, there *is* no debate on the Scientific Revolution in the sense of a debate carried on in full awareness of the full scope and the full depth the proliferation of scholarly views has meanwhile acquired. The formulation of viewpoints goes on and on, yet the 'debate' is conducted in the form of essentially separate monologues; the topic is compartmentalized in remarkably watertight niches, each inhabited by students of the Scientific Revolution in one or another of its many aspects.

In short, the aggregate of the literature pointed at above is still far from constituting a full-fledged history of the historiography on the nature and causes of the Scientific Revolution. But why should we care to have one?

Getting farther than where we are now.   As historians of science we perhaps take it too much for granted that time sifts out whatever is worthwhile in our historical studies, with the worthwhile portions being added more or less automatically to the accumulating stockpile of 'valid' historiography. But I believe that no such mechanism functions in the historiography of science in any measure comparable to how this mostly happens in the sciences of which we study the history. Hence, we must actively construct some collective memory to ensure optimal cumulation of research findings. One suspicion with which I started the present inquiry has found considerable confirmation as I went on: that much of value tends to get lost if no one takes care to draw the lines together. It is true that certain interpretations of the Scientific Revolution have become obsolete beyond repair. But it is no less true that there exist, for example, attempts at explaining the Scientific Revolution which have scarcely ever been recognized for what they are. Studies by Leonardo Olschki, Joseph Ben-David, and R. Hooykaas are cases in point. I found that ideas of immense scope and potential for further exploitation were once put forward that remain thoroughly underexploited (Alexandre Koyré's work, in particular, provides striking examples). I discovered that Joseph Needham's key problem—why

was modern science born in the West rather than in China?—if suitably transformed contains grandiose opportunities for students of Western science, too.
I found, in short, that a systematic study of what past masters had to say on the
Scientific Revolution may serve as an ongoing, and not soon to be exhausted,
source of inspiration for getting farther than where we are now.

In undertaking this study I am also guided by a sense of sheer *gratitude*. I
have myself had the privilege of being introduced to the history of science by
one of its pioneers: R. Hooykaas. If we consider the present-day flourishing of
the field, we have every reason to think back to those bold thinkers who made
it flourish in the first place and to squeeze out of their works every ounce of
insight that may still have been left untapped.

Approaches taken.   One way to achieve these aims has been systematically to
compare views on the Scientific Revolution with one another. Thoughts of one
historian may shed considerable, and often quite unexpected, light on thoughts
of another—a light that, in its turn, helps to elucidate standing issues of controversy. Examples are such vexed questions as when the Scientific Revolution
may be taken to have started; what the respective parts in the event were of
astronomy, of mechanics, of chemistry, of the life sciences, and of medicine;
what precisely we mean by 'mechanization'; and scores of other bones of ongoing contention.

Great thinkers are almost never of a piece. Much of the tension that is so
instrumental in keeping their works alive after they themselves die often derives, so I believe, from certain unresolved contradictions at the very bottom
of their thought. Tracing down and delving up such contradictions is a method
I have frequently employed, not indeed for purposes of cheap exposure but
rather because such inner contradictions may often be made surprisingly productive for future analysis and research.

It is these two modes of operation in particular—those of comparison and
of the search for stimulating inconsistencies—that make possible what I regard
as perhaps the principal crop to be yielded by the present study. Ideas of previous thinkers, however innovative and comprehensive, can rarely be adopted
wholesale anymore, because in a real sense they have become obsolete. For
example, in section 2.3.3 I shall seek to show that much in Alexandre Koyré's
thought has remained extremely stimulating and fruitful despite much present-
day neglect—yet the neglect is not at all incomprehensible, because new historical facts (particularly on Galileo) have since come to light that render it
impossible to hold on to his views in their pristine entirety.[14] What Koyré's
conceptions as well as those of many others need in my view is neither wholesale adoption nor sweeping neglect, but rather *judicious combination and
cross-fertilization* leading to *consciously applied transformation*. Many of the
ideas studied in the present book are just too good to die due to the partial
revision of the empirical facts originally invoked to sustain them. One of my

9

purposes here is to open ways toward their viable resurrection. This is not to say that the revision of empirical facts is an unfortunate accident reluctantly to be accepted. Surely the empirical fact must be the ultimate judge in any dispute over historical interpretation. But historical facts rarely make themselves heard unambiguously. When they speak, we had better listen. Yet often they do not begin to speak at all clearly until they are exposed to the searching torchlight of ideas, and it is the ongoing transformation of ideas that I am addressing here as a process as vital for progress in the historiography of science as I.B. Cohen has shown it to be for the progress of science itself.[15]

Boundary lines. It follows from the objectives and approaches just defined that this book is not meant to survey all that has been written from the 17th century onward about what happened in science from Copernicus to Newton—to trace down the centuries how, for example, the Copernican hypothesis or the meetings of the Royal Society have been handled by historians. Rather, my concern throughout is with the birth of early modern science *as a historical problem,* and this is why I focus on historiography from the twenties of the present century onward. No disparagement is meant of such very erudite works (from which present-day historians can still learn a great deal) written, for example, around the turn of the 19th century by Montucla on the history of the mathematical sciences up to his own time, or around the turn of the 20th by Lasswitz on the history of atomism up to Newton. Such works taken together yield a history of their own, of which N.M. Swerdlow has recently drawn a pioneering outline as far as the mathematical sciences are concerned.[16] Naturally this vast and, as yet, unwritten history enters mine at many places; but the one is not to be taken for the other.

The present book is meant to be comprehensive; it is not meant to be exhaustive. I have surely done my best to take up the most important and interesting and influential among the studies that over time have been devoted specifically to the what, when, and why of the Scientific Revolution. Equally surely, I must have missed a number of those, and I apologize beforehand for whatever is missing but should not be. Yet I have *not* aimed for exhaustive coverage—I have preferred in-depth treatment of the most significant contributions to our understanding of the birth of modern science over superficial sketches of everything that has ever been said about it regardless of inner value.

My focus on historians' ideas about the Scientific Revolution as a historical problem has determined throughout the kind of literature I have used. My primary material naturally consists of works (mostly books, but also articles, papers, and chapters of books) in which these ideas have found expression. More than one such work has taken shape as a textbook for students, and it is only fair to warn the reader in advance that some of the ideas of historians of the Scientific Revolution I am to discuss were inevitably expressed by their authors in a simpler, less sophisticated vein than would have been the case if they had been addressing none but their scholarly peers. On the other hand, I focus very much on such features of their leading ideas, of the arrangement of their expositions, etc., as are touched relatively little by didactic concerns *per se*.

What secondary literature goes into the present book consists in part of a good deal of background reading of the most varied kind, but mostly of books and articles on the historians whose ideas form my primary material. Since the middle eighties, while I was working on this book, such literature has been proliferating, and I have striven to keep up with it. I have not, however, found it my duty to take the whole of the literature on every single figure in, or aspect of, the Scientific Revolution as my province. My aims are historiographical in the sense indicated, not bibliographical, and if I were to consider, and refer the reader to, the latest literature on all those numerous topics that come up tangentially, this book could never be completed. Moreover, such studies as I did wish to examine here will not cease to come out, and I had to draw the line at some point. Barring a few exceptions no literature has been considered that came out after 1990.

One more respect in which the present book may be considered incomplete is that it focuses on *ideas* about the Scientific Revolution rather than on the intellectual and social *backgrounds* out of which those ideas emerged in their turn. For example, one might study not only what Koyré meant with his fresh concept of the Scientific Revolution, but also how, in framing it, he focused certain ideas of Cassirer, Meyerson, Bachelard, and many others upon the problem of the origins of modern science. Or one might set out to discover how Koyré's thought is related to his somewhat low position, in pre-war France, on the status ladder of Parisian academic life.[17] In such a way I might easily have doubled or trebled the content of this book by turning it into a vast 'intellectual and social history of the historiography of the Scientific Revolution'. I might well read such a book with great pleasure—only, I have chosen not to write it myself. As it stands, the present volume should offer enough food for thought. Also, much backgrounding in this fashion has a way of *reducing* ideas to their intellectual and social origins, whereas I prefer to study ideas in their own right before finding out where they came from.[18] All this is not to say that I have missed the opportunity to make occasional observations on the background of certain ideas discussed here—Pierre Duhem's inspired exaggerations, the 'Hermeticist' current, certain Marxist approaches, and Joseph Needham's thought, in particular, proved irresistible in this respect—but I have made no *systematic* effort in this direction at all.

Acts of mediation.   My general approach to so many seminal studies that together have shaped what understanding we now possess of the Scientific Revolution is marked by a spirit of critical sympathy. I have not hesitated to lay bare flaws where I see them, but I have done my utmost (despite one or two black sheep, whom the reader will not fail to discover) to do so in a constructive vein, with opportunities for future scholarship uppermost in mind. I have not failed to articulate my deeply felt admiration wherever I could. I once found the spirit that pervades the present investigation eloquently expressed in a book by E.A. Burtt (whose ideas on the birth of modern science are discussed in section 2.3.4), *In Search of Philosophic Understanding* (1965). His topic here is systems of philosophy rather than the ideas on the origins of modern science that form my subject, yet the general approach is quite the same:

> Our aim has been to expound each philosophy in terms that reveal the grave limitations haunting it, but also—and more especially—in terms that bring out clearly its promising possibilities and enable us to glimpse how they might win their most significant fulfillment.[19]

Thus I have aimed likewise at acting as a mediator between the past and a possible future of the historiography of the Scientific Revolution. But the present book also forms an act of mediation through the adoption of a role perhaps more customarily associated with a Dutchman: that of a mediator between Continental and Anglo-Saxon currents of thought. Much inspiring thought on the origins of modern science has become virtually inaccessible to many present-day members of the English and American worlds of scholarship as a consequence of rising language barriers. There, however, the center of the historiography of science has been located ever since the Second World War, as has happened in so many areas of scholarship. Translations, which appear in increasing quantities, certainly make up for resulting losses to some extent. Yet more than enough room seems left for passing on in concentrated form pertinent ideas originally expressed in French or German or Dutch.

A third and final act of mediation has to do with the opposition between the specialist and the scholar who aims for synthesis. Many a thinker in the past and the present has expressed dismay at the ongoing specialization of scholarly work. Specialization is being felt everywhere, and the history of science is no exception. The specialist is indispensable—it would be sheer obscurantism to regret the spectacular increase in our knowledge of the Scientific Revolution that has been achieved over the past half-century, and I stand in awe of the unusual variety of personal resources that must be brought to bear on the successful transmutation of raw facts from the past of science into nuggets of historical understanding.

Nevertheless, the very act of painting details of the overall picture in ever more sharply drawn lines renders the picture itself more and more opaque. Specialization goes on relentlessly. Everyone knows it, everyone professes to regret it, yet few do something about it. And among those rare efforts consciously to alleviate the unfortunate consequences of specialization, one has to stand on guard all the time against grandiose speculations devoid of even a modicum of sustenance from empirical fact. The perennial challenge as I see it is to strike a balance between the relative security of research into minutiae on the one hand and a sense of complacency with hasty and sweeping unification on the other. Thus I offer the present book as one contribution in the valiant struggle against the dragon of ultraspecialization. J. Huizinga once articulated "that fatal question that menaces all humanities which are expanding rather than turning inwards: Where is all this effort to get us?"[20] Empirically based efforts at recovering a sense of the wholeness of things should therefore be seen, I think, as adding an inherently valuable 'plus' to our understanding of what the indispensable specialist has revealed, with some measure of indulgence to be granted perhaps for the inevitable 'minuses'. Variety can only be enjoyed to the full in my view when interspersed with an accompanying awareness of overarching unity.

The secret treasure.   In the present case, the chief consequence of specialization to be combated has already been touched upon at the beginning of this Introduction. I mean the curious position of the Scientific Revolution as a key juncture in world history that finds itself being studied in its own right but out of its principal context by historians of science, whereas the very same event constitutes the chief missing link in studies of the origins of the modern world and how it came about in the West in the first place. The vagaries of history writing in the 19th and 20th centuries have put historians of science in possession of an almost secret treasure. It has remained secret at least to virtually all those scholars ('general' historians, sociologists, economists, and others) who want to find out about the rise of the West. (The few exceptional cases are discussed in the course of the present book.) It has equally remained hidden to most students of the very many aspects of human thought and activity that have been touched and transformed since the 17th century by the magic wand of modern science.[21]

Such a situation is not without its risks. A major task of integration still lies ahead of us, with the present book being offered as just one stepping-stone toward that perhaps distant aim. Having become the sole guardians of the treasure, historians of science bear a heavy responsibility for keeping it in good shape pending the arrival of the day when full integration has finally been achieved. Yet they cannot be blamed for handling the concept in accordance with the uses that come naturally to them. And the internal logic of the changes the concept of the Scientific Revolution has undergone over the past half-century, as I am concerned to demonstrate at length in this book, goes far to undermine it. It is at least conceivable that the concept of the Scientific Revolution may evaporate entirely before the day of integration has come. There are already tangible signs of such a tendency, a topic to which I will return in the Conclusion to this book.

Fifty years and more of ongoing and increasingly specialized research have made it very hard to construct a unitary picture of the Scientific Revolution that remains in keeping with the historical facts. Half a century ago, such pictures were put forward with great boldness and confidence. Nowadays, when we take a bird's-eye view of what has meanwhile happened to the concept of the Scientific Revolution, we face either sharp definition in a quite restricted setting or loss of conceptual clarity in a widened setting or structural difficulties encountered in the two-current approach to the Scientific Revolution that has been set up as an effort to salvage both virtues. Not by chance much of our day-to-day usage of the concept of the Scientific Revolution has become almost entirely vacuous. Starting with the 16th century, we are in the habit of cutting up the history of science in didactic pieces of one century each, calling the 16th century 'Renaissance'; the next, 'Scientific Revolution'; followed by '18th-century' (or 'Enlightenment'), '19th-century', and '20th-century' science. In this common parlance, the concept has deteriorated into an empty label, suitable for purposes of facile periodization only. For the rest, we seem ready to give to the concept just about any meaning that suits our purpose, with little community of content left. From there to proclaim the entire concept 'superfluous', as Einstein did with the aether, seems a small step indeed.

Altogether, the Scientific Revolution as a tool of historical dissection has become rather blunt, so that it may well be put aside before it has had a chance to render its most-needed services. Needless to say, I would regard the evaporation of the concept of the Scientific Revolution within half a century of its existence as a major intellectual disaster. A considerable deal of my effort at this somewhat critical stage in its eventful career is therefore devoted not only to showing in some detail how the concept could land in such a crisis but also, and most importantly, to indicate lines along which it may be rescued and, after suitable transformation, be restored to its previous, robust health.

Outline of the book.   The building plan of the book that follows is devised in accordance with the purposes just outlined. The chief division is into three parts. Part III draws together the lines laid bare in the body of the book. Central to the division of that body into two is the distinction between 'what' questions and 'why' questions concerning the Scientific Revolution. Historians' attempts to find out what was characteristically new about modern science as distinct from previous systems of natural philosophy fill Part I, whereas Part II is devoted to efforts undertaken to *explain* the birth of modern science. I am aware that the distinction, although logically satisfying and surely pertinent, is not watertight in practice. Its chief merit as a principle of organization is that it keeps apart questions that are definitely distinct and thus makes it possible to lay bare mutual connections between the various answers given to these questions as, in my presentation, they make their successive appearances. That this can be done at all is due to the fact that the questions of 'what' and of 'why' have mostly been taken up by distinct sets of persons. The chief drawback comes from the fact that this has not been entirely the case. Some students of the Scientific Revolution—I am thinking primarily of Rupert Hall, Robert Merton, Frances Yates, and Joseph Ben-David—have presented in roughly equal measure views on the nature of the Scientific Revolution and on its causes, or have blended the two issues inextricably. With these authors I had to face a choice between either giving them a dual appearance or pigeonholing their views in somewhat arbitrary fashion. In neither case could the solution adopted be altogether satisfactory, but I have done my best to minimize the resulting discomforts for the reader. Further, when the discussion is about Galileo in particular, the issues of what and of why seem almost to merge. Again, this could not be repaired, but damage has been minimized as best I could.

Part I is made up of two chapters. Chapter 2 takes up in by and large chronological order what I like to call 'the Great Tradition'. I mean by that the movement of thought through which the concept of the Scientific Revolution was brought into being and went through its subsequent development as an analytical tool meant to clarify what the novel science of the 17th century ultimately was about. This movement was largely concerned with scientific ideas: with their content and with their philosophical ramifications. Among the themes that run through this chapter are the following:

—the birth of the history of science as a discipline, not out of 'general' history but rather out of the philosophy of science;
—the framing of the concept of the Scientific Revolution;
—the shifting relations between the specific notion of the unique Scientific Revolution and the general idea of a plurality of scientific revolutions;
—the question of whether the overwhelming fact about the birth of modern science is a marked continuity with preceding conceptions of nature or rather a more or less sudden break with the past;
—attempts at finding the 'key' to the Scientific Revolution in unprejudiced observation, in setting up experiments, in the mathematical treatment of natural phenomena, or perhaps in 'mechanization' in any of the meanings given to that versatile term;
—difficulties that have arisen in presenting an integrated picture of the Scientific Revolution because of problems in dating it, in assigning a proper role to Descartes in particular, and in finding a fitting place for contemporary developments in chemistry, the life sciences, and medicine.

Chapter 3 addresses clusters of themes in the historiography of the Scientific Revolution pursued mostly outside the framework of the 'Great Tradition'. Here such novel developments originating in the Scientific Revolution come to the fore as the attitude of 17th-century scientists toward authority, progress, and the future completion of science; scientific method; scientific instruments; ideas on the possible exploitation of nature through knowledge of its laws; the organization and institutionalization of the scientific movement and the roles of the universities and scientific societies therein. Another, fairly recent current in historiography has been concerned with attempts to reconstruct elements of the mental world in which 17th-century scientists lived, and with the repercussions thereof upon their ways of doing science—it has become customary to cover this topic with the umbrella term 'Hermeticism'. Finally, I discuss efforts by historians to study the Scientific Revolution as embedded in wider events and historical processes: a 17th-century crisis in the social order, the breaking up of feudal Europe, and the entire course of Western civilization. All this is rounded off with a conclusion in which I show that one developmental line running through the historiography of the Scientific Revolution has been an ongoing effort to turn the apparent inevitability of the birth and subsequent development of modern science into an array of contingencies.

Part II can be outlined more briefly. Historians' efforts at explaining how modern science could come about are spread over three chapters. Chapter 4 discusses the emergence of modern science out of previous thought on nature in the West as exemplified in Greek, medieval, and Renaissance thought. In chapter 5 causal chains are treated that have been suggested to connect the birth of modern science with both previous and contemporary events in the history of western Europe. Here such causal agents come to the fore as Puritanism and the Reformation, technology and the arts and crafts, the rise of capitalism, the Voyages of Discovery, the printing press, and the rise of a scientistic movement. Chapter 6 discusses answers given to one variant or another of Needham's famous question: What, if anything, can we learn about the causes of the Scientific Revolution by considering the 'failure' of modern science to emerge in any of the other great civilizations of the past?

In the concluding Part III, chapter 7 draws our findings together in the form of an essay entitled "The Scientific Revolution: Fifty Years in the Life of a Concept." In Chapter 8, finally, I exchange my chosen role as a historian of the historiography on the nature and causes of the Scientific Revolution for that of a historian proper in offering a sketch of the Scientific Revolution as I have meanwhile come to see it. This final chapter is by all means intended to be an open-ended one. Throughout this book I have done all I could to ensure that a reader who might utterly disagree with my views may still benefit from the argument that precedes—the argument does not *depend* upon the conclusion.

Hence, it is only in the final chapter that opinions stated are without exception my own. Insofar as, in the main body of the book, I take my job to be the analysis of other historians' views on the Scientific Revolution, I gladly accept responsibility for a correct rendition of those views, but I do so for their content and the viability of their factual basis only at those occasions when I join in with explicit agreement. A fair appraisal of the effort undertaken here is possible only if the author's ideas are not mixed up with those of his subjects.

Practical matters and conventions. I do not think that a reader can come to this book entirely unprepared. I have tried to explain a great deal, but I do not explain what Galileo's principal exploits were or what the Glorious Revolution was all about. I do hope the book to be accessible to anyone familiar with at least one good introduction to the history of 17th-century science and one good textbook of modern European history. Examples might be Butterfield's, Dijksterhuis', Hall's, or Westfall's books about the Scientific Revolution, and Palmer and Colton's *A History of the Modern World.*[22]

Next, a few matters of terminology. I have used the label 'historian of the Scientific Revolution' for everyone who has addressed the topic in depth, irrespective of his or her own disciplinary origins, which may well have been in, for instance, philosophy, sociology, or economics. I have grouped historians of subjects other than science under the umbrella term 'general historians', acknowledging their respective domains of specialization when relevant. More importantly, I had to cut several knots in order to arrive at an unambiguous vocabulary for distinguishing one stage in the history of science from another. The science of ancient Greece has, besides 'Greek', sometimes been called 'ancient', sometimes 'classical'. The Middle Ages and the Renaissance are above confusion, but science between c. 1550 and c. 1900 has been called by some 'classical', by others 'early modern', by most 'modern', whereas 20th-century science is invariably called 'modern'. From here on I shall without exception call Western science in antiquity either 'Greek' or 'ancient', the science of Galileo and Kepler and Newton 'early modern', and science since Einstein and Bohr 'modern'. Thus the term 'classical' is avoided altogether, and if it occurs in quotations I shall indicate the author's meaning whenever confusion might still arise.

All quotations in the main text are rendered in English, with a note indicating provenance. If the quoted text is a translation, the original is printed in the accompanying note, with the exception of texts originally in Dutch. With preexistent translations I have felt free to make changes wherever I thought fit and have indicated the fact of alteration in the first corresponding note.

The notes consist almost exclusively of references. As a rule the first note to a section provides a brief survey of the available literature on the historian whose ideas are being discussed in the main text. The nonspecialist reader may safely ignore all notes without missing anything indispensable to the argument.

So much for preliminaries; let us now go ahead and face the rise, in the early 18th century, of the idea that science advances in revolutionary steps.

PART ONE

DEFINING THE NATURE OF THE SCIENTIFIC REVOLUTION

✍ Two ✑

THE GREAT TRADITION

2.1. 'SCIENTIFIC REVOLUTIONS' AND THE SCIENTIFIC REVOLUTION

THE TERM 'SCIENTIFIC REVOLUTIONS' is *generic*. It stands for a *philo-sophical* idea about the ongoing process of science. It signifies the idea that scientific discovery generally proceeds in a convulsive sort of way. Advances in science, in this view, take place in leaps rather than by small, incremental additions. Scientific revolutions are taken to occur with a certain frequency, or even regularity; there is nothing unique about them. Both the term and the idea have made an enormous impact on the intellectual world since the publication, in 1962, of Thomas S. Kuhn's *The Structure of Scientific Revolutions,* but as I.B. Cohen has shown, both the term and the idea have been around ever since the early 18th century.

The term 'Scientific Revolution', in contrast, is *specific*. It stands for a *historical* idea about one episode in the past of science. It signifies the idea that there has been a period in history, which is hard to date with precision but which almost always is meant to include the first decades of the 17th century, when a dramatic upheaval occurred in science. This upheaval was unique. The term 'Scientific Revolution' gained considerable currency for the first time through Herbert Butterfield's series of lectures *The Origins of Modern Science,* which came out as a book in 1949. As I.B. Cohen has shown, the 'Scientific Revolution' as a term had been incidentally employed since 1913.[1] As a con-ceptual tool for understanding the birth of early modern science, it was created by Alexandre Koyré in the thirties.

The notions of scientific revolutions and of the Scientific Revolution have often been conflated. Sometimes this has been due to sheer misunderstanding because the terms are—not by chance, of course—so very similar. For ex-ample, after I.B. Cohen published his findings on the origins of the term 'scien-tific revolutions', some inferred without more ado that the term the 'Scientific Revolution' equally goes back to the early 18th century.[2] But there are also more substantial grounds for conflation. On the one hand, some of those who uphold that there has been a unique historical event called the 'Scientific Revolution' have distinguished one or several later 'Scientific Revolutions' as well: a Second Scientific Revolution in the late 19th and 20th centuries, and so on.[3] If such Revolutions are allowed to proliferate, the two originally distinct notions naturally come close to effective merger.

On the other hand, one may strike at the heart of many conceptual mixups by asking—for some reason, it is rarely asked—whether the Scientific Revolution should be counted as one of those scientific revolutions that are distinguished by those who uphold the revolutionary advance of science over time.[4] If the answer is an unqualified yes, the Scientific Revolution stands to lose the unique character on which the historian insists. But if no, it is very odd that the most incisive upheaval that has ever taken place in science is not counted as a revolution. There is a paradox here, which is hard to unravel. Scientific revolutions may peacefully flow on like waves on the surface of the river of science all across the landscape of the past, only to mesh, and often to clash, with the Scientific Revolution when crossing the 17th century. Or is it rather the case that science did not acquire a revolutionary character until the emergence of early modern science, so that the Scientific Revolution stands at the origin of an untold number of subsequent scientific revolutions? We shall see later in this chapter in what interesting ways the two notions have sometimes complemented, sometimes stood athwart, one another in the minds of Kant, Whewell, and Kuhn, in particular.

Enlightenment thinkers on the birth of early modern science.   One early example of the philosophically general and the historically unique meshing in the term 'scientific revolution' is to be found in the *Encyclopédie*—that celebrated document of the Enlightenment.

As demonstrated by I.B. Cohen, the 17th century contributed decisively in two different ways to making possible the subsequent coinage of the term 'scientific revolutions'. Largely through the Glorious Revolution of 1688 the *political* term 'revolution' lost its cyclical connotation (borrowed in its turn from the *planetary* revolutions studied by astronomers) and came increasingly to stand for radical advance in the sense of progress toward a better future. In simultaneously giving birth to early modern science, the 17th century made possible the rise of the idea that just like political events, science, too, might display progress attained through a revolt against the past. The dual sense in which this new term 'scientific revolutions' was employed in the 18th century comes out clearly in d'Alembert's four-page entry "Expérimental" in volume 6 of the *Encyclopédie* (1756).[5]

Here d'Alembert presented an outline of the "rebirth of Philosophy properly speaking,"[6] which he identifies with the appearance on the scene of Bacon and Descartes. In advocating an experimental approach, these two men put an end to the "vague and obscure method of philosophizing" characteristic of the "dark times," or "centuries of ignorance," when Aristotelianism reigned.[7] However, with Bacon and Descartes and the academies devoted to experimental research following in their wake the revolution was not over yet, nor could it be, since "one should not expect the mind to rid itself so promptly from all prejudices. Newton appeared."[8] And now follows a characteristic twist. On the one hand, d'Alembert describes Newton's achievement in glowing terms:

> [He] was the first to show what his predecessors had no more than envisaged, namely, the art of introducing Geometry into Physics and, by uniting experiment with calculation, of forming an exact, profound, luminous, and novel science. . . . the light has finally prevailed.[9]

And the remainder of the entry confirms the impression that, in d'Alembert's view, Newton's achievement in his *Principia* and *Opticks* was unique in yielding "truths that are today the basis and, as it were, the principles of modern physics."[10] On the other hand, after celebrating Newton's achievement, he goes on to observe that such a two-stage course of a scientific revolution is nothing uncommon: "once the foundations of a revolution have been cast, it is almost always in the next generation that the revolution is completed."[11]

Such a mixture of uniqueness and generality in describing the scientific achievement of the 17th century appears to have been rather the rule in Enlightenment days. We should hardly expect an independent interest in the past of science to have arisen in the 18th century—too pervasive was the dichotomy felt to obtain between the dark ages of pre-17th-century science and the sudden appearance of the light of scientific truth; too exhilarating was the sense of participation in what Newton and his immediate predecessors had begun. I.B. Cohen's chapter "Eighteenth Century Conceptions of Scientific Revolution" strongly suggests that although an interesting study might certainly be devoted to Enlightenment views on 17th-century science as a mirror of the eighteenth century's own preoccupations, no comprehensive, truly historical conception is to be expected. One finds brief outlines like d'Alembert's entry, penetrating remarks on the achievement of individual 17th-century scientists, and discussions of 'revolutions' in one or another of the mathematical sciences, each of which found its first historian in the 18th century (e.g., Bailly, the author of a history of astronomy, on the 'Copernican revolution'). One even finds, in J.-E. Montucla's four-volume *Histoire des mathématiques* (2 vols., 1758; 2nd ed., 4 vols., 1799–1802), an extensive and still most instructive survey of all mathematical sciences ('pure' as well as 'mixed') from the earliest times to his own day, with one whole volume dedicated to the 17th century. The driving force behind the effort was surely Montucla's exceptional consuming interest in the past almost entirely for its own sake. With so many learned books devoted to the history of mankind's slaughtering sprees, he felt it was about time to record the history of mankind's steady progress in invention and discovery, marked above all by the mathematical sciences with their in-built certainty.[12] To their progress, so he felt sure, no century (for Montucla thought very much in terms of centuries) had contributed more than the one just passed. Nevertheless, the 17th century (fed in its turn by a "happy revolution in the minds of men" in the second half of the 16th century)[13] marks in large part the big flight ahead made by the mathematical sciences. He also distinguished sharply between the recovery, the limited extension, and the wholesale transformation of one or another portion of the Greek heritage in the mathematical sciences. But this is about as far as Montucla's conceptualization of the birth of early modern science went. For the rest his volumes are filled with extensive, very knowledge-able discussions of the writings of individual authors in pure mathematics, mechanics, astronomy, and optics, these subjects being arranged from 1550 onward by half-century and their history punctuated with manifold 'revolution-ary epochs'.

One also finds, in Voltaire's work in particular, science being taken up as one constitutive element of a history of culture, a genre soon virtually to die out and not to be resurrected until the twentieth century, but then with the history of science left out almost entirely.[14]

In considering the historiographical landscape of the French Enlighten-ment, and the kind of historical consciousness then prevailing, it appears that one would search in vain for a conception of the birth of early modern science that is at once systematic, analytical, broadly interpretive, and concerned with the past in its own right. The road toward such a conception was traveled be-tween the end of the Enlightenment and the beginning of the present century. Names that mark it are Kant, Whewell, Mach, and Duhem. Each of these think-ers took an intermediate position in at least one of these respects. The first to be discussed here is Kant. Despite his concern for the general forms of human thought rather than for the details of its history, he combined in an interesting way the notion of scientific revolutions with a remarkably precise idea of the birth of early modern science.

2.2. FIRST APPROACHES TO UNDERSTANDING THE BIRTH OF EARLY MODERN SCIENCE

2.2.1. Kant's "Revolution in Mode of Thought"

Kant's *Kritik der reinen Vernunft* (Critique of Pure Reason) was regarded by its author as an effort to turn metaphysics into a science. In the preface to the second edition (dated April 1787) he made his case for the viability of such an effort. His argument was by analogy: He wished to show how and when other 'concerns of Reason' had succeeded in becoming truly scientific.

Kant's point of departure in the preface is that metaphysics, unlike some other 'preoccupations of Reason', has never reached the status of a secure and indubitable science.[15] How has this sorry state of affairs come about, and what can be done about it? The answer must follow from an inquiry into how sciences come into being generally. The various disciplines into which the concerns of Reason may be subdivided, Kant says, fall into two distinct classes. Some are successful; that is, at some point in their development they acquire "the secure pace of a science."[16] Others, in contrast, tend to bog down after a more or less promising start, or pick one path after another in an unceasing search for the right one, or display an ongoing lack of unanimity among their practitioners on how to proceed. In all such cases we witness "a mere groping in the dark."[17] So far, only three fields where mankind applies its reason have achieved the status of a secure science: logic, mathematics, and the empirical sciences. Mathematics presents an interesting case. For a long time, in particular when cultivated by the Egyptians, mathematics displayed the "groping in the dark" characteristic of the first stage. Not until the Greeks turned to mathematics was the secure, the 'royal' path found. To find it required no small effort. How it all happened has got lost in the darkness of the past, yet we may be sure that the decisive turn was due to

> a revolution, brought about in one effort by the happy thought of one individual man. From this point on the course to take could no longer be missed. The secure pace of a science had now been adopted and established for good, for all times, and into infinite vistas. The history of this revolution in mode of thought . . . , and of the person fortunate enough to bring it about, has not been preserved for us.[18]

This account of how mathematics was turned into an established and indubitably secure science sets the stage for the empirical sciences. For here, Kant assures us, the pattern is quite the same, only the empirical sciences needed much more time to "reach the highway of science."[19] Here Francis Bacon was the central figure. His role was twofold: to make the key discovery that enabled the empirical sciences to move on from the groping stage to the secure pace of mature science and, equally importantly, to inspire others who were already on the right track. What happened must once again be characterized as a "revolution in mode of thought." The experiments carried out by men like Galileo, Torricelli, and Stahl (the originator of 'phlogiston' chemistry) suddenly enlightened their fellow students of nature. Nature, so they now understood, must be examined in an active manner, instead of the researcher allowing nature to keep him on a leash. In the "groping in the dark" stage of the empirical sciences observations had indeed been made, but these had been arbitrary and unorganized. Before the revolution no one had aimed to organize observations systematically into laws. But this can only be done if the researcher approaches nature, not as a pupil who is passively told by his teacher what he has to learn, but rather as a judge who actively interrogates witnesses so as to find out the truth. It is this revolution in the conventional mode of thought that accounts for the physical sciences having finally reached the status of a true science, after hundreds and hundreds of years of aimless groping.

Kant goes on to argue that for metaphysics, too, the time has now come to undergo a similar "revolution in mode of thought," so as to be put on the high road of a secure and indubitable science as well. (The radical change in point of view necessary to pull off this feat is what, somewhat improperly, is wont to be called 'Kant's Copernican revolution'.)[20] But for our purposes we have already learned enough.

What Kant presented in his preface in summary fashion is essentially the same combination of the philosophically general and the historically unique that characterizes much Enlightenment thinking on 'scientific revolutions'. But what was rather an indefinite mixture with, for example, d'Alembert turns here into a real theory of the birth of early modern science, embedded in a general theory of scientific revolutions.

Kant agrees with his predecessors that science is a revolutionary affair. Only, for every science, or set of sciences, no more than one revolution takes place. This is what allows the emergence of early modern science to appear here as a unique phenomenon. Every such revolution is marked by a specific turnabout. In the case of the empirical sciences, this is the shift from aimless observations to conscious experimentation. Here, in Kant's view, lies the key to the birth of early modern science.[21]

The transformation brought about by Kant in Enlightenment thinking on the topic is therefore twofold. On the one hand, he provides a 'structure of scientific revolutions' with a special place reserved for the Scientific Revolution (to employ present-day terminology). Here we recognize not a little resemblance to Kuhn's later schema, in particular. Kant's markers for the "groping in the dark" stage have surprisingly much in common with Kuhn's criteria for the 'pre-paradigmatic state' of a given discipline (more on this in section 2.4.4). To phrase the point in a more general vein: Kant presents here a reconciliation, in the form of a structured argument, between the revolutionary nature of science in general and the unique upheaval of the birth of early modern science.

On the other hand, he offers a clear-cut definition of what made this new science so different from previous conceptions of nature. The active, experimental interrogation of nature is the constitutive feature of the new science, in contrast to the passive, merely observational method of earlier attempts at coming to grips with nature. With Kant, this criterion is a largely philosophical *a priori* construction, in keeping with his overall idea of what science is about, and the criterion is imposed on history rather than derived from it. In this feature, too, Kant sets the stage for much later thought on the characteristic properties of the Scientific Revolution.

Unlike later thinkers, though, Kant makes scarcely an effort to underpin his distinguishing criterion with historical facts. He is so much more concerned with the fixed forms of human thought than with its shifts over time that he seems to regard the facts of the history of science as impenetrable anyway. A note attached to the passage on Galileo, Torricelli, and Stahl even states: "I am not following here in a precise manner the line of the history of the experimental method, the origins of which are not well known in any case."[22]

We turn now to a thinker who fully shared Kant's concern for generalized insights into the nature of science, but who believed in addition that such insights must to some extent be derived from, and must certainly find their ultimate confirmation in, as broad a survey as can be attained of the entire course of the history of science. That thinker was William Whewell.

2.2.2. *Beginning to Learn from the Past of Science: William Whewell*

Like Kant and the other Enlightenment philosophers, the British philosopher of science William Whewell (1794–1866) was not interested in the history of science for its own sake. His primary concern, too, was to account in a general way for how science progresses. But what turned Whewell into the man who may rightly be considered the 'father', or perhaps more fittingly the 'grandfather', of the historiography of science was his belief that, in order to define in any precise way what these patterns of scientific advance are, one must turn to history.[23]

And turn to history is what he did, in the really grand manner. His *History of the Inductive Sciences,* the first edition of which came out in 1837, comprises three volumes—altogether 1,447 pages in the third, and final, 1857 edition. No more than three years after the first edition of the *History,* two accompanying volumes, *Philosophy of the Inductive Sciences,* came out. In their final form these take up another 1,387 pages, which serve as the "moral to the story."[24]

Whewell was well aware of the novelty of such an enterprise. His principal motive, as he explains in the introductory chapter to the *Philosophy,* was that he found the method of 'case studies' (as we nowadays call them) insufficient for the formation of a truly satisfactory philosophy of science:

> For the conclusions stated in the present work, respecting knowledge and discovery, are drawn from *a connected and systematic survey of the whole range of Physical Science and its History;* whereas hitherto, philosophers have contented themselves with adducing detached examples of scientific doctrines, drawn from one or two departments of science. So long as we select our examples in this arbitrary and limited manner, we lose the best part of that philosophical instruction, which the sciences are fitted to afford when we consider them as all members of one series, and as governed by rules which are the same for all. . . . When our conclusions assume so comprehensive a shape that they apply to a range of subjects so vast and varied as these, we may feel some confidence that they represent the genuine form of universal and permanent truth.[25]

One other motive for Whewell to examine the history of science in considerable detail was his strong sense of gratitude toward all those great thinkers from the past who have made it possible for us, moderns, to stand on such a huge mountain of solid scientific achievement. From that vantage point, not only should we look forward toward all those still higher slopes that lie ready to be climbed by our own and by following generations, but we should also look backward, in order both to acknowledge our intellectual debt and to learn how science is done. This is not to say that Whewell believed that the investigation of the past of science can yield to us an 'Art of Discovery'—he was far too strongly convinced of the element of the rationally unaccountable in the process of scientific discovery.

Yet we can learn from the history of science. Two features in particular define, in Whewell's view, successful science. In their presence, science flowers, whereas in their absence we find periods of barren infertility. Whewell named his two defining characteristics "the Explication of Conceptions" and "the Colligation of Facts." Together, these two mental operations enable the scientist to turn the two basic ingredients he has at his disposal, Ideas and Facts, into Scientific Knowledge.

It is essential for an understanding of Whewell's entire work to note that he does not employ these two defining features as a Procrustean bed onto which the past of science is squeezed so as to conform to his preconceptions or as pure and untinged generalizations ostensibly derived from a preceding, philosophically neutral overview of the whole of history. Rather, they provide the key elements around which to organize his historical research. This is what gives his five solid volumes such a strong inner unity. In the *History* Whewell's general conception of science is employed as the leading idea around which to unfold the tale of the history of the sciences; in the *Philosophy* he constantly turns and returns to history in order now to draw in a more explicit fashion the general lessons it has to teach us. It is this procedure, together with a most impressive erudition, a genuine sense for the historical, and great care in handling his views on the process of science in an undogmatic way,[26] that turns these five volumes into a model for what a few historically minded philosophers of science in our own day are making great efforts to achieve: history turned by philosophy to another and perhaps grander purpose than the reconstruction of the past in its own right, and philosophy enriched by history so as to remain in living contact with the actual practice of science.

What, then, were these key ideas Whewell entertained about science and put to productive usage in organizing his account of its history? The term 'inductive', which seems to serve such a programmatic function in the title of his books, should not put us on the wrong track. Even though Whewell often employs the well-known imagery of inductivism in asserting that the progress of science consists in ever wider generalizations attained by inductive reasoning from a continuously growing body of facts, his actual view of scientific advance is much richer than such terminology suggests. Whewell was convinced that discovery involves bold speculations and happy guesses that often defy rational interpretation:

> Advances in knowledge are not commonly made without the previous exercise of some boldness and license in guessing. The discovery of new truths requires, undoubtedly, minds careful and fertile in examining what is suggested; but it requires, no less, such as are quick and fertile in suggesting. What is Invention, except the talent of rapidly calling before us many possibilities, and selecting the appropriate one? It is true, that when we have rejected all the inadmissible suppositions, they are quickly forgotten by most persons; and few think it necessary to dwell on these discarded hypotheses, and on the process by which they were condemned, as Kepler has done. But all who discover truths must have reasoned upon many errours, to obtain each truth; every accepted doctrine must have been one selected out of many candidates. . . . Discovery is not a 'cautious' or 'rigorous' process, in the sense of abstaining from such suppositions. But there are great differences in different cases, in the facility with which guesses are proved to be errours, and in the degree of attention with which the errour and the proof are afterwards dwelt on. . . .
>
> [In short:] This is the spirit in which the pursuit of knowledge is generally carried on with success: those men arrive at truth who eagerly endeavour to connect remote points of their knowledge, not those who stop cautiously at each point till something compels them to go beyond it.[27]

Hence we should not expect Whewell's key notions of the Explication of Conceptions and the Colligation of Facts to be of a superficially inductivist nature. In the brief analysis of the two notions that follows I reluctantly confine myself to what is indispensable for understanding Whewell's views on the origins of early modern science—so rich is the content of his five volumes, and so great is the temptation to go on and on in quoting and discussing the many insights this eminent expert in the past of science formed about the general nature of science.

Whewell's key principles of scientific advance. Whewell's Explication of Conceptions comes down to the unfolding of certain Ideas that are basic to scientific discovery. Among these Ideas, those of Time, Number, and Space are among the most fundamental, although far from the only ones (Cause and Resemblance are others). In being unfolded in greater detail, the Ideas in their turn give rise to such more specific notions, or Conceptions, as proportionality, pressure, gravitation, and the like. As a rule such key notions are somewhat vague when they are first employed to bind certain observed facts together— their rigorous definition and axiomatization are *a posteriori* acts.

Hence the twin process of the Colligation of Facts is really the other side of the same coin: Observed facts can be turned into scientific knowledge only by binding them together by means of such "clear and appropriate Conceptions." To determine precisely what conception is the appropriate one in any particular case is not something that can ever be done mechanically or as a matter of routine. Here is where individual sagacity or even genius comes in: No fixed *rule* could tell Archimedes that he might "refer the conditions of equilibrium on the lever to the Conception of *pressure,* while Aristotle could not see in them anything more than the results of the strangeness of the properties of the circle."[28] However, it is not worthless or vacuous to say that doing science involves these two fundamental thought processes. One reason is that Whewell was indeed somewhat more specific about their operation. Usually, he says, it consists of three steps, namely, the Selection of the Idea, the Construction of the Conception, and, finally, the Determination of the Magnitudes. Of the first two steps, in particular, it can be said that they can be carried out with success only by simply trying.

Even though it is true that these overall insights do not teach us how henceforward to achieve success in any particular scientific enterprise, they do go a long way in accounting for manifold errors, dead ends, and periods of general infertility such as can be found aplenty in the history of science. (In sections 4.2.1 and 4.3 I shall discuss Whewell's pertinent account of the overall stagnation he perceived in Greek and in medieval science.)

Still, the broad picture of the progress of science is one of ever increasing generality. An ever larger number of facts is colligated at every new turn, and progress occurs whenever a new conception, or a new application of an earlier one, is used to colligate a greater number of facts (including earlier theories of smaller scope that owing to their having been found to represent truth now also have acquired the status of facts) than any preceding explication had succeeded in binding together. The resulting theoretical construct, however, is turned into a new scientific fact in its own right if, and only if, it stands up to the most-rigorous experimental testing.

This entire cycle, explained by Whewell at great length, is what scientific induction comes down to. One prime example is the conception of universal gravitation, by means of which Newton colligated the facts of Galileo's and Kepler's laws into "the first example of the formation of a perfect Inductive Science."[29] Thus, whereas science displays unilinear progress in the sense of ever ascending inductions of ever higher levels of generality, we can also discern, and, as it were, superimpose upon the overall line of scientific progress, certain continuous cycles of successful Colligations of Facts. These cycles, Whewell believes, are of a revolutionary nature.

> The great changes which thus take place in the history of science, the revolutions of the intellectual world, have, as a usual and leading character, this, that they are steps of *generalization.* . . . Among [successive steps of generalization], we shall find some of eminent and decisive importance, which have more peculiarly

influenced the fortunes of physical philosophy, and to which we may consider the rest as subordinate and auxiliary.[30]

Since such revolutions in science are ultimately dependent on the performance of gifted individuals, we can distinguish in the history of science certain Inductive Epochs closely connected to prominent individual achievement in scientific discovery, for example, the Inductive Epoch of Hipparchus or of Newton or of Fresnel.

Whewell's views on the revolutionary nature of these Epochs were not fully consistent. When carried away by his enthusiasm in relating them, he often called attention to what made them so pathbreaking and essentially discontinuous with preceding views. In this vein, Whewell wrote about Newton's Induction of Universal Gravitation

> that any one of the five steps into which we have separated the doctrine, would, of itself, have been considered as an important advance . . . All the five steps made at once, formed not a leap, but a flight,—not an improvement merely, but a metamorphosis,—not an epoch, but a termination.[31]

But Whewell also believed that certain results of a previous Epoch could remain valuable, and indeed valid, for a later Epoch, and thus could survive fundamentally unscathed. In summing up his overall view on the matter, he strikes the balance between continuity and break in scientific revolutions as follows:

> The principles which constituted the triumph of the preceding stages of the science, may appear to be subverted and ejected by the later discoveries, but in fact they are, (so far as they were true), taken up into the subsequent doctrines and included in them. They continue to be an essential part of the science. The earlier truths are not expelled but absorbed, not contradicted but extended; and the history of each science, which may thus appear like a succession of revolutions, is, in reality, a series of developments.[32]

The organization of Whewell's History. The general view of the process of science we have just sketched determined the overall organization Whewell gave to his five volumes. Understanding how he organized them will enable us to see how it came about that, despite Whewell's full awareness that something quite unprecedented happened to science in the 17th century, there is no full-fledged, distinct treatment of the birth of early modern science to be found in his works. One good reason for dwelling on this quite conspicuous absence is that, in my view, Whewell had moved himself into a far better position than any other historian of science for a long time to come to set forth his views on the origins of early modern science. If the arrangement of the five volumes had not barred the way, the historiography of the Scientific Revolution might well have taken an altogether different course and, more specifically, might have been damaged far less severely by the onslaughts of positivism, which had already begun during Whewell's own lifetime. I shall return presently to this little exercise in if-history.

The overall organization of the *History,* then, is according to subject matter. Each separate science receives its own chronological treatment. Where available in Whewell's view, his Inductive Epochs are made the focal points in his account of the history of the science in question, with each individual Epoch divided into three stages: a Prelude, during which the Epoch is prepared; the Epoch itself; and a Sequel, during which its results are systematized and brought to perfection.

Despite this overall arrangement, not all unity is lost. In confining Greek science to astronomy and physics and in virtually denying the existence of any medieval science, Whewell manages to give some unitary account of these periods. However, right at the start of the Scientific Revolution (a term never employed by Whewell), the various sciences begin to branch out, and thus any coherent treatment of the period of most interest for our present inquiry inevitably eludes us.

The Revolution in Science according to Whewell.   Whewell was very much aware of the uniqueness and the revolutionary nature of science during the period between Copernicus and Newton. For example, he calls Copernicus one of the "practical discoverers to whom, in reality, the revolution in science, and consequently in the philosophy of science, was due." [33] This remark is taken from Book XII of the *Philosophy of the Inductive Sciences,* where Whewell presents a systematic exposition of conceptions previous to his own of scientific knowledge and method. Although this book is therefore *not* a systematic overview of those events in science that, together, constituted the birth of early modern science, it did give occasion for Whewell to express his views on who played the key roles in the process and how they did it.

In Whewell's original view, Francis Bacon was the all-important figure here. True, in the 16th century various visionary thinkers had announced that it was necessary to cut loose from Aristotle's doctrines and that an overall renewal of science was in order. Others, in a more practical vein, had gone ahead and actually tried to achieve such a renewal. Examples of the class of visionaries are Telesio, Bruno, Ramus; of the class of actual innovators, Leonardo, Benedetti, Copernicus, Galileo, and Kepler. Whewell then goes on to compare Bacon to his fellow visionaries:

> Bacon stands far above the herd of loose and visionary speculators who, before and about this time, spoke of the establishment of new philosophies. If we must select some one philosopher as the Hero of the revolution in scientific method, beyond all doubt Francis Bacon must occupy the place of honour. [34]

Bacon deserves this place for "his confident and emphatic announcement of a *New Era* in the progress of science"[35] and for his deep, though not indeed perfect or complete, insight into the nature of scientific method: general induction.

In a later edition of the *History,* Whewell acknowledged that those "practical discoverers" of the 16th and early 17th centuries had achieved so much on their own that it would be misleading to regard Bacon and Descartes as the only, or even the first, great opponents of Aristotelianism. Still, it remained Bacon's outstanding merit to have provided such an "august image" of a future philosophy that "he thus converted the Insurrection into a Revolution."[36] And this is how Bacon's work continued to convey, for Whewell, the essence of the Revolution in Science of which he was the herald:

> this prompt and vigorous recognition of the supreme authority of observation as a ground of belief; this bold estimate of the probable worthlessness of traditional knowledge; and this plain assertion of the reality of theory founded upon experience.[37]

The above is more or less all Whewell had to say, at various places scattered across his five volumes, about the what and the how of the origins of early modern science. On the one hand, Whewell senses the uniqueness of the event. On the other, his conviction that there is a uniform pattern of progress in science, coupled to the overall organization of his inquiry in accordance with that view, precluded his going deeper into an analysis of what might be regarded as defining characteristics of the new science. In this respect we are left to infer from Whewell's treatment, somewhat unsatisfactorily, that the Inductive Epochs of Copernicus, Kepler, Galileo, and Newton were so many examples of successful Colligations of Facts. We do not learn whether, besides the overthrow of Aristotelianism and the new foundation of theory upon experience, other generalized statements might be made about this dense succession of Inductive Epochs.

If, therefore, we wish to determine Whewell's views on the nature of the new science of the 17th century, we are thrown back on his views on the nature of science in general. In discussing the Explication of Conceptions, Whewell expressly denied that these conceptions are exhausted by the ideas of Number, Space, and Time. He readily admitted that, because of their unique precision, such notions are the most obvious ones to give rise to scientific knowledge. Yet, throughout the history of science other, quantitatively less precise conceptions have always played an important part of their own in binding certain facts together. Examples are the ideas of Resemblance and of Natural Classification, which, in being unfolded, gave rise to such eminently pertinent scientific conceptions as Affinity, Polarity, Genus, and so on. This is why Whewell explicitly rejects a view of science as uniquely defined by "mathematical reasoning as [its] most essential part."[38] In later stages of our inquiry we shall find that the very view contested here by Whewell was to stand at the origin of the modern historiography of science in the twenties and thirties of our own century. We shall also have occasion to observe that it was to be qualified in a similar way by those modern historians of science who, just as Whewell did in the *History,* gave an important place to chemistry and the life sciences in their overview of 17th-century science.

Whewell on continuity and discontinuity.   Let us now sum up what we have learned so far about Whewell's views on the progress of science over time as compared to those of his predecessors, and let us do so from the point of view of continuity and discontinuity in scientific advance. We find that Whewell adopted both the notion of revolutions in science and the notion of the Revolution in Science that were part and parcel of the heritage of the Enlightenment. However, he was the first to apply these notions on the grand scale and to unfold them into something much more intricate and sophisticated than before.

Concerning the overall advance of science through the ages, we find the Enlightenment idea of progress to be pervasive, thoroughly informing Whewell's leading conception of unilinearity, which in its turn is rendered more explicit by his personal, very elaborated version of inductivism. The element of the discontinuous and the revolutionary finds its expression in the various Inductive Epochs scattered throughout the past of science but is at the same time toned down by Whewell's awareness that the ingredients of truth inherent in scientific discovery tend to live on in successor theories, thus providing a key element of continuity throughout the overall advance of science. Whewell elaborated this latter insight in an essay "On the Transformation of Hypotheses in the History of Science" (1851), which has deservedly been called "Whewell's best historical-philosophical analysis" and "a literary gem."[39] All taken together, we find that in Whewell's view, although Inductive Epochs are revolutionary in their own right, their successive concatenation over time leads to an overarching continuity that is embodied in the unilinear progress of science marked by general induction.

If from this vantage point we now ask how 17th-century science can be fitted into the overall schema just sketched, we find that such an admixture of continuity and discontinuity simply leaves no systematic place for the one grand and unique event of the Revolution in Science Whewell was so keen to perceive. In Whewell's work we find the first instance of a feature we shall come across again: the notion of revolutions in science being allowed ultimately to overrule the conception of the Seventeenth-Century Revolution in Science.

Finally, we have found that Whewell's ideas on the nature of 17th-century science as distinguished from preceding philosophies of nature (chiefly Aristotle's) were somewhat vague. In this respect Kant's ideas had been much more specific. For Kant, the one defining feature of modern science was guided and active experiment as opposed to loose and passive observation. Whewell was certainly aware of the distinction, but he did not assign it a prominent place in his conception of science. Nor did he mark any feature other than the foundation of theory in experience as particularly constitutive of the new science. As we have seen, this was in part because there was something inherently vague in his key ideas of the Explication of Conceptions and the Colligation of Facts. But an equally important reason was that, in studying the histories of all natural sciences, including geology and the life sciences, Whewell was far too much impressed by their great variety to elevate just one or two prominent features of early modern science to The Key to what distinguished it in an essential way from previous attempts at doing science. In particular, Whewell rejected in advance what Burtt, Dijksterhuis, and Koyré were to nail down in the 1920s and 1930s as the essential distinctive feature of early modern science: the mathematization of nature. Whewell was certainly aware of the outlines of this process, too. Yet he steadfastly refused to take either this or any other element of 17th-century science as uniquely distinctive.

Whewell and Comte.   Whewell completed his general survey of theories of scientific method previous to his own with an elaborate attack on a contemporary account of the overall progress of science that, from a formal point of view, was remarkably similar to his own effort as embodied in the *History* and the *Philosophy* taken together. This rival account was Auguste Comte's *Cours de philosophie positive* (written 1830–42), of which Whewell discussed, and tore to pieces in his own quiet, civilized, and deadpan manner, the first two volumes.

Comte thought that the history of each science could be split up in three successive stages: the theological, the metaphysical, and the positive stage. What distinguishes the first two stages from the final one is that they are utterly worthless. That is to say, in the overall developmental process the theological and the metaphysical have a function of their own, which, however, comes down to little more than to ushering in the positive stage. And this positive stage, to which every science should aspire, is marked by laws that, operating with complete certainty, bind together facts in a quantitative fashion. Such things, however, as causes, or hypotheses, are just elements of metaphysics, which, therefore, can have no place in positive science.

In Comte's view the empirical sciences of nature had already attained the positive stage, so that a similar revolution in the social sciences was the only thing still needed to complete the process. To usher in that revolution personally was Comte's ultimate goal:

> This general revolution of the human mind has today been almost entirely accomplished. As I have explained, nothing else remains but to complete positive philosophy by including in it the study of social phenomena, and subsequently to sum it up in one single body of homogeneous doctrine. Once that dual effort

has advanced sufficiently, the definitive triumph of positive philosophy will spontaneously take place, and order in society will thereby be restored.[40]

When had the empirical sciences of nature undergone their positive revolution? Although astronomy and physics and chemistry and the other natural sciences had not quite simultaneously reached the positive stage, Comte identified the 17th century as the period when "this revolution" took place:

> I shall point at this [epoch] of the grand movement impressed upon the human mind two centuries ago by the combined action of Bacon's precepts, Descartes' conceptions, and Galileo's discoveries, as the moment when the spirit of positive philosophy began to pronounce itself in the world, in evident opposition to the spirit of theology and of metaphysics. It is at that time that positive conceptions were effectively and with precision set free from the superstitious and scholastic alloy that more or less disguised the veritable character of all previous efforts. Since that memorable epoch the ascending movement of positive philosophy, and the decaying movement of theological and metaphysical philosophy, have been marked with extreme clarity.[41]

So much for Comte's conception of the 17th-century Revolution in Science as marking the onset of the third, positive stage of science as opposed to the theological and metaphysical stages, when causes and hypotheses, among other things, had still plagued science. We can easily imagine how repugnant such views were to Whewell and how deeply irreconcilable he found them to his own conception of both the history and the philosophy of science. Indeed, his final conclusion stated this in no uncertain terms: "Thus Mr. Comte's arrangement of the progress of science as successively metaphysical and positive, is contrary to history in fact, and contrary to sound philosophy in principle."[42]

Philosophically, Whewell had an easy time in showing that the exclusion of hypotheses and causes and the like from science would preclude for good any meaningful progress in science: "To exclude such inquiries, would be to secure ourselves from the poison of errour by abstaining from the banquet of truth."[43] Historically, Whewell had no difficulty at all in pointing out the crude distortions to which Comte's conception of science would inevitably lead, and had already in Comte's own works wherever he ventured into the past. As Whewell insisted, "there is, in Galileo, Kepler, Gassendi, and the other fathers of mechanical philosophy, as much of metaphysics as in their adversaries. The main difference is, that the metaphysics is of a better kind; it is more conformable to metaphysical truth."[44] And elsewhere, slightly less daring but even more pointedly:

> Physical discoverers have differed from barren speculators, not by having *no* metaphysics in their heads, but by having *good* metaphysics while their adversar-

ies had bad; and by binding their metaphysics to their physics, instead of keeping the two asunder.[45]

Not only was Comte wrong in his interpretation of the past of science, but it entailed an even more disastrous consequence:

> Metaphysical discussions have been essential steps in the progress of each science. If we arbitrarily reject all these portions of scientific history as useless trifling, belonging to the first rude attempts at knowledge, we shall not only distort the progress of things, but [the following italics are mine] *pervert the plainest facts.*[46]

Whewell then illustrates the point by tearing to pieces Comte's account of how Kepler derived his area law as the resultant of accelerations caused by an external force. The faithful historian that Whewell was dutifully indicated why such a procedure was inconceivable from the point of view of Kepler's actual dynamical views and came up with the dry but pertinent remark that "there is not a trace of the above propositions in the work *De stella Martis,* which contains Kepler's discovery of his law, nor, I am convinced, in any other of Kepler's works."[47]

There is an important point to this amusing yet seemingly innocent incident in the historiography of science. It is that the crude distortions of Comte and countless later positivists were to become the dominant feature of history-of-science writing for a long time to come. Whewell was as careful and conscientious a historian as one can imagine for his time. In writing the history of whatever given science, he referred scrupulously and as a matter of course to the sources he had at his disposal, nor did he suffer from any positivist tendency to focus on past successes and ignore the many errors, dead ends, and failures the history of science is littered with. A beautiful remark in the *History* may even be read as an antipositivist manifesto *avant la lettre:* "This ex-post-facto obviousness of discoveries is a delusion to which we are liable with regard to many of the most important principles."[48]

Whewell's limitations as a historian of science. Our esteem for Whewell's historical sensibilities should not blind us to a number of limitations inherent in his mode of operation. Essentially there are two. The one is a consequence of his being a philosopher over and above a historian. The other is due to the limited availability of source material at his time of writing. One obvious limitation to the *History* is that Whewell relied almost exclusively on source material that had been published during the lifetime of his scientists, or shortly thereafter. His command of all this material—the books and articles written and published by the scientists themselves—was impressive both in scope and in depth, yet he never suspected that we may learn more about past scientific discoveries by investigating the huge numbers of manuscripts that, with only a handful of exceptions, still lay dormant all over Europe's archival collections.

Other, more profound limitations are rather of a philosophical nature. I mean this in two distinct senses. One is that the idea of progress dominated his thought as it did that of virtually all of his contemporaries. Inevitably, Whewell turned the full corpus of science and its gradual advance into considerably more of a resounding success story than is quite palatable for us, more sceptical inhabitants of the 20th century that we are. Also, Whewell's mode of history writing was marked by a feature common to the historiography of science until, in the twenties and thirties of our century, a rather sudden shift in historical consciousness began to occur among historians of science. This shift, which will occupy us tangentially in later pages, has often been noted (even though as such it awaits its historian). What I mean is this. Whewell's history is almost never contextual in the sense of inquiring into the thought of a given scientist of the past as one unified whole. For example, Galileo's contribution to the science of dynamics is discussed by Whewell in the format of a step-by-step retracing of Newton's successive laws of motion. The way Whewell handled the sources at his disposal was generally fair enough, yet at the end we are left with a range of unconnected statements regarding the extent to which Galileo, as part of a chain of 16th- and 17th-century scientists, participated in the discovery of (1) the principle of inertia, (2) the principle that forces accelerate, rather than cause, motion, and so on. What we, with our sensibilities reared in a different mode of historical understanding, tend in retrospect to find missing from his treatment is an integrated examination of the thought of a scientist in its own right, irrespective of whatever place it might be assigned on some evolutionary line from falsehood to truth.

An even more profound limitation to Whewell's historiography is equally inherent in his philosophy. However undogmatic Whewell was in applying his philosophy to his history, in the last analysis the *philosophy* was what really counted for him. His unquestioned belief that one pattern of progress in science can in the end be distilled from history was never to leave him. As we have seen exemplified by his handling of the birth of early modern science, the historically unique tended to evaporate from his work, subservient as he ultimately made it to those grander purposes he had in mind.

A follow-up that failed to occur.   We conclude that, on the one hand, Whewell's brand of philosophy went a far longer way than any previous philosophy had ever done to make genuine historiography viable in the first place. But we must also be aware of the inevitable limitations of his approach. In order to become really historical, the historiography of science had to liberate itself from its foster father, the philosophy of science. Conceivably, this might have been accomplished by the generation after Whewell. After all, theirs was the time when all over Europe historians—in particular, those concerned with political history—began to develop a craft of their own, with refined methods of criticism of sources and with an eye to writing history "wie es eigentlich gewesen" (Ranke: "as it actually happened"). In thus fantasizing for a moment on what might have been, one can easily imagine how the generation after Whewell might have discovered that such a thing as medieval science had in fact existed and had played a part of its own in bringing about the 17th-century Revolution in Science. Taking Whewell's five pioneer volumes as a starting point, one can without difficulty imagine gradual progress in the historiography of science along lines such as these.

But this was not to be. Whewell's historically minded philosophy was replaced, not by an even more historically oriented conception of the past of science, but rather by what is (at least from a historical point of view) a remarkably crude philosophy. It is ironical that, in taking on Comte, Whewell himself had already laid bare all the shortcomings of this philosophy, which was to dominate the historiography of science for more than half a century to come. The philosophy I am alluding to was, of course, positivism.

In the next section we shall see how remarkable a primitivization of the historiography of science was brought about by the onslaught of positivism.[49] One thing it managed to achieve was the virtually complete elimination of Whewell's own work from the collective memory of historians of science. When, in the first two decades of the 20th century, the existence of a flourishing and, to some extent, original science of the Middle Ages was finally discovered, its importance had to be assessed, not in the sophisticated context provided by Whewell's work, but rather against the background of far less worthy opponents, among whom the Austrian physicist and philosopher Ernst Mach (1838–1916) holds pride of place. In the next section we shall discuss Mach's contribution to the historiography of 17th-century science as one example of positivism applied.

2.2.3. The Positivist Picture of the Birth of Early Modern Science as Exemplified by Ernst Mach

What has made positivism so appealing to scientists with an interest in the past of science is that this doctrine fits in beautifully with working scientists' virtually inborn prejudices regarding the achievement of their predecessors. Almost inevitably, successful scientists tend to regard the truths they have discovered as necessary truths. In retrospect, the achievements of earlier generations seem self-evident, and it requires a real effort of the mind to place ourselves in the position of those to whom 'truth' had not yet become manifest.

This particular lesson has for decades provided the stuff historians of science must fill their opening classes with if they wish to entertain any hope of ever getting anywhere. To get rid of the positivist inside us is precisely what we have to learn in order to become viable historians of science. Evidently, positivism itself is not a good school to go through in order to achieve such an aim. It was even less so in the 19th century—the Era of Progress. And thus, fitting in both with scientists' natural prejudices regarding the past of science and with the general conception of unilinear and remorseless progress that was so natural to the times, 19th-century positivism took what historiography of science there was by storm. The only professional people who might have prevented this from happening (or who at least might have blocked its success in dominating the field entirely) were the historians proper—but they had other worries. With few exceptions, 'general historians' have never considered the past of science as a proper field of historical study, for reasons that would carry me too far from my proper subject to do more than deplore in these pages.

This, then, is how the field was left to Ernst Mach and his ilk. Let us see what he did with it.[50]

Mach's philosophy of science.   At the end of the previous section, when discussing Whewell's various limitations as a historian, we found that he tended to write history as an exercise in topically tracing back the origins of known laws (really inductions, in his view), not as a treatment of the collected works of scientists as entities to be studied in their own right. This same feature strikes us when we open Mach's *Science of Mechanics,* with the one difference already mentioned that Mach's handling of the sources was far cruder than Whewell's. Now what was this physicist's purpose in turning to the history of science?

This is not so easy to pin down. Mach's philosophical views are quite complex and there is by no means agreement among historians over their actual content and meaning.[51] For our present purpose there is no need to delve very deeply into those views. Unlike the case of Whewell, where the history and the philosophy of science helped to shape one another in a productive manner, Mach used the history of science primarily for illustrating and applying a philosophical position already arrived at before. As far as my understanding goes, Mach's aim was to eradicate metaphysics from science by showing that the foundations of science ultimately go back to some basic notions of the scientist at work—notions characterized by the following two features: (1) these notions represent an 'economy of thought' in that they are the simplest and most self-evident ones conceivable and (2) they derive their simplicity from an unprejudiced, even 'instinctive', look at the world around us, from an intuitive capability of distinguishing the essential *facts* of nature. Mach's conception of science thus represents a rather extreme form of empiricism. It is this empiricism that he imposed on the history of mechanics in his book *Die Mechanik in ihrer Entwickelung historisch-kritisch dargestellt* (1883; translated in 1893 as *The Science of Mechanics: A Critical and Historical Account of Its Development).*

The first-mentioned idea—the principle of the 'economy of thought'—was what caused Mach to turn to history in the first place. His basic interest in mechanics was its axiomatics: Somewhat crudely formulated, the simpler one's axioms, the better. Here is where history could come in usefully: It could provide us with alternative axiomatizations. Let me give one example of this procedure. In Mach's time, it was a point of debate whether or not the principle of inertia can serve as an axiom to mechanics, or at least as a necessary corollary of the law of causality. Mach denied this, arguing, among other things, that "a principle that has been universally recognized for so short a time only cannot be regarded as *a priori* self-evident."[52] It is enlightenment of such a nature that Mach seeks in the history of mechanics, and he admits freely that he is not so much interested in history for its own sake.

Galileo at the center.   The main effect of Mach's book in the various editions it went through between 1883 and 1912 was that it pinpointed Galileo as the central figure in the birth of modern physics. In one sense this had of course been a commonplace all along, celebrated as it was in many biographies and popular accounts. But what was new about Mach's book was that it purported to discuss the entire science of mechanics from a historical point of view (which is not the same as providing a history of mechanics—that was not Mach's objective). Biographies, by their very nature, focus almost exclusively on the importance of the person whose life and works are being studied. By contrast, Galileo was now, in a seemingly well documented manner, made the central focus of the first stage of the new physics. At the same time, this focus received a quite distinct coloring through Mach's peculiar conception of science—a conception that has continued to fit in well with the prejudices of many philosophically and historically untrained scientists regarding the nature of their craft. Some time ago Dudley Shapere collected a hilarious sampling of what 'historical' introductions to present-day physics textbooks have to say on Galileo. The striking thing about his survey is that virtually all these perfunctory pronouncements still go back to what Mach had to say about Galileo in his *Science of Mechanics*.[53]

In briefly overviewing Mach's interpretation of Galileo's achievement, we should note first that on the continuity versus discontinuity scale where Whewell occupies such a sophisticated middle-ground position Mach appears as a propounder of radical discontinuity. Dynamics, he claims,

> is entirely a modern science. . . . Dynamics was founded by Galileo. We shall readily recognise the correctness of this assertion if we but consider a moment a few propositions held by the Aristotelians of Galileo's time.[54]

Similarly, on the discovery of the relation of time and space in free fall:

> We must first remark that no part of the knowledge and concepts with which we are now so familiar existed in Galileo's time, but that Galileo had to develop this knowledge and these concepts for us.[55]

And again:

> An entirely new notion to which Galileo was led is the idea of *acceleration*.[56]

What was it that enabled Galileo to put on such a one-man show? Evidently, his keen application of the true scientific method. But this was a topic on which Mach entertained views of his own:

> We see thus . . . that Galileo did not supply us with a *theory* of the falling of bodies, but investigated and established, wholly without prejudice, the *actual facts* of falling.[57]

Of Galileo's method it can be said that

> there is no method of procedure more surely calculated to lead to that comprehension of all natural phenomena which is the *simplest* and also attainable with the least expenditure of both feeling and the intellect.[58]

In the following statement on method, Mach and Galileo seem almost to merge:

> If we are to understand fully Galileo's train of thought, we must bear in mind that he was already in possession of instinctive experiences prior to his resorting to experiment. . . . But for scientific purposes our mental representations of the facts of sensual experience must be submitted to *conceptual* formulation. Only thus may they be used for discovering by abstract mathematical rules unknown properties conceived to be dependent on certain initial properties having definite and assignable arithmetic values; or, for completing what has been only partly given. This formulation is effected by isolating and emphasizing what is deemed of importance, by neglecting what is subsidiary, by *abstracting,* by *idealizing.* The experiment determines whether the formulation chosen is adequate to the facts. Without some preconceived view the experiment is impossible, because its form is determined by the view in question. For how and on what could we experiment if we did not previously have some suspicion of what we were about?[59]

It is not so easy to infer from this statement of method precisely what respective parts, in this view, are played in scientific discovery by rough observation, by conceptualization, by experimental testing, by mathematical treatment, or by theory formation in general. However, two interesting consequences appear to follow. One is that mathematics, in Mach's view, comes in only through the backdoor, as it were, of scientific discovery: It is just part of the formalization process *after* the discovery has been made. Second, Mach was well prepared to take Galileo's own lively accounts of how he arrived at his results quite literally, as indeed appears from various places in the *Science of Mechanics* where he cites them without questioning their reliability for a moment.

In summing up Galileo's achievement in the field of dynamics, Mach coined a phrase destined to become famous:

> The modern spirit proclaimed by Galileo is expressed here, at the very outset, by the fact that he does not ask *why* heavy bodies fall, but poses to himself the question, *How* do heavy bodies fall? In accordance with what *law* do freely falling bodies move?[60]

Instead of dismissing too easily Mach's heavily tinged account of Galileo's scientific method, we should bear in mind that, in allowing his own methodological views to enmesh with his account of Galileo's scientific method, Mach set a pattern that is still with us. If one believes that Galileo was somehow central to the 17th-century Revolution in Science, and if one also fosters ideas of one's own about the nature of scientific method, then it is almost inevitable to project those ideas onto the historical reconstruction one makes of Galileo's achievement. This became so striking a feature of modern historiography precisely because of the countless ambiguities Galileo's own works and notes and letters display as regards what this method of his actually was. Because this is so, and because everybody entertains notions on the true nature of science—whether consciously or not—the predicament of the Galileo scholar, and, by implication, of the student of the Scientific Revolution, is impossible to avoid.[61]

Mach and the historian's craft.   The historical material adduced by Mach in establishing his account of Galileo's pioneering work in dynamics was ostensibly taken from Galileo's own works—in particular, from the *Dialogo* and the *Discorsi*. Some of Mach's observations do betray a keen study and understanding of these works. One example is his insistence that Galileo's discovery of free fall being uniformly accelerated was not derived by Galileo as a consequence of considering the effect of a constant force:

> It would be an anachronism and utterly unhistorical to attempt, as is sometimes done, to derive the uniformly accelerated motion of falling bodies from the constant action of the force of gravity. 'Gravity is a constant force; *consequently* it generates in equal elements of time equal increments of velocity; thus, the motion produced is uniformly accelerated.' Any exposition such as this would be unhistorical, and would put the whole discovery in a false light, for the reason that the notion of force as we hold it today was first created by Galileo.[62]

But for the factual error made in the very last clause, this is quite a convincing refutation of a persistent historical myth created by Newton.

However, later historians of science have not failed to point out that Mach was not always so felicitous in handling his source material. Whewell's anticipation of the aptness of positivism to "pervert the plainest facts" was well borne out by such feats as Mach's account of how Galileo derived the principle of inertia along a chain of argument of which there is literally no trace in Galileo's own works. The derivation is as simple as it is ingenious, but as Dijksterhuis was to remark dryly in 1924, "the only person to deserve credit for

this find is Mach himself."[63] Another example would be Galileo's purported
discovery of the parallelogram of forces, or Mach's wholly unfounded statement that, after Galileo found the parabolic trajectory of a horizontally projected body, "oblique projection no longer presented any essential difficulty"
for him.[64] Finally, we may mention Mach's free introduction of mathematical
formulae wherever it suited him to make a point, thus obliterating the nota-
tional habits of Galileo's own time and thereby seriously distorting some of
their conceptual consequences. In this obnoxious habit, too, Mach has proved
the predecessor to more than one historian of science.

A few later concessions to continuity. The overall picture thus created in
Mach's *Science of Mechanics* is one of a complete break, in the fundamental
science of dynamics, with the past. This break was single-handedly effected
by Galileo, who was enabled to do so by his pioneer application of the true,
empiricist method of scientific discovery. This method relies heavily on unprejudiced observation, on discerning the essential elements among the facts nature displays before us, on experimental testing, and also, in a somewhat undefined manner, on the application of mathematics. It was not, however, given to
Mach to adhere to the letter to this picture throughout the seven editions his
book went through between the years 1883 and 1912. In the meantime an increasing number of facts about the history of mechanics came to light that
seemed to clash with the simple but highly persuasive picture of discontinuity
sketched by Mach.

One major source of these new facts lay in the major headway that was
made in the later decades of the 19th century in publishing editions of various
scientists' notebooks, unpublished works, letters, and so on. One example is
the scientific Notebooks of Leonardo da Vinci, which began to come out in
1881. These appeared to contain a great many ideas which hitherto had been
ascribed solely to Galileo. Mach still felt free to dismiss their historical relevance because, not having been published until 1881, they could not have influenced the course of science. But the critical edition of Galileo's works,
notes, and letters in the Edizione Nazionale, and increasing attention being
paid to such 'precursors' as Benedetti, led to a number of insights Mach could
no longer entirely ignore. Thus in later editions of the *Science of Mechanics*
Mach sought to come to terms with the view, put forward notably by the German Galileo historian Emil Wohlwill, that Galileo had had many prominent
predecessors, in particular, Leonardo and Benedetti, and also, more fundamentally, that

> the predecessors and contemporaries of Galileo, nay, even Galileo himself, only
> *very gradually* abandoned the Aristotelian conceptions for the acceptance of the
> law of inertia.[65]

Mach happily went along with this view:

> Wohlwill's researches deserve our gratitude and show that Galileo had not attained perfect clarity in his own pathbreaking ideas and was liable to frequent reversion to the old views, as might have been expected.[66]

And thus Mach succeeded to his own satisfaction in accommodating views that soon were to lead to a very different picture of the 17th-century Revolution in Science, as if they were nothing but welcome support for his original discontinuity view, which he continued to uphold in the main body of the text of the *Science of Mechanics.*

In fact, the decisive blow to this original view—the discontinuity conception of the birth of early modern science in its pristine, empiricist form—had already been delivered at the time of Mach's death in 1916. The culprit was another physicist, philosopher, and historian of science, but in this case a Frenchman. Rather than dismissing Leonardo's Notebooks as uninfluential by definition, this man was inspired by these very Notebooks to embark upon a systematic survey of Leonardo's sources and followers. In so doing, Pierre Duhem was to transform the core issues of the historiography of science almost beyond recognition.

2.2.4. *The Duhem Thesis*

The historical thesis which, in 1913, Pierre Duhem put before the world can scarcely be summarized more briefly and succinctly than in his own words:

> When we see the science of a Galileo triumph over the hard-headed Peripateticism of a Cremonini [one of Galileo's adversaries in Padua], we believe, badly informed as we are about the history of human thought, that we are witnessing the victory of young modern Science over medieval Philosophy and its obstinate parrotry. But in reality, we are watching the triumph, prepared long in advance, of the science which was born in Paris in the 14th century over the doctrines of Aristotle and Averroes, which in the meantime had been restored to honor by the Italian Renaissance.[67]

In other words, Duhem arrived at broadly the following picture of the rise of early modern science. When we read both Galileo's works and the things his contemporary adversaries had to say in defense of their views, we get the impression that Galileo was doing battle with a form of Aristotelianism that had reigned ever since its reception in 13th-century Europe. (Such, as we have seen, is indeed the picture that arises from reading Ernst Mach's *Science of Mechanics*; Whewell's view, too, although far more nuanced, was not fundamentally at variance with it.) However, in adopting this overall conception of the birth of early modern science, one overlooks entirely that Aristotle had already come under attack way before Galileo began his celebrated onslaught. In the 14th century, a number of natural philosophers at the University of Paris had laid the groundwork for early modern science. Their teachings were gradually received in Italy, where, in the 16th century, they came under attack from orthodox Aristotelians with a strongly Averroist inspiration. It is this reinforced Aristotelianism that Galileo engaged in battle. In this battle, he did nothing but recover and, at a later stage, elaborate the science already created at Paris two and a half centuries before.

One does not go too far in ascribing to Duhem the view that, in reality, the Revolution in Science took place in the 14th rather than in the 17th century:

> Thus, from William of Ockham to Dominicus Soto, we see how the physicists of the Parisian School laid all the foundations of the Mechanics which Galileo, his contemporaries, and his disciples would develop.[68]

And in a discussion of one particular idea of one of the main propounders of this new science, Jean Buridan, Duhem even went so far as to write:

> If one wishes to separate through one precise line the reign of ancient Science from the reign of modern Science, we believe that it has to be drawn at the moment when Jean Buridan conceived of this theory, that is, at the moment when one ceased to regard the heavenly bodies as moved by divine beings, when it was stated that the heavenly and the sublunary motions depended on one single Mechanics.[69]

Now these, to say the least, were startling claims. How did Duhem arrive at them?

Duhem's early Leonardo studies. Pierre Duhem was born in 1861. He died in the same year as Ernst Mach—1916. Just like Mach, he was not only an important physicist but also a philosopher of science of considerable originality and influence. Just like Mach's, his general views on science were connected with his interest in its history in a manner that is not easy to define. Yet there seems to be one important common source for both Duhem's philosophy and his history: namely, faith. Among philosophers of science Duhem is chiefly known for his advocacy of the view that scientific theories can never claim to represent reality. The ideal of science is, as the ancient expression has it, 'to save the phenomena', that is, to find the most economical model that can account for them. The fundamental view behind this conception of science, at least in Duhem's mind, was that we can know the truth only through faith.

Faith, for Duhem, meant Catholicism of a particularly orthodox nature, tinged with an extremist French nationalism. We shall see that these strongly held sentiments were certainly driving forces in Duhem's historical research, which began when he became intrigued by a couple of references he found in Leonardo's Notebooks to previous scholars from the University of Paris (of which Duhem himself was an alumnus).[70] Looking up these references in a wealth of scholastic manuscripts preserved in the Bibliothèque Nationale at Paris and tracing these ideas ever farther backward, he eventually uncovered a comprehensive school of Parisian scholastics who appeared to have achieved two most noteworthy exploits: They had broken the backbone of Aristotelianism, and, in so doing, they had laid the foundations of early modern mechanics and astronomy. Among the claims Duhem made on their behalf, the following are paramount:

—Against Aristotle's idea that projectile motion is sustained by the surrounding air, Jean Buridan elaborated earlier objections into a full-fledged impetus theory, that is, the idea that, rather, an inner motive force sustains the motion. This impetus was set proportional to the mass and velocity of the body and thus adumbrated the early modern concepts of quantity of motion and momentum. Through its property of indestructibility, impetus also served as a predecessor to Galileo's principle of inertia. Finally, in being applied to the motions of the planets, impetus provided a key link between earthly and heavenly dynamics, thus turning Buridan into the predecessor of both Copernicus and Galileo.

—Nicole Oresme, in giving a geometrical representation to certain arithmetical values, anticipated Descartes in creating analytic geometry and Galileo in deriving from his graphs the fundamental laws of falling bodies, even though he failed to apply these to the actual process of falling. (But this shortcoming was made good by Parisian scholastics of the 16th century, as appears from the works of Dominicus Soto.)

—Finally, in favorably discussing the possibility of the daily rotation of the Earth around its own axis, and in underpinning this hypothesis with a plausible dynamics, Oresme anticipated, and to some extent even surpassed, Copernicus.

It is a fascinating and very rewarding effort to follow Duhem through his three volumes of Leonardo studies in which these discoveries are successively reported. The first two volumes, which appeared in 1906 and 1909, respectively, still focus on Leonardo, in that each separate essay attempts to connect ideas in Leonardo's Notebooks to one or two earlier or later thinkers. The volumes are appropriately entitled *Études sur Léonard de Vinci: Ceux qu'il a lus et ceux qui l'ont lu* (Leonardo Studies: Those Whom He Read and Those Who Read Him). But in the third and final volume, which came out in 1913, Leonardo was gradually allowed to phase out, and the Paris Terminists, as they were called, came to occupy the center of attention, properly announced in the subtitle Duhem gave to this volume: *Les précurseurs parisiens de Galilée* (Galileo's Parisian Forerunners).[71]

The continuity of the history of science. Duhem's thought was heavily imbued with the notion of 'forerunners'. Right at the beginning of the preface to volume 1 we find an eloquent confession of faith in continuity as the essence of scientific advance:

> The history of the sciences is falsified by two prejudices which are so much alike that one might conflate them into one only. It is commonly thought that progress in science is made by a succession of sudden and unexpected discoveries and thus, so one believes, is the work of men of genius who have no precursors at all.
>
> It is a useful effort, and one worth insisting on, to mark the point where these ideas are erroneous, the point where the history of scientific development is subject to the law of continuity. Great discoveries are almost always the fruit of slow and complex preparation, which is pursued in the course of the centuries. The doctrines which are propounded by the most powerful thinkers result from a multitude of efforts which are accumulated by a crowd of obscure workers. The very same persons whom it is fashionable to call creators—the Galileos, the Descartes, the Newtons—did not formulate one doctrine which is not attached through innumerable links with the teachings of those who preceded them. A too simplistic history leads us to admire them as colossi born through spontaneous generation—incomprehensible and monstrous in their isolation. A better-informed history retraces for us the long filiation from which they came forth.[72]

There is no doubt at all that Duhem was indeed 'better-informed'. His enduring claim to fame in the historiography of science is this single-handed rediscovery of an important body of 14th-century scholastic manuscript texts which had almost completely been lost from sight since the late 16th century. Also, Duhem was the first to connect the subsequent fate of those Parisian ideas he found most worthy of his attention, through an unbroken line of subsequent scholastic efforts, with the onset of the 17th century, thus confirming his strongly held belief that "no more than Nature does Science make any sudden jumps."[73]

The key problem with Duhem's thesis.   However, Duhem's status as the originator of what was to turn eventually into an area of specialization in the historiography of science in its own right—the history of medieval science—should be carefully distinguished from the larger question of the extent to which his broader claims were indeed justified by the wealth of source material he employed to underpin them with. After all, there was something inherently absurd about these claims. From a formal point of view, the treatises of the Paris School were in no way different from conventional commentaries on Aristotle's works. With the one exception of Oresme's treatise on the *latitudines formarum*, such issues as the impetus theory of projectile motion were put forward in the format of Quaestiones on Aristotle's *Physics*. On the surface, to say the least, it is not so readily apparent that, generally speaking, the Parisian thinkers did more than question certain separate statements in the received body of Aristotelian doctrine, or that they ceased to work inside the overall

Aristotelian framework. Throughout the three volumes of *Études sur Léonard de Vinci,* Duhem continued to regard the achievements of Galileo, Descartes, Kepler, Newton, *et al.,* as the yardstick against which to measure the performance of his Parisian heroes.[74] Was it not more reasonable, then, to leave the origins of early modern science where they had always been thought to be—in the 17th century—and be content with having pointed to some interesting adumbrations three centuries earlier which for various reasons had aborted?

One thing that makes reading Duhem so fascinating is that these dilemmas were to some extent present in his own mind as well—there is an inner tension which makes its presence felt throughout his volumes. On the one hand, Duhem stretched his claims as far as he could, and we shall discuss presently his motives for doing so. But on another, less conscious level he knew very well that in a sense the Parisian revolution—if indeed it was one—had aborted. Duhem even knew why: because the Parisian tracts had been written far too early to benefit from the invention of printing.[75]

The unresolved tension thus present in the Duhem thesis from the outset was soon to discharge in, and in that very act to dominate for decades to come, an entirely new Grand Debate over the nature of the Revolution in Science. Most of our present-day thinking on the topic was shaped by the debate. Even though virtually no one has ever adopted Duhem's thesis in its pristine radicality,[76] opinions on the relative degree of its validity varied widely, from slightly modified continuity conceptions to attempts at incorporating Duhem's findings in restatements of sharp discontinuity. This debate will occupy us throughout later sections of the present chapter. First, however, we have to return to certain features of Duhem's thought which enabled him to put forward a thesis that, on the surface, seemed so paradoxical as to defy plausibility.

Three such features seem to me particularly pertinent: Duhem's pride in his discovery, his views on the role of mathematics in science, and his nationalist brand of Catholicism.

The background to the thesis: Duhem as a proud discoverer. Neither the *Études sur Léonard de Vinci* nor Duhem's concomitant *Système du monde* (The System of the World: History of Cosmological Doctrines from Plato to Copernicus; 10 vols.) is written in the format of sustained arguments. They run up and down the centuries and turn and return to the same thinkers of the past in order to expose freshly discovered scholastic doctrines as they come in the order of Duhem's loan slips at the Bibliothèque Nationale. In this way Duhem makes his readers share in his sense of excitement over his ongoing discoveries. His finely chiseled, powerful, and rousing prose style adds much to the effect. And indeed, we may well ask whose attention would ever have been seized by a dull, scholarly article in which some hitherto neglected scholastic ideas were cautiously weighed as to their degree of originality? In cases such as the discovery of the Paris Terminist School (and we shall come across a comparable case in Yates' calling forth the Hermetic tradition from the dung-heap of history), overstatement is of the essence in gaining oneself an audience in the first place. The careful weighing and toning down of the original thesis can safely be left to later generations, thus ensuring a certain degree of continuity, not so much in the history as in the historiography of science.

The background to the thesis: Duhem's conception of science. Besides rhetorical overstatement, there was also much in Duhem's conception of science that made for aggrandizement of his claims. It is not easy to specify this conception exactly, because throughout the *Études* one seeks in vain for a clear-cut statement of what particular features, in Duhem's view, constituted the birth of early modern science. If we take his claims literally, an argument on the daily rotation of the Earth anticipating Copernicus, a principle of inertia of sorts and the two most fundamental laws of free fall anticipating Galileo, the invention of analytic geometry and the concept of quantity of motion anticipating Descartes, and, finally, handling on a par the motive forces of planets and projectiles anticipating Newton, taken together, were sufficient for constituting the birth of early modern science. Even if we leave aside the fundamental question (to be addressed in later sections in some detail) of whether these claims were factually sound, that is, of whether they were tenable interpretations of the evidence, we still have to ask if, granted for the sake of argument their basic correctness, this is enough to be called the birth of early modern science. After all, there is no trace, in this picture, of the annual revolution of the Earth around the Sun, of the elliptic orbits of the planets, of the parabolic trajectory of projectiles, of a particle theory of matter, of the employment of scientific instruments, and so on and so forth. Clearly there is some gap to be filled here.

This is not to say that Duhem presented the case so frankly. But filling the gap is what he did, in various ways. For one thing, separate findings of his Parisians were almost invariably blown out of proportion with respect to their original context. For example, the impetus theory of projectile motion as conceived by Buridan (again, leaving aside whether Duhem interpreted it in an acceptable manner) was really no more than one bright idea brought up to replace one specific notion of Aristotle's and then applied to two other cases. For Duhem this was enough to allow him to make constant reference to Buridan's 'Impetus Dynamics', as if the term 'dynamics' does not stand for something far more intricate, far more elaborate, and far more structured that what Buridan's, or anyone else's, loose statements could possibly be made to represent.

Another case in point is the jumps Duhem made in attributing to certain non-Aristotelian, or anti-Aristotelian, statements the status of forebodings of "a true scientific spirit,"[77] without ever taking the trouble of defining what this spirit should be taken to be. Was it just opposing Aristotle? Was it doing experiments? Was it applying mathematics to natural philosophy?

Certainly not the last. What little we find in the *Études* on the 'true spirit of science' leaves only a small place for mathematics. Time and again we find that, in Duhem's view, the chief contribution Galileo and his contemporaries made to science is that they elaborated, in a more precise fashion, insights that had been arrived at earlier in a purely qualitative vein. Thus Duhem could state that for two and a half centuries nothing essential was added to Buridan's impetus conception, after which period the theory "was divested of its purely qualitative form in order to be clothed now in a more precise, quantitative form."[78] Another very revealing passage is the following, which also displays in a nice manner the tension mentioned earlier. After having found such a bad reception with the Averroists in Italy, Buridan's 'mechanics'

> was to find there adherents [notably, Leonardo and Benedetti] who, through their reading of the ancients, had been trained in the subtle procedures of Geometry. They translated [Buridan's mechanics] into mathematical language; they thus made explicit the truths it potentially contained and, in the process, guided it so as to produce modern Science.[79]

Two comments are in order. The first is that Duhem continuously shifts early modern science back and forth to where it suits him. Now Buridan founds it; then it is a 17th-century acquisition; then its origin is ascribed to Leonardo and Benedetti.[80] All of which betrays that at heart Duhem knew he was badly overstating his case.

The second comment has to do with the role of mathematics in science. For Duhem its role is a second-order phenomenon. Mathematics comes in after the fundamental discovery has been made; it helps to make the discovery more precise, to explicate its meaning, to bring out into the open what is potentially there; yet the core of the discovery lies in the qualitative statement itself. It appears from a careful perusal of the *Études* that for Duhem scientific discovery is made in the format of separate statements containing loose and generally unconnected insights. To combine them, to make them mutually coherent and consistent, to quantify them, to axiomatize them—all such operations, although certainly relevant to science, are, at bottom, of a secondary nature.[81]

It is this implicit conception of scientific discovery that enabled Duhem to burden his favorite scholastic statements with such heavy loads. This is what made it possible for him to argue from the similarity of certain basic statements in works of a 17th-century thinker and a 14th-century thinker to the essential originality and modernity of the latter. Partly, this fundamental feature of Duhem's mode of historical thought may be attributed to the state of the art of the historiography of science. No 'contextual' method had been developed yet—a successor generation, in grappling with Duhem's intellectual legacy, was to create the method in the next two decades. But we should also recognize that, to some extent, Duhem shared a rather scholastic bent of mind with his beloved Parisians: For both him and them, single affirmations were the very stuff of science and scholarship. What counted were opinions rather than co-herent bodies of theories and conceptions.

This, then, leads us quite naturally to the third defining feature of Duhem's historiographical habits.

The background to the thesis: Duhem's chauvinist Catholicism. One other powerful stimulus toward overstatement lay hidden in Duhem's extrascientific beliefs. Clearly Duhem felt close to French scholasticism. It is not by chance that the non-Christian philosophers Aristotle and Averroes (the latter in partic-ular) were among his *bêtes noires.* In his own time, Duhem stood for about everything that is summed up in the notions of *revanchisme* and *anti-Dreyfusisme.* He was anti-Enlightenment, anti-German, and (so it seems) anti-Semitic; the other side to the same coin being his fierce nationalism and his extremely orthodox Catholicism.[82] His eagerness to demonstrate that French Catholicism provides a particularly fertile soil for scientific progress shines through his *Études.* The most remarkable feat in this connection was his inter-pretation of the celebrated decree issued in 1277 by the bishop of Paris, Étienne Tempier. This decree was not uncovered for the first time by Duhem—it had been known all along among medievalist scholars—yet he exploited it to support his Parisian claims. In Duhem's interpretation of the decree, Tempier opened the way for a fresh approach to both Aristotle and nature by forbidding an array of statements which came down to restrictions of God's omnipotence. In forbidding, for example, the proposition "that the first cause cannot make several worlds,"[83] Tempier, as it were, created room for scholars' entertaining the idea of the plurality of worlds. Not, Duhem hastened to add, that they actu-ally went ahead and did so—the point is rather that they were now free to do so if they pleased.

Duhem was so impressed with the new liberty thus demanded for science by the bishop of Paris that, at the end of volume 2 of the *Études,* he situated the birth of early modern science at this event:

> If we had to assign a date to the birth of modern Science, without doubt we would choose this year 1277 when the Bishop of Paris solemnly proclaimed that several Worlds could exist, and that the entirety of the heavenly spheres could, without contradiction, be animated by rectilinear motion.[84]

There is more at stake here than Duhem's customary enthusiasm run rampant. Witness the following characteristic remark made in the context of a discussion of Leonardo's and Benedetti's achievements:

> It is from the Logic and the Physics of the Parisians that, in Italy, the initiators of modern Science borrow arms in order to combat the out-of-date teachings of the Philosopher and the Commentator [Aristotle and Averroes, both of them pagans]. Those who exert themselves in shaking off the yoke of tyrannical rou-

tine have their eyes fixed upon Paris, with its nominalist Scholasticism, which has for centuries possessed intellectual liberty.[85]

In the presence of so much missionary zeal coupled with such fervent discoverer's enthusiasm, historical accuracy and a modicum of internal consistency may easily give way. It is in the above spirit that Duhem ends the preface to the third volume of his *Études*, which contains the summing up of his fundamental thesis, with the following, truly enchanting exclamation:

> [The] substitution of Aristotle's Physics by modern Physics resulted from an effort of long duration and extraordinary power.
>
> This effort took its support from the most ancient and the most splendid of the medieval Universities: the University of Paris. How could a Parisian fail to be proud of this?
>
> Its most eminent promoters were Jean Buridan from Picardy and Nicole Oresme from Normandy. How could a Frenchman fail to entertain a legitimate sense of pride in this?
>
> It resulted from the obstinate struggle which the University of Paris—at that time the true guardian of Catholic orthodoxy—waged against peripatetic and neoplatonic paganism. How could a Christian fail to acknowledge his gratitude for this to God?[86]

Conclusion. The quotations given above are not meant primarily to make Duhem look ridiculous. It is of importance to acknowledge the spiritual climate in which the Duhem thesis, with its revolutionary consequences for our present-day conception of the Scientific Revolution and for the historiography of science generally, came to be enunciated in the first place. Swept on and on by a highly personal, even unique set of motives, Duhem *ipso facto* invited those of his readers who did not share his world-view to divest his message from its more outlandish embellishments and to inspect in a more sober vein to what extent his far-reaching claims possessed a reasonable core suitable for adoption by all. This, then, was the challenge Pierre Duhem put before those scholars in the next generation who cared at all for the history of science.

2.3. SHAPING THE CONCEPT OF THE SCIENTIFIC REVOLUTION

We must now find out how the concept of the Scientific Revolution has grown out of a debate that finds its origin in issues put on the agenda by Pierre Duhem. Here we must distinguish between two strands: the general reform of the historiography of science that lies implied in Duhem's work and the particular thesis he put forward on how early modern science had come into being. In what guise did these two issues present themselves to Duhem's successors?

From philosophy to history: One lasting step forward.   Duhem had certainly taken a major step on the path leading to mature historiography of science. True, his concern had not been to present one coherent picture of one or another scientist's thought. This indispensable goal of the present-day historian of science was not to be fulfilled until the generation that succeeded Duhem. Like Whewell, Duhem had, rather, written history by tracing across the centuries separate statements in the work of individual scientists. Yet in so doing—and this is our point for now—he had been guided by an acute sense of what in the *Études* he once called "the chain of scientific tradition."[87]

In Kant's, Comte's, and Mach's conceptions of the birth of early modern science we have found no trace of such an awareness. For each of these philosophers, the 17th century simply marked the demarcation between wrong science and right science, in accordance with the version adopted in each case of what 'wrong' and 'right' amounted to: empirical versus experimental in Kant's view, metaphysical versus positive in Comte's, prejudiced versus unprejudiced in Mach's. In each case the transition was presented as having been accomplished quite suddenly, without any preparation or intermediate stages. For the philosopher, as distinct from the historian, there is little that is inherently implausible in such a view. Before Duhem, only Whewell had been able to overcome it, but the wave of positivism had swept over his sophisticated approach and had powerfully reinforced the philosopher's natural tendencies when addressing the history of science. Half a century later, Duhem's work marks the definitive breakthrough, achieved by a philosopher of science, toward a view of history characterized by continuity rather than by one or more sudden and unprepared breaks with the past. One might say that Duhem's work marks the point where an ahistorical view of *absolute discontinuity* in the birth of early modern science is left behind more or less for good. What takes its place with Duhem might be called *absolute continuity*—the radical conviction, that is, that no breaks occur in "the chain of scientific tradition" at all. We shall see that other conceptions of the advance of history are possible and have been adopted, in fact—conceptions that restore the room needed for positing fairly radical breaks inside the tradition. But never were such 'discontinuity' interpretations of the history of the birth of early modern science to return to absolute discontinuity in the ahistorical sense of Kant, Comte, or Mach. To strike the proper balance between a perception of historical events as *relatively continuous* or *relatively discontinuous* has been the historian's task ever since the craft attained maturity in the course of the 19th century. Duhem's work made it possible to conceive of the task of the historian of science in similar terms. No subsequent treatment of the emergence of early modern science was to lose again the ground gained here by Duhem. What Duhem once eloquently observed about Leonardo in particular has, in a more general fashion, become the implicit guiding line, not to say the commonplace point of departure, of later generations of historians of science:

[He] does not therefore appear to us any longer as a genius isolated in time, without connection with either past or future, devoid of intellectual ancestors or of scientific posterity. We see how his thought feeds itself on the juices of the science of preceding centuries so as to fertilize in its turn the science of future ages. [Thus] he resumes his place as an admirably solid and brilliant link in the chain of scientific tradition.[88]

New questions.   One major issue of the ensuing debate over the Duhem thesis was, simply, to assess the use Duhem had made of his sources. His consistent habit in both the *Études* and the *Système du monde* was to present relevant fragments of the Latin original (only Oresme had written some of his treatises in the vernacular) in French translation. Now did he have his facts straight? Or had he been carried away too much by his own enthusiasm? And also, had he not torn his beloved Parisians out of their scholastic context to such an extent as to distort seriously their meaning and, consequently, their significance?

Obviously, these were issues that struck at the heart of the Duhem thesis. And they were *new*. No one had asked such questions before in connection with a piece of history-of-science writing. It was precisely the peculiar nature of the thesis as Duhem had shaped it that now gave rise to such questions in the first place.

In the next sections I shall address the pertinent works of those three historians of science from the generation after Duhem who, in respect to our problem of the origins of early modern science, made the most productive usage of the challenge the Duhem thesis had laid before them. These three are Anneliese Maier, E. J. Dijksterhuis, and Alexandre Koyré. Their work covers roughly the twenties through the sixties of the present century.

Dijksterhuis' first relevant publication dates back to 1924, but because it was written in Dutch, it had very little impact (Koyré cites it a couple of times, but it seems highly unlikely that he actually read it). Maier began to publish her findings in 1939, and did so in German (not until 1982 did a small selection of her papers appear in English translation). Koyré's epoch-making *Études Galiléennes* (Galileo Studies) also dates from 1939, and it had an almost immediate impact. Between the three there were also some interesting comments upon one another's works concerning the issues raised by the Duhem thesis, which had been their common source of inspiration.

The upshot of Dijksterhuis' contribution to the debate was that the process of the mathematization of nature, which was the key to early modern science, had been a very gradual and continuous affair, to which Duhem's Parisians had contributed in a major way. The upshot of Koyré's contribution was that the process of the mathematization of nature, which was the key to early modern science, signified an unprecedentedly new and revolutionary mutation in human thought, which to a large extent had been achieved by overcoming the modified Aristotelianism of the Parisians. The upshot of Maier's contribution was that a more scholarly responsible study of late scholastic manuscript texts than had been carried out by their discoverer revealed that the world-view of Duhem's Parisian heroes and other 14th-century scholastics represented a major, but not the decisive, step necessary in overcoming Aristotelianism. Her studies ushered in the field of medieval science as an area of scholarly specialization in its own right. We shall discuss Maier's views first.

2.3.1. *Reviewing the Sources: Anneliese Maier*

Impetus and inertia. Of all the claims Duhem had made on behalf of his Parisians, none was quite so explosive as the one that directly linked Buridan's impetus theory with the principle of inertia. After all, the principle appeared to lie at the heart of Galileo's mechanics. It was the chief weapon in his defense of the Copernican system as physically plausible. Further, the principle of inertia served as a cornerstone in Descartes' corpuscular conception of the material world, and it figured prominently as Newton's first law of motion. Thus, if it could be shown not only that the impetus theory was one major step on the road toward the discovery of this principle in Galileo's own work but also that, in fact, the impetus theory already *embodied* the very principle of inertia, then indeed a powerful, although surely not yet decisive, argument had been found for shifting the birth of early modern science from the 17th to the 14th century.

It is important to distinguish carefully between the two issues at hand. There is no doubt at all that, in the early notes from his Pisan days, *De motu* (On Motion), Galileo employed a concept of 'vis impressa' (impressed force) that is similar in many respects to the medieval impetus concept. The narrower question is: What role did the concept play in Galileo's subsequent thought? The far wider problem is (and this is how Duhem presented it to the world): Was impetus itself a kind of inertia? It was this wider question that Maier subjected to extensive, critical scrutiny in a number of essays starting in 1940.[89]

All these and a great many more essays by Maier were based upon the careful study of an enormous mass of manuscripts collected in the Vatican Library, where Maier spent virtually the whole of her scholarly life. Born in 1905, she had left Germany in 1933, "when it began to become intolerable there."[90] She had been educated as a philosopher and a manuscript specialist but had also taken some courses from Planck in physics—not at all sufficient to qualify her as a physicist but, as she disarmingly wrote to Dijksterhuis in 1948, she could recall just so much of this as was needed for her studies in late medieval scholasticism.[91] Altogether she published eight volumes of essays on late scholastic thought, which appeared between 1949 and 1977 (the final one appeared posthumously; she died in 1971). Four of her essays were devoted to the impetus/inertia issue. The most succinct among these is her article "Impetustheorie und Trägheitsprinzip" (Impetus Theory and Principle of Inertia). Significantly, the collection to which it belongs was entitled by her *Die*

Vorläufer Galileis im 14. Jahrhundert: Studien zur Naturphilosophie der Spätscholastik (Galileo's Predecessors in the 14th Century: Studies in the Natural Philosophy of Late Scholasticism; 1949).[92]

Here as in her other pertinent essays Maier showed that impetus and inertia are two fundamentally different answers to Aristotle's original question of what sustains the motion of a projectile after it has left the agent that put it into motion. This is a problem in Aristotelian natural philosophy, because 'everything that moves is necessarily moved by something'. Now what was this 'something' in the case under discussion? Aristotle attributed a moving force to the ambient air (communicated to it by the moving agent; for example, by a hand that throws a stone). Buridan replaced this external moving force of the air by an internal moving force originally communicated to the body by the same agent that put it into motion. From the point of view of early modern science, however, no special explanation of the process of projectile motion is required, since bodies tend to persist in uniform, rectilinear motion unless some external force intervenes. Thus ontologically the two concepts are entirely at variance with one another: The one explains something for which the other denies the need of an explanation. Physically, too, they are wide apart. Buridan and his pupil Albert of Saxony were the only ones ever to regard the impetus as a self-sustaining force, that is, as one that does not peter out just by itself. But this does not imply that it cannot be destroyed. Even if all kinds of opposing forces such as gravity and air friction are abstracted away, the impetus would still be destroyed. There always remains, in the sublunary sphere, an innate resistance to further motion:

> The impetus is gradually destroyed by the very resistances which are indispensable for bringing about motion in the first place. . . . In inertial motion there is a moment which can simply not be abstracted away: the inertia of the moved mass point. One may ignore external impediments and forces, but not the mass of the projectile. And the difference between scholastic and classical mechanics is precisely that the latter conceives of this inertia as the real factor that sustains the motion, whereas the former regards it as resistance against this motion and against the impetus that causes it.[93]

This is why Buridan wisely confined his hypothesis of an unimpeded, and therefore eternal, impetus motion to the heavens—"an analogous application to Earthly mechanics did not occur to him, nor could it have."[94]

Thus Galileo's reasons for giving up the impetus theory must be sought elsewhere. Just as it had seemed to Buridan that Aristotle's account was not good enough to save the appearances, so was Galileo concerned with certain phenomena that the impetus account, in its turn, was incapable of saving. Ironically, the phenomenon in question was one that might well have been made the subject of theoretical reflection by the scholastics: It is the rolling movement of a spherical body over a virtually frictionless, horizontal plane. In attempting to explain this persisting movement, so Maier argued, Galileo was gradually led toward his principle of inertia, even though in his case the principle always remained valid for horizontal (i.e., in Galileo's view, circular) motion only.

The conclusion Maier drew from her analysis is as follows:

> The transition from the impetus theory to the principle of inertial motion has nothing to do with any transformation in the picture of the world at large. Just one physical theory is replaced by another, which matches better with the phenomena to be explained. And this transition takes place in a way quite similar to the corresponding one in the 14th century. In the one as in the other case one starts with no other purpose in mind than to stick to the traditional explanation. One wishes to modify these only in some details, in order to save certain appearances. But later on it appears that these details have such a fundamental significance that they make the old building collapse in its entirety. In the 14th century the correction in question concerned the 'subiectum' [subject] of the moving force. In the 17th century the role of the vis inertiae [inertial force] was at stake. In either case these seemingly insignificant modifications ushered in a new epoch in the history of physical thought.[95]

Two revolutions in science rather than one.   It is clear from the above quotation from Maier's essay on impetus and inertia that a conception of the Revolution in Science lurked behind her analysis that differed substantially from the one dominating, albeit not entirely without qualifications, Duhem's thesis. In the introduction to her 1949 volume Maier made her conception of a dual revolution in science explicit.

In one fundamental respect, she believed, Duhem had been right. It was indeed true that "the conception of nature to be found in the 14th century [could be seen] as a preliminary stage to, and a preparation for, classical physics."[96] Duhem's major mistake in this respect had been "often [to] have interpreted the details of the scholastic doctrines in too modern a sense and [to] have read more in them than is actually there."[97] (An error of lesser significance was that, besides the Parisians, other 14th-century scholastics, such as the Oxford School, had also contributed to the ongoing renewal.) These shortcomings, which were shared by many of Duhem's critics and followers, could only be overcome by attempting as Maier herself had done, to "make as objective and as exact a picture as possible of the physical theories of the 14th century *and of the overall world-view of which they are a part*" [my italics].[98]

In so doing, Maier urged, it was crucial to know what to look for. One should not allow oneself to be deceived by superficial analogies between efforts to explain physical phenomena which, after all, were usually the same in the 14th as in the 17th century. The goal was to find out whether and, if so, to what extent concepts, principles, and theories match: "ever since Duhem that has been the question at issue."[99]

The overall conclusion to which a study guided by such principles ultimately led, Maier asserted, was that the entire history of the sciences in Christendom, right from the 13th century up to the 18th, should be seen as

> a history of the gradual overcoming of Aristotelianism. This overcoming did not take place in one single, grand revolution, as has been thought for a long time. On the other hand, neither was it a stable and continuous emancipation process evenly spread over the centuries. Rather, it displays a development executed in two grand stages, of which the first finds its culmination point in the 14th, and the second in the 17th, century.[100]

Yet the two stages were not quite of equal importance. Seen as a whole, Maier granted, it is certainly permissible to regard the new conception of nature embodied in late scholasticism as a precursor to early modern physics, "that is, in the sense of the first stage of a grand developmental process of which the second *and decisive* [my italics] stage falls in the 17th century."[101]

The six-page introduction just summarized is the only place where Maier ever expressed herself in general terms on the nature of the Revolution in Science. Consequently, even though she fully deserves to be called the 'mother' of the historiography of medieval science, her middle-of-the-road view was largely ignored.[102] This was not only because it remained locked in the German language until quite recently. More important perhaps is that, in contrast to Duhem, she did not apply it as the guiding principle of all her essays, nor did she make it the culmination point of a chronologically arranged, fully historical overview set up in such a way that the conclusion finally arrived at acquires something of the inevitable for the reader who so far has followed the historical argument. Two historians of science who were masters in pushing this skill to a high degree of persuasiveness were Dijksterhuis and Koyré. We take on Dijksterhuis first.

2.3.2. Dijksterhuis and the Mathematization of Nature

Eduard Jan Dijksterhuis, who lived from 1892 to 1965, was a mathematics and physics high school teacher for by far the longer part of his life. (He was appointed to a newly created chair in the history of science at Utrecht University only in 1953.)[103] By training, he was a mathematician. His first book, which has never been translated, was *Val en worp* (Free Fall and Projectile Motion; 1924).[104] Many of its 466 pages testify to the profound impression Duhem's *Études* had made upon Dijksterhuis. At a later point in the present section we shall try to find out whether his own strong belief in the continuity of science over time was due to the lasting influence Duhem exerted upon him or whether, the other way round, his reading of Duhem's work served to strengthen a conception of the advance of science over time that had already taken hold of him. In any case, the view that science grows by bits and pieces was never to leave Dijksterhuis, even though we shall have occasion to note that, increasingly, it

came into conflict with his maturing insights into the nature of what he steadfastly refused to call the Scientific Revolution.

These insights found their expression in his 1950 book, which finally secured him a lasting place among the important historians of science in our
century when it appeared in 1961 in English translation under the title *The
Mechanization of the World Picture*. However, it may come as somewhat of a
surprise to the non-Dutch reader to learn how many of these mature views
appeared in his early masterpiece, *Val en worp*. One major difference is that
the earlier book, although an independent study in its own right, was at the
same time an ongoing confrontation with the Duhem thesis, a feature much
less conspicuous in his later book. But overall it is true to say that many major
themes that mark *The Mechanization of the World Picture* already present
themselves in *Val en worp* (subtitled *A Contribution to the History of Mechanics from Aristotle to Newton*). The preface presents *Val en worp* as

> a contribution to our knowledge of how the foundations of modern physical
> thought, such as formulated in Newton's Leges Motus, have come into being. It
> finds it origin in the conviction that, in order to acquire a clear understanding of
> the meaning and significance of these foundations, it is indispensable to follow
> as accurately as possible *the gradual development* as a result of which they ap-
> pear. [My italics][105]

These introductory words point to the past rather than to the future of history-
of-science writing, at the crossroad of which *Val en worp* is in fact to be situated in more than one sense. On their basis alone the reader might well be
misled into anticipating a book that would do little more than blend Mach's
preoccupation with the foundations of mechanics with a dash of Duhemian
continuity. However, three distinctively new features helped to turn *Val en
worp* into a great deal more than such an uninspiring mixture.

New features.   One of the new features is announced in the preface. It is the
methodical determination to stick throughout to the original sources, coupled
with an equally methodical distrust in translations of the originals. Obviously,
the thrust of the first point is directed against Mach, the thrust of the second
against Duhem, who, in the course of *Val en worp*, is indeed frequently found
guilty of distorting, mainly by means of far too free translations, the meaning
of the texts he used to underpin his far-reaching claims.

Two other, even more important characteristics of Dijksterhuis' mode of
history writing come to the fore as one ventures beyond the preface. One is
closely connected with the overall organization of the book. *Val en worp* is not
written as an exercise in tracing the origins of Newton's laws of motion. Rather,
it is an attempt to record the development over time of those particular ideas
in the history of mechanics that ultimately were incorporated into Newtonian
mechanics. In other words, rather than tracing 'primitive' forms of individual

statements down the centuries, we meet here with a systematic, truly contextual treatment of the history of a science. *Val en worp* aims at presenting an overview of theories of free fall and projectile motion, from their original status as fitting, although slightly irregular, elements of Aristotelian natural philosophy, through their becoming the highly problematical focus of increasingly unorthodox thought about nature, up to and including their final codification in Newton's laws. Dijksterhuis concentrated on free fall and projectile motion because in his view these were the two key phenomena around which the transition from Aristotelian natural philosophy to early modern mechanics had taken place. His account is *historical* in the sense of being devoted to an understanding of the pertinent theories in the context of thought during the times in which they were put forward—no philosophical preconceptions are allowed to lord it over the chronological account.

This is not to say that Dijksterhuis had no such philosophical preconceptions. Of course he had some, and they influenced his work, as could not be otherwise. We shall discuss them presently, but before we embark on such an exposition it should first be made absolutely clear that in *Val en worp* a decisive process of emancipation of history of science from philosophy of science has set in. No longer is the history made subservient to some other cause that lies outside the domain of the past and the ideal of its faithful reconstruction. Such a reconstruction requires certain guidelines, which may well be philosophical, as, naturally, they have often been in the historiography of science. But the point is that they are no longer employed to *prove* something. The exploration of "wie es eigentlich gewesen" (an ideal to which Dijksterhuis happily subscribed) has finally begun for the history of science.[106] The method employed is a patient and painstaking reconstruction of the thought of those past scientists whose works and other recorded thoughts were relevant to the historical topic under discussion. Together with Burtt's *The Metaphysical Foundations of Modern Physical Science* (to be dealt with in section 2.3.4), *Val en worp* seems indeed to be the earliest book in the historiography of early modern science that lends itself to be read, not only as a historical document, but also as if the author were still among us and one could engage throughout in lively debate with him over the tenability of his conclusions.

Mathematics in science. Dijksterhuis' views on scientific method, although originally clothed in inductivist language (which he changed when, in the thirties, he became familiar with Popper's early work), are of the hypothetico-deductive variety. As such they are, although remarkably early, not particularly exciting. What made them inform his history writing in a decisive way was the *mathematical* turn he gave to them. In Dijksterhuis' view, the only true scientific knowledge is knowledge of the quantitative relations between entities unknown as such. It is not given to us to perceive the essential nature of things; the most we can aim for is to determine, as precisely as possible and, hence, in mathematical language, their functional interrelationships. For Dijksterhuis, scientific method

> requires an exact description of the phenomena of nature in the only language which is subtle and fine enough to make such exactness possible: the language of mathematics. Not until our knowledge of the facts has thus been established does the method bear attempts at an explanation. It is true that this explanation by itself is again a description, but now made from a broader point of view—it is not, as unscientific thought often tends to believe, a penetration into the essence of things. But this does not diminish in any way the value and the justification of the method. Even though we have to acquiesce in the opaqueness of the essence, we sense the explanatory effect of the broader viewpoint. This is explanation in the literal sense of the phenomena becoming plainer—it does help to enlighten our thinking about nature.[107]

In roughly one-third of *Val en worp*, when dealing with Leonardo (whom, because of his unmathematical vagueness, he did not value nearly so highly as Duhem had), Dijksterhuis allowed this particular view of scientific knowledge and method to shine through his historical exposition for the first time. Here it is, once more in Dijksterhuis' own, inimitable, beautifully sculptured (yet not so easily translatable) style of writing:

> Moreover, the widespread idea that the roots of modern science are always there where we find more distinct attention being paid to the phenomena of nature and a higher estimation of experience as a source of knowledge than occurred in Scholasticism is generally erroneous. Precisely in the field of mechanics, which after all is the foundation of modern physics, the above opinion loses its force. We saw already how in the very depths of the most abstract scholastic dialectics the first kinematic results were reached and the outlines were sketched of the most important dynamical concepts. As our inquiry continues, we shall learn how, time and again, the same influences as were operative there continue to bring about every important advance in mechanics—abstraction made from the disturbances which, in the reality of nature, obscure the ideal phenomenon; the deepening and sharpening of those vague and intuitive conceptions which already from the earliest times had been called forth by experience; in one word: the introduction of natural science into the sphere of mathematics.[108]

From this point on in his book, Dijksterhuis' view of the mathematization of nature as the defining feature of early modern science began to inform his entire historical account. The quotation itself already shows that Dijksterhuis' assessment of the contribution of the Parisians to the birth of early modern science could not but be tinged by rather different colors than Duhem's. Duhem had been hunting for adumbrations of 17th-century views; Dijksterhuis cared for kernels of mathematization. To the extent that medieval science had produced these, it entered the history of mechanics on the positive side, and

Dijksterhuis never wavered in his conviction that, to some extent, it had, and therefore provided a vital link in the long and unbroken chain whose history he was writing.

And yet, it proved impossible for Dijksterhuis to maintain that the impulses toward mathematization had made themselves felt in as gradual a fashion as his continuity view required. Right after the above quotation he goes on to show that in the 16th century the Parisian impulse had run its course and that, as a result, "mechanics has got stuck" in the mud of Aristotelian orthodoxy—"a powerful impulse is needed to get it going again."[109] This new impulse is then identified as the publication, in the 16th century, of the works of a number of Greek mathematicians, in particular of "the most miraculous genius of the entire history of Greek mathematics . . . Archimedes":

> The rebirth of mechanics dates from the moment when his work begins to penetrate, and the incalculable impact which the mathematization of natural science has exerted upon the European mind is principally to be traced back to his influence as a point of departure.[110]

Dijksterhuis consistently opposed the use of the term 'revolution' in connection with the history of the sciences, and we shall presently delve somewhat deeper in his motives for banning the term from his private vocabulary. Yet for Archimedes he is apparently willing to make an exception, for the above passage continues: "The revolution brought about by Archimedes' works becoming known occurred at two distinct points."[111] First, his hydrostatic theory destroyed the fundamental Aristotelian distinction between light and heavy. In so doing, one of the cornerstones, and thus the inner consistency, of the Aristotelian building was seriously undercut. "Secondly, however, the geometric method [Archimedes] had employed for his work—never enough to be admired—proved that it was possible to apply to the study of nature an instrument which was infinitely finer and purer than the most subtle scholastic logic."[112] The passage then ends with the observation that the Archimedean influence did not fully reveal itself until the advent of Galileo.

My purpose in the above was not so much to catch Dijksterhuis in the act of inadvertently employing a term he was loath to use as a matter of principle. Rather, by highlighting his inconsistency over the fitness of the concept of revolution in connection with the origins of early modern science, I wish to draw attention to a deep ambiguity in Dijksterhuis' thought that marks virtually everything he wrote on the subject up to the time of his death, and that also goes a long way toward explaining some major peculiarities of the thesis he was to advance in *The Mechanization of the World Picture.*

The continuity of the history of science.   One important link between Dijksterhuis and earlier historians of mechanics (in particular, Mach) is the fundamental importance he attached to Newton's first two laws of motion: the principle

of inertia and the conception of a constant force producing uniform acceleration rather than motion *per se*. In his view, early modern mechanics may be
considered to have been born at the moment when the latter insight appears as
self-evident. In *Val en worp* he goes to great lengths to show that in this sense
Galileo never made the final step over the threshold, close as he had come with
his circular principle of inertia. It is not possible for the historian, Dijksterhuis
maintains, to pinpoint this crucial transition exactly. Galileo stuck to the Aristotelian conception of force, even though this is masked in his work by the fact
that he applied dynamic arguments only in those cases where 'Aristotelian'
and (to put the argument anachronistically) 'Newtonian' forces have the same
effect—namely, when the time periods over which they are applied are
equal.[113] One generation later, in Huygens' *Horologium oscillatorium,* the transition is a foregone fact. Dijksterhuis then goes on to observe that for mechanics in the second half of the 17th century the proposition 'a constant force
produces uniformly accelerated motion' can be *proven.*

> Thus here we are confronted with a curious phenomenon. We have here an idea
> which, in the Aristotelian conception of things, had simply been absurd. It had
> been so even for those thinkers who in physical matters resisted the Stagirite
> [Aristotle]. Now this very same idea—namely, *accelerated* motion as the effect
> of a *constant* force—is already at the end of the 17th century, when it has
> scarcely been born, conceived as being demonstrable by a very simple argument.
> Does this phenomenon point to a sudden new light, flashed upon the human
> mind by the redemption from the Aristotelian jail, as a result of which centuries-
> old riddles, once they had through major efforts been resolved, suddenly ap-
> peared as not being riddles at all, but only as problems which could be resolved
> through an obvious and quite elementary argument? Or are we confronted here,
> as has happened so often in the history of science, with a younger generation
> that has not taken part in the laborious development of which it reaps the fruit,
> and thus tends to underestimate the effort needed to arrive at a seemingly self-
> evident insight? In other words, is not the simplicity of the logical argument
> which makes that insight appear as almost a necessity of our thought (and thus
> giving rise to surprise at the failure of earlier generations to achieve it) no more
> than an illusion?[114]

The reader need not guess twice for which horn of his self-styled rhetorical
dilemma Dijksterhuis opts:

> The historical overview we gave of the theory of free fall already makes it clear
> that the first supposition is to be rejected completely. Throughout the develop-
> ment of every single part of the knowledge we find now to have matured in
> Huygens' work, we have been able to observe all the time such a large measure
> of continuity that, when we take a quick hindsight survey of the entire course of
> the development [from Aristotle, through the Parisians, Galileo and his disciples,
> up to and including Huygens], it is completely impossible to point, at whatever

place, toward an advance that suddenly created entirely new ideas and opened entirely new vistas.[115]

These are strong words, and unusually vehement for such a cautious balancer as Dijksterhuis was. But their very outspokenness enables us to identify the sources of his lifelong belief in the continuity of scientific advance. Continuity for Dijksterhuis means: Scientific discoveries do not fall like pennies from heaven; it is a sin against a proper sense of history to consider the past of science in that light.

Here, it seems to me, we see the lasting influence of Machian positivism at work. In Dijksterhuis' view, apparently, Mach's discontinuity conception and his positivist distortion of the historical record are fruits of the same tree. To take a genuinely historical view of things, Dijksterhuis seems to be saying in the above quotation, implies the rejection of discontinuity. History teaches us as a general rule that, in science, Schopenhauer's famous dictum fully applies: "Truth will be allotted only a brief feast of triumph in between its condemnation as paradoxical and its being dismissed as trivial." This is where our tendency to distortion by hindsight of the true historical picture comes from. We can guard against it only by showing again and again the overall continuity of the true development of science over time.

It is my firm conviction that, with the very best of intentions, Dijksterhuis had been mixing up two distinct issues. The confusion is due to his urge— born from a genuine sense of history—to find weapons against Mach, who was still the leading authority on the history of science in Dijksterhuis' early day.[116] Here we find confirmation of what we said earlier about the pernicious effect of positivism: In being combated it still managed to force its adversary into a position that was in fact at odds with his true views. It is perfectly possible to maintain that, yes, what is paradoxical for one generation may appear as self-evident to the next, whereas, no, this does not necessarily exclude the occurrence of sudden breaks in the development of science over time—new orientations of so radical a kind that indeed 'break' and 'revolution' are the only fitting labels if these words are to have any meaning at all. Curiously enough, Dijksterhuis, as we have already indicated, would in one sense be the last to disagree. Here is his summing up of Galileo's achievement:

Indeed, the Third Day of the *Discorsi* opens, as a completely original work, a new period in the history of science: the classical period of geometrical mechanics. . . . With Galileo, a new period begins in the history of man's thinking about nature.[117]

Enough, for now, of our paradox. We may perhaps sum it up in the following statement: In opposition to Mach and his fellow positivists, Dijksterhuis felt it to be his historian's duty to regard the advance of science as an essentially continuous affair, whereas his equally firmly held conviction that the mathematical treatment of natural phenomena constitutes the essence of scientific method almost forced him to conceive of the origins of early modern science as a decisive break with the past. The inner tension that resulted from this unresolved dilemma is palpable in Dijksterhuis' pioneering *Val en worp.* Yet in his magisterial *The Mechanization of the World Picture,* written a quarter-century later, it is present in no lesser degree, although hidden much more deeply under the surface.

Dijksterhuis' later views on the nature of early modern science. When, in 1950, *The Mechanization of the World Picture* appeared in the original Dutch, it was not quite the first comprehensive history of the Scientific Revolution. One minor reason is that Butterfield's *The Origins of Modern Science* (on which more in section 2.4.3.) had appeared one year earlier. But more important is that *The Mechanization of the World Picture* is not really set up as such a history. Rather, the book is organized as an answer (given in the format of history, to be sure) to a question about the meaning of a certain term: the very word 'mechanization' that appears in the title. In his introduction Dijksterhuis calls attention to the enormous impact which "the emergence and the development of the conception of the world usually called mechanical or mechanistic"[118] has exerted not only upon the methods proper to science itself but also on technology and, hence, industrialization, on philosophical views of man's place in the world, and on many branches of scholarship which originally were far removed from natural science. This is why, regardless of value judgments on these wider, cultural effects of mechanization, history of science is an important part of cultural history. Now, Dijksterhuis continues, it is not at all *a priori* clear what 'mechanization' means:

> In one sense the whole of this book is really an attempt to discover what it means to speak of a mechanistic world-picture. Are we in doing so thinking of the meaning 'tool' or 'machine', implied in the Greek term *mechane,* and are we thus considering the world—the mind of man included or not—as a machine? Or is the term used to express the view that natural events can be described by means of concepts, and handled by means of methods, which come from a branch of science that is called mechanics—the word here being used in quite a different sense from the original, namely, the science of motion? For the present this question has to be left pending.[119]

Dijksterhuis then goes on to argue that the answer can be found only by turning to the historical roots of what the term 'mechanization of the world-picture' stands for. This is so, he maintains, because a gradual and unbroken line connects its definitive formulation in Newton's *Principia* to antiquity:

> Classical physical science gradually arose from ancient science and, as has always happened in the development of thought about nature, it did not need to repudiate or ignore several of the results obtained in the preceding period, but could adopt them after suitable reshaping. In so doing it continued to move along

the paths already broadly indicated by the ancients. If we wish to understand its genesis, we shall therefore have to go back from effect to cause, from pupil to master, from follower to predecessor; so that we are bound in the end to arrive at Greek Antiquity—the true nursery of European culture in this domain as in so many others.[120]

And thus Dijksterhuis embarks on a 485-page-long exposition of the history of the physical sciences from the Greeks up to and including Newton (covering all but the life sciences), in order, ostensibly, to find out what 'mechanization' means.

For those of his readers who were already familiar with *Val en worp*, the answer Dijksterhuis eventually arrives at in his "Epilogue" hardly came as a surprise. In seven powerful pages Dijksterhuis disposes of several possible current meanings of the term (to which we come back shortly) in order, finally, to identify 'mechanization' with the discipline of mechanics in one particular mathematical sense:

> A complete characterization is not reached, nor is the true contrast between classical and medieval science made perfectly clear, until the definition of mechanics as the science of motion is made to include the feature of mathematical treatment as well. And this statement must be made even more precise. Not only is classical mechanics mathematical in the sense that is makes use of the tools of mathematics for the sake of convenience in abbreviating arguments which, if necessary, could also be expressed in everyday speech, but it is so in the far stricter sense that its basic concepts are mathematical concepts, that mechanics itself is a mathematics. In fact, it is only thus that the cardinal difference from medieval physics is revealed; as has been shown, at one stage of its development the latter, too, liked to make use of mathematical methods, but only a few of its representatives—Oresme, in particular—tried, as a precursor to a principle later to be formulated by Galileo, to use mathematics as the language of physics.[121]

Dijksterhuis then goes on to show that "not until the seventeenth century . . . [were] the content and scope of this principle gradually realized,"[122] which he amply illustrates by recalling examples from his preceding account. Thus we find that 'mechanization' really comes down to mathematization in the sense just defined. And in this sense, modern science, that is, the science that started with Planck and Einstein and Bohr, should be seen as a continuation of early modern science (which Dijksterhuis calls 'classical'):

> This is why the relation between classical and modern science is quite different from that between ancient and classical science. Whereas the latter had to repudiate ancient science on important points and had frequently to struggle in order to liberate itself from it, it lives on in modern science as a first approximation which is even sufficiently accurate for large domains of science.[123]

The truly "wide chasm," [124] then, is not to be found in the transition that took place at the turn of the 20th century but rather in the earlier transition around 1600:

> In those days an entirely new perspective on nature had to be conquered: 'substantial' thinking, which sought the essence of things, had to be exchanged for 'functional' thinking, which wishes to ascertain the behaviour of things in their interdependence; the treatment of natural phenomena in words had to be abandoned in favour of a mathematical formulation of the relations observed between them. In the present century, however, functional thinking with its essentially mathematical mode of expression has not only been maintained, but has even come fully to dominate science. [125]

It is hard indeed to leave this "Epilogue" with an impression other than that the transition from Aristotelian to early modern science constituted the one and only fundamental break in the history of science. Once again we see how Dijksterhuis' conviction of the mathematization of nature being the true core of the 'mechanization of the world-picture' easily got the upper hand over the overall continuity he had urged upon his readers at the beginning of the book. Earlier signs of this can be spotted throughout the *Mechanization*: Whenever Dijksterhuis discusses the achievement of men such as Galileo or Kepler, he points out how fundamentally different their approaches and results were in many, though not of course all, respects from their predecessors. True, he displays their many links with the past (which is particularly rewarding in the case of Galileo) and always aims at striking a careful balance between what is old and what is new in their works; yet the overall impression gained by the reader is quite clearly aimed to prepare him or her for the final conclusion of the book.

The dating of the Scientific Revolution. At what precise moment in time did the true break occur in Dijksterhuis' view? The answer is clear enough: The crucial date is 1543, the year of publication of Copernicus' *De revolutionibus orbium coelestium*. Already in 1924, in *Val en worp*, Dijksterhuis had liberated himself from the Duhem thesis in its extreme form—medieval science had to some extent helped prepare the new mechanics and astronomy but could not be said to have started it. The "new era" [126] starts with Copernicus. Why Copernicus? Dijksterhuis gives no clear answer to that question; he simply imposes it on his reader. In fact, his account seems to emphasize the grounds for questioning his choice. The author goes out of his way to show in how many respects Copernicus' own work, *as distinguished from its lasting effects,* still belonged to the tradition of Greek mathematical astronomy, with its fictional wheels and epicycles and its ideal of 'saving the appearances' regardless of the true state of affairs in the heavens. Nowhere in the treatment of Copernicus do we come across such enumerations of new features as we find when Dijksterhuis discusses Galileo and Kepler, where he focuses in particular on their independence from authority and their mathematical handling of the phenomena of nature.[127] In short, despite the author's protestations of continuity and despite organizing his book so that the period of early modern science is made to begin with Copernicus, the attentive reader leaves the book with the impression that in the past of science we find one unique break, which was occasioned in the main by the Archimedean Galileo and by Kepler, the Platonist and Pythagorean—the first two scientists truly to mathematize nature.[128]

Mathematization and mechanization. Let us ask now how Dijksterhuis managed to reconcile his identification of mechanization and mathematization with other conceptions of 'mechanization', 'mechanical', and 'mechanistic'. First, we quickly take stock of the alternative meanings of these terms that Dijksterhuis explicitly rejected in his "Epilogue." Next we compare Dijksterhuis' usage of the terms with 'mechanicism' in the sense of the corpuscularian world-view devised by Beeckman, Gassendi, Descartes, Boyle, and others.

The first possible meaning to be attached to the word 'mechanicism' is the well-known 'machine' analogy which goes back to the Greek word *mechane*. Dijksterhuis rejects this interpretation, because the picture of the universe as a giant machine presupposes a Maker and a purpose, attributes that do not belong to early modern science proper, even though the image of the Divine clockmaker was occasionally used by 17th-century scientists to make their views more palatable to a Christian audience (and to themselves as well). Nor will 'mechanization' do in the cognate sense of 'picturable, machinelike, by means of a graphic representation—something one can visually imagine.' Now this 'picturability' is the primary property extolled by such contemporary mechanicists as Huygens and Leibniz, and, remarkably, this is precisely why Dijksterhuis would have none of it: If these two scientists could reject Newtonian forces as unmechanical, as indeed they did, and if it is also true that Newton's laws embody the essence of what we mean when we talk about the 'mechanization of the world picture', then the criterion of picturability is wrong. After all, not ever the most skillful instrument builder is capable of constructing machines which represent to us the movement of material bodies as resulting from their mutual gravitation; yet gravitation is commonly regarded as being of the essence of mechanicism.

One final possible meaning of the term 'mechanicism' comes in for quick dismissal. It is to take 'mechanistic' in a sense contrary to 'animistic,' or 'organic.' In this sense of the word the lifeless aspect of the machine is stressed. The consequence of such a view, Dijksterhuis maintains, is that in early modern science all change would have to be accounted for by means of external influences, whereas Aristotelian physics would then be taken as explaining change by means of some inner principle. But in practice this criterion does not work. For example, inertia can easily be accounted for as an internal principle of motion. Also, "the definition of mechanistic as non-animistic has far too meagre a content to pinpoint the very pronounced character of classical science."[129] And after this massacre the path has finally been cleared to usher in the true, mathematical meaning of this versatile term, 'mechanicism.'

Unfortunately, the path was now also cleared for multiplying the confusion the entire "Epilogue" was intended to eradicate for good.[130] It is one thing to maintain that the mathematization of nature was the key feature that distinguished early modern science from preceding natural philosophies. But it is another thing to go ahead from there and tell the world that the true content of the term 'mechanization' is really 'mathematization'. Why not leave the term 'mechanization' to its chief contemporary usage and underscore only that this is not where the essence of the Scientific Revolution is to be sought? This is the road traveled by other historians of science. It is a far more productive road, as we shall see illustrated in particular when, in section 2.4.5., we come to discuss Richard Westfall's conception of the Scientific Revolution, for along such lines one can satisfy Dijksterhuis' major objection by showing that mechanicism in the sense of the 'mechanical philosophy' itself underwent a drastic change in the hands of Newton—thus replacing a debate over the meaning of words by a debate over the significance of things.

Naturally, Dijksterhuis did not ignore that 17th-century science had witnessed something called the 'mechanical philosophy', of which, in his view, Descartes was the main originator, Boyle an important contributor, and Huygens the culmination point. In order to understand why, according to Dijksterhuis, the corpuscularian view of things was only a secondary characteristic of early modern science compared to the mathematization of nature, we must recall Dijksterhuis' views on the essence, and the limits, of scientific knowledge. The very same view he had expressed in *Val en worp* is to be found in *The Mechanization of the World Picture*. But the earlier book contains one thing that is left out of the later one: an assessment of the corpuscularian worldview in the light of Dijksterhuis' mathematical epistemology. In *Val en worp* Dijksterhuis contrasts Galileo's achievement in the area of free fall and projectile motion with the well-known contempt in which his results were held by Descartes, since after all Galileo "had built without foundation."[131] Here, Dijksterhuis says, we find one stage in

> a battle repeated so often in history: the battle between the modesty of the mathematical-physical method, which, after careful study, wishes to describe the phenomena of nature in ever more embracing schemata by means of the exact language of mathematics, and the arrogance of philosophical thought, which, in one leap of genius, wants to embrace the entire world and aims at once for an insight into the essence of things from the conviction that this is the only road toward valuable knowledge of the phenomena. One can be convinced of the superiority of the former method, without therefore allowing oneself to be robbed of the

charm exerted by the execution of the latter in the hands of a Descartes or a Schopenhauer.[132]

This is why "mechanical causality," as Dijksterhuis calls it elsewhere in *Val en worp*,[133] is subordinate to the mathematical approach to nature. It also accounts for why, in general, Dijksterhuis treated Descartes as little representative of the scientific upheaval of the 17th century. Even though Descartes rejected the content of Aristotelianism, he was no less of a system-builder than Aristotle had been, and to the extent that system-building is what he sought his strength in, he was doomed to remain outside the mainstream of the actual renewal of science as effected by his contemporaries and immediate successors. And this feature, Dijksterhuis maintained, was considerably strengthened by the fact that Descartes' system was much less empirical than Aristotle's. This is not to say that Dijksterhuis believed in the key importance for science of the empirical observation of nature; we have already seen him assert the contrary earlier in this section. But it does mean that, in Dijksterhuis' view, Descartes failed to establish the one door through which the 'facts' of nature really enter: experiment.

The place of experiment in the history of science.   The role of experiment, too, follows with inexorable logic from Dijksterhuis' views on matters scientific—his mathematical version of hypothetico-deductivism.[134] He does not believe in deductivism run rampant, which is what happened to Descartes. But the heuristic value to be assigned to experiment is also low. True, the history of science shows that experiment—that is, in true Kantian fashion, the active interrogation of nature—may serve as a direct source of scientific knowledge. But far more important is the conception that gave rise to such active interrogation in the first place. And here we come once more across our good old friend "mathematical formulation as a means of description," which now also serves to yield "mathematical deduction as the guiding principle in the search for new phenomena, to be checked by experiment."[135] Thus the main function of experiments is to verify whether those phenomena whose existence has been predicted by mathematical deduction do actually take place in accordance with the prediction.

   One corollary of this view is that Dijksterhuis was not at all favorably disposed toward taking at face value Galileo's accounts of his experiments insofar as Galileo presented them as heuristic—one other respect in which Dijksterhuis distanced himself from Mach's brand of empiricism. One of the key elements of Galileo's greatness, in Dijksterhuis' view, was precisely that

he did not experiment in order to find a law of nature, but in order to verify *a posteriori* a relation he had deduced by mathematical reasoning from suppositions that appeared more or less self-evident.[136]

71

In linking so closely his appreciation for Galileo with his own view of what constitutes science, Dijksterhuis obviously put his view at risk from the moment on when documentary or experimental evidence might show that Galileo had perhaps been far more of a heuristic experimenter than Dijksterhuis was willing to allow. This issue will be pursued in further depth in connection with the fate of Koyré's conception of the Scientific Revolution, which in this respect was almost indistinguishable from Dijksterhuis'.

Chemistry and the life sciences.   *Val en worp* was about mechanics only, whereas *The Mechanization of the World Picture* aimed to cover all physical sciences in the broad sense of the term. The addition of astronomy was of course most welcome, since in particular the contributions of Kepler and of Newton in Book III of the *Principia* were so much grist for Dijksterhuis' mathematical mill. But chemistry was a different matter. The chemists' contribution to the renewal of science in the 17th century is partly subsumed by Dijksterhuis in a treatment of Boyle's corpuscular theories, and for the rest it is confined in the main to theories on the structure of matter, in particular, on the issue of the extent to which 'elements' continue to exist in a compound. In other words, 'philosophical' chemistry takes heavy precedence, in Dijksterhuis' treatment, over practical chemistry. Nor do we hear much about such nonmathematical branches of physics as electricity, magnetism, and the like. Astronomy, mechanics, and some optics are the subjects around which the story is made to turn and return. No productive role was left for those sciences which did not contribute directly to the mathematization of nature. They were there, and their development over time could of course be described in a sympathetic way, but they did not take part in the main action. The same state of affairs that had made William Whewell abstain from assigning to mathematization the key role in bringing about the new science forced Dijksterhuis into a relative neglect of those branches of science which did not fit into his overall conception. The dilemma was certainly not one that Dijksterhuis alone had to confront. The more one wishes to do justice to all the sciences of the 17th century, the more one tends to lose a clear view of what it was that distinguished early modern science from its predecessors. In a slightly modified form, this dilemma continues to plague present-day historians of science.

Conclusion.   There is little left for conclusion that has not already been said in the foregoing, but two cautionary remarks are perhaps in order. In highlighting certain conspicuous features in Dijksterhuis' mode of history writing, it has not been my intention to present my late countryman as their single-handed inventor. To the extent that such key topics are to be found in *Val en worp*, Dijksterhuis may indeed be regarded as their originator, often together with, though entirely independent of, E. A. Burtt. Here I am thinking in particular of Dijksterhuis' methodical criticism of sources, his contextual approach, his curious brand of 'continuism', his elevation of the mathematization of nature into the sole key to early modern science and his identification of the spread of Archimedes' works as the principal new element in the origin of early modern mechanics.[137] But other things with which Dijksterhuis grappled in constructing the grandiose building of *The Mechanization of the World Picture* were, and are, no less of a problem for quite a few fellow historians of science, notably, the vexed problem of what to do with chemistry and the life sciences, the proper place of Descartes in the story, and the problem of how to assign a suitable starting date to the Scientific Revolution.

In the foregoing I have repeatedly alluded to certain inner tensions to be found in the works of both Duhem and Dijksterhuis, mainly in connection with their respective views on the continuity of the advance of science. I shall continue to point out such tensions in discussing certain later historians of science. At this place I wish to make it clear once again that I do not mean this in a pejorative sense at all. I do not think that authors should aim for inconsistencies. I also do not think that perfect consistency should be an author's highest ideal—to the contrary, few books are quite so boring as the fully consistent ones. One key element, in my view, that makes for a truly living book is precisely the inner tension that comes from a beautiful and high-minded, overriding thesis of which the author has never been able fully and completely and unrestrictedly to convince himself.

2.3.3. *"From the World of the 'More-or-Less' to the Universe of Precision":* *Koyré's Conception of the Scientific Revolution*

The previous section was entitled Dijksterhuis and the Mathematization of Nature, and the present one might just as well have been called Koyré and the Mathematization of Nature. For the similarities in the views of these twin brothers in the historiography of science are nothing but striking. Yet, in order to call attention to the equally striking fact that they were twins of the nonidentical variety, I opted rather for a title that brings out something of their differences as well. (As a matter of fact, the next section, which is about their step-twin-brother, might well receive the exciting title 'Burtt and the mathematization of nature.')

One similarity between Dijksterhuis and Koyré is that they were very close contemporaries. They were born in the same year, 1892, and died within one year of each other (Koyré in 1964, Dijksterhuis in 1965). One major difference is that, while the Dutchman spent by far the largest part of his active life as a quiet, provincial schoolteacher who was nowhere quite so happy as at the various desks in his own study, Koyré, who was born in Russia, lived the life of a true cosmopolitan. He served on a voluntary basis in both world wars, studied in Germany, settled in France, spent several years in the Middle East, and split up his teaching between France and the United States after World

War II while learning to write in English when he sensed that that was the scholars' language of the future. In short, Koyré was as brilliant and flamboyant and outgoing as Dijksterhuis was brilliant and introverted and fundamentally a loner.[138]

Another similarity is that both were to stick for the rest of their lives to the conception of the origins of early modern science which they unfolded in their first published writings on the subject. Another difference is that, over the years, Koyré appeared a little more flexible in allowing some slight modifications, and some more-than-slight extensions, to merge with his original views. One more similarity is that both had been educated as mathematicians (Koyré had also, and in fact primarily, been raised as a philosopher) and that they shared largely, though not fully, the same conception of scientific method. One more difference is that Dijksterhuis habitually expressed himself in the format of book-length arguments, whereas Koyré's mind was predisposed to the essay—if these displayed sufficient mutual coherence, he put them in covers; if not, so did his posthumous editors.[139]

One final similarity is that both were deeply convinced of the mathematization of nature as the key to early modern science. One final difference is that Dijksterhuis thought he could combine this view with an equally strongly held belief in the continuity of scientific advance over time, whereas right from the start Koyré proclaimed the revolutionary nature of early modern science. This start was taken in 1940, with the publication of three essays collected together under the title *Études Galiléennes* (Galileo Studies).

Revolution and revolutions. The framing of the term 'the Scientific Revolution' in more or less its current meaning was largely the work of Koyré. Throughout his writings, Koyré continued to distinguish several revolutions in science: the Greek creation of the idea of the Cosmos; the revolution of Galileo and Descartes, which as he went on he tended more and more to extend over time up to and including Newton's achievement; an unspecified 'revolution' in the 19th century; and, finally, the 20th-century revolution of Einstein and Bohr. Thus, somewhat paradoxically, we find that Dijksterhuis, the great upholder of overall continuity, in reality distinguished only one sharp break in the history of science—the one carried out principally by Galileo and Kepler—whereas for Koyré, who focused so emphatically on discontinuity as a feature of scientific advance, the events of the 17th century constituted only one revolutionary stage in the history of science beside at least three others. Yet, because as a historian of science Koyré was to write exclusively about 16th- and 17th-century science, the unique character of this particular period in the past of science stands out very outspokenly in his life's work.

Koyré, then, believed in revolutions in science. His principal argument for this can be rendered in one brief quotation: "A well-prepared revolution is nevertheless a revolution."[140] Which, given Koyré's performance as a truly contextual historian of science, may be taken to mean: True, as a historian one has a duty to find out what a given great scientist from the past owed to certain predecessors; to reconstruct as faithfully as one can that scientist's own thought in the context of the thought of the time; to regard errors, dead ends, and the like as at least equally instructive as major exploits; in other words, to over-come the positivist approach to history; and yet, there is nothing in this cata-logue of sound methodical prescriptions that enforces upon us a picture of the growth of science as gradual. Rather than let his strong sense of discontinuity creep in through the backdoor, as Dijksterhuis had in fact done, Koyré in one sentence cut right through the Gordian knot his twin brother had managed to entangle himself in for his entire scholarly life.

The 'intellectual mutation'.   Why was it so important for Koyré to insist that what happened in science around 1600 indeed constituted a revolution? The reason is that, in his view, the transition signaled a decisive change in world-view all by itself, "of which modern physical science . . . was at once the ex-pression and the fruit."[141] Earlier concepts and theories lost their meaning be-cause they no longer made sense in the context of the new world-view; the new concepts and theories at once began to look seductively self-evident for the same reason.

Thus the birth of early modern science was not just the emergence of a number of new statements about nature; not even such fundamental proposi-tions as the principle of inertia or Newton's second law marked the transition in their own right. Their very discovery and subsequent adoption could only be accounted for in the framework of this larger transition, which Koyré described as a fundamentally new overall conception of *motion.* And this transition, in its turn, could only take place in the even wider framework of a new conception of the universe at large. Movement ceased to be regarded as a goal-oriented *process* in our heterogeneous, finite Cosmos; largely through the work of Galileo and Descartes, movement henceforth came to be regarded as a value-neutral *state* of bodies on their way through the homogeneous infinity of Euclidean, geometrized space. This sentence sums up the core of Koyré's views; this is why he regarded the origins of early modern science as no less than an 'intellectual mutation'; this is the basic conception to which all his other views—on the mathematization of nature, on the significance of the impetus theory, on the role of experiment in Galileo's work, on the relevance of philosophy for the history of science, on his fellow historians of science, and so on and so forth—were ultimately subsidiary. In one sense, after the reader has made himself thoroughly famil-iar with this core sentence and has repeated it as frequently as Koyré was to do in his own writings (I counted its scarcely varied occurrence at least a dozen times), I could end the section on Koyré here and leave it to the reader to think up the proper Koyréan answers to all these and a great number of related sticky issues such as we have seen come to the fore when discussing Dijksterhuis' contribution.

However, I shall not let the reader down. In the following I shall first describe how, in his *Études Galiléennes*, Koyré worked out his core conception. Next, I discuss in what ways Koyré was later to repeat, elaborate, modify, and extend the consequences of his core view, as he drew on this inexhaustible source both for the purpose of polemics and to make observations on a great many other relevant historical phenomena.

The 'Galileo Studies'.   The first essay of the *Études Galiléennes* is "À l'aube de la science classique" (At the Dawn of Classical Science) originally published in 1935/36. Its main purpose was to demonstrate two historical propositions: (1) The history of medieval and Renaissance science is to be divided into three stages, which correspond to three distinct modes of thought, even though these do not always obey strict chronological order. These stages are constituted by the Aristotelian, the impetus, and the Archimedean modes of thought, respectively. (2) These three stages were traversed in condensed form yet in due order by Galileo in the period before his physical thought reached maturity.

Two consequences of this categorization *cum* periodization are important for our purpose. We note in the first place that right from the start Koyré ties up his views on the upheaval in the science of the 17th century with the work of one of its main propounders: Galileo. Even though an unbridgeable chasm separates Koyré's conception of science from Mach's, he nevertheless shares, on a far more sophisticated and historically responsible level to be sure, not only the latter's discontinuity view but also his focus on Galileo as the central figure. Mach's discontinuity view, however, had been 'absolute', whereas Koyré's full awareness that the Scientific Revolution had been "prepared" turns his conception into one of 'relative discontinuity.'

It also followed for Koyré that impetus physics was relevant to the advance of science mainly in its *failure* to bring about the revolution. Galileo's early thought, as embodied in *De motu,* demonstrated the ultimate futility of the effort made there to treat impetus physics in the Archimedean way—impetus physics had to be given up before a genuine mathematization of nature could become feasible. And this process of mathematization was right at the heart of the overall upheaval in science. It provided the framework for the ideal, abstract world of free fall in the void, of perfectly smooth planes, of completely frictionless motion, in which Galileo had already begun to delve in *De motu,* and which was to remain his hallmark for the rest of his life.

> One senses: motion begins to emancipate itself; the Cosmos is breaking up; space is being geometrized.[142]

76

Throughout the two other essays of the *Études Galiléennes* Koyré traced the further unfolding of this process through its mature stages in the work of Galileo and Descartes. In "La loi de la chute des corps: Descartes et Galilée" (The Law of Falling Bodies: Descartes and Galileo) Koyré took as his point of departure the mistake these two men had made in their first attempts at finding out how speed, time, and distance are connected in free fall. This mistake was particularly instructive because, without knowing each other, both men had erred in exactly the same way, thus highlighting the fact that the proportionality of speed and time in free fall is not nearly so self-evident as it may seem in retrospect. This is so because the law is embedded in a whole system of presuppositions regarding the nature of space, motion, and action of bodies in general that are interdependent in a quite complex way. Their gradual disentanglement results from asking a new kind of question:

> Nature responds only to questions posed in mathematical language, because nature is the domain of measure and order.[143]

And this is the only way for experience, thus guided by mathematics, to penetrate into the essence of nature.

In the third and final essay, "Galilée et la loi d'inertie" (Galileo and the Law of Inertia), Koyré really means business, for the principle of inertia is the central event in the revolution he is describing precisely because it stands for and, as it were, embodies the very geometrization of space that was at the heart of the revolution. This is so for two different, although interrelated, reasons. On the one hand, the principle served Galileo, in the *Dialogo*, as the principal weapon to defend Copernicanism, that is, the one doctrine that did more than any other to usher in the destruction of the Greek Cosmos. On the other, the principle of inertia, in stating the persistence of rectilinear motion in empty space, asserts something that is not there in actual reality: It explains what is, by means of what is not, and cannot even be.[144] This accomplishment of the principle could be achieved only through the conversion of the empirical into a mathematical world, whereas at the same time this conversion could not be a complete one, because otherwise no link with reality would remain at all, and thus nothing would have been explained.

The conversion in question involves the exclusion of purpose from the world of motion, the homogenization of space and abstraction made from anything else but bodies moving in inertial motion through infinite space, which is the space of Euclidean geometry. The person to prepare this conversion to a very large extent without actually completing it was Galileo. The men to state the principle fully without actually seeing its larger implications were his disciples, Torricelli among others. The person to execute the full conversion was Descartes, who was the first to think up, in a systematic and thorough fashion, a world consisting of nothing but bodies moving through infinite space in accordance with the principle of inertia. Together these two men, Galileo and

Descartes, thus carried out the Scientific Revolution of the 17th century. It is true that Galileo did not travel the road down to the end. Since he never fully made the transition from the finite, ordered Cosmos to the homogeneous, infinite Universe, he never arrived at the formulation of the principle of inertia in such a form that it equipped a material body for infinity. Yet the traditional view of Galileo as the father of early modern science is right, because he was the first scientist to realize the idea of a mathematized physics. In so doing, he was heavily influenced by the thought of Plato, as is testified principally by the Platonic leanings of the *Dialogo*. In the classical antithesis regarding the role of mathematics in physics, Galileo sided with Platonian 'mathematicism' against Aristotelian empiricism with its claim that real physical bodies have no geometrical forms. However, the response Galileo gave to this traditional, and by itself quite justified, objection was wholly new, for, unlike Plato, he did not believe that mathematical forms find expression in the Idea only. Rather, Galileo denied the premise common to both Greek philosophers;[145] in Koyré's words:

> The geometrical form can be realized by matter. More than that: it always is. . . .
> The geometrical form is homogeneous with matter, and this is why geometrical
> laws have a real significance, and dominate physics.[146]

This is what enabled Galileo to go further than Plato in aiming for a true mathematization of nature. However, Koyré insists, it was not enough for Galileo to *say* so. Galileo's greatest merit was his ability to act upon his own insight that the only way to prove that physics can be treated in a mathematical fashion is to go ahead and actually *do* it. Koyré then concludes as follows:

> Motion obeys a mathematical law. Time and space are linked by the law of number. Galileo's discovery transforms the failure of Platonism into victory. His science is the vindication [*revanche*] of Plato.[147]

This vindication, he finally observes, was still incomplete: Descartes is to draw the final conclusions. Even though, unlike Galileo, Descartes did not regard himself as a Platonist, both share the same idea of science as the reduction of physics to mathematics. But this new Platonism is far removed from the original, because space has now lost the cosmic value it possessed for Plato. The new science has won. "But never has a victory been bought at a higher price."[148]

An interim assessment. This, then, was Koyré's conception of what he called 'the Scientific Revolution of the 17th century'. Note that the scope of the term in its usage by Koyré is confined here to the activity of no one else but Galileo and Descartes. Note also the many similarities, and the equally numerous differences, with the interpretation put forward by Dijksterhuis fifteen years before in *Val en worp*. Rather than spelling these out here in detail, I wish to call attention to what seems to me the deepest distinguishing feature in the thought of these two early, truly contextual historians of science—the underlying distinction from which both their similarities and their differences can be seen to follow. Dijksterhuis held mathematics to be a language by means of which one can describe reality without actually penetrating into its essence; for Koyré physics was truly the incarnation of mathematics, which is why the language of mathematics does indeed express the essence of reality.

One other striking feature is the relentless inner logic of the argument, which in the case of both Dijksterhuis and Koyré betrays the mathematician's mind at work. Trains of thought so compulsively carried on have their assets as well as their liabilities. The main liability was to reveal itself shortly after they died: The inner coherence of their thought-building turned it into a fortress of great strength and resistance, yet as soon as an adversary had managed to turn one stone, however small, upside down, the very same inner coherence was quite likely to make the entire building collapse. But in the shorter run the fortresslike structure of the argument was nothing but an asset. For Koyré, at any rate, it served as an ideal starting point to drive his message home, to elaborate it, to extend it somewhat, to connect it to other aspects of the history of 17th-century science, and to use it as a home base when out to beat opponents. We shall now trace in rough outline these varied uses Koyré made of his fortress.

Repetition.   Let there be no mistake about it: Koyré had a most powerful message. It had all the strengths of a unitary account in which, through the magnetic action of the core conception, a huge number of hitherto unrelated historical facts were now arranged, like so many iron filings, along neat lines of force. (Such, I think, may well be the ultimate in what historical analysis is capable of achieving.)

But Koyré also knew how to sell his message. This was the more admirable since his *Études Galiléennes*, although dated 1939, actually came out in 1940—the year when France was invaded by Nazi Germany, and most people, including Koyré himself, had other concerns than the interpretation of the thought of certain long-defunct scientists. Koyré had written his book while in Cairo (not omitting, on arrival, to take his personal copies of the twenty volumes of the Edizione Nazionale of Galileo's works through customs). When war broke out, he returned at once to Paris, only to find that there was nothing useful for him to do but go back to Cairo. Having managed to arrive there, he made contact with the one remaining center of French resistance, General de Gaulle. The general thought that it might be useful to enhance his tenuous credentials with his allies *in spe* by sending an intellectual to the United States. Whether Koyré's arrival in the United States did much to endear President Roosevelt to the cause of the Free French is perhaps to be doubted. It is certain, however, that it did a lot of good to the cause of the historiography of science, which, *as a profession,* is often and rightly said to have been born from the marriage, in the early forties, of American opportunity (prepared above all by the émigré Belgian Sarton) with the thought of Koyré. But more on this in later sections; for the moment I wish to return to the manner in which, once having arrived in the United States, Koyré worked to get the message from his *Études Galiléennes* across. His basic techniques were condensation, translation, and repetition. Substantial chunks of the *Études Galiléennes* appeared as articles in English-language journals, the most famous being "Galileo and Plato," which appeared in 1943 in the *Journal of the History of Ideas* and sums up Part I, and "Galileo and the Scientific Revolution of the XVIIth Century," which appeared in the same year in the *Philosophical Review* and is in fact little more than a translated condensation of Part III of the *Études Galiléennes.* However, Koyré never tired of putting at the head of his numerous articles, as well as of his later books, a most emphatic restatement of his core view. Here it is, in its mature form, as taken from his last book, the *Newtonian Studies* (1965):

> I shall therefore characterize this revolution by two closely connected and even complementary features: (a) the destruction of the cosmos, and therefore the disappearance from science—at least in principle, if not always in fact—of all considerations based on this concept, and (b) the geometrization of space, that is, the substitution of the homogeneous and abstract—however now considered as real—dimension space of the Euclidean geometry for the concrete and differentiated place-continuum of pre-Galilean physics and astronomy.
>
> As a matter of fact, this characterization is very nearly equivalent to the mathematization (geometrization) of nature and therefore the mathematization (geometrization) of science.
>
> The disappearance—or destruction—of the cosmos means that the world of science, the real world, is no more seen, or conceived, as a finite and hierarchically ordered, therefore qualitatively and ontologically differentiated, whole, but as an open, indefinite, and even infinite universe, united not by its immanent structure but only by the identity of its fundamental contents and laws; a universe in which, in contradistinction to the traditional conception with its separation and opposition of the two worlds of becoming and being, that is, of the heavens and the earth, all its components appear as placed on the same ontological level; a universe in which the *physica coelestis* and *physica terrestris* are identified and unified, in which astronomy and physics become interdependent and united because of their common subjection to geometry.
>
> This, in turn, implies the disappearance—or the violent expulsion—from scientific thought of all considerations based on value, perfection, harmony, meaning, and aim, because these concepts, from now on *merely subjective,* cannot have a place in the new ontology.[149]

Elaboration.   A careful look at this lengthy quotation reveals that in a quarter-century a few things had been added by Koyré to the original message. This was mainly the result of the fact that the message itself not only was meant to be spread among the heathen (in this case, the positivist historians of science) but also functioned as a research program for its author. Implicit in the message of the *Études Galiléennes* are in-depth inquiries to be made into the actual process of the 'destruction of the cosmos'. These Koyré carried out in his two books *La révolution astronomique* (here Copernicus, Kepler, and Borelli are discussed mainly in relation to their dynamical views) and *From the Closed World to the Infinite Universe*. Also implicit in the program was the study of Newton, whose work signals the final unification of terrestrial and heavenly physics; this was the chief occupation of Koyré's final years. As one result, his concept of the Scientific Revolution drew ever wider circles. In 1940, it covered only the performance of Galileo and Descartes in geometrizing nature. Gradually, the concept began to encompass Copernicus' work as well, with whose *De revolutionibus*, in Koyré's subsequent opinion, the revolution might henceforward be considered to have started.[150] And when, in 1950, he embarked on his Newtonian studies, we find that the time scope of the concept has now been extended to encompass all that historians have since been wont to call (in the wake of Butterfield and Hall besides Koyré himself) the Scientific Revolution of the 17th century:

> ... the deepest meaning and aim of Newtonianism, or rather, of the whole scientific revolution of the seventeenth century, of which Newton is the heir and the highest expression ... [151]

One other elaboration of a thought already present in the *Études Galiléennes* concerns the role of *experiment* in Galileo's work. It follows in a straightforward manner from Koyré's mathematical view of natural science that there was little room here for experience to come in all by itself. To begin with, reality itself is mathematical. This is not to say that no other link between reality and our mathematical deductions is necessary; Descartes, so Koyré was quite ready to admit, had turned his mathematical deductivism into a "géométrisation à outrance," a "geometrization run rampant." In such a case, where geometrization "dissolves the real entity into the geometrical," "reality takes its revenge."[152] Thus a certain balance between the mathematically abstract and the world of experience is indeed necessary, and Galileo was the man to strike that balance.

This was not, however, to be done by introducing experience as such. Koyré never tired of emphasizing that naive, daily experience of the Aristotelian kind cannot lead to knowledge of a type that may count as science proper in the new universe of precision. Central in the process of linking the ideal to the real is the active interrogation of nature along guiding lines supplied by previous mathematical deduction.[153] The confirmation of reality comes in through experimental testing only. There is no heuristic value in experiment, either—"good physics is done *a priori*."[154] And Koyré then goes on to show that what Galileo himself had to say on the topic in the *Dialogo* and the *Discorsi* fully fits in with this view. All of which does not imply that Koyré was not interested in 17th-century experiments—to the contrary, he devoted a number of fascinating articles to detailed analyses of some famous cases. But he invariably did so in order to underscore the pitiful state of the actual means 17th-century scientists were forced to employ compared to the infinitely subtle, elegant, and refined nature of their abstract, mathematical tools. As a natural result, they were often compelled to confine the introduction of experience to experiments of a particular type in which Galileo had been the first to excel: the imaginary, or thought-experiment.[155]

Extension. We noted that, taken as a research program, Koyré's core conception pointed in three different, although interrelated, directions. One of these was terrestrial physics; this he handled in the *Études Galiléennes*. Another was astronomy in the sense of the science which formed the permanent background to the blowing-up into infinity of the Greek Cosmos. And the third was the ultimate synthesis of these two domains—which had already been linked by Galileo through the principle of inertia—into one, unified dynamics. We have also noticed how the more intricate study of these two last areas of 17th-century science led to some adaptation of Koyré's conception of the Scientific Revolution. Another thing he now had to confront was Kepler's place in that revolution. Galileo and Descartes had been Koyré's principal characters in 1940—now what to do with Kepler?

The problem with Kepler was that, despite his marvelous performance in the emergence of early modern science, he scored badly on all but one of the criteria implicit in Koyré's core conception. Kepler regarded the universe as finite. As a Copernican familiar with the parallax problem he was forced to assign a considerable measure to the universe, yet he was very emphatic in his refusal to go any further. And Kepler's dynamics, too, never ceased to be of the Aristotelian variety; although the very author of the expression 'inertia', for Kepler this meant the tendency of a body to return to rest unless there was a force to keep it going. The result of this was that Kepler's original attempt to unify terrestrial and heavenly physics, although instrumental in the establishment of this three laws, ultimately got him nowhere near the envisaged goal.

Thus, although Koyré greatly admired Kepler's achievement, and in particular his mathematical conception of nature, he had little use for him in his overall schema of things, for reasons remarkably different from those of many of his co-professionals at the time, who tended to praise Kepler for his three laws but condemn him for his 'mystical' leanings—after all, the no less 'mystical' Giordano Bruno fares far better under Koyré's treatment because he indeed proclaimed the infinity of the universe:

One stands confused when confronted with the boldness, and the radicalism, of Bruno's thought. Bruno wrought a transformation—a veritable revolution—of the traditional image of the world and of physical reality.[156]

But no such revolution for Johannes Kepler. And Koyré's historiographical habits were such that, once having assigned a scientist his fitting place as derived from the core conception, he was willing to go to considerable lengths in interpreting the data accordingly. Take, for example, Kepler's own celebrated claim to fame when, in rejecting his *hypothesis vicaria* because of a divergence of 8 minutes of arc between the predicted value and the actual value as found in Tycho's observations, he said that this was what had ultimately enabled him to build a new astronomy. Most historians of science—for example, Dijksterhuis—have indeed seen here one of those watersheds that mark the onset of early modern science. Not so Koyré. He coolly observes that the important thing for Kepler was not so much his possession of Tycho's observations but rather his determination, already established years before he gained access to the observations, to handle astronomy as *aitiologetos,* as founded upon causes—in other words, as celestial physics. Koyré is apparently not willing to consider that Kepler's construction of such a heavenly physics would have remained stuck in the atmosphere of beautiful but ultimately barren speculation but for those despised hard facts such as accumulated by Tycho, the patient empiricist.[157]

Koyré's core conception dealt exclusively with physics and astronomy. Chemistry and the life sciences had no place in it. Dijksterhuis, when forced to deal with the issue in his *Mechanization,* had left out the life sciences and had in effect left chemistry pending in midair. Koyré, who never wrote an overview of the scope of the *Mechanization,* disposed of the issue in an even more summary fashion. He simply excluded the life sciences from the Scientific Revolution *as conceived by him* because, being inherently qualitative, their concepts were doomed to remain the nonmathematical ones of Aristotelian logic. And although he admitted chemistry was a quantifiable science, he had nothing else to say about it.[158]

Modification.   There is only one substantial modification of earlier core views to be found in Koyré's writings on the history of science. Even here one might perhaps rather speak of an extension than an actual modification. Koyré's participation in the Gassendi Tricentennial of 1955 provided him with an opportunity to acknowledge that, if it were true that the Scientific Revolution constituted the 'vindication of Plato', it was also true that Plato had had an ally in defeating Aristotelian empiricism—a rather unnatural ally, to be sure, yet an ally all the same, namely, Democritus. By introducing Democritean atomism as the fitting ontology for the new science, Gassendi, more than anybody else, had broken up the Aristotelian ontology of substance and attribute.[159]

In an earlier essay (later to be reprinted in the *Newtonian Studies*), Koyré had worked out these thoughts in another, very promising direction. Besides the physico-mathematical current in early modern science, which no doubt was by far its most original and important ingredient, another, parallel current could be discerned, so he wrote. This side-current was more empirical and experimental; it aimed at a much smaller scope than the mathematical generalizations it regarded with such distrust; it was satisfied with the discovery of new facts and the construction of partial theories in order to explain them; it was not inspired by the Platonic idea of a mathematical construction of reality but rather by the Democritean conception of its atomic structure. The principal members of this current were Gassendi, Boyle (the best of the group), and Hooke. These men generally preferred a cautious and safe, corpuscularian philosophy over the panmathematism of Galileo and Descartes. The man to unify the two currents was Isaac Newton. He of course created a synthesis between terrestrial and heavenly physics, but, in so doing, he also synthesized these two major currents of the Scientific Revolution. For Newton, Koyré said in a striking image, the Book of Nature was written in corpuscular characters, which are kept together by a mathematical syntax that invests these characters with meaning.[160]

It is indicative of the interpretational controversies Descartes' place in the Scientific Revolution could, and still can, elicit, to find him here assigned to the mathematical, rather than the corpuscularian, tradition in 17th-century science. This is a topic that shall occupy us again and again throughout the present chapter. For now I wish to call attention to another important aspect of Koyré's discussion: the very fact that *a two-current interpretation of the Scientific Revolution* has been put forward here. True, this particular interpretation remained as yet confined to one introductory page in one article on Isaac Newton. Yet here was a ready-made source of inspiration for two of Koyré's most prominent disciples, Thomas Kuhn and Richard Westfall, for their subsequent creation of the two most fruitful reinterpretations of the Scientific Revolution that have been advanced since.

The role of philosophy in the history of science.   Throughout his numerous books and articles written after the pathbreaking *Études Galiléennes*, Koyré used the core conception developed there as a source of observations on related issues. These may be subsumed under two headings: explanations of the Scientific Revolution and the general question of the relevance of philosophy for the history of science.

We postpone a treatment of Koyré's views on the causes of the Scientific Revolution to its proper place in the present book (section 5.2.1), thus conforming literally to Koyré's conviction that the *analysis* of what actually constituted the Scientific Revolution ought to be prior to an attempt at *explanation,* since only the former can tell us what there is to be explained.[161]

Koyré's ideas on the role of philosophy in the history of science are scattered over six of his articles and books. They consist partly of the reaffirmation of his own philosophical position, which at one place he very aptly labels "mathematical realism." [162] The other part comes down to a sustained confrontation with two rival views, namely, positivism and the related view that, in order to come into being, science had to emancipate itself from philosophy in the first place.

For positivism Koyré felt little but scorn. It is hard to decide what he found more pernicious—its influence on the history or on the historiography of science. Positivist historiography of science could elicit remarkably vehement language from this usually courteous polemicist:

> ... a historiography infected by the virus of the empiricist and positivist epistemology which caused, and still causes, so many ravages among historians of scientific thought.[163]

Just as positivism distorts in a systematic fashion the reconstruction of past scientific thinking, so has it acted as a brake on the progress of science itself. In positivist historiography of science the legalist, the phenomenal, in short, the 'positive' side of 17th-century science tends to be opposed to the realist deductivism of antiquity and the Middle Ages. But this opposition is doubly mistaken. Not only does it ignore that in establishing mathematical relations science does not connect phenomena but rather abstract notions. But it also neglects that the positivist posture goes way back to antiquity and that, as such, it signaled nothing but premature resignation. The methodological limitations inherent in the 'saving the appearances' attitude, which had been traced back by Duhem to Hipparchean astronomy and beyond, indeed made it possible to predict without explanation. Yet this attitude is really one of resignation and despair; it is regressive and defeatist. (It was left to the 19th-century variety of positivism to replace "resignation by boastfulness.")[164] True science becomes possible only by overcoming the positivist posture:

> It was in a revolt against this traditional defeatism that modern science, from Copernicus ... and up to Galileo, and Newton, accomplished its revolution against the shallow empiricism of the Aristotelians, a revolution based upon a deep conviction that mathematics was much more than a formal device for ordering data, in fact, the very key for the understanding of Nature.
>
> ... It is not by renouncing the apparently impossible and unattainable goal of knowing the real, but, on the contrary, by boldly pursuing it that science progresses on its endless path towards truth.[165]

Time and again scientists have resigned in the face of this challenge, but time and again there have been other scientists to take up afresh "the allegedly unprofitable, or impossible, solution of problems which had been declared to be devoid of sense."[166]

Thus philosophy has been of the utmost importance throughout the entire history of science, and its historians had better recognize this. As yet, Koyré observed in 1954, there were very few historians of science to do so. The great majority still adhered to the crude and fundamentally mistaken picture that 17th-century science meant the liberation from the yoke of metaphysics, some regrettable remnants of which are still to be found in the works of Descartes and Newton, but were happily removed in the 19th century. In fact, so Koyré went on in this tirade for the benefit of his co-professionals, he was aware of only one historian besides himself who had indeed drawn at least some of the consequences of the fact that metaphysical views had been absolutely indispensable for bringing about the Scientific Revolution. This historian was the American philosopher Edwin Arthur Burtt. We shall discuss his book *The Metaphysical Foundations of Modern Physical Science* in the next section. But first I wish to bring to a close the present section by presenting a final summing-up of Koyré's own achievement.

A summing-up in Koyré's own words. In 1948 Koyré wrote an article in French that purported to be no more than a review of five books on various aspects of the history of technology but which in reality contains an exposition of some of the wider vistas opened up by his core conception. He entitled it with a phrase that was to linger on in his work ever since. The phrase sums up very beautifully the core conception: "from the world of the 'more-or-less' to the universe of precision."

This was the formulation in a nutshell of the fundamental split in Western civilization effected around the turn of the 17th century. The Greeks had indeed sought for precision, but they had found it only in abstract geometry and in the heavens: "never would [Greek thought] have admitted that exactitude can be of this world." [167] Our world, in this view, which was shared by Platonism and even by Archimedes, is the world of the more-or-less. True, both Pythagorean number metaphysics and the Bible verse which says that God created the world according to number, weight, and measure, pointed toward such precision, yet it was not until Galileo that these injunctions were taken seriously. Up to Galileo's time, descriptions of, for example, technological apparatus of great ingenuity are striking for their really "approximative character." [168] True, many calculations were made all the time, but "no one ever sought to go beyond the practical usage of number, weight, and measure in the imprecision of daily life—to count months and beasts, to measure distances and fields, to weigh gold and corn—in order to turn it into an element of precise knowledge." [169]

The reason is not so much that the proper instruments were lacking, nor that there was no language to express any results in. True, these things were generally deficient, but the true reason for the deficiency is to be sought rather in an underlying, quite characteristic *mentality*—"the general structure of the 'world of the more-or-less'."[170] No better witness for this than the case of alchemy. For hundreds of years this was the only science of earthly matters which had succeeded in acquiring a vocabulary, a notation, and a collection of apparatus which withstood the centuries and were indeed to pass on eventually into chemistry. Treasures of observations were accumulated; important discoveries were made; experiments were carried out. However, descriptions of alchemical operations read like family recipes: The same style of 'take a little bit of this and a teaspoon of that' reigns here. Never did alchemy succeed in carrying out one precision experiment, for the simple reason that it was never tried—not for any lack of precision instruments, to be sure, given the availability of quite precise jewelers' balances.[171] The same is true of thermometry: "It is not so much the thermometer which is lacking, but rather the idea that heat is susceptible of exact measurement."[172] The same is true of the early history of optics. After the invention of spectacles in the 13th century, it did not occur to anybody for four full centuries to grind a lens with slightly smaller dimensions and slightly higher curvature, thus inventing the microscope. The same is true of the measurement of time, which was doomed to remain relatively imprecise as long as it stayed in the domain of the artisans (generally, mechanical clocks before Huygens' invention of the pendulum clock had still to be checked regularly against sundials or hourglasses).

Here, Koyré insisted, is the real point. As long as these things continued to be situated in the realm of the artisan, of the engineer, of those unskilled in theory, they were doomed to partake in the properties of the world of the more-or-less. The application of 17th-century science to these and similar areas drew them at once into the new universe of precision. This, then, was the fundamental transition. Nobody else could have carried it out but those who knew, or discovered, how to apply mathematics, that is, the embodiment of precision, to our physical world, thus turning our world into one small part of the new universe of precision. Henceforth, this universe was to absorb ever larger domains of human life and thought into its realm.

There are a few sentences to be found in Koyré's work which indicate that he was not altogether happy with life and thought as these have manifested themselves in the universe of precision. The closing line of the *Études Galiléennes* ("But never has a victory been bought at a higher price") serves as one such signpost. The most elaborate statement of Koyré's pertinent views is to be found right at the end of his 1950 article "The Significance of the Newtonian Synthesis":

Modern science broke down the barriers that separated the heavens and the earth. . . . it did this by substituting for our world of quality and sense perception, the world in which we live, and love, and die, another world—the world of quantity, of reified geometry, a world in which, though there is place for everything, there is no place for man. Thus the world of science—the real world—became

estranged and utterly divorced from the world of life, which science has been
unable to explain—not even to explain away by calling it "subjective." . . . This
is the tragedy of the modern mind which "solved the riddle of the universe," but
only to replace it by another riddle: the riddle of itself.[173]

It is most proper that Koyré attached to this brief reflection a footnote with a
reference to E.A. Burtt. For Burtt was the man who, a quarter of a century
before, had developed this very thought, not only into an extensive and radical
philosophical critique but also into a systematic, truly historical treatment of
key elements of the Scientific Revolution itself, and of this vital aspect in particular.

2.3.4. *Burtt and the Mathematization of Nature*

The year 1892 was a great year for the mathematization of nature—not only
Eduard Jan Dijksterhuis and Alexandre Koyré were born in that fruitful year
but also, on the other side of the Atlantic Ocean, Edwin Arthur Burtt. He was
educated in theology and philosophy, and he devoted only one book to the
history of science: his 1925 doctoral thesis at Columbia University, *The Metaphysics of Sir Isaac Newton,* which in April 1924 was published under the
definitive title *The Metaphysical Foundations of Modern Physical Science: A
Historical and Critical Essay.*[174] A revised edition appeared in 1932, and from
then on its author left the book to its own fate. In bibliographical lists at the
end of current overviews of the history of 17th-century science it is often mentioned as one other such overview. In reality, it is at the same time far more
than that, and something quite different. It is a book that starts from, and is
pervaded by, a particular philosophical thesis of Burtt's own making—a thesis
about the status of mind and spirit in modern thought. The author turned to
history, not in order to illustrate a ready-made thesis by means of historical
examples squeezed into a preset pattern, but because the nature of the thesis itself compelled the author to seek clarification in the past. The thesis is
handled throughout the book as a guiding-line; it helps to select and organize
the material but does not impose itself upon it. Together with Dijksterhuis' *Val
en worp,* which also came out in 1924, Burtt's book constitutes the first truly
contextual historical treatment of vital aspects of the origins of early modern
science. In sharp contrast to Dijksterhuis' book, though, it did not take its starting point in the Duhem thesis. Although both the *Études* and the *Système du
monde* are mentioned in the bibliography, Burtt was apparently not impressed.
His handling of the Middle Ages as providing one unified conception of man
and nature is not disturbed by any doubts of whether, in the light of Duhem's
revelations, such a unitary treatment could still be upheld.

Altogether, *The Metaphysical Foundations of Modern Physical Science*
had a somewhat odd fate. It did not come out of any mainstream; nor did it
create one for itself or was it ever made to fit any other. In its location beyond

philosophical or historical currents or fashions it just represents that priceless thing: the individual thought of an individual thinker.[175] The book need not be rescued from oblivion, though, for it continues to be reprinted and read to the present day. Still, I do not think that the significance of the book is widely recognized.[176] At a later point in the present book we shall find that there is a sense in which lonely Burtt's account is still very relevant to the course of present-day discussions of the Scientific Revolution. And what is even more: It is from Burtt's book that Koyré's concept of the Scientific Revolution probably found its origin.

Burtt's key problem. The Metaphysical Foundations of Modern Physical Science *is striking for the combination it displays of an intense and deeply felt admiration for the achievement of 17th-century science as an unprecedented intellectual adventure, with an equally intense conviction that the resulting overall world-view, which has come to pervade all our modern thinking, is extremely one-sided, depressing, demonstrably untenable, and fundamentally irresponsible from the point of view of logical, philosophical analysis. Burtt's view may be summed up by the paradox that concurrent with, and as a direct result of, one of the highest achievements of the human mind, the autonomy of that very same mind was downgraded and banished from the supposedly real universe of atoms moving according to mathematical laws across geometrical space.*

Before the advent of early modern science, throughout the Greek and the medieval eras of Western civilization, the reigning conception of man's place in nature, Burtt observed, had been fundamentally different:

> The prevailing world-view of the period was marked by a deep and persistent assurance that man, with his hopes and ideals, was the all-important, even controlling fact in the universe. . . . The entire world of nature was held not only to exist for man's sake, but to be likewise immediately present and fully intelligible to his mind. Hence the categories in terms of which it was interpreted were not those of time, space, mass, energy, and the like; but substance, essence, matter, form, quality, quantity—categories developed in the attempt to throw into scientific form the facts and relations observed in man's unaided sense-experience of the world and the main uses which he made it serve. . . .
>
> . . . just as it was thoroughly natural for medieval thinkers to view nature as subservient to man's knowledge, purpose, and destiny; so now it has become natural to view her as existing and operating in her own self-contained independence, and so far as man's ultimate relation to her is clear at all, to consider his knowledge and purpose somehow produced by her, and his destiny wholly dependent on her.[177]

For Burtt the primary task of philosophy was to reinstate "man with his high spiritual claims"[178] to a more fitting position than that of an entity reducible to the atomic categories of modern science. He surely did not recommend anything like a return to the Middle Ages. Rather, the new philosophy should be built upon the achievements of science; it is its metaphysical substructure that is to be repudiated. In order to do so, Burtt firmly believed, one must turn to history. Not, however, to the history of philosophy. The reason is that ever since Newton philosophers have struggled largely in vain to accomplish such a desired reinstatement. The very failure of their attempts testifies to the extraordinarily powerful grip the metaphysical substructure of early modern science has exerted upon intellectual thought ever since. It prevented philosophers from "rethink[ing] a correct philosophy of man in the medium of this altered terminology." [179] Thus, in order to clear the way for reaching a point of departure from which a new, fully modern, and more satisfactory philosophy of man and nature becomes at all attainable, a historical survey is needed of how this metaphysical substructure with its consequent dominance over our intellectual habits came into being in the first place. The question then is:

> Just how did it come about that men began to think about the universe in terms of atoms of matter in space and time instead of the scholastic categories? . . . What was happening between the years 1500 and 1700 to accomplish this revolution? And then, what ultimate metaphysical implications were carried over into general philosophy in the course of the transformation? . . .
> . . . we are proposing . . . a rather neglected type of historical inquiry, that is, an analysis of the philosophy of early modern science, and in particular of the metaphysics of Sir Isaac Newton. . . . We must grasp the essential contrast between the whole modern world-view and that of previous thought, and use that clearly conceived contrast as a guiding clue to pick out for criticism and evaluation, in the light of their historical development, every one of our significant modern presuppositions. . . . It is the creative period of modern science, then, in the seventeenth century chiefly, to which we must turn for the main answer to our problem.[180]

Burtt was well aware that his own generation, which had witnessed the first successful overthrow of Newton's virtually unlimited authority in the realm of science, was in a better position than any previous one for now turning to Newtonian metaphysics. Newton, so Burtt insisted, had been a far greater scientist than a philosopher. Large chunks of the new world-view that had been created by the previous generations of Kepler, Galileo, Descartes, Boyle, and others were adopted fairly uncritically by Newton, who added little but his own positivist leanings to what he had found readily available. The resulting, essentially metaphysical conceptions of space, time, causality, "the relations of man with the objects of his knowledge," [181] and so on were then uncritically carried along with his scientific achievement and adopted by the European intellect:

> None of these keen and critical minds, however—and this is the major instructive lesson for students of philosophy in the twentieth century—directed their

critical guns on the work of the man who stood in the center of the whole significant transformation. No one in the learned world could be found to save the
brilliant mathematical victories over the realm of physical motion, and at the
same time lay bare the big problems involved in the new doctrine of causality,
and the inherent ambiguities in the tentative, compromising, and rationally in-
construable form of the Cartesian dualism that had been dragged along like a
tribal deity in the course of the campaign. For the claim of absolute and irrefut-
able demonstration in Newton's name had swept over Europe, and almost everybody had succumbed to its authoritative sway. Wherever was taught as truth the
universal formula of gravitation, there was also insinuated as a nimbus of surrounding belief that man is but the puny and local spectator, nay irrelevant product of an infinite self-moving engine, which existed eternally before him and will
be eternally after him, enshrining the rigor of mathematical relationships while
banishing into impotence all ideal imaginations; an engine which consists of raw
masses wandering to no purpose in an undiscoverable time and space, and is in
general wholly devoid of any qualities that might spell satisfaction for the major
interests of human nature, save solely the central aim of the mathematical physicist. Indeed, that this aim itself should be rewarded appeared inconsistent and
impossible when subjected to the light of clear epistemological analysis.

But if they had directed intelligent criticism in his direction, what radical
conclusions would they have been likely to reach?[182]

Perhaps most impressive about Burtt's radical historical critique is that he carried it out in the name of an as yet nonexistent philosophy of which, beyond
its barest outlines (tentatively sketched in the "Conclusion" to his book), he
freely acknowledged to know nothing. His book aimed at clearing the path for
that hoped-for reader who, better than the author himself, would be able to
construct upon both the achievement of modern science and the ruins of its
attendant metaphysics a new philosophical building which would better satisfy
man's spiritual needs and would be more in conformity with our keen experience of the unique place of the human spirit in the surrounding cosmos.
Hitherto, such a hoped-for reader seems not to have come forward.

It is clear that the above position, although not necessarily religious in
character[183] might easily lead to a religious outcome. Burtt's subsequent career,
which was devoted "to construct[ing] a convincing and encouraging philosophy of man," [184] shows that this is indeed the course he took. Here is the most
probable reason for why, from 1932 until his death in 1989, he no longer cared
to elaborate, or to bring up to date, the historical and philosophical views put
forward in *The Metaphysical Foundations of Modern Physical Science*. In
1987, when he was ninety-five years old and I had a chance to talk with him
for a couple of minutes, he told me that he had often been admonished by
friends and colleagues to pursue the history of science further, but that, for
better or worse, his intellectual concerns had shifted irrevocably. I believe

nonetheless that there is an underlying unity connecting *The Metaphysical Foundations* with such later books as *In Search of Philosophic Understanding* (1965) and *The Human Journey* (1981), and I hope one day to be able to devote a deeper study to a reconstruction of that unity in Burtt's overall historical and philosophical and religious thought.

In *The Metaphysical Foundations* religion made its appearance chiefly where, in Burtt's opinion, both scientists and philosophers of the 17th century sought an all-too-easy way out of their self-created metaphysical dilemmas by pointing to God and by making Him responsible for whatever remained inexplicable in their new, scientific world-view. What, then, were the essential features of this new world-view as Burtt saw it?

The mathematical world-view.   Let it first be made clear that Burtt's aim was not to write a historical overview of scientific ideas in the course of the 17th-century Revolution in Science, such as was to be done later by Dijksterhuis, Butterfield, Hall, and others. Rather, he wished to focus on the philosophical underpinnings and implications of these scientific ideas. For those who deny, in a positivist vein, that Kepler, Galileo, Newton, *et al.* had been philosophers besides being scientists, Burtt's book was, and is, necessarily devoid of substance. Indeed, it was argued in 1936 by E. W. Strong that what 16th- and 17th-century scientists had to say on the structure of the world at large had really been irrelevant, if not downright damaging, to their positive, scientific achievement.

Yet what makes *The Metaphysical Foundations* so important a contribution to the historiography of the Scientific Revolution is precisely the careful link Burtt provided throughout between the actual performance of such men as Galileo or Kepler and the metaphysical underpinnings and implications of their scientific accomplishments. His book relies heavily on the printed works of the scientists he studied. Although Burtt found generally that these men had been far better scientists than philosophers, the key point was that so much of their heritage, surrounded as it was by the aura of the new science, went into the mainstream of modern thought. This thorough grounding in the historical sources viewed in a new light is what enabled this most unusual of historians of science to contribute in a major way to our understanding of the origins of early modern science.

The historical overview took its point of departure in the question of what it was that, in the face of so much plausible and empirical evidence to the contrary, made Copernicus and Kepler adopt the heliocentric conception of the universe. The answer is that both found such a conception more satisfactory than geocentrism because it "threw the facts of astronomy into a simpler and more harmonious mathematical order." [185] Burtt then points to some features of the wider intellectual framework of 16th-century thought that provided a fertile soil for such a new, mathematical treatment of things empirical (among such currents of the time he briefly mentions neo-Platonism) and goes on to show how, in the course of the 17th century, this new mathematical approach drew ever wider circles.

In Copernicus' work little more than the first vestiges are to be encountered. Nevertheless, the new step in astronomy would have been inconceivable but for the simultaneous emergence of an alternative world-view. Its formulation was taken a substantial step further by Kepler, who regarded

> the underlying mathematical harmony discoverable in the observed facts as the cause of the latter, the reason, as he usually puts it, why they are as they are. This notion of causality is substantially the Aristotelian formal cause reinterpreted in terms of exact mathematics.[186]

Kepler's mathematical conception of cause is linked to the empirical through his views on scientific hypotheses:

> A true hypothesis is always a more inclusive conception, binding facts which had hitherto been regarded as distinct; it reveals a mathematical order and harmony where before there had been unexplained diversity . . . this more inclusive mathematical order is something discovered *in the facts* themselves.[187]

This view leads Kepler to a fundamental distinction between two different levels of reality:

> Thus we have in Kepler the position clearly stated that the real world is the mathematical harmony discoverable in things. The changeable, surface qualities which do not fit into this underlying harmony are on a lower level of reality; they do not so truly exist.[188]

Such an overall geometrical conception of reality was expanded further by Galileo. The conception binds together his method, his results, and his underlying world-view. The mathematical structure of the world implies that "the world of the senses is not its own explanation."[189] The senses provide us with the world which is to be explained, but they fail to provide the rational order which makes it explicable in the first place. From this follows Galileo's peculiar experimental method:

> Viewed as a whole, Galileo's method then can be analysed into three steps, intuition or resolution, demonstration, and experiment; using in each case his own favorite terms. Facing the world of sensible experience, we isolate and examine as fully as possible a certain typical phenomenon, in order first to intuit those simple, absolute elements in terms of which the phenomenon can be most easily and completely translated into mathematical form; which amounts (putting the matter in another way) to a resolution of the sensed fact into such elements in quantitative combinations. Have we performed this step properly, we need the sensible facts no more; the elements thus reached are their real constituents, and deductive demonstrations from them by pure mathematics (second step) must always

be true of similar instances of the phenomenon, even though at times it should be impossible to confirm them empirically. This explains the bold tone of his more *a priori* passages. For the sake of more certain results, however, and especially to convince by sensible illustrations those who do not have such implicit confidence in the universal applicability of mathematics, it is well to develop where possible demonstrations whose conclusions are susceptible of verification by experiments. Then with the principles and truths thus acquired we can proceed to more complex related phenomena and discover what additional mathematical laws are there implicated.[190]

Swept on by the underlying mathematical world-view, Galileo, like Kepler, but in a much more outspoken fashion, arrives at a radical distinction between what have since been called the 'primary' and the 'secondary' qualities: "the clear distinction between that in the world which is absolute, objective, immutable, and mathematical; and that which is relative, subjective, fluctuating, and sensible."[191]

This is the historical process which lies at the heart of Burtt's philosophical concerns:

Now, in the course of translating this distinction of primary and secondary into terms suited to the new mathematical interpretation of nature, we have the first stage in the reading of man quite out of the real and primary realm.[192]

Precisely because that which is essential to the life of man cannot be handled in a quantitative fashion, the only domain of true reality has to be the world outside man: "The only thing in common between man and this real world was his ability to discover it."[193] And this raises for the first time the knotty question of how the human mind can know a world it is not even part of.

The stage is now fully set for Cartesian dualism:

Man begins to appear for the first time in the history of thought as an irrelevant spectator and insignificant effect of the great mathematical system which is the substance of reality.[194]

From this point on Burtt's book shifts into an extensive, brilliant account of how the new world-view was subsequently expanded, systematized, and amended. The full power and comprehensiveness acquired at once by the new world-view is also demonstrated by Burtt *a contrario* by showing that even its critics, such as the Cambridge Platonists, felt compelled to accept it, albeit with qualifications. Unable to resist the pull of the bandwagon, these men failed to develop conceptions of their own—conceptions truly independent of the framework the originators of the new, scientific world-view had set before them.

From this perspective Descartes, Hobbes, More, Barrow, Gilbert, and Boyle come in for in-depth treatment of their views on space, time, sense perception, the mind-body problem, and so on, until finally the whole movement is crowned by the specific direction Newton gave to it. However great the temptation (for, as must have become clear by now, Burtt was a powerful and rousing author), it would carry us much too far away from our own purposes to go any further into all this. We have seen almost enough to acquire a clear idea of what Burtt took the key features of the origins of early modern science to be.

The distinguishing features of early modern science.   The decisive point about Burtt's view of what the distinguishing features of early modern science were, is that they all hang closely together. The mathematization of terrestrial and heavenly physics, the geometrization of time and space, the distinction between the primary and the secondary qualities, the analysis of matter into material corpuscles, the role of experiment, the new conception of causality, the epistemological paradox of mind grasping the world without partaking in the world: all these and other features of 17th-century science were shown by Burtt to follow inexorably as soon as the view developed that reality is fundamentally mathematical.

Inside this overall scientific and metaphysical upheaval Burtt distinguished two main currents. The predominant current was almost exclusively mathematical. Here the mathematical world-view entailed the establishment of exact quantitative laws of diverse motions. However,

> back in the days of Kepler and Galileo, besides the exact mathematical movement in science, so powerfully advanced by their achievements and bringing in its train the remarkable metaphysical revolution it seemed to imply, there was another scientific current under way, flowing by slower and more tentative steps, but none the less scientific in interest and fruitfulness. Its method was wholly empirical and experimental rather than mathematical, and it was primarily in connection with this other current that attempts to give science a correct metaphysical groundwork made a quite positive and definite appeal to this 'spirit of nature' or, as it was more commonly called, 'ethereal spirit'.[195]

This spirit, in Burtt's account, fulfilled various functions. For such a critic of the new metaphysics as More, the Cambridge Platonist, it served to provide the human mind with a place of its own beyond the "pitifully meagre"[196] location inside the brain allowed to it by Cartesian dualism. For such scientists as Gilbert and Boyle 'spirit' served rather as the medium which could account for all those material processes, such as magnetic and electric action, that remained inexplicable because they could not readily be subjected to mathematical treatment.

This very notion of 'mathematical treatment' was not, however, entirely unambiguous. In its general sense, it came down to the adoption of the overall world-picture that had grown out of the conviction that the underlying reality of our world is mathematical. But only a few of the chief contributors to this intellectual revolution "had caught the full vision of Galileo, that motion is to be expressed in exact mathematical terms."[197]

Evidently, Newton was among these men. Yet his work incorporated the other current in scientific thought as well:

> For Newton . . . mathematics must be continually modelled on experience. . . . Newton was thus the common heir of the two important and fruitful movements in the preceding development of science, the empirical and experimental as well as the deductive and mathematical. He was the follower of Bacon, Gilbert, Harvey, and Boyle, just as truly as the successor of Copernicus, Kepler, Galileo, and Descartes; and if it were possible wholly to separate the two aspects of his method, it would have to be said that Newton's ultimate criterion was more empirical than mathematical. Despite the title of his great work, he had far less confidence in deductive reasoning as applied to physical problems than the average modern scientist. Continually he called in experimental verification, even for the solution of questions whose answers would seem to be involved in the very meanings of his terms. . . . It is not too much to say that for Newton mathematics was solely a method for the solution of problems posed by sensible experience.[198]

And this is how Burtt succeeded in combining an, at bottom, unitary conception of the 'intellectual revolution of the 17th century' with his equally pioneering two-current account of the advance of the revolution over time.

Conclusion.   On a scale of continuity versus break in the transformation that brought early modern science into being, Burtt is to be located squarely on the side of the adherents of 'relative discontinuity'. He may well have been the first to adopt such a position. His lack of concern for Duhem's ideas on medieval science, perhaps mercifully, spared him the need to consider whether his own unitary account of medieval thought about man and nature could still be upheld. It is unlikely that a deeper acquaintance with the Duhem thesis would have made much difference for him, given the criteria he employed for distinguishing the break in the first place. These criteria were deeply rooted in the consequences for our world-view of the idea of the underlying mathematical nature of the world, which made possible early "modern science, the most successful movement of thought that history so far records."[199]

The fundamental problem with this mathematical conception of the world, Burtt believed, was its pervasiveness. Whatever the new philosophy he hoped for was to look like, it should begin to confine the mathematical to its proper realm. Rather than reducing the 'secondary' qualities to manifestations of the 'primary', mathematical ones, the domain of the primary qualities should be restricted to those qualities which indeed lend themselves naturally to mathematical treatment. Only along such lines can the human mind be restored to its proper dignity, and only thus can one begin to understand how mind is capable of acquiring knowledge of our world, which must remain incomprehensible if the knowing mind is not even part of it.[200]

Thus we find how the deploration of certain consequences of the Scientific Revolution which Koyré was to mark in only a few closing lines constituted for Burtt the central issue of his inquiry. In this respect, too, Burtt stands, largely unrecognized, at the head of an array of historians of the Scientific Revolution who, in the sixties, began to look upon this signal event in European intellectual history as having brought, besides impressive gains in our knowledge, some no less consequential losses.

2.4. THE CONCEPT WIDENS

2.4.1. *The Great Four: Views Compared and Views Exchanged*

Let us review what had been accomplished since Duhem, in 1913, put forward his startling thesis.

Between the four of them, Burtt, Dijksterhuis, Koyré, and (to a lesser extent) Maier created a picture of the Scientific Revolution displaying virtually all those core features with which we are still familiar. Emphases have since shifted; foci of attention have wandered to and fro; mounds of new source materials have been brought to bear on fresh interpretations; new venues have been explored—all this is true, and we shall presently chart these new directions. Yet, insofar as we are talking about the analysis of scientific ideas, the principal themes are all to be found, whether more or less explicitly, in the clear-cut pictures drawn in the twenties to early sixties by these four scholars of the first post-Duhem generation.

Nor is this all they accomplished. Before, writing history of science had largely meant either illustrating preconceived, philosophical viewpoints with examples from the past or tracing received scientific views down the centuries. The Great Four, together with a number of their contemporaries to be sure,[201] turned history-of-science writing into the art of analyzing and understanding a scientist's thought as an integrated whole. That is to say, in their hands the job of the historian of science became, above all, the art of interpreting any given thought of any given scientist in the context of both his own work and the thought of his contemporaries and predecessors. Philosophical conceptions might well continue to serve as guidelines, but the principal purpose was henceforward the faithful reconstruction of 'how things had really happened'. Whatever has since changed, this mode of operation has remained the primary standard for measuring fresh contributions to the historiography of science.

The effects of the new approach were drastic. Key positions previously taken for granted were now left behind for good. Thinking about continuity and discontinuity in absolute terms gave way to historiographically more-fruitful considerations of the extent to which, in the ongoing flow of history, one or more revolutionary convulsions might nevertheless be distinguished. The issue of continuity versus break thus remained as alive as ever, yet the crude and unreservedly empiricist guise in which Mach had clothed it was destined henceforth to live on outside the domain of responsible historiography of science. The radical counterthesis that the Scientific Revolution had really begun in the 14th century was also repudiated, although opinions continued to differ considerably over the extent to which 14th-century scholasticism had contributed to the rise of early modern science.

Likewise, a number of fresh interpretations of the Scientific Revolution had been put forward by the Great Four. Foremost among these is that the birth of early modern science was now subjected for the first time to treatment as a historiographical issue in its own right, and in the process had received a fitting label all of its own. In the meaning originally given by Koyré to the concept, the 'Scientific Revolution' stood for the transformation wrought by Galileo and Descartes in our conception of the world. When Koyré went on to widen his concept somewhat in time so as to make it span the period from Copernicus through Newton, the net result was for the new concept to coincide largely with what Burtt had meant all along by the 'intellectual revolution of the 17th century'. This implied, among other things, that Burtt and Koyré, but also Dijksterhuis, agreed that the Scientific Revolution had begun with Copernicus' *De revolutionibus,* even though Copernicus' links with the past were strongly emphasized.

Further, the general idea that, in the *Principia,* Newton had achieved a synthesis of previously distinct elements in the description of nature was now made more precise in that partly overlapping two-current interpretations of the Scientific Revolution were put forward by Burtt and Koyré, respectively. Precisely along what lines the two currents should be distinguished from one another had not yet been resolved—the topic was still no more than a side issue, though fraught with promising kernels of further analysis.

Next, the ambiguous nature of the place occupied by René Descartes in the emergence of early modern science had been put into high relief. Should his contribution be taken, as Burtt had upheld, to be marked above all by the radical excision of mind from matter that had been prepared by Galileo? Should Descartes be seen, in Koyré's similar vein, as the man who finished what Galileo had started—the geometrization of space? Or should Descartes' ambition to put a new, complete *system* of nature in the place of its Aristotelian predecessor rather be taken for a regressive enterprise, fundamentally at odds with the true spirit of the new science, as Dijksterhuis strongly implied?

From Enlightenment days onward there had been universal agreement that Aristotle's doctrines had provided the particular natural philosophy in opposition to which the new science had taken shape in the first place. The Great Four adopted this view without reservation. The nature of the new science, however, was now defined in a much more precise fashion than had been done before. To Burtt and Dijksterhuis as well as to Koyré it was clear that the central distinguishing feature of the new science was the mathematization of nature. They differed somewhat among themselves as to what exactly this meant, and how deep the process went, yet they agreed that here was the one true key to what made the new science essentially distinct from preceding conceptions of nature. They adopted Kant's distinguishing criterion of the active, experimental interrogation of nature versus the loose observation characteristic of earlier times, but they made the process of interrogation follow from the mathematical theorizing to which, in their view, it was subservient. In so doing, Koyré and Dijksterhuis, in particular, narrowed the role of experiment down to the *a posteriori* verification of mathematical laws deductively established. Both sought key evidence for this view of the role of experiment in Galileo's work. Some of their wording even suggests that their entire conception of the birth of early modern science, which in fact had far wider implications than the role played in it by experiment, hinged critically on the validity of their strongly antiempiricist Galileo interpretation.

Finally, the preponderant part inevitably assigned to astronomy and mechanics by the mathematical view of the origins of early modern science left wide open the question of what, if any, role nonmathematical physics, chemistry, and the life sciences had played in the birth of early modern science. Dijksterhuis had virtually ducked the issue. So had Koyré. True, his original concept of the Scientific Revolution had left no room for these sciences, but it was not automatically clear whether, once he widened his concept, silence on the issue was still justified. Burtt, in contrast, had recognized in the domain of the nonmathematical physical sciences one particular manifestation of the principal shortcomings inherent in the mathematization of the secondary qualities. In his account these neglected areas of 17th-century science served as signposts to those very deficiencies.

Such are some key elements in what the Great Four accomplished. But what stands out most in all they did is the sense of sheer *excitement* with which the birth of early modern science was surrounded in their treatment. Burtt had glowingly spoken of "the most stupendous single achievement of the human mind." [202] Koyré had waxed eloquent over this "veritable 'mutation' of the human intellect . . . —one of the most important, if not the most important, since the invention of the Cosmos by Greek thought." [203] Dijksterhuis had expressed himself in equally captivating terms. All three possessed in high degree the ability to communicate their intellectual enthusiasm to their readers. Let us now find out what happened to the ideas of the Great Four when thrown into a historiographical debate elicited by themselves in the first place. I shall first say a few words about the institutional framework that came into being during, and partly as a result of, the creative period of these authors. Next, I discuss the illuminating comments the Great Four directed to one another's views. That done, I shall outline a number of fresh developments that have manifested themselves during the past quarter-century, after the contributions of the Great Four had come to an end owing either to death (Koyré, Dijksterhuis) or to major shifts of interest (Maier and Burtt).

The professionalization of the history of science. As indicated in the section devoted to Koyré, the professionalization of the history of science as an academic field crystallized intellectually around his account of the Scientific Revolution. This was the case in particular in the United States and, to a somewhat smaller extent, in the United Kingdom as well. Before the Second World War, the history of science, although taught as a side issue in quite a few science departments, had been studied in a truly creative way by no more than a few such loners as Charles Singer in the United Kingdom and Lynn Thorndike (the author of eight volumes on the *History of Magic and Experimental Science*), E. A. Burtt, and George Sarton in the United States. What singled out Sarton was his determination to create an institutional framework for the history of science as a profession in its own right, to be equipped, besides departments and chairs, with such tools as a society, a journal, bibliographies, introductions to the field, and the like. All this was set up around Sarton's core vision of the history of science as the most viable vehicle on the path of humanity toward a 'new humanism' that was to embody more or less everything worthwhile in our cultural heritage. The vision has largely gone, but the institutional groundwork so patiently laid by Sarton in the decades before the Second World War began to bear spectacular fruit in its immediate aftermath.[204]

It is not difficult to understand why Koyré's single-minded and incisive writings were much more germane to filling in this institutional framework with intellectual content than was the work of Sarton's stimulating, yet strongly encyclopedic and taxonomic turn of mind. It is less easy to explain how it is that rather than the, as it were, native treatment of the 17th-century mathematization of nature to be found in Burtt's book, it was the Frenchman's handling of the same process that was singled out to serve as an intellectual crystallization point in the United States. Be that as it may, the sense of excitement over the discovery of Koyré's work appears vividly in the following account of one of the participants in the new, professional movement, Charles C. Gillispie:

A new generation of historians of science, the first to conceive of the subject in a fully professional way, was just then finding an opportunity in the expanding American university system, which more than made up in flexibility and enthusiasm for science whatever it may have lacked in scholarly sophistication and philosophical depth. Casting about through the literature in search of materials, they came upon *Études Galiléennes* as upon a revelation of what exciting intellectual interest their newly found subject might hold, a book which was no arid tally of discoveries and obsolete technicalities, nor a sentimental glorification of the wonders of the scientific spirit, nor yet (despite the author's Platonism) a stalking horse for some philosophical system, whether referring to science like the positivist outlook or to history like the Marxist.

Instead, they found a patient, analytical, and still a tremendously exciting history of the battle of ideas.[205]

This new generation was to do more than just seize upon the Koyré treasure. Over time the treasure was to be used as a starting point for fresh interpretations of the Scientific Revolution. Before discussing those, we close our extensive treatment of the achievement of the key members of the pioneer generation with an examination of their own exchanges. It will appear convenient to do so by focusing on Koyré as the central figure.

Koyré and Burtt.   In his published writings Koyré had precious little to say on Burtt's book. From the three or four somewhat disparaging footnotes and peripheral remarks he altogether devoted to *The Metaphysical Foundations,* one would be hard put to infer whether Koyré really did not care that much for it, or whether he was in effect covering up a substantial indebtedness of his own. However, Henry Guerlac reports that "in personal conversation with this writer he once remarked that his reading of E.A. Burtt's remarkable book, the *Metaphysical Foundations of Modern Physical Science . . .* played an essential role. . . . As Burtt abandoned the history of science for the rest of his career to teach the philosophy of religion . . . , so Koyré moved in the opposite direction."[206]

If this is what happened, Koyré's comments certainly appear in a somewhat odd light. True, he praised Burtt for his very rare, nonpositivist approach to the history of science. But he went on to blame him for his failure to distinguish between two forms of neo-Platonism: the productive neo-Platonism that helped to create 'the vindication of Plato', embodied in the Scientific Revolution, and the unproductive variety that led only to pointless number speculations.[207] His principal criticism meanwhile aimed at the very heart of Burtt's book. True, Burtt was aware like no one else but Koyré himself that the Scientific Revolution constituted a break in man's thought of a far deeper nature than the mere transition from certain scientific propositions to certain others. Nevertheless, Koyré wrote, Burtt regarded this deeper, metaphysical level as one that the scientists in question had felt free to discard after they had arrived at what really mattered to them, namely, their scientific insights—as scaffolding that could be removed when the building was done.[208]

This was a most curious criticism. Not only is such a message absent from *The Metaphysical Foundations,* but it would go flatly against the very point Burtt was laboring so hard to get across. He did not at all state or imply that the scaffolding had become useless once the true, positive work had been done, but rather that after three centuries the time had come to replace the metaphysical foundations of science taken for granted for so long with a more adequate metaphysics.

Despite this scarcely explicable misunderstanding,[209] Koyré did briefly defend Burtt against an attack in E.W. Strong's book of 1936, *Procedures and Metaphysics*. Dealing with science in the Italian Renaissance and culminating with Galileo, this book was intended as a sustained critique of *The Metaphysical Foundations*. Strong's book, which belongs much more to the ill-reputed 'preface-history' genre than Burtt's, was written in a fairly positivist vein. It strongly emphasized "the operational autonomy of science and the irrelevance of the metaphysical tradition" and asserted flatly that "mechanical knowledge marches by method, not by metaphysics."[210] After all that we have found out about Koyré's views, it need not perhaps be spelled out in what precise wording he marked his vigorous disagreement.[211]

Koyré, Maier, and Dijksterhuis.   How important it was for Koyré to defend his interpretation of the Scientific Revolution as the outcome of a change in world-view appears also from a review he wrote in 1951 of Maier's first book on late scholasticism—the very *Vorläufer Galileis* we discussed in section 2.3.1. To the extent that Maier's detailed investigations of 14th-century scholastic thought tended to undermine the hypothesis of straightforward continuity with 17th-century science, they of course provided Koyré with so much grist for his mill. Thus, in discussing the impetus/inertia issue, he began by taking up arms for Maier against Dijksterhuis, for the latter, in a previous review of the same book, had attempted to show that Newton's concept of 'inertial force' was virtually identical with the medieval notion of impetus.

The little debate that ensued brings out in a nutshell the similarities as well as the differences between its three participants. Koyré and Dijksterhuis shared a great admiration for Maier's painstaking research. Dijksterhuis even ascribed to her "the rare faculty of carrying her mind back to the period which she is studying and to participate in the reasonings of the Scholastics, so to say from within."[212] He accepted to a large extent her corrections of Duhem's interpretations of late scholastic texts, which—lacking access to most of the original manuscripts—he had been compelled to adopt in *Val en worp*. Also, the somewhat toned-down continuism in Dijksterhuis' *Mechanization*, compared with *Val en worp*, was equally due to Maier's influence in the first place. Nevertheless, Dijksterhuis always felt that she went too far in denying certain conceptual continuities over time, in particular where the concept of impetus was concerned. The debate between Maier and Dijksterhuis took place partly in public (in mutual reviews and in the *Mechanization* itself) and partly in the few, highly respectful letters exchanged by these two touching, somewhat otherworldly elderly people who never met despite their rather urgent wish to do so. Thus in 1960 Maier wrote to Dijksterhuis: "One cannot alter the fact that in some fundamental points we have diverging views, and will continue to have forever."[213]

Almost to the end of his writing days in the early sixties, Dijksterhuis was to uphold the continuity between Newton's inertial force and Buridan's impetus.[214] This was the critical thrust in his review of Maier's *Vorläufer Galileis*. And here Koyré, in his own review, came to Maier's rescue by pointing out that Newton's language in his First Law, of bodies persevering in a *state* of motion, was inconceivable from the scholastic perspective, in which motion was rather a *process*.

Koyré's own difficulties with Maier's treatment of the inertia issue came at another point. It will be recalled that in Maier's view Galileo had more or less stumbled upon inertia when he found that the experience with the rolling ball could not be accounted for through the impetus conception, and thus the discovery had nothing to do with an underlying change in world-view, which was rather to *result* from the accidental discovery. Koyré, in his review, expressed his "vivid surprise" at Maier's thus ignoring what really were the most precious results of her own investigations.[215] No experiment could ever establish, or even lead to, the principle of inertia. Rather, the principle signaled the radical change in world-view we have seen Koyré denote so very often by the terms 'geometrization of space' and 'destruction of the Cosmos', with all that these twin thought processes implied. Thus Koyré had little time for the 'two revolutions' conception Maier had developed regarding the origins of early modern science.

Interestingly enough, in later years Maier's views indeed shifted somewhat in the direction courteously recommended to her by her self-styled French *chevalier*. Two passages from her correspondence with Dijksterhuis mark the transformation in a curious manner. In 1949 Maier, in thanking Dijksterhuis for his review in *Isis* of her *Vorläufer,* went on to refer to their common interest in "these problems in natural philosophy, which really constitute the most interesting area of scholasticism. Again and again one finds something unexpected that, in one form or another, lives on in early modern science."[216] Eleven years later, in the course of their ongoing debate over the value of certain specific attributions of modernity to scholastic ideas, Dijksterhuis confronted Maier with her past willingness to make such attributions herself. But now she responded quite differently:

> When you put before me that I myself more than once established 'anticipations' of future insights, you are undoubtedly right; yet I did not do that often, and whenever I did I did it very reluctantly. But when I began to take an interest in these problems, this mode of putting the question had been customary for a generation, and then little else remains for the individual but to conform, even while protesting from time to time.[217]

Leaving aside the intriguing question of what may turn very good historians of other people's past thought into such atrocious recorders of their own thought, we obviously have to ask what had changed in the meantime. Maier's later essays display a markedly increased interest in the metaphysical level of late scholastic thought. However, this shift away from more readily tangible, concrete issues was accompanied by a growing annoyance at the persistent interest shown by the scholarly world in the extent to which her researches tended to confirm or call into question the Duhem thesis or parts thereof. In 1960, eleven years before she died, she published an essay "'Ergebnisse' der spätscholastischen Naturphilosophie" (The Achievements of Late Scholastic Natural Philosophy), in which she confronted the issue for the last time. The tone which pervades the piece makes it clear that meanwhile she had become fed up with treating late scholastic thought as if it had no value in its own right but could serve only to decide an issue inherently foreign to it: the origins of early modern science.[218] Her essay centered nonetheless around a quite explicit assertion that is indeed most relevant to this vexed issue, viz., that a small but highly significant set of Aristotelian principles was never challenged before the 17th century, thus effectively precluding the coming about of more than incidental results of such a nature as to satisfy early modern scientific thought. One of the examples she gave in the paper brings her position remarkably close to Koyré's:

> Scholastic philosophers not only did not carry out any measurements in practice and did not establish any theory of measurement, but over and above this, they declared really exact measurement to be impossible for reasons of principle. The ground for this conclusion was, in the last analysis, a matter of world-view. . . . whereas for God the whole world is numbered and measured in every detail, for man this is not so: He has not been granted the possibility of knowing the exact measures of things, or, so to say, of verifying them.[219]

Perhaps equally remarkably, Maier does not mention Koyré's name in this or any other context.[220]

Dijksterhuis and Koyré. Not only Maier, but Dijksterhuis, too, was in the habit of occasionally criticizing Koyré without mentioning his name. In 1957, on the initiative of the historian of medieval science Marshall Clagett, an eleven-day international conference of historians of science was held at the University of Wisconsin. Here Dijksterhuis gave a paper with the title "The Origins of Classical Mechanics from Aristotle to Newton." In this twenty-one page *précis* of his—as yet untranslated—*Mechanization* he took on pet ideas of Koyré twice and, as the reader will not be surprised to learn, the continuity issue provided the bone of contention. At one point in his argument Dijksterhuis complained about a tendency of historians of science to treat certain discoveries of Galileo lightly, now that they had been found to have been made already by Oresme and other scholastics, but which, "only a few decades" ago,

> were considered original and brilliant contributions of Galileo. . . . That a truly fundamental discovery has had to be dated about 300 years earlier is, after all, a fact of great historical importance, and it is by no means clear why the appreciation of this discovery should now become smaller. Or can this be an after-effect

of the belief held so persistently in the nineteenth century, namely, that the Middle Ages were devoid of all significance for the evolution of science?[221]

This interesting attempt to associate Koyré with a dash of Machian discontinuism was preceded by a more fundamental attack on one of Koyré's core views. Again, without mentioning his name, Dijksterhuis referred to a "view [which] is also met with," namely, that attempts to link the impetus concept to any other concept of early modern mechanics are doomed at the outset because of the fundamental difference between two distinct conceptions of motion: as a process in Aristotelian science and as a state in early modern science. And here Dijksterhuis indeed made an important point. He made it in order to salvage the right to maintain conceptual continuity from medieval impetus to quantity of motion in Newtonian mechanics. His point was that

> the antithesis in question between the two conceptions of motion exists only so long as uniform rectilinear motions are considered; all other motions are processes in classical mechanics as well, that is to say, they equally require the constant action of an external force, even though its effect is evaluated quite differently from the way it was done in Antiquity.[222]

For the rest, Dijksterhuis asserted, the question of whether the distinction is indeed so fundamental depends rather on the philosophical point of view one wishes to take (and thus, he seems to imply here, is quite arbitrary). He did not elaborate the point further, but it seems fair to assume that it was meant to strike at the heart of Koyré's celebrated core conception, for it implies that the very transition deemed so fundamental by Koyré was indeed important but not nearly so crucial, or so strongly linked to a new world-view, as Koyré would have it. It would be interesting to know how Koyré, if present at the meeting, would have responded to this challenge. However, other than his acting as the Tertius Interveniens (the third man in between) on behalf of Anneliese Maier in 1951, I am not aware of any occasion where Koyré expressed himself on the achievement of the man with whom he shared so many of his basic insights into the origins of early modern science.

Koyré, Duhem, and Crombie. Koyré was much more eager to take on another proponent of the continuity view. In 1953 the Australian/English historian of science Alistair C. Crombie came to the fore with a new thesis which, although quite different in content from the Duhem thesis, had the same effect of pushing substantial and, according to the author, essential portions of early modern science way back into the Middle Ages. Crombie's book was entitled *Robert Grosseteste and the Origins of Experimental Science, 1100–1700*. This book centers on methodological discussions to be found in the works of Robert Grosseteste (1168–1253) and Roger Bacon. Unlike Duhem, Crombie did not claim that major scientific breakthroughs were actually yielded by these medieval discussions at the time. Yet, Crombie maintained, in setting forth for the first time what a truly experimental and mathematical science could accomplish, these two authors form key links in a chain of methodological argument that was to continue through the 17th century. After all, it was not by chance that so many protagonists of the Scientific Revolution wrote extensively, and in a vein quite similar to that of Grosseteste and Bacon, about these matters; witness Francis Bacon's works, Descartes' *Discours de la méthode,* Galileo's and Newton's disquisitions on method, and so on. In starting this tradition, Crombie claimed, the two 13th-century Englishmen had indeed prepared the ground for the Scientific Revolution in a major way.[223]

Koyré's critique of Crombie's thesis appeared in a lengthy review published in 1956. It is a virtuoso performance, in which openness for what the opponent has to say combined in a most fruitful way with a reaffirmation and elaboration of Koyré's own point of view. The piece also brings out Koyré's favorite debating style, which was to describe sympathetically the opponent's view, to go out of his way to agree with him or her wherever he could, and then to show why the data adduced by his opponent really pointed to conclusions fundamentally at variance with those of the opponent—the true conclusion to be drawn belonging rather to Koyré's own core conception of the origins of early modern science.

Koyré courteously began his essay review by stating that the continuity conception had now found its most eloquent defender in Crombie. It may not be amiss to regard this opening statement as a sly dig in the direction of Duhem. Even though many key ideas in Koyré's own work—however much they run counter to the Duhem thesis—are inconceivable but for Duhem's previous discoveries, it does not seem from what little Koyré had to say about him in his books and articles that he held him in particularly high esteem. In a piece on ideas about infinity and the void in the 14th century Koyré made fun of Duhem's handling of Tempier's 1277 condemnation—in so doing, Duhem had "put the proclamation, by the bishop of Paris, of two absurdities at the origin of modern science."[224] But no such lack of philosophical subtlety occurs in Crombie's book. Yet, Koyré wrote, the latter's thesis suffered from a serious overestimation of the importance of methodological discussions. Taken by themselves, all those attempts at defining the true role, in scientific discovery, of theory and hypothesis, of analysis and synthesis, of 'resolution and composition,' of experience and experiments, were really little more than variations on Aristotelian themes with little inherent significance unless connected in a straightforward manner with actual scientific discovery. But this, on Crombie's own admission, was not to take place on any significant scale until the 17th century. Thus the very lack of real scientific achievement in the works of Grosseteste, Bacon, and all those other keen epistemologists showed rather that something else was needed to get the Scientific Revolution going. Altogether, Koyré believed, the efforts of Grosseteste and Bacon as analyzed by Crombie testified rather to the fact that purely methodological disquisitions have little importance for science: "the place of methodology is not at the beginning of scientific development, but, we might say, in the middle of it."[225]

Crombie's thesis has gained considerable respect rather than acceptance. The author turned it into the centerpiece of his historical survey *Augustine to Galileo,* but his thesis has failed to be incorporated into later accounts of the Scientific Revolution. The particular episode of Koyré's attack on Crombie's thesis serves for us to highlight an approach to history which distinguishes Koyré from a great many other historians of science. The more common approach is to pick up certain elements found important in 17th-century science, and then, when such an element is found back in earlier times, a certain continuity with early modern science is claimed to have been established thereby. Koyré, in contrast, tended rather to see in those previous elements so many reasons for why early modern science had *not* been produced earlier: The sheer fact of their existence was insufficient for bringing it forth. The distinction is to occupy us again when, in Part II, we begin to discuss historians' accounts of the causes of the rise of early modern science.

2.4.2. New Problems and a New Generation

We have now reached the end of our treatment of the generation which put the problem of the rise of early modern science on an almost wholly new footing. The four members of this generation on whom we focused our attention did more than just frame a highly outspoken set of conceptions about the origins of early modern science; they managed to turn this set of conceptions into the central issue of the debate that their work helped so much to unleash. Take, as one example, the first few hundred pages of the collection of essays that came from Clagett's 1957 Wisconsin conference, which was entitled *Critical Problems in the History of Science.* Here one can see the extent to which all this stuff from the twenties and thirties still formed the framework of the discussions held at that occasion. But there is also something stale about it all. The debate was highly loaded with philosophical notions and filled with 'medievalisms'; it tended to center on the principle of inertia to the virtual exclusion of any other element of the Scientific Revolution; it paid scant attention to domains of early modern science other than astronomy and mathematical physics; and, most importantly, when one leafs through the collection, it seems to cry out for some fresh air to be let in by a new generation of historians of science—the first to have benefited both educationally and professionally from the wave of institutionalization after the Second World War.

The fresh air was to come in from several sides, and some of it had already been in the making from the early fifties onward. Here is a list of some key elements, which come under five headings: the dropping out of the medievalists, the Galileo debate, philosophy on its way out, the Hermeticist challenge, and, last but not least, coming to grips with the nonmathematical sciences.

The dropping out of the medievalists.   The Duhem thesis greatly stimulated research in science during the Middle Ages—a thing deemed virtually nonexistent before. (As Lynn White was to sum up the situation in 1974: "As an undergraduate fifty years ago I learned two firm facts about medieval science: (1) there wasn't any, and (2) Roger Bacon was persecuted by the church for working at it.")[226] For decades, much of this research centered around those elements in medieval thought claimed to establish links of the 'forerunner' type with 17th-century science, with Marshall Clagett, in particular, documenting his continuity conception in impressive manner in *The Science of Mechanics in the Middle Ages* (1959). But we have seen how the most authoritative pioneer of these studies, Anneliese Maier, whose early work had been undertaken under the aegis of the Duhem thesis, in the late fifties turned increasingly to studying the period purely on its own terms and wished rather to emancipate such research from the original impulse that had got it going. What enabled her to adopt this posture in the first place was her rejection of the Duhem thesis. Once having reached the conclusion that the Scientific Revolution did not take its origin in late scholastic thought, she became increasingly interested in studying the period for its own sake.

During the period of vast, institutionalized expansion of the historiography of science, Maier's later conception of the task of the medievalist was equally adopted by the rapidly growing number of historians of medieval science, who equally rapidly turned the field into an area of specialization in its own right. Whereas in the fifties such medievalists as Clagett and Moody still occasionally set out to establish fresh links between medieval and early modern science, the connection seems to have been all but severed since.

However, now that as a result of the efforts of Maier, Koyré, and others, the Duhem thesis had been found untenable, a new question could be asked, or rather, an old question could be asked afresh, namely, what it was that had *prevented* medieval science from ushering in the Scientific Revolution. As a result of the very specialization just indicated, though, only very few medievalists have since felt called upon to pronounce their opinions on this subject, Maier herself being one and Edward Grant another. Their views on the matter are to be discussed in section 4.3, which deals with the issue of why the Scientific Revolution, having 'failed' to emerge in ancient Greece, did not emerge in medieval western Europe either.

The Galileo debate.   When, in 1961, Thomas Settle (then a graduate student) let a reconstructed ball roll over a reconstructed inclined plane in very precise accordance with the instructions given by Galileo in the *Discorsi* and acquired an outcome very close to the one claimed by Galileo, in that very act he ushered in a debate over the true role played by experiment in Galileo's work that is still with us today, and which is united over just one thing, namely, that Koyré had been wrong. Following up on Settle's "Experiment in the History of Science" a renewed interest was taken in Galileo's scrap papers, and evidence was sought for and against his having found at least some of his quantitative laws through actual experiment, not thought-experiment, employed heuristically, not for verification purposes only. The ensuing debate, in which Stillman Drake became the central figure,[227] has over time become exceedingly technical and is hard to follow for outsiders who do not possess instant recall of 'folio 116 verso' and like pet items of the aficionados. It is out of the question to deal with the debate in the present book. Nor would such an exercise really be called for. For our purposes it suffices to note four significant features of the debate in question.

First, the very technicality of the debate has tended to rob the issue over which it is fought of the core importance it had rather uncritically been assigned from Mach down to Koyré and Dijksterhuis. Somehow, and strongly supported by Koyré's (and Dijksterhuis') own wording, the issue of the nature of early modern science had become closely linked up with the question of whether Galileo had actually carried out his experiments and, if so, to what purpose. But, even if we fully accept the central position Galileo holds in the Scientific Revolution however defined, this simply puts more weight on the narrower issue than it can possibly bear.[228] With this heavy load removed, the wider debate on the nature of the Scientific Revolution could now be continued in a less overcharged and perhaps healthier atmosphere.

However, and this is the second point, Koyré himself, who had given so much occasion for virtually identifying the two issues, lost much when they were, in effect if not on purpose, disentangled. Koyré appeared to stand refuted, and although there is much in his vision of the Scientific Revolution that seems to me to have lost nothing of its original vitality and analytical fertility, this vision of his has become marginal to present-day thinking on the Scientific Revolution to a far greater extent than, at least in my view, it deserves by its own merits. Indications of the undiminished fertility of much in Koyré's key ideas will surface throughout the present book.

Another consequence of the new approach to Galileo was that one particular subdebate not mentioned so far also lost much of its original significance. This was the question of whether or not Galileo had himself been familiar with the late scholastic writings which, according to Duhem and later continuists, had foreshadowed or actually preempted important discoveries of Galileo. In the final pages of his *Études sur Léonard de Vinci* Duhem had confidently established such a direct link, but his particular claim (through Galileo's "Iuvenilia" was quickly (though, as it appeared later, too rashly) disproved. The issue was exacerbated by Galileo's evident lack of familiarity with the writings of his most immediate and most obvious 'forerunner,' Benedetti. Part of the riddle was finally solved when, in the early eighties, William Wallace (roughly simultaneously with Adriano Carugo and Alistair Crombie) established a direct link between Galileo and previous thought on nature through the Jesuit

Collegio Romano. This continues to be an intriguing line of inquiry. It remains of importance for the *explanation* of the Scientific Revolution; hence it is to receive due treatment in Part II (section 4.3.3). Owing to the historiographical developments sketched above, however, the issue no longer carries the special significance for the nature of early modern science once allotted to it by Duhem and his followers.

Finally, now that, since the sixties, the Galileo debate had become so technical, it was also largely divested of the highly philosophical clothing it had been clad in by the original proponents of the mathematization of nature. (As one result, many crudely empiricist notions were at once allowed to creep in again through the backdoor.) This was part of a larger process in contemporary historiography of science and is therefore taken up separately under the next heading.

Philosophy on its way out. In the hands of the pioneer generation the historiography of science was still heavily tinged with philosophical conceptions. True, no longer was the writing of history made subservient to ultimately philosophical purposes, as had been largely the case throughout the preceding century. Yet a further emancipation from philosophy seemed possible and was in fact carried out by the post-Koyréan generation. This was particularly true of developments in England. Crombie's views on the origins of early modern science are somewhat of a borderline case here, but of the historical interpretations given in the late forties through the early eighties by, successively, Butterfield, Hall, Boas (later Boas-Hall), and Hall again, it can safely be said that they tended strongly to give the facts and put these in historical perspective without caring much for their possible wider philosophical ramifications. As a result, such clear-cut yet one-sided pictures as drawn by Koyré, Dijksterhuis, and Burtt gave way to more inclusive, but not so sharply drawn, accounts in a more narrowly historical vein. We shall trace this new development in the next section under the title Butterfield and the Halls: The View from England.

The Hermeticist challenge. An entirely new perspective on the Scientific Revolution, which was unfolded in the sixties and, in the next decade, took the profession by storm, also issued from England. It did as much as anything else to make the ongoing debates over impetus and the principle of inertia, and similar bones of contention, so obsolete in so short a time. This new perspective was the view that there were intimate, but as yet largely ignored or misunderstood, connections between the Scientific Revolution and preceding Renaissance thought of a magical, broadly neo-Platonic kind that began to flower through the coming to light, in the 15th century, of the Hermetic corpus of writing. Not so much polar opposites (as had hitherto been taken for granted), magic and science now appeared to be connected through links whose nature was quite variously described, but whose existence seemed undeniable.

To a considerable extent, the debate over the precise connection to be established between 17th-century science and 16th/17th-century Hermetic thought took over in the late sixties and seventies the vital focus once assigned to the details of the transition from Aristotelianism to early modern science. Besides obvious differences in content there is one major analytical difference between the two debates. Whereas the latter went far to shape our present-day conceptions of how the scientific achievement of the Scientific Revolution should be defined, the former debate has tended to take these results mostly for granted and to focus rather on the wider intellectual context of early modern science as well as on the question of whether and, if so, to what extent, Hermetic thought was instrumental in *bringing about* the Scientific Revolution. For the organization of the present book this means that the subject of Hermeticism is to be split up between the next chapter, which is about the wider ramifications of the Scientific Revolution, and Part II, which deals with its causes.

Coming to grips with the other sciences.   One major reason for reducing somewhat the virtually exclusive attention given to the mathematization of nature by the pioneers of the twenties and thirties was a felt need to reinsert other sciences than just astronomy and mathematical physics into the story. The chief impulse toward feeling a need for the inclusion of the organic sciences, in particular, came from historians of science with a background in chemistry, such as Pagel, Debus, and Hooykaas. The first two showed how, under the inspiration of Paracelsus, something like a 'chemical world-view' appeared in the late 16th century, which, during the first decades of the 17th century, competed powerfully with mechanical modes of explaining the operations of nature.[229] Hooykaas insisted on the contribution made by chemists in reuniting intellectual and manual efforts in the 16th and 17th centuries. All these historians emphasized the importance of chemistry for the Scientific Revolution, but none of them went on to write a full-fledged history of the Scientific Revolution primarily from the point of view of chemistry rather than of astronomy and mathematical physics.

Whether that can be done at all is still an open question. What has been done is to include chemistry, but also experimental optics, electricity, magnetism, and the life sciences into the story of the Scientific Revolution. In the United Kingdom this was done in a rather literal fashion, that is, by inserting them in a story. In the United States, in contrast, we find a couple of attempts to draw up new organizing schemata which, while preserving major insights gained by the previous generation, were also equipped to embrace on equal footing the nonmathematical physical sciences as well as chemistry and the life sciences and to fit all these into a wider yet equally coherent picture. Major efforts in this direction were undertaken by Kuhn and by Westfall, both of whom elaborated hints by Burtt and Koyré into full-fledged two-current accounts of the Scientific Revolution. Before discussing these we fix our gaze on what happened in the United Kingdom when, in 1948, the Cambridge historian Herbert Butterfield was invited to give a series of lectures on what he, having read but also having slightly misread Koyré, decided, at a momentous instant, to call the Scientific Revolution.

2.4.3. *Butterfield and the Halls: The View from England*

The book that came out of these lectures was published in 1949 under the title *The Origins of Modern Science, 1300–1800*. Herbert Butterfield (1900–1979) was a 'general' historian, not a historian of science, and the book was to remain his only contribution to the field. He never claimed to have done original research in the history of science or to have based his book on any but secondary material. It is for this reason chiefly that one admirer of Burtt's *Metaphysical Foundations* was in the habit of calling *The Origins of Modern Science* a 'poor man's Burtt.'[230] However, Butterfield selected his secondary material in a masterful way and added to it some noteworthy features uniquely his own which helped greatly to turn his book into a landmark in the historiography of the Scientific Revolution. Its principal accomplishment was threefold.

To begin with, Butterfield's usage of the term 'Scientific Revolution' as a central category in his historical survey gave wide "currency to the name" for the first time, even though he had not been the first to use it, as he was well aware. In I.B. Cohen's words, "Butterfield was largely responsible for making the Scientific Revolution a central issue in the mind of every reader."[231] He lent the concept a sense of historical urgency precisely because, as a 'general' historian, he was in a position to articulate in a persuasive manner the idea that the Scientific Revolution set Western civilization on a new, exciting, and so far entirely uncharted course.

What did Butterfield take the 'Scientific Revolution' to mean? At the surface level, the meaning is plain enough: It is the upheaval in science that took place between 1300 and 1800. A remarkably long period compared to the few decades originally assigned to the event by Koyré or to the 140 years Burtt and Dijksterhuis and the later Koyré were talking about![232] Butterfield's chief reason for starting so early is the impetus theory, which, taken in conjunction with its wider applications, is said by Butterfield, in true Duhemian fashion, to "represent the first stage in the history of the scientific revolution."[233] The story subsequently centers on 16th- and 17th-century achievements in astronomy and mechanics, "which hold the strategic place in the whole movement."[234] Anatomy and physiology come in for independent treatment in a chapter that is set up as a parallel treatment to the one on Copernicus and deals mainly with Vesalius and Harvey, after which these subjects are dropped. The other life sciences are handled as an example of how the corpuscularian world-view (represented chiefly by Boyle) was applied to natural phenomena. The same goes for chemistry. Yet chemistry is one of the reasons why the book runs through 1800. Having dealt with the Newtonian synthesis and having shown how the results, and the new thinking habits, of the Scientific Revolution could usher in the era of the Enlightenment, Butterfield introduces Lavoisier as the originator of a "Postponed Scientific Revolution in Chemistry"—a revolution, apparently, that ought to have taken place in the 17th century but somehow didn't.[235]

Despite the very long duration thus assigned to the Scientific Revolution, the bulk of the historical survey centers on the period between Copernicus and Newton. This is the other sense in which Butterfield employs the concept of the Scientific Revolution. At the back of his mind seems to reside some sort of 'inner' Scientific Revolution inside that larger Scientific Revolution that encompassed the full five-hundred-years period. It is hard indeed to leave *The Origins of Modern Science* without gaining the distinct impression of a 'Revolution in the Revolution' (to borrow a once-famous phrase from that forgotten political ideologue of the sixties Régis Debray). Here and there in the book the terms adopted by Butterfield reveal his dual usage of the concept. For example, when discussing the effects, during the Renaissance, of the Voyages of Discovery, he remarks:

> The discovery of the new world, and the beginning of a close acquaintance with tropical countries, released a flood of new data and a mass of descriptive literature which itself was to have stimulating effects. *The essential structure of the sciences was not changed—the scientific revolution was still far off—*but the Renaissance. . . . [my italics].[236]

Here a much narrower time span to be covered by the term 'Scientific Revolution' is obviously intended. But the discrepancy is not brought to the surface or made the subject of further historical analysis. Behind the discrepancy lurks a tension equally present in much later writing on the Scientific Revolution in that it seems so very hard to reconcile the long list of excitingly novel things that happened in European science from at least the 16th century onward (Paracelsus' intellectual revolt, Vesalius' and Tycho's observations, Gilbert's and Harvey's discoveries, Boyle's sceptical chemistry, Leeuwenhoek's findings, and so on) with the very specific upheaval in 'the essential structure of the sciences' wrought in the decades around 1600 by Galileo and Kepler and Descartes and later completed by Newton. How did these Scientific-Revolutions-in-two-different-senses hang together? To bring them together harmoniously in one coherently structured picture constitutes, so I think, a key challenge to future historians of the achievement of the Scientific Revolution.

Butterfield not only gave wide academic currency and a sense of intellectual urgency to the concept of the Scientific Revolution however defined but also had influential things to say on what constituted the nature of the transformation. Right on the first page of the book he points to what he regards as the essential feature of scientific change during the revolution:

> We shall find that in both celestial and terrestrial physics—which hold the strategic place in the whole movement—change is brought about, not by new observations or additional evidence in the first instance, but by transpositions that were taking place inside the minds of the scientists themselves. . . . Of all forms of mental activity the most difficult to induce . . . is the art of handling the same bundle of data as before, but placing them in a new system of relations with one another by giving them a different framework, all of which virtually means putting on a different kind of thinking-cap for the moment.[237]

Throughout his book Butterfield points to such "mental transpositions," which, for him, are at the heart of historical transitions generally:

> It is never easy—if it is possible at all—to feel that one has reached the bottom of a matter, or touched the last limit of explanation, when dealing with an historical transition. It would appear that the most fundamental changes in outlook, the most remarkable turns in the current of intellectual fashion, may be referable in the last resort to an alteration in men's feeling for things, an alteration at once so subtle and so generally pervasive that it cannot be attributed to any particular writers or any influence of academic thought as such. . . . Subtle changes like this—the result not of any book but of the new texture of human experience in a new age—are apparent behind the story of the scientific revolution, a revolution which some have tried to explain by a change in man's feeling for matter itself.
>
> It is fairly clear in the sixteenth century, and it is certain in the seventeenth, that through changes in the habitual use of words, certain things in the natural philosophy of Aristotle had now acquired a coarsened meaning or were actually misunderstood. It may not be easy to say why such a thing should have happened, but men unconsciously betray the fact that a certain Aristotelian thesis simply has no meaning for them any longer.[238]

Two distinct sides to this conception of what was basic to the Scientific Revolution were to put their stamp on later thinking on the subject. For one, the idea of the new "thinking-cap" was to act as one formative influence on an American physicist who, after turning to history and writing a book on the Copernican revolution, began to intensify his reflections on the nature of scientific change: Thomas S. Kuhn. The passage also betrays a keen sense for what remains intangible in historical transitions of such magnitude as the Scientific Revolution—a sense that was to live on in the historical surveys of the Scientific Revolution written by Butterfield's pupil A.R. Hall, to whom we now turn.

Indeterminacy as a Hallmark. Although I have not carried out a word-count in Hall's successive histories of the Scientific Revolution, I am sure 'complexity' is among the most prominent terms to occur in them. Throughout his scholarly career, A. Rupert Hall (born 1920) has been a vigorous opponent of attempts at monocausal explanation and one-sided interpretation. For him the pleasure of history lies in its endless variety, and the attendant bane in its equally endless complexity. The unique event is what counts for him, and whenever he ventures at a statement of some level of generality, one can be virtually certain that it will be qualified in the very next sentence, just as one may reasonably expect that, one more sentence further down, the qualification will be qualified in its turn.

Perhaps it is fair to state at the outset that this is not the style of history writing I feel personally most at ease with, yet I am quite ready to admit that it may serve as a welcome antidote to all-too-ambitious attempts at throwing the facts of history into overly fanciful, or too tenaciously held, patterns or theories. In Part II of the present book we shall see some typical Hall-antidotes in full operation.

Unlike Butterfield, Hall is a full-blown historian of science, who, besides Butterfield, recognizes Joseph Needham and Alexandre Koyré, in particular, as the original guides toward his craft. Whereas *The Origins of Modern Science, 1300–1800* is a fairly brief, introductory text which requires no expertise in science at all, Hall's *The Scientific Revolution, 1500–1800: The Formation of the Modern Scientific Attitude,* which came out five years later (in 1954), is far more in the nature of a historical overview, albeit concise, and is also somewhat more technical. As such, it has served as the standard text for a generation of English-language students of the Scientific Revolution.

In discussing, as we are about to do, not only this book of Hall's but also a number of sequels written by him and his later wife, Marie Boas, it is useful to note at the outset that these surveys were written in the course of scholarly careers almost wholly devoted to an understanding of the Scientific Revolution. An ongoing stream of more-specialized and detailed contributions forms the ever-present background to their surveys, to which I refer readers in search of the complete scholarly basis on which the Halls erected their successive monographs. Here we confine ourselves to gleaning from these the authors' leading ideas on what the Scientific Revolution was all about.

Hall's core approach to the history of science, and of the Scientific Revolution in particular, is stated right at the outset, in the introduction to his first book on the subject:

> The cumulative growth of science, arising from the employment of methods of investigation and reasoning which have been justified by their fruits and their resistance to the corrosion of criticism, cannot be reduced to any single theme. We cannot say . . . why some men can perceive a truth, or a technical trick, which has eluded others. From the bewildering variety of experience in its social, economic and psychological aspects it is possible to extract only a few factors, here

and there, which have had a bearing on the development of science. *At present at least, we can only describe, and begin to analyse, where we should like to understand* [my italics]. The difficulty is the greater because the history of science is not, and cannot be, a tight unity. The different branches of science are themselves unlike in complexity, in techniques, and in their philosophy. They are not all affected equally, or at the same time, by the same historical factors, whether internal or external. It is not even possible to trace the development of a single scientific method, some formulation of principles and rules of operating which might be imagined as applicable to every scientific inquiry, for there is no such thing.[239]

This statement of negative principle sets the tone for the entire account. Hall makes the Scientific Revolution start around 1500 because that is when, in his view, European science began to cut loose from the Greek heritage for the first time. He recognizes impetus theorizing as an important account of motion in its own right, yet, since it was developed inside an overall Aristotelian framework, it cannot be regarded as more than a precursor to the Scientific Revolution.[240]

Nowhere in the story do we come across anything in the nature of breaks or revolutions; differential rates of renewal is what characterizes the period throughout. Yet at the end we do find that something fundamental has changed indeed. European science has now developed some features uniquely its own. True, all civilized societies had to respond to "the ubiquitous challenge of Nature; but [outside Europe] the organized, conscious, rational response to it that we call science was of minor importance."[241] What made the science that developed in Western Europe between 1500 and 1800 so unique was this overall rational character which marks it throughout. This is the one thread that runs through Hall's account and holds it together: the rational nature of the Scientific Revolution as opposed to the various brands of mysticism, magic, superstition, and the like, which early modern science conquered and gradually outgrew. Here, for Hall, is the one and only true distinguishing criterion. Thus, in the case of many a scientist of the 16th or 17th century we find Hall weighing carefully whether his work can be called predominantly rational or whether, because of its irrationality, it should rather be taken as harmful, or at best as irrelevant, to the gradually evolving Scientific Revolution. What Hall has to say on alchemy seems to sum up nicely his attitude toward the entire Scientific Revolution: His search is for "the grain of real knowledge . . . concealed in a vast deal of esoteric chaff."[242]

Having made this fundamental distinction, Hall displays little interest in defining in a more narrow sense what the new science came down to. Mathematics, experiment, and 'mechanism' all have their part to play, yet "we cannot rely only upon appeal to experiment, observation, measurement, or any other over-simplification of the complex processes of science.[243]

One particular feature seems nonetheless to stand out a bit more clearly than others. It is what Hall calls 'mechanism,' a term which he does not subject to rigorous definition, but which gradually comes to stand for 'matter in motion' as the basic set of categories according to which natural phenomena were handled in the course of the Scientific Revolution. But again, this description fits the physical sciences better than the life sciences, and thus is not used without proper qualification. Altogether, the treatment of the fate of the various sciences throughout the period of the Scientific Revolution is such that they are handled essentially on a par, as separate strands in one, not very closely knit, fabric. And however closely Hall follows Koyré in his interpretation of Galileo, in particular, nothing at all remains, in his overall treatment, of the core conception from which Koyré had drawn his Galileo picture in the first place.

We have seen that, for the benefit of the Enlightenment and of the revolution in chemistry, Butterfield had stretched the Scientific Revolution so as to include the 18th century. So did Hall, but for partly different reasons. His tactical reason was that, besides chemistry, he felt a need to include the early history of electricity and magnetism in the story but could not really do so if he left them bungling in the 17th century, when these fields of study had scarcely reached the status of independent subjects in their own right. His rationale for the procedure is to be found in his statement that, since 1800, "though profound changes in scientific thought have occurred . . . , and though the growth of complexity in both theory and experimental practice has been prodigious, the processes, the tactics and the forms by which modern science evolves have not changed." [244] What has been revised since in science has more to do with content than with overall structure. But because, as a matter of principle, Hall had left it so vague what this structure really should be taken to be, it is not made clear by him why the dividing line between three centuries of 'preparation' and the years of subsequent 'accomplishment' should be drawn at the year 1800.

The first sequel.   Hall intended his book to be a "character-study" rather than a "biography" of the Scientific Revolution.[245] Eight years later he participated in an effort that was more in the nature of such a biography. Under the general title *The Rise of Modern Science* two volumes appeared, in 1962 and 1963, which together covered the whole event. The first volume was written by Marie Boas Hall under the title *The Scientific Renaissance, 1450–1630*; its sequel was written by Hall and called *From Galileo to Newton, 1630–1720*. In most respects, these two volumes are quite similar in organization, emphasis, and tone to *The Scientific Revolution*. But there are some notable differences as well. The revival of learning in western Europe is now taken to start already in the middle of the 15th century. The years around 1630 are now taken to separate the early, preparatory stages of the Scientific Revolution from its "central, decisive stages,"[246] so that here we find the same implicit distinction of a core Scientific Revolution inside a larger one that we encountered with Butterfield. The story is now brought to an end with the third and final reedition of the *Principia* still authorized by Newton himself rather than with the year 1800. Yet such shifts of emphasis are marginal to the overall setting. Again, we find the telling of a story rather than a historical argument; again, the one vital distinguishing feature of the Scientific Revolution appears to be its rational character. Boas' contribution focuses very much on the magical side of natural philosophy in the 15th and 16th centuries, in a tone that the following quotation represents quite well:

> The area to which magic could be and was applied in the sixteenth century was still very great; it is fascinating to observe the way in which, out of the muddled mysticism of sixteenth-century thought and practice, the scientifically valid problems were gradually sifted out to leave only the dry chaff of superstition.[247]

In this view, the emergence of early modern science comes down to a general process of purification, to which the three undistinct sisters 'magic, mysticism, and superstition' contributed in an essentially negative way by allowing themselves to be gradually eliminated from science:

> [Around 1630] the sheer success of science and the steady advance of rationalism generally meant the end of the magical tradition. Mathematician no longer meant astrologer; the word chemistry replaced alchemy as a new science was born; the number mysticism Kepler loved [*sic*] gave way to number theory . . . ; natural magic was about to be replaced by experimental science and the mechanical philosophy. Science and rationalism were to become synonymous, cemented together by Descartes's *Discourse on Method* (1637).[248]

The same view of the Scientific Revolution as the unfolding of rationalism pervades Hall's sequel volume. There is a "mainstream of rational scientific development,"[249] and this is what prevailed during the second half of the 17th century, whereas "the first thirty years of the seventeenth century had shaken the old order of things but by no means disrupted it."[250]

Another changing feature of early modern science was the role of experiment. "Even as late as 1630," Hall contended, "systematic observation had proved a far more constructive method in science than had experiment." This was true of the observations on the human body which revolutionized anatomy; of Tycho's observations; of the new data enabling botany and zoology to "assume organized form."[251] Hall even went so far as to state that

> observation had brought cosmological theory to the point of crisis. However much the establishment of a new science was indebted to the experimental inquiry later, as yet . . . mere observation of what is happening all the time was almost enough to destroy the formal rigidity of the old. To have the use of one's eyes—if one knew how to direct them—was to see that what Aristotle or Ptol-

emy had described was false and to find reason for a fresh idea of nature. And it was in this idea that the explosive force of the scientific revolution lay.[252]

This seems a clear-cut commitment to a view of the nature of the Scientific Revolution that had been abandoned by *avant-garde* historians of science ever since Mach. But seventy pages later we find the view properly qualified:

> The scientific revolution was not effected by empirical methods only nor indeed by the use of any single new method of science. It was the replacement of one system of thought by another, and this process had begun long before the time of Bacon, or Descartes or even Galileo. The critical feature of seventeenth-century science was that it embraced new or revived ideas—an infinite universe in which the sun was but one star like countless others and life perhaps a commonplace, atomism, the mathematisation and mechanization of natural processes, the idea of law and regularity—which were not and could not be proved experimentally and were related only incidentally to the rise of empiricism.[253]

Thus, after all, we find that a number of distinguishing features of the Scientific Revolution are once more being handled on a par. One of these features—the process of mechanization—finds further clarification in a later chapter in the book. We have seen before how variously this term was being used by historians of the Scientific Revolution. In his earlier book, Hall had defined 'mechanism' as the treatment of natural phenomena by means of the categories of matter and motion; now he identifies the 'mechanical philosophy' squarely with the corpuscular view of nature, that is, with the *more specific* conception that natural phenomena can be reduced to the motions of invisibly small material particles of varying sizes, shapes, and distributions over space.[254] But again, in a more generalizing mood Hall had already indicated how,

> if the phenomena of nature arise from the motion of particles, the science of motion provides the essential key to physics and beyond that to chemistry and even physiology—since these sciences depend on the physical properties of matter. Thus the mechanistic explanation of things could develop as an endless chain, beginning with the laws of motion and stretching further and further into successively more complex departments of science; always however traceable back to firm roots in the laws of motion that could not be false. Here, potentially at least, lay the unshakable foundation for all scientific knowledge.[255]

The net conclusion is that, although at places more specific about the defining features of the Scientific Revolution than in his earlier book, Hall ultimately declined again to commit himself to anything else but his, and Boas', idea of its overriding rationality. At the time of writing, he cannot have expected that, when a quarter-century later he was to bring his *Scientific Revolution* up to date with all the detailed research that had since been carried out by himself and hosts of other scholars, it was this core view of his and his wife's that now had to be defended against the attacks under which it had come from the viewpoint that 'magic, mysticism, and superstition' had contributed in a more productive sense to the emergence of early modern science.

The Scientific Revolution *brought up to date.*   The revision was a thorough one, and Hall marked the fact by giving his almost wholly rewritten book a new title, *The Revolution in Science, 1500–1750.* It came out in 1983, which makes it the most recent book-length survey of the Scientific Revolution currently available. The updating may be exemplified by the new treatment of Galileo, which takes into account all those discoveries about Galileo's experimental work made since by Drake and others. Hall readily grants that his own, and many others', earlier interpretation of Galileo's thought had been "wrong," because it was too "'idealistic'."[256] The treatment of Kepler is also considerably modified. Kepler's 'mystical' side so severely treated in earlier volumes is now acknowledged to have been more than mere "dross."[257] Altogether the urge to distinguish between 'rationalist' and 'irrationalist' thought in the 16th and 17th centuries, with the Scientific Revolution yielding the gradually emerging victory of the latter over the former, is considerably toned down in the new version. True, in a strident passage in his introduction Hall still defends his right to focus on the "victor's side of the battle,"[258] but in the actual telling of the story the dichotomy hardly serves any longer as the chief guiding-line.

No other organizing idea or principle takes its place. Even more so than in earlier volumes Hall's views on defining characteristics and structural elements of the Scientific Revolution must be gleaned from occasional remarks scattered throughout the book. The start of the Scientific Revolution in 1500 is justified by an appeal to "a shift in the cultural base" that took place at the time.[259] It consisted of a "shift in the form of explanation from the verbal to the mathematical"[260] as well as of the new idea European man acquired of himself "as the great practical inventor to whom the combination of common sense, manual handiness and natural knowledge had brought power and wealth."[261] The Scientific Revolution is now taken to have come to an end around 1750 rather than in 1800, because "when Newton died the great creative phase of the scientific revolution was already finished, though its acceptance and assimilation were still incomplete."[262] Chemistry, geology, and the life sciences deserve no less a legitimate place in the Scientific Revolution than the physical sciences, even though it is true that there are great "historical asymmetries" between them[263]—a "general contrast" in that changes in the former took place largely on the level of phenomena, whereas the latter signal "shifts of metaphysical perspective."[264] In the biological sciences the revolutionary situation, even though equally present, "was much less successfully exploited."[265] Hence, although "the study of living things did participate in the scientific revolution, [its] typical core was certainly in the mathematical sciences."[266]

The Scientific Revolution, so Hall makes clear at one point, may be broadly divided into two stages. The first is the "century of confusion,"[267] when all is in gestation and flux, but no clear program of research has yet emerged. "By contrast, if we leap a century ahead to the years near 1640, a far more positive situation presents itself and the labours of the next couple of generations—the period of the scientific revolution *par excellence*—seem to follow in almost logical succession."[268] Once again we come across a core Scientific Revolution set inside a larger one. The new stage is said to take over from the "century of confusion" in 1642, with Galileo's death and Newton's birth. The "defining characteristics" of the "century of fulfillment" that follows are juxtaposed as the "questioning of mere authority, acceptance of Copernicanism and mechanism, faith in empirico-rational arguments, and especially mathematics."[269] Thus, the 'Scientific Revolution *par excellence*' appears closely linked with the mathematical sciences, in particular. Once again we are left to wonder how precisely the inner Scientific Revolution, which betrays remnants of Koyré's original conception, is connected to the outer Scientific Revolution— the one that had been introduced by Butterfield and Hall to take into account all novelty in European scientific thought in all the various sciences cultivated at the time.

Conclusion.   We may distinguish between two radically different conceptions of the Scientific Revolution. On the one extreme, the revolution comes down to a fundamental upheaval in cosmology and mechanics—an upheaval which inevitably affected other areas of inquiry into nature and substantially speeded up their autonomous rate of change, without however itself being affected by them. This extreme position had almost without qualification been taken by our pioneers of the twenties. Inside this overall framework the position still left room for varying definitions of what the essence of the upheaval should be taken to be.

At the other extreme, the Scientific Revolution can be regarded as an overthrow of earlier patterns of thought about nature that essentially affected all sciences alike. All disciplines at the time were caught in a process of accelerated change and growth, with only the rate of change varying across the various domains of inquiry involved. In this view, which is exemplified by Butterfield and more outspokenly so by the Halls, no sharper distinguishing criterion between early modern science and its Aristotelian and magical predecessors was found than its newly won rationalism. Leaving aside for now how satisfactory a description of the essential distinction this is, the salient point is that, at this extreme, the unique events in cosmology and in the emerging science of mechanics can only be brought out by giving them some added emphasis in the course of the overall story, and the question remains of whether that is enough to do them justice.

Ultimately, the issue comes down to whether the 'Scientific Revolution' is to be seen as a shorthand label for 16th- and 17th-century science in its entirety or rather as an analytical concept that stands for whatever it was that, inside this wider domain of 17th-century science, distinguished vital parts of it so sharply from preceding conceptions of nature. For the Britons in our account—Whewell, Butterfield, Hall—the former position had been the more attractive, and perhaps even the only natural one to take. The Continentals and the one American who dominated our historical overview in between Whewell and Butterfield had implicitly committed themselves to the latter, analytical conception of the Scientific Revolution.[270] We shall now devote the final two sections of this chapter to those two Americans who, in following Koyré and adopting a similarly analytical view of the Scientific Revolution, nevertheless made an attempt to salvage more of the surrounding sciences than their Continental predecessors, and Koyré in particular, had thought possible.

2.4.4. *Kuhn and the Scientific Revolution*

The title of the present section is less self-evident than it seems. The primary thing with which to associate the name of Thomas S. Kuhn (born 1922) is the scientific revolutions whose structure he studied in that famous book that came out in 1962. So popular has his book become that, in the minds of many academics outside the discipline of the history of science, the generic concept of scientific revolutions and the specific concept of the Scientific Revolution have conflated to the point of complete, although unconscious, identification. It is ironic that this happened only fifteen years after the concept of the Scientific Revolution had managed to emancipate itself from the centuries-old, general notion of scientific revolutions. It is even more ironic that this was an unintended side effect of the work of a disciple of the very two historians of science to create the term and the concept of the Scientific Revolution in much of its present-day usage: Koyré and Butterfield.

For Kuhn himself the two distinct notions—the unique Scientific Revolution and scientific revolutions in general—have always remained apart to a surprising extent. Although he has written extensively on both topics, he has never to my knowledge examined how they are connected either conceptually or in the reality of the past. Such an examination seems the more worthwhile because his views on the one and on the other appear hard to reconcile. Although it cannot be the task of the present section to carry out such an examination in full, we shall come across vital aspects of the problem. First, we examine what happened to the Scientific Revolution in *The Structure of Scientific Revolutions*. Next, we discuss a fascinating contribution Kuhn began to make in 1972 toward a resolution of virtually all the sticky issues that we have found to come to the fore in the Great Tradition of the historiography of the Scientific Revolution. We conclude by pointing at the possible origin of the relative incompatibility of Kuhn's views on scientific revolutions and on the Scientific Revolution.

The historical origin of The Structure of Scientific Revolutions.   It is remarkable how many of the ideas on the origins of early modern science that have
been discussed so far in the present chapter found their way into certain key
notions that make up the argument presented in *The Structure of Scientific
Revolutions.*[271] This is not to say that the book lacks originality—on the contrary, its originality lies precisely in the refined manner in which Kuhn brought
together all kinds of previously wide-apart ideas on the historical growth of
science and made them merge both with one another and with a number of
ingredients of his own making so as to yield a smoothly appealing account
of the revolutionary character of science. As far as the prehistory of the notion
of scientific revolutions is concerned, Kuhn's debt to the past has been demonstrated extensively by I. Bernard Cohen in his search for 18th-century usages
of the idea of the revolutionary growth of science. But it is no less true for
accounts of the Scientific Revolution. Substantial bits and pieces of the conceptual apparatus that from d'Alembert to the Halls went into the making of
historical representations of the Scientific Revolution have equally found their
way into *The Structure of Scientific Revolutions.* In the first pages of his book
Kuhn has himself called attention to this fact.

He does so in direct reference to the period, in the twenties through forties
of the present century, when the historiography of science reached maturity.[272]
He mentions Koyré's name, in particular, and goes on to say that

> seen through the works that result [from the new historiography of science as
> exemplified by Koyré] . . . science does not seem altogether the same enterprise
> as the one discussed by writers in the older historiographic tradition. By implica-
> tion, at least, these historical studies suggest the possibility of a new image of
> science. This essay aims to delineate that image by making explicit some of the
> new historiography's implications.[273]

In other words, *The Structure of Scientific Revolutions* can be read as a generalized account of what lessons lay hidden in the new historiography of the Scientific Revolution that started in the twenties and thirties.

These lessons may easily be identified by following up Kuhn's own frequent invocation of the names of Koyré and Butterfield. When, in 1947, the
twenty-five-year-old graduate student of physics Thomas S. Kuhn—already
greatly interested in the nature of science—decided to switch to the history of
science, he soon came to recognize Alexandre Koyré as his true "maître."[274]
We may assume that Koyré's core conviction of the revolutionary nature of the
origins of early modern science made some lasting impression on his student.
After all, the idea that the growth of science is not a gradual affair but involves
veritable convulsions in human thought about nature lies at the heart of the
works of both master and disciple.

In Butterfield's *The Origins of Modern Science* Kuhn found a key idea
about how transformations in science actually take place: the notion of mental

transposition, of the 'new thinking-cap,' of the new way of looking at a set of well-known natural phenomena for which the current standard interpretation no longer appears to make sense. Kuhn went ahead and found that this somewhat informal concept (who, after all, ever wears a thinking-cap, let alone exchanges one for another?) is expressed in a more sophisticated fashion by psychologists in the theoretical framework of the 'gestalt-switch', which came to play such an important part in Kuhn's overall schema of revolutionary science. Generally speaking, this schema is dominated by the discontinuity view of scientific growth, which Kuhn found ready in Koyré's core conception of the Scientific Revolution.

However, according to the structure sketched in the book of the same name, revolution is not all there is to scientific advance. Although revolutions are the principal agents of truly creative conceptual and theoretic growth, they are the exception. 'Normal' science is the rule, and this comes down to the quiet and successive resolution of remaining puzzles after the revolution has led to the establishment of a fresh 'paradigm'—a set of principles and *modi operandi* accepted as valid, for the time being, by the community of scientists organized around the paradigm. (If this sounds slightly circular, I cannot help it: The tinge of circularity has been there all along in the definition of this wonderfully versatile concept—the Kuhnian paradigm.) It is the perception, at first by isolated geniuses and then by the profession at large, of a crisis in the current paradigm that, eventually, through the mechanism of the gestalt-switch, leads to its revolutionary overthrow and the establishment of a new paradigm, after which a fresh cycle of the quiet cultivation of normal science possibly followed by a new perception of crisis and a new paradigm sets in all over again.

Kuhn has insisted that such a schema does not apply to all branches of scholarly endeavor. It is true only of those disciplines which at some point in their existence have found a single shared framework of thought, or 'paradigm', in the first place. This has generally become true of the natural sciences, whereas the feature of a shared paradigm is equally conspicuously *lacking* in, for example, most or all of the social sciences up to the present. As long as lengthy but unresolved debates over fundamentals continue to be the order of the day in a given branch of scholarship, this serves as a sure sign that the field in question still finds itself in the pre-paradigm state. It is precisely the arrival at such a state, one after another, of the various natural sciences that makes these fields of inquiry so different from all the others. As soon as the paradigmatic stage has been reached, specialization sets in; the language used becomes incomprehensible to all but the initiated, and the field becomes cumulative in the sense that, in handling a certain puzzle, its practitioner can take for granted the underlying structure of concepts, theories, theorems, and laws— unless, that is, the anomaly that gave rise to the puzzle appears to be entirely insoluble inside the given framework, and thus may lead to the perception of crisis that stands at the origin of a new revolution.

Kuhn's basic distinction between pre-paradigmatic and paradigmatic sciences thus displays a striking structural analogy with Kant's idea. In one sense, Kuhn's book is nothing but the detailed elaboration of this view of Kant's that at a certain stage in its development, after a longer or shorter period of "groping in the dark," one discipline or another attains the "secure pace of a science." This was Kant's account of the origin of early modern science, and we have to find out now whether or not this is true in Kuhn's book as well. In other words, is it or is it not Kuhn's opinion that the attainment of the paradigmatic state in the natural sciences coincides with, or even defines, the Scientific Revolution?

Two different answers to this question can be dug up from Kuhn's work. If one confines oneself to a reading of *The Structure of Scientific Revolutions,* the answer is no. But a later essay of Kuhn's gives another impression—it yields a 'yes and no' of a type that helps to enhance our insight into the nature of the Scientific Revolution. We shall take up these two different answers in turn.

The Scientific Revolution dissolved.   Our outline of the main structural elements of Kuhn's theory of scientific growth may have reminded the reader of the somewhat similar endeavor undertaken more than a century earlier by William Whewell. Although the operating mechanisms employed by the two authors are quite different, the resulting picture is not. In both cases we find unilinear progress made up of revolutionary cycles which contain strong elements of continuity over time (although Kuhn is much more sceptical of whether this indeed constitutes 'progress' in the sense of arriving at the *truth*). Interestingly, we find that this shared overall structure of the growth of science in general entails in both cases exactly the same consequences for the Scientific Revolution: It simply dissolves, because the general picture leaves no room for the unique event.

We know how this came about in Whewell's case. A perusal of the relevant chapters in *The Structure of Scientific Revolutions* quickly shows that Kuhn does not at all handle the Scientific Revolution as one key example of those scientific revolutions the book is about. Rather, at least four distinct portions of the Scientific Revolution come in for treatment as cases of revolutionary paradigm shifts all by themselves: Kepler's work on planetary orbits, Galileo's new view of pendular motion, the mechanical philosophy, and 17th-century optics. Elsewhere I have argued that this failure, or refusal, to take the Scientific Revolution as constituting a 'paradigm shift' in its own right has unfortunate consequences for Kuhn's account of theory replacement in general.[275] Here I wish to confine myself to my job as a historian of historiography in pointing out the strange irony that the very conception which gave rise to the writing of *The Structure of Scientific Revolutions* in the first place appears dissolved at the end of that great, yet endlessly controversial, book.

The Scientific Revolution restored. When the same question, namely, of whether the arrival of the natural sciences at the paradigmatic state is indeed what defines the Scientific Revolution, is put to other portions of Kuhn's work, we find a far more nuanced answer than the negative one implicit—though scarcely elaborated—in *The Structure of Scientific Revolutions.* This answer is to be found in his article "Mathematical versus Experimental Traditions in the Development of Physical Science," which started life as a lecture in 1972, appeared in a journal in 1976, and was codified one year later in Kuhn's collection of essays *The Essential Tension.* For the sake of convenience, I shall refer to this momentous article by the shorthand title "Traditions."

The article begins with an illuminating historiographical introduction. Kuhn distinguishes between two currents in the historiography of science, taking as his starting point the question of whether the sciences are one or many. The historiographical current that takes the sciences as many tends to retrace the prehistory of scientific ideas in a given discipline as defined now, that is, in its present demarcated state among adjacent fields of academic concern. The other current, in handling the sciences as one, necessarily confines itself to their extrascientific environment while neglecting its evolving substantive content. Since neither alternative is really acceptable from a truly historical point of view,

> historians who wish to illuminate actual scientific development will need to occupy a difficult middle ground between the two traditional alternatives. They may not, that is, assume science to be one, for it clearly is not. But neither may they take for granted the subdivisions of subject matter embodied in contemporary science texts and in the organization of contemporary university departments.[276]

By way of an example of how to overcome this dilemma, Kuhn then offers a "schematic presentation"[277] of an arrangement of scientific disciplines in the past that aims to do justice to their past demarcations while at the same time preserving something of what they had in common and what made them interact. In order to achieve this aim, Kuhn makes a fundamental distinction between two categories of sciences: the 'classical physical sciences' and the 'Baconian sciences.'

The classical physical sciences are defined by Kuhn as those "which . . . were already in antiquity foci for the continuing activity of specialists."[278] The "extremely short list" Kuhn comes up with comprises five such fields, namely, astronomy, geometrical optics, statics (including hydrostatics), mathematics, and 'harmonics'. What these disciplines, as cultivated in ancient Greece, had in common is that they were domains of technical expertise inaccessible to the layman, they witnessed "the accumulation of concrete and apparently permanent problem solutions,"[279] and together they formed a cluster in the sense that scientists who contributed to one of them were most likely to contribute to other members of the cluster, too (examples are Euclid, Archimedes, and Ptolemy). It is interesting to note that Kuhn might well have replaced, but did not replace, this whole set of characteristics by one key concept to be taken from his *Structure*, by saying that these were the only ancient sciences that had reached the paradigmatic state.

One other basic thing the sciences in this mathematical cluster had in common was that

> their considerable ancient development required little refined observation and even less experiment. For a person schooled to find geometry in nature, a few relatively accessible and mostly qualitative observations . . . provided an empirical basis sufficient for the elaboration of often powerful theories.[280]

This is what enabled the practitioners of the classical physical sciences to make rapid progress of a type that eluded students of such other subject matters as electricity, magnetism, and the like. On their long journey from ancient Greece to Renaissance Europe the classical sciences continued to be cultivated on a high level of technical proficiency, while some new features were added to them by contributions from other civilizations. The most notable contribution, though, came from medieval western Europe. While originally the mathematical cluster "was subordinated to a dominantly philosophical-theological tradition," gradually a sixth topic was added in that,

> partly as a result of fourteenth-century scholastic analysis, the subject of local motion was separated from the traditional philosophic problem of general qualitative change, becoming a subject of study in its own right.[281]

Since this new field—the study of motion—shared the characteristic of needing only few observational data for being formulated in largely mathematical terms, it could easily fit into the closely knit cluster.

Thus constituted—so Kuhn continues his schematic account—all the sciences that made up the cluster, with only one exception, underwent a process of drastic transformation in the 16th and 17th centuries. This exception, in Kuhn's view, was 'harmonics'. All the other sciences were radically reconstructed: mathematics, astronomy, and the study of motion in ways too obvious to dwell upon here; statics in the sense of the extension of hydrostatics to include pneumatics in the hands of Torricelli, Pascal, and Boyle; and optics through its "gaining a new theory of vision, the first acceptable solution to the classical problem of refraction, and a drastically altered theory of colors."[282]

Having thus arrived at the radical scientific upheaval that forms our proper subject, Kuhn goes on to outline some of its key features. To begin with, these were the *only* physical sciences to undergo, in the 16th and 17th centuries, such a transformation, a conclusion elaborated in the following key passage:

> These conceptual transformations of the classical sciences are the events through which the physical sciences participated in a more general revolution of Western thought. If, therefore, one thinks of the Scientific Revolution as a revolution of *ideas,* it is the changes in these traditional, quasi-mathematical fields which one must seek to understand.[283]

So much for the classical sciences. At this point Kuhn introduces the emergence, early in the 17th century, of a different cluster, which he labels the Baconian sciences. These were experimental in a sense fundamentally different from the empirical basis common to the classical sciences:

> The participants in the new experimental movement, often called Baconian after its principal publicist, did not simply expand and elaborate the empirical elements present in the tradition of classical physical science. Instead they created a different sort of empirical science, one that for a time existed side by side with, rather than supplanting, its predecessor.[284]

One key difference between the two clusters thus concerns the nature of experiment. Experiments in the classical sciences were mainly carried out for the sake of confirmation; those in the Baconian tradition were largely heuristic:

> Men like Gilbert, Boyle, and Hooke . . . wished to see how nature would behave under previously unobserved, often previously nonexistent, circumstances. [The resulting collections of data may often look quite loose and unstructured to us, but, under closer examination, such collections] often prove less random in choice and arrangement of experiments than their authors supposed. From 1650 at the latest, the men who produced them were usually guided by one or another form of the atomic or corpuscular philosophy. Their preference was thus for experiments likely to reveal the shape, arrangement, and motion of corpuscles.[285]

However, for a long time there would remain a gap here, since the corpuscular view of things did not in fact lend itself naturally to experimental testing.

Another feature of the new experimentalism inside the Baconian cluster is that the experiments, rather than taking nature and its phenomena for granted, were set up in such a way as to force nature to "exhibit [itself] under conditions it could never have attained without the forceful intervention of man."[286] This is how the Baconian cluster added a characteristic to science it had never possessed before: A substantial set of scientific instruments was invented and employed by means of which nature could thus be challenged to display phenomena not otherwise observable. In the Baconian tradition, too, the reporting of experiments was as careful and circumstantial as it was loose, 'idealized,' and ambiguous on fundamentals in the case of the classical sciences, in particular with Galileo but also with Pascal.

Having thus characterized the Baconian cluster, Kuhn goes on to ask what effect the cluster exerted on the development of the classical sciences. His basic view is that "to the conceptual transformations of the classical sciences, the contributions of Baconianism were very small."[287] This is illustrated by a brief account of the quite limited importance experiments possessed for these transformations. To the extent that instruments played a significant role in the transformations in question, they go largely back to the tradition of the classical sciences itself. Also, "all owe their special effectiveness to the closeness with which they could confront the evolving theories of classical science, which had called them forth."[288]

Two principal conclusions are made by Kuhn to follow from the above argument. First, he says, the conception of the Scientific Revolution put forward by Koyré and Butterfield, although in some need of qualification, is fundamentally right:

> The transformation of the classical sciences during the Scientific Revolution is more accurately ascribed to new ways of looking at old phenomena than to a series of unanticipated experimental discoveries.[288]

But this is not to say that such experimental discoveries were insignificant to the whole process:

> If Baconianism contributed little to the development of the classical sciences, it did give rise to a large number of scientific fields, often with their roots in prior crafts.[290]

Examples of such new fields are the study of electricity, of magnetism, and of heat; also, chemistry, which in a sense had existed long before the Scientific Revolution, was elevated from a craft to the status of a regular science as a result of its absorption, through the revision of its experimental methods, into the Baconian cluster.

It would be too much to expect from these "new foci for scientific activity in the seventeenth century"[291] that they at once display the cumulative characteristics of the classical sciences:

> If the possession of a body of consistent theory capable of producing refined predictions is the mark of a developed scientific field, the Baconian sciences remained underdeveloped throughout the seventeenth and much of the eighteenth centuries. Both their research literature and their patterns of growth were less like those of the contemporary classical sciences than like those discoverable in a number of the social sciences today.[292]

Translated into another terminology, remarkably eschewed by Kuhn in the article under discussion, this means that the Baconian sciences did not achieve paradigmatic status until the middle of the 18th century. In fact, Kuhn continues, they were not to merge with the classical tradition until the 19th century, when the barriers between the mathematical and the experimental approaches finally began to break down.

These barriers had existed in full bloom in the 17th century. The mathematical approach traditional to the classical sciences had scarcely any influence on the Baconian cluster, whereas the emergence of the Baconian sciences produced "a gradual refinement rather than a substantial change in the nature of the classical sciences"—their transformation had other sources, and a substantial "ideological gap" remained.[293]

At this point Kuhn anticipates two objections likely to be directed against the picture of the Scientific Revolution thus arising, namely, that neither Galileo nor Newton seems to fit it. Regarding Galileo, Kuhn argues that, although indeed he took part in an engineering and instrumental tradition, his "dominant attitude toward that aspect of science remained within the classical mode . . . —he could and did make epochal contributions to the classical sciences but not, except through instrumental design, to the Baconian."[294]

Newton, however, fully participated in both traditions. But as such, "with the partial exception . . . of his continental contemporaries, Huyghens and Mariotte, Newton's example is unique."[295] The *Principia* fully belongs to the mathematical tradition, whereas the *Opticks* straddles both, in that it fuses a transformation of geometrical optics with new experimental data on light (interference, diffraction, polarization) which the classical tradition was powerless to handle. Also, the speculations in the "Queries" at the end of the *Opticks* all stem from the Baconian tradition.

Attractive features of the schema.　We break off the account of Kuhn's argument at this point. Having arrived so far, Kuhn himself pursues three paths, along which we shall not now follow him. We have to ask first in what sense his overall schema of the events that together made up the Scientific Revolution helps us to understand it better.

Part of our answer has to be delayed until Part II of the present book. One of the major claims Kuhn makes further on in "Traditions" on behalf of the analytical fertility of his account is that it contributes to a resolution of some long-standing debates regarding the *causes* of the Scientific Revolution, in particular the debate over the Merton thesis. For the moment we confine ourselves to the question of whether Kuhn's account helps to clarify the issues we have pursued in the present chapter. The answer cannot be other than mixed. Let me begin by enumerating what, in my view, are the major strengths of the schematic overview of the Scientific Revolution thus arrived at.

Perhaps the remarkable, largely intuitive sense of recognition that the article induces in the reader is what accounts for its singular appeal in the first place. One principal strong point is that here we find a coherent picture of the Scientific Revolution which, while focusing on the mathematical side of the revolution, also provides a fitting place for nonmathematical physics and for chemistry, while at the same time taking into account the corpuscularian view of things to some extent as well. Anticipated by no more than the brief hints we found in the almost entirely mathematizing accounts of Burtt and Koyré, Kuhn's schema elaborates their hints into a full-fledged two-current interpretation of the Scientific Revolution which attempts to do justice to all the sciences that took part in the process. Kuhn's case is even stronger than he himself seems to have realized. Recall that, in his account of the 16th/17th-century transformation of the classical sciences, he takes 'harmonics' to be an exception, in that this particular discipline failed to undergo transformation. In Kuhn's view, 'harmonics' was just allowed to fade out of the mathematical tradition, save for some acoustical portions which at the end of the 17th century began to merge with the Baconian tradition in the area of sound. Inspired in part by this brief remark of Kuhn's, I have since gone ahead and found what some historians of musical theory (Walker, Palisca, Kassler) knew already, namely, that ever since Pythagoras a lively tradition of scientific endeavor centered on the problem of musical consonance has been pursued. More than that, I found that the science of music fits Kuhn's schema in that this most ancient of all scientific disciplines underwent a fundamental transformation during the very same period. Pythagoras-like calculations on the proportions of interval ratios were rapidly replaced, at the hands of Kepler, Galileo, Beeckman, Mersenne, Descartes, and others, by a mathematico-physical account of when and why certain musical intervals turn consonant. Thus, what happened in this neglected area seems reasonably analogous to what, according to Kuhn, happened in the other classical physical sciences.[296]

Take further Kuhn's introduction of scientific instruments into the debate over the nature of the Scientific Revolution. So far these had been either dismissed as too crude to count (Koyré) or mentioned as another thing you find in 17th-century science without being assigned a structural part in it.

One other strong point of Kuhn's two-current picture seems to be that it provides a neat resolution of the controversial 'Parisian' issue. The 14th-century scholastic theories of motion now acquire a relative independence of their own without being forced in the process to bear the full brunt of having to foreshadow, in one way or another, Galileo's revolutionary results.

Finally, there is Kuhn's handling of the Baconian sciences as a relatively independent cluster in its own right. In particular, the distinction between heuristic experimentation and experiments aimed rather at confirmation of results mathematically predicted becomes readily understandable if split up over two largely independent currents of the Scientific Revolution. It is necessary to bear these strong and quite substantial points of Kuhn's interpretive schema in mind when we turn now to some of its weaker points.

Problems with the schema. These points come under four headings: the life sciences, the revolution in mechanics, the role of the protagonists of the Scientific Revolution, and, finally, a host of persistent anomalies, in particular regarding optics and the corpuscularian view of things.

First, Kuhn left the life sciences entirely out of his account. He explains to have done so on purpose, in order to avoid "excessive complexity" and "on grounds of competence." [297] However, he thinks that a distinction between the medical and the nonmedical life sciences might lead to an equally fruitful account of their respective developments in the 16th and 17th centuries and, in particular, to an explanation of why anatomy and physiology—the classical life sciences—were the only ones to be transformed during this crucial period in a way similar to the classical physical sciences. But in the absence of any elaboration of the three footnotes to which these cursory remarks are relegated, we can only conclude that in this respect the schema was left incomplete by its author.

Second, there are a number of issues about which questions of factual and interpretive accuracy may be raised. I confine myself to the most important case of all: the transformation of Kuhn's sixth classical science, local motion. Granting as we did that Kuhn's peculiar handling of this domain provides a quite sophisticated middle-ground view, nevertheless some questions remain to be asked. To begin with, is it true to say that the impetus theory and concomitant proposed changes in Aristotle's account of some cases of local motion were sufficiently 'mathematical' in character to justify their conceptual inclusion in the range of the 'classical sciences'? Next, can one really distinguish these non-Aristotelian theories so thoroughly from the overall Aristotelian framework in which they were put forward as to grant them sufficient autonomy for being counted as a separate science? And, finally, is sufficient justice done to Galileo's creative genius in implying that he transformed an existing, rather than created a fundamentally novel, science? These are queries, not statements. Even though my own reading of the evidence is such that I tend to answer them all in the negative, my point is that these are highly debatable issues—in fact, these topics, as we have seen, have been debated for almost a century. Therefore, one may expect even a schematic overview of the Scientific Revolution to provide some more factual underpinnings for its peculiar interpretation of such a highly controversial issue as the origin of the science of mechanics, which, after all, is so close to the heart of the revolution.

Third, we have to address the issue of the protagonists of the Scientific Revolution. Here again, a variety of problems turns up. To begin with, it is doubtful whether what Kuhn has to say on the almost exclusive extent to which Galileo belongs to the mathematical tradition can continue to stand when put under the searchlight of later Galileo research, which has shown him to have been more empirically minded than was generally believed by the Great Four and their followers. Perhaps more importantly, one may wonder whether it is really illuminating first to proclaim a 'split' between the classical and the Baconian sciences, then to state that Newton belongs to both, to go on by proclaiming a corresponding split in Newton's own work, and finally to let the split run right across the *Opticks*. All of which is compounded by the fact that Mariotte and, more significantly, a figure of the stature of Huygens are regarded by Kuhn as just other exceptions. No attempt is made to determine in a more constructive vein the admixture of the 'mathematical' and the 'Baconian' in the work of Huygens, whose specific place in the Scientific Revolution has been equally passed over in most other interpretations of this signal event to which he contributed in so many and important ways.[298] Even worse, it is hard to see how Descartes, whose very questionable position in the Scientific Revolution we have stressed before, could be made to fit Kuhn's schema. Kuhn mentions him once, in connection with the classical tradition. But if we recall that Descartes was also one of the principal originators of the mechanical philosophy, then the logic of Kuhn's own argument, which takes this to be the underlying, corpuscularian metaphysics of the Baconian tradition, would compel Kuhn to stick Descartes in the Baconian pigeonhole as well—which would surely be an odd place to put him in! It is at this point that one begins to wonder whether the original distinction does not raise more difficulties than it solves.

This sense of wonder is compounded when, fourth, we take a quick look at another host of exceptions and anomalies, some of which are acknowledged by Kuhn himself. Such, for example, is statics, which in fact was a branch of science already largely completed by the Greeks (it is the only discipline besides pure geometry they got that far in). Whether one prefers to regard what Stevin did to the field as a transformation or as an elaboration, in either case it makes little sense to ignore his contribution entirely and then to handle the revolutionary work of Torricelli, Pascal, and Boyle on the void as the 17th-century transformation of a nonexistent subcluster to be called 'statics and hydrostatics'. (Although the two were connected through the analogical use made in hydrostatics of the balance and the law of the lever, the topics remain distinct.) Next, the significance of scientific instruments for the 'transformation' of the classical sciences seems so great that I do not find it particularly convincing to take this for no more than the subordinate contribution of the Baconian to the classical sciences. Again, Huygens' life's work appears to belie such a construction if nothing else does. There is the problem of optics, which is also made by Kuhn to straddle the distinction between the two clusters and thus helps to undermine the distinction further. In the view of one historian of scientists' conceptions of the nature of light, Hakfoort, only a threefold distinction in which the tradition of 'natural philosophy' acquires an independent position beside the two traditions distinguished by Kuhn can do justice to the entire development of 17th-century science, and to the history of optics in particular.[299] Finally, one may well wonder whether the corpuscularian world-view is done sufficient justice if it receives no other place than that of an underlying substructure of the Baconian sciences—quite apart from the odd results to which such a construction is bound to lead in the cases of not only Descartes but also Beeckman and Gassendi, who are not mentioned by Kuhn at all. But here may also lie the clue to a useful amendment. A close look at Kuhn's schema reveals that the Baconian sciences are linked up with corpuscularianism only from 1650 on. As a result, the Continental originators of the corpuscularian world-view are, in Kuhn's scheme of things, left hanging in midair. Does this not strongly suggest that a new stage in the course of the Scientific Revolution was reached when corpuscularianism moved across the Channel and, in blending with the Baconian sciences as cultivated in England, gave rise to a new mixture—a mixture which, through its experimental bent, became much more fruitful than the pure and untinged corpuscularianism hitherto cultivated by Descartes and Gassendi and Beeckman? We shall pursue this suggestion further in the next section, where the significance of the corpuscularian view of things for the unfolding of the Scientific Revolution is discussed in greater detail. But first we bring the present section on Kuhn to a fitting end.

The Structure *versus the "Traditions": Kuhn's dual vocation.*   By way of conclusion to Thomas S. Kuhn's handling of the Scientific Revolution amid a multitude of scientific revolutions, let us make more explicit a question we have asked before: How is it that the Scientific Revolution appears dissolved in *The Structure of Scientific Revolutions,* whereas it is made to operate as a key concept in "Traditions"? Two distinct lines of argument lead to the resolution of this question. Part of the answer has to do with Kuhn's celebrated 'paradigm' concept; another part with Kuhn's general views on, and his experiences in, research traditions in distinct academic disciplines.

If, in accordance with the conception of Alexandre Koyré, the Scientific Revolution had been handled in the *Structure* as one, probably the major, case of a revolution in science, then, according to Kuhn's schema of scientific revolutions, it would have had to stand for either of two events. Either the Scientific Revolution would have marked the collective entrance of the natural sciences into their paradigmatic state (in which case Kuhn's account would have coincided with Kant's in this respect), or it would have stood for a collective paradigm *shift* of those sciences that, in Kuhn's view, made up the Scientific Revolution. However, according to the view presented in the *Structure,* various natural sciences—or certain problem areas inside them—entered the paradigmatic state at various times, and they underwent paradigm shifts at various times as well.[300] But in "Traditions" Kuhn takes a somewhat different view. Here the Scientific Revolution is characterized by simultaneous paradigm shifts in the six classical sciences, accompanied by the emergence of a cluster of new sciences which were not to enter paradigmatic state until roughly the middle of the 18th century. Clearly, these two views, although conceptually related, are incompatible. How could this remarkable incompatibility of views arise in the unretracted works of one undivided individual, Thomas S. Kuhn?

Throughout his scholarly career Kuhn has been preoccupied almost to the point of obsession with the different questions asked, and the different approaches taken, in different areas of scholarly endeavor. Some of his finest pieces owe their existence to this preoccupation; Kuhn has a very keen eye for the consequences of scholarly specialization.[301] His own work displays many efforts spent in overcoming the limitations inherent in working inside such established research traditions. And yet, in the final analysis it seems to me that Kuhn overrates the height of the boundary fences erected by tradition between established academic fields. As is demonstrated in particular in his essay "The Relations between the History and the Philosophy of Science,"[302] Kuhn himself, when practicing now the philosophy of science (as he did in the *Structure*), then its history (as he did in "Traditions"), wears somewhat different 'thinking-caps.' He is so convinced of the different natures of these two disciplines that he seems to regard the possibility of a unitary outcome of a topic taken up in both as elusive from the start. His account of the different approaches current in these two disciplines is so striking, and partly so persuasive, that one almost forgets that to a considerable extent the differences can also be overcome.

Let me conclude. The reader should not leave the foregoing account of pertinent portions of Kuhn's work with the impression that I do not like it at all—I really do (to borrow Holden Caulfield's favorite expression). I do believe indeed that, in both the *Structure* and many of his historiographical essays, together with a great deal of overschematizing Kuhn has asked a generous number of fruitful questions and created some truly useful concepts, in more or less close connection to the Scientific Revolution. I feel less sure, though, that the same is true of the overall schema of the Scientific Revolution he presented in his "Traditions"; too many exceptions and anomalies seem to present themselves there. But the proof of the pudding can only be in the eating, and the only way for the schema to show its mettle would be for someone to write a history of the Scientific Revolution in accordance with it. Only then would it appear whether Kuhn's imaginative schema can provide an account of the revolution capable of satisfying the increasingly sophisticated requirements that ought to be put to such an account after more than half a century of historiographical work already spent on coming to grips with the event. So far, this has not been done. However, some competing rearrangements of the structural elements of the Scientific Revolution have been presented in the past two decades, the most important of which was indeed written right at the outset in the format of a full-fledged historical survey. Its author is R.S. Westfall, and it is his account that I take up now before bringing the present chapter to a close.

2.4.5. The Scientific Revolution as a Process: Westfall's Conception of the Origins of Early Modern Science

The American historian of science Richard S. Westfall (born in 1924) has devoted the major part of his scholarly career to the study of Isaac Newton's science and life. In 1980 his efforts culminated in the publication of *Never at Rest: A Biography of Isaac Newton.* Ten years earlier, Westfall had interrupted his Newtonian studies in order to write, for the benefit of undergraduate students, a 165-page survey of the Scientific Revolution, which appeared in 1971 under the title *The Construction of Modern Science: Mechanisms and Mechanics.* Besides serving students' needs, the book explicitly "aspires to present a coherent interpretation of the scientific revolution that will have more than ephemeral value."[303]

It achieves this aim through the smooth fusion of all kinds of elements taken from previous historiography into a coherent picture largely Westfall's own. The interpretation put forward in what I shall henceforth abbreviate as the *Construction* naturally betrays the effects of his immersion, at the time, in Newton's life's work. In one sense, the *Construction* can be read as the extension beyond the science of dynamics of the basic organizing theme employed in Westfall's *Force in Newton's Physics: The Science of Dynamics in the Seventeenth Century.*[304] At the same time, the *Construction* also represents the elaboration at book length of one particular idea of Alexandre Koyré's: the juxtaposition of what Koyré had termed the Platonic and the Democritean currents as the main constitutive elements of the Scientific Revolution. Altogether, Westfall's contribution to our understanding of the Scientific Revolution can best be summed up by stating that he elaborated a sketchy juxtaposition into a developed account of a dynamical process of interaction between these two currents.

Two currents in conflict.   The most concise way to get at Westfall's interpretation of the Scientific Revolution is to quote almost in full his own introduction to the *Construction.*

> Two major themes dominated the scientific revolution of the 17th century—the Platonic-Pythagorean tradition, which looked on nature in geometric terms, convinced that the cosmos was constructed according to the principles of mathematical order, and the mechanical philosophy, which conceived of nature as a huge machine and sought to explain the hidden mechanisms behind phenomena. This book explores the founding of modern science under the combined influence of the two dominant trends. The two did not always mesh harmoniously. The Pythagorean tradition approached phenomena in terms of order and was satisfied to discover an exact mathematical description, which it understood as an expression of the ultimate structure of the universe. The mechanical philosophy, in contrast, concerned itself with the causation of individual phenomena. The

Cartesians at least were committed to the proposition that nature is transparent to human reason, and mechanical philosophers in general endeavored to eliminate every vestige of obscurity from natural philosophy and to demonstrate that natural phenomena are caused by invisible mechanisms entirely similar to the mechanisms familiar in everyday life. *Pursuing different goals, the two movements of thought tended to conflict with each other,* and more than the obviously mathematical sciences were affected. Since they proposed conflicting ideals of science and differing methods of procedure, sciences as far removed from the Pythagorean tradition of geometrization as chemistry and the life sciences were influenced by the conflict. The explication of mechanical causation frequently stood athwart the path that led toward exact description, and *the full fruition of the scientific revolution required a resolution of the tension between the two dominant trends.* [My italics][305]

The distinction thus made appears to be aimed at two distinct objectives. For one thing, it served as a clue to clearing the path for inserting into the account of the Scientific Revolution, besides cosmology and mathematical physics, which, in this respect, had never presented a problem, nonmathematical physics, chemistry, and the life sciences. In this sense, Westfall's arrangement, although in an important respect different from the one Kuhn was to sketch one year later, served essentially the same purpose. We shall presently inquire, by means of a brief survey of the eight chapters the *Construction* consists of, how well it achieved this aim and also compare it with what, on this score, had been accomplished by Kuhn.

The distinction Westfall made at the outset served another purpose as well. Through the idea that the two currents not just coexisted but rather interacted and, through their interaction, provided the dynamics of the development of 17th-century scientific ideas over time, Westfall added a new dimension that so far had been lacking in accounts of the Scientific Revolution. Briefly said, and I shall come back to this, Westfall was the first to conceive of the Scientific Revolution as a structured *process* rather than as either an unstructured advance over time or as a static structure. In discussing the successive chapters of the book, I hope to clarify what I mean by this.

Conflict and harmony in the Scientific Revolution. Unlike the other 20th-century historians of the Scientific Revolution we have met so far, Westfall takes his start around 1600. This is largely so because the *Construction* is part of a series on the history of science, with treatment of other centuries being assigned to other authors. But also, chapter 1 makes it very clear that, for Westfall, Galileo and Kepler placed the fundamental landmarks for the Scientific Revolution through their work in astronomy and mathematical physics. The developments of the 16th century, including Copernicus' work, although setting the stage for the Scientific Revolution in a major way, are thus not taken to belong to the process of the Scientific Revolution proper; until the advent of

Galileo and Kepler it was an open question indeed whether or not Copernicus' proposed reform in astronomy was to usher in a revolution in science at all.[306] Of 14th- through 16th-century impetus theorizing it is said, in true Koyréan fashion, that it had provided a natural starting point for Galileo's thought, which, nonetheless, had to be given up before he could make any further advances toward the resolution of the problems of motion set by his adoption of Copernicanism as a true representation of the world. In making those advances Galileo, together with Kepler, inaugurated the mathematization of nature which is characteristic of the 'Platonic-Pythagorean' tradition in which both men stand.

Chapter 2 starts with an exposition of Gilbert's work on the magnet as an example of the empirical side of Renaissance theorizing on nature and then goes on to introduce the 'mechanical philosophy.' The term is made to stand for the corpuscularian view of things—for the conviction, that is, that nature is fully transparent to human thought, and that its manifestations can ultimately be reduced to the movements of particles of different shapes and sizes.

> Suggested in Galileo and Kepler, it assumed full proportions in the writing of such men as Mersenne, Gassendi, and Hobbes. . . . Nevertheless René Descartes . . . exerted a greater influence toward a mechanical philosophy of nature than any other man, and for all his excesses, he gave to its statement a degree of philosophic rigor it sorely needed, and obtained nowhere else. . . . The effect of Cartesian dualism . . . was to excise every trace of the psychic from material nature with surgical precision, leaving it a lifeless field knowing only the brute blows of inert chunks of matter. It was a conception of nature startling in its bleakness—but admirably contrived for the purposes of modern science.[307]

Although not the only one to contribute to, or to work out wildly incredible details of, the explanation of the world contained in the mechanical philosophy, Descartes holds a central position, which indeed makes him "fit into the work of the scientific revolution."[308] The term 'mechanical philosophy' itself, West-fall acknowledges, comes not from Descartes but from Boyle, who defined it in terms of " 'two catholic principles', matter and motion."[309] The chapter ends with a sketch of the overwhelming importance of the mechanical philosophy for 17th-century scientific thought:

> In the 17th century, the mechanical philosophy defined the framework in which nearly all creative scientific work was conducted. In its language questions were formulated; in its language answers were given. Since the mechanisms of 17th century thought were relatively crude, areas of science to which they were inappropriate were probably frustrated more than encouraged by its influence. The search for ultimate mechanisms, or perhaps the presumption to imagine them, diverted attention continually from potentially fruitful enquiries and hampered the acceptance of more than one discovery. Above all, the demand for mechanical explanations stood in the way of the other fundamental current of 17th century

science, the Pythagorean conviction that nature can be described in exact mathematical terms. *Despite its rejection of a qualitative philosophy of nature, the mechanical philosophy in its original form was an obstacle to the full mathematization of nature, and the incompatibility of the two themes of 17th century science was not resolved before the work of Isaac Newton.* [My italics][310]

The sentence italicized in the above quotation contains the program for all remaining chapters of the *Construction,* with the sole exception of chapter 6, which deals with matters of scientific method and organization that do not concern us here. In chapter 3, entitled "Mechanical Science," the theories and experiments of Torricelli, Pascal, and Boyle on the void and on air pressure, followed by discussions of new explanations of light and colors by Kepler, Descartes, Newton, and Huygens, are all "attributed to the rise of the mechanical philosophy and mechanical modes of explanation."[311] Westfall explains the exemplary nature the debate over the barometer acquired in the middle of the 17th century through

> the golden opportunity which the barometer offered to the mechanical philosophy. A simple phenomenon with a quantitative factor, it presented the most favorable ground on which the mechanical philosophy might attack animistic conceptions. The quantitative factor, moreover, made the question ideally suited for experimental investigation.[312]

Thus we have here the exact reverse of what we found in Kuhn's two-current interpretation: Experimentation is here regarded as a consequence of the mechanical philosophy, carried out in its name, rather than the other way round. Of 17th-century achievements in optics Westfall equally asserts that they were "profoundly influenced by the mechanical philosophy."[313] At least, this philosophy "encouraged the generation of mechanical conceptions of light which might account for the known phenomena," and although it did not lead to any discoveries and "may even have frustrated the comprehension of some," it certainly "provided the idiom in which the 17th century discussed optics."[314] All this notwithstanding,

> by the end of the 17th century, the mechanical philosophy . . . had become an obstacle to its further progress. Experimentation had discovered three properties or phenomena which were frankly unintelligible to either of the mechanical models in use.[315]

The subordination of experiments to the mechanical philosophy as thus related in chapter 3, however, is not left unqualified. For one thing, Westfall expressly denies that Kepler's contribution to optics should be seen in the tradition of the mechanical philosophy. More troubling is perhaps another feature of this chapter: When dealing with Torricelli's and Pascal's barometric experiments in particular, the word 'mechanical' is plainly made to stand for something else than for 'belonging to the corpuscularian world-view', which elsewhere in the book defines the mechanical philosophy. In stating that such experiments made clear the "superiority of the mechanical [explanation]," Westfall means not so much an explanation in terms of motions of material particles as, rather, an explanation in terms of elementary mechanics—in this case, "the basic relations of the lever and balance."[316] In other words, at least in this chapter the word 'mechanical', already employed in so many different senses by successive historians of science, once more acquires a dual meaning. Another sign of trouble is that it is not made clear how, *in the given context,* the 'quantitative factor' could acquire such a crucial importance.

Before attempting to point out the common source of these oddities in an otherwise very straightforward and clear account, I pursue Westfall's argument to its culmination in the Newtonian synthesis. As might be expected from the program enunciated at the end of chapter 2, chapters 4 and 5 are devoted to "Mechanical Chemistry" and to "Biology and the Mechanical Philosophy," respectively. In both chapters we deal almost all the time with 'mechanical' in the sense of corpuscularian, and in both the pattern of the argument is the same: The mechanical philosophy is said to have provided the language in which findings in these areas were expressed. In the case of chemistry, the great factual gains made in the 17th century through experimentation had not led to substantial progress in theory, the chief reason being that,

> since there were no criteria by which to judge the superiority of one imagined mechanism over another, the mechanical philosophy itself dissolved into as many versions as there were chemists.[317]

The main service rendered to chemistry by the mechanical philosophy was that it helped to turn the field into a respectable, relatively autonomous part of the scientific enterprise.

Something similar was true of the life sciences. Here Westfall reaffirms that "the concept of a scientific revolution has validity for the organic sciences as well as the inorganic."[318] After describing the immense increase in factual knowledge resulting from both microscopic discoveries and the importation of tropical plants and beasts, Westfall discusses "the reconsideration of the nature of life as the mechanical philosophy extended its influence over the last stronghold of Aristotelianism."[319] However, but for Harvey's discovery of the circulation of the blood, mechanical modes of thought had no appreciable positive effect, since the mechanisms imagined in this framework were much too crude to be adequate to the "subtlety of biological processes."[320]

So far, we have witnessed how Westfall's two-current interpretation served principally as a framework for discussing major developments in 17th-century scientific thought. In the two final chapters, it acquires another function as well. Focusing from this point onward on "The Science of Mechanics" (the title of ch. 7), Westfall goes on to show that, increasingly, the two dominant currents ran into conflict with one another, and that this was a consequence of the way the concept of force was treated in the mechanical philosophy. In this philosophy, force is the effect, not the cause, of motion. Thus,

> the inability of mechanical philosophers to consider any conception of force except the 'force of a moving body' became an obstacle to the development of a mathematical dynamics and tended to confine mechanics within kinematic problems, in which motions were described without reference to the forces that cause them.[321]

The resulting conflict is illustrated for several cases. In the first place, Cartesian corpuscularianism proved incapable of finding a mechanism such that the free fall of bodies could not only be explained (both Descartes and Huygens succeeded to their own satisfaction in doing just that), but, what is more, be explained so that Galileo's law of the uniform acceleration of falling bodies could be derived from it. Something similar holds for Kepler's laws: Cartesian vortices could indeed be manipulated in such a way as to make the planets revolve around the Sun in closed curves, but there was no conceivable way to derive the exact, elliptical shape of their orbits from whatever corpuscularian mechanism could be imagined. Nor would it have been possible to derive Kepler's two other laws from vortical mechanisms.

Westfall goes on to show that Huygens, who was "the heir of Galileo" and, "in many ways . . . also remained the disciple of Descartes," continued to adhere to the latter's mechanistic concept of force.[322] However, his pupil in mathematics, Leibniz, not only became aware of the relative incompatibility of the mathematical and the mechanical traditions but also began to see that the road toward the successful resolution of the conflict lay in a reform of the concept of force. In its Cartesian form, Leibniz showed, it could mathematically be demonstrated to lead to consequences irreconcilable with Galileo's quantitative kinematics. Thus,

> in Leibniz, the conflict between the two [currents] began to resolve itself by modifications of the mechanical philosophy. . . . The development of a conception of force as action on a body to change its state of motion, a conception that contributed greatly to the further elaboration of mathematical mechanics, was inhibited by the mechanical philosophy during the century. . . . It remained for Isaac Newton to pick up that conception again and to use it both to extend mechanics and to revise the mechanical philosophy.[323]

In the final chapter of the book, "Newtonian Dynamics," Westfall shows how Newton became aware of the conflict and resolved it through a redefinition of the concept of force in the sense of an active principle, acting on bodies from outside and causing them to change their motions. Right from the start of his immersion in the ongoing scientific upheaval Newton had displayed a strong interest in precisely those natural phenomena which the mechanical philosophy was least equipped to handle. Not only did it fail to provide an explanatory framework from which Galileo's and Kepler's laws could be derived, but for such phenomena as magnetism, the generation of heat in chemical reactions, and the cohesion of bodies, the usual invisible mechanisms which made up the standard inventory of the mechanical philosophy simply did not do. The key to reconcile the ideal of an exact, mathematical science with the mechanical philosophy was the reformation of the latter through a new concept of force:

> Newton's admission of forces acting between particles of matter constituted a major break with the prevailing mechanical philosophy of nature. . . . Newton himself considered forces between particles, not as a denial of the mechanical philosophy, but as the conception needed to perfect it. By adding a third category, force, to matter and motion, he sought to reconcile mathematical mechanics to the mechanical philosophy. Force to him was never an obscure qualitative action [as it had been regarded in the Renaissance]. He set it in a precise mechanical context in which force was measured by the quantity of motion it could generate.[324]

Newton's life's work must be considered fragmentary to the extent that the actual measurement of most of those forces continued to elude him. Only in the case of the unification of earthly and heavenly movements did he succeed—but in such a resounding way that the *Principia* may indeed be regarded, not only as the synthesis between these two previously distinct areas of scientific inquiry, but also as the embodiment of

> the reconciliation of the tradition of mathematical description, represented by Galileo, with the tradition of mechanical philosophy, represented by Descartes. By uniting the two, Newton carried the scientific work of the 17th century to that plane of achievement which has led historians to speak of a scientific revolution.[325]

An assessment of Westfall's two-current account. In assessing the principal novel features of Westfall's interpretation of the Scientific Revolution we have two distinct topics to address. One is the insights gained by describing the Scientific Revolution as a process which went through stages of conflict and reconciliation between the major trends it was made up of. The other topic, which we take up first, is the performance of Westfall's particular two-current interpretation in accounting for what he himself calls "the work of the Scientific Revolution."

Let us begin to look at terminology. The labeling of the mathematical strand of the fabric of the Scientific Revolution is unequivocal, although not perhaps altogether fortunate. As has been suggested by Aiton, 'Archimedean' might be a better term. (My reason for the change would be that Pythagoreanism was too strongly tinged with numerology to make it fit the Galilean tradition in 17th-century science.)[326] But this is quite a secondary matter compared to the various usages to which the word 'mechanical' is put at various places in the book. To be sure, the broad usage of the mechanical philosophy as an overall framework for considerable chunks of revolutionary science in the 17th century works admirably up to a point. No more natural way to make chemistry and the life sciences take an active part in the making of the Scientific Revolution had been found before, or has been found since, for that matter. Also, since Westfall traces the original inspiration for Newton's new conception of force back to the latter's own work in alchemy and chemistry, in particular, the inclusion of these areas into an account of the Scientific Revolution is made more urgent than it has ever been before.[327]

It is chapter 3, however, the one on "Mechanical Science," where, as already suggested in the course of my summary, trouble comes in. Old trouble, to be sure, for in the works of both Dijksterhuis and Hall—the two historians of the Scientific Revolution most concerned to define what 'mechanical', 'mechanistic', and like derivations of the Greek word *mechane* should be taken to mean—we have seen confusion to be rampant.

In Westfall's chapter 3, then, 'mechanical' is rather indiscriminately employed in the senses of 'belonging to the (mathematical) field of mechanics' and of 'pertaining to the mechanical philosophy'. Thus Pascal's barometrical experiments are interpreted in the framework of the mechanical philosophy, even though their mechanical element derives from the science of mechanics, and in spite of the fact that, according to the logic of his own argument, the element of precision, which Westfall stresses as crucial, cannot but be attributed to the mathematical, as opposed to the mechanistic, current. The resulting problem would not have arisen if the Platonic-Pythagorean and the mechanistic traditions did not indeed have some quite important features in common (the very features, in fact, that had led Burtt to derive the latter from the former). We know from Westfall's opening statement how he defines the connection: Mechanical philosophers sought to demonstrate "that natural phenomena are caused by invisible mechanisms entirely similar to the mechanisms familiar in everyday life." However, these mechanisms are *not,* or at least far from exclusively, the mechanisms of mechanics. In the introductory chapter to *Never at Rest,* which is a survey of the revolution in science up to the moment when young Isaac Newton became aware of it, Westfall exposes the connection as well as the distinction between the two strands in a slightly different manner:

Aspects of the mechanical philosophy suggest an inherent harmony with the developments I have surveyed in astronomy, mechanics, and optics [note that, in the *Construction,* optics had been subsumed under the mechanical philosophy!]. The very word 'mechanical' seems to incorporate the science of mechanics, and the program of the mechanical philosophy, to trace all phenomena to particles in motion, seems to demand the same. The assertions that quantity alone is real and that qualities are only sensations recall the mathematically formulated laws of physics. The harmony may be more apparent than real, however. . . . The words

'quantitative' and 'mechanical' should not mislead us into seeing the two dominant trends in seventeenth century science as different facets of one program.[328]

Thus the mechanical philosophy is linked up with the mathematical mode of
thought (as Burtt and Dijksterhuis, in particular, had shown) in that the 'primary' properties attributed to the particles of which matter is made up are the
'geometrical' ones of size, shape, and location.

One reason for stressing the discrepancies in Westfall's chapter 3 is that,
in my view, the very content of the chapter really points toward a somewhat
different conclusion from the one he is, by the logic of his own argument,
compelled to draw. To me it seems that large parts of the early history of the
Scientific Revolution, up to at least 1655 or thereabouts, may fruitfully be described as the fairly harmonious flowing together of the various currents that
made up the Scientific Revolution. Perhaps one might even say that to some
extent they reinforced one another in that a number of discoveries were made
under their combined inspiration (one example of the harmony between the
two components would be Huygens' laws of impact). Only in the course of a
subsequent stage did the latent sources of conflict become apparent, with the
conflict finding its resolution during the final stage, in the Newtonian synthesis.

Such a modification of Westfall's organizing principle, however, would
only partially resolve the structural unclarity of chapter 3. Much of the trouble
comes from the usage, here as elsewhere, of 'mechanical' in at least three
distinct senses. First, as noted, 'mechanical' is made to connote 'mechanics'
as well as the mechanical philosophy, resulting in the continuation of precisely
the sort of terminological confusion Dijksterhuis had in vain tried to eradicate through his equally confusing *identification* of 'mechanization' with
'mathematization-in-the-sense-of-the-discipline-of-mechanics.' But also, the
mechanical philosophy is made to stand for two things which are not quite
identical, namely, the corpuscularian view of things and the *more general* idea
that natural phenomena are to be explained through Boyle's two 'catholick
principles': matter in motion. That this is not merely a verbal quibble may be
shown by pointing at the example of Marin Mersenne. Even though Mersenne
was determined to explain natural phenomena—in particular, those pertaining
to music and sound—through the principle of matter in motion, he would have
none of the corpuscularian explanations put forward by his friends Descartes,
Gassendi, and Beeckman, who together stand at the origin of the mechanical
philosophy.[329] One generation later Pascal, who certainly aimed for an account
of things through the categories of matter and motion, quantitatively applied,
equally certainly rejected corpuscularian mechanisms as undemonstrable constructions of an overimaginative mind (namely, Descartes').[330] Thus Westfall's
statement that "no figure of importance in European science in the second half
of the seventeenth century stood outside the precincts of the mechanical phi-

losophy of nature"[331] rests more heavily on the qualifier 'second half' than transpires from the overall picture presented in the *Construction*; it would apply to the earlier period only if 'mechanical philosophy' is defined in the wider sense of the term, not in the strict sense of corpuscularianism. The problem, of course, is that the latter was indeed a pervasive, although not an all-pervasive, feature of the Scientific Revolution, as has been demonstrated by no one more clearly than by Westfall himself.

The best way to get at the heart of all these dilemmas may be to ask next what happens with *experiments* in Westfall's account. In this area Kuhn, in his "Traditions" article, seems to be on somewhat more solid ground. After all, Kuhn's identification of the Baconian sciences as a cluster in its own right led to a number of useful further distinctions—regarding, in particular, the distinct roles of experiments in the classical and the Baconian traditions—which appear more blurred in Westfall's account.[332] Once again, the key event here may have occurred when, during the middle decades of the century, the corpuscularian view of things crossed the Channel and resettled in a new, Baconian environment which tinged it with an experimental bent not, or far less, applied to its legacy on the Continent. If left to themselves, both the Baconian program and corpuscularianism had a strong potential for sterility, which soon became apparent wherever they continued to be cultivated in isolation from one another. Their blending, however, yielded a somewhat muddled yet highly fertile mixture that, in its turn, could be married to the mathematical tradition in a harmonious way not within reach of either Cartesian mechanicism or Baconian experimentation proper.[333]

What I sketch here is little more than a suggestion (to be elaborated somewhat in the final chapter of the present book). It is not easy to capture all this in a neat picture, and it may well be that one is bound to encounter difficulties here one way or another. Ways out of the difficulties have been sought in more than one direction. For example, K. van Berkel has suggested that, although it does make sense to distinguish two currents, these should be redefined somewhat in accordance with the protagonists' own motives at the time. In van Berkel's view the category of 'mathematization' is imposed by historians upon the period rather than derived from what actually motivated the men engaged in a two-pronged battle with both Aristotelianism and a variety of neo-Platonic and vitalist programs in natural philosophy. Some, like Galileo and Mersenne, opposed to the comprehensive claims of natural philosophy a programmatic limitation to positive, partial insights (often, but not necessarily, of a mathematical nature). Others, like Beeckman and Descartes, opposed a corpuscularian natural philosophy, which offers a concretely picturable conception of the world at large, to the opaque essences and forces of the Aristotelians and the neo-Platonic vitalists. The two currents of the new science—the 'positive' and the mechanicist one—shared a predilection for arguments taken from the domain of mechanics, which explains how they could generally operate in harmony, until Newton's reform of mechanicism blew up the temporary connection.[334]

Proposals have also been made to expand two-current approaches further. C. Hakfoort suggested the addition of a third category—natural philosophy—to the mathematical and the Baconian traditions distinguished by Kuhn.[335] In *Quantifying Music* I attempted to distribute my protagonists over a mathematical, an experimental, and a mechanicist approach, with the last being taken only in the strict sense of corpuscularianism.[336] All such suggestions may work, or have worked, up to a point—none has so far proved its mettle.

And now for the moral. The moral of all our criticizing and vacillating seems to be that the organization of the substantive content of the Scientific Revolution might be attempted most successfully in the future in either of two basic directions. One possibility is to rescue the insights gained through Westfall's and Kuhn's categorizations either by redefining them or by introducing a third, relatively independent strand. If so, it remains as yet unclear in what way the strands, other than the almost undisputed mathematical one, should then be defined. The other possibility is a return to the position adopted most radically by Burtt, who made both experimental practice and the corpuscularian view of things follow from the new mathematical metaphysics of Kepler, Galileo, and Descartes. However, in the process one almost inevitably loses nonmathematical physics, chemistry, and the life sciences, whereas their inclusion in the story is precisely the strongest point of both Kuhn's and Westfall's two-current accounts of the Scientific Revolution. Or should we perhaps return to the Koyréan conception that the Scientific Revolution was an event in cosmology and mechanics exclusively, and consequently regard everything else that happened in 17th-century thought on nature (also on human nature) as the absorption into the new universe of precision of all other sciences as well as of technology and of so many other manifestations and products of the human mind?

The dynamical approach to the Scientific Revolution. A final consideration concerns the dynamical approach pioneered by Westfall. He added a new dimension to the study of the Scientific Revolution in that its advance over time was now being made problematical. This had not been done before; it always seemed that, once the first, decisive steps in the direction of the new science had been made, the event could move forward under its own steam and reach its manifest destiny in the Newtonian synthesis. Kuhn's two-current account, too, is essentially static in that the sum total of scientific achievement at the time is neatly aggregated and split up over his two categories. In Westfall's account, in contrast, his two currents acquire a life of their own. Neither the one current nor the other could have completed 'the work of the Scientific

Revolution,' but together, in an intricate pattern of harmony and conflict and reform and reconciliation, they could.

I have argued why I do not think that, with all the illumination it presents, the resulting picture of the Scientific Revolution is wholly satisfactory. An analysis internal to Westfall's argument itself already displays discrepancies, and additional incentives toward further modification are to turn up as we proceed. But I do think that the conception of the Scientific Revolution as a dynamic process is quite as fertile as it is underexploited. At the very least, the idea may appear to be a helpful tool in resolving the various dilemmas that still beset our notions of the Scientific Revolution. Barring Hall's 1983 revision of his original *The Scientific Revolution,* Westfall's is still the most recent, book-length attempt to come to grips with the Scientific Revolution, and there is good reason to consider his survey the most sophisticated attempt yet undertaken to resolve the intricate problems that came to the fore as a by-product of the marvelous harvest yielded by the Great Tradition. It is time now to round off our inquiry into what the Great Tradition has brought. In a brief concluding section we restate its achievement in terms that bring out the key controversy over the nature of the birth of early modern science. That is to say, we sum up what has happened, over more than two centuries, to the sense of an absolute break with the past that was expressed, at the end of the 18th century, by those thinkers who, in retrospect, set the Great Tradition in motion.

2.5. CONCLUSION: CONTINUITY AND BREAK WEIGHED
IN THE BALANCE

There are no absolute discontinuities in history. Nothing happens entirely out of the blue; no event, however unexpected, is without its prior preparation. To us, these statements are, or should be, truisms. To an earlier generation they were not. The truth of these statements is a lesson 'general' historians learned, or rather taught themselves, in the course of the 19th century. Together with the urge to check the authenticity of their sources, this particular insight into the connectedness of all events over time may be said to have marked the birth of history as a scholarly discipline.

For the time being, the lesson passed by the history of science. Historians in the 19th century first devoted themselves to the history of states and their diplomatic relations; social and economic history followed with more or less hesitation; but, whatever further aspects of life and thought came under the scrutiny of the historian, the history of science was not among them, 'and this has remained so to the present day'. To the extent that the history of scientific thought was cultivated at all, it was almost exclusively the province of the philosopher. For the philosopher, however, there is nothing extraordinary in think-

CHAPTER TWO

ing in terms of absolute discontinuities. And so the lesson learned by 'general' historians in the course of the 19th century had to be learned later and in a different manner by those philosophers who had an interest in the history of science.

It is to the immortal merit of Pierre Duhem to have brought the lesson home to himself and his fellow philosophers with a sledgehammer. True, we may speculate that if, rather than an Italian Catholic and a German Protestant among others, such orthodox French Roman Catholics as Buridan and Oresme had inaugurated early modern science at the turn of the 17th century, Duhem would have been less eager to insist upon the continuity of science across the ages. But no matter what his motives were, he was the man to break through a thinking habit that seemed to preclude for good the examination of the past of science in a fashion at all in line with the new historiographical standards.

Even though the free room thus created by Duhem was almost closed at the same stroke in that, after demolishing absolute discontinuity, Duhem proclaimed instead a virtually absolute continuity of scientific thought over time, not all members of the successor generations allowed themselves to be bound by the new dogma. Dijksterhuis provides a fairly spectacular example of one who did. In his mind the notion that there are no absolute beginnings became equated so indissolubly with the idea of the overall continuity of science over time that no room was left for his scarcely less vivid awareness of the radical novelty brought about by early modern science. One conseqence for Dijksterhuis was his steadfast *rejection* of the concept of the Scientific Revolution. In a lecture "Renaissance en natuurwetenschap" (Renaissance and Science, 1956), Dijksterhuis criticized for their lack of precision such images for the advance of science over time as 'development', 'heritage', and 'flourishing', and then went on as follows:

> And yet such imagery is by far to be preferred over the usage of such terms as "renaissance" or "rebirth," or of the 'revolution' terminology dear to English authors, which has already led to distinguishing a considerable number of revolutions in the course of the history of science. The reason is that such expressions suggest images which, after all, conflict with the essence of the process of the history of science, continuity in its development being one of its essential features.[337]

Another outspoken continuist, Hall, has been less consistent in this respect, in that he has always combined his usage of the term 'Scientific Revolution' with an argument directed above all toward showing how unrevolutionary this revolution really was. Such usage is rather in keeping with the gradual devaluation the word 'revolution' has undergone in our time in domains far removed from the history of science.[338]

The more fruitful position, in my view, has been the one adopted for the first time by Burtt and more outspokenly so by Koyré: the view of 'relative discontinuity'. This is the recognition that one may be well aware of all that went into the making of a scientific thinker and yet acknowledge his achievement as constituting a revolution, or a break, with all that had gone before. Judging by the terminology and the overall setting of their accounts, both Kuhn and Westfall concur in Koyré's judgment on the issue. Meanwhile Herbert Butterfield was the man to express succinctly the general conception of history that underlies the position of 'relative discontinuity':

> There is just the unbroken web of history, the unceasing march of generations which themselves overlap with one another and interpenetrate, so that even the history of science is part of a continuous story of mankind. . . . Though everything comes by antecedents and mediations—and these may always be traced farther and farther back without the mind ever coming to rest—still, we can speak of certain epochs of crucial transition, when the subterranean movements come above ground, and new things are palpably born, and the very face of the earth can be seen to be changing.[339]

This is a very balanced view, far removed from the absolute terms in which debates over the continuity or discontinuity of 17th-century science in respect to preceding times have been carried on for too long. One may or may not have a preference for 'relative discontinuity' over 'relative continuity' in the sense here expressed—the point I should like to urge is that the debate be held in none but such relative terms. It should be possible to uphold the revolutionary nature of 17th-century science without being taken to task for believing in absolute beginnings, just as it should be possible to deny the revolutionary nature of 17th-century science without committing oneself *ipso facto* to the view that science always advances by bits and pieces.

This does not imply at all that discussions of the revolutionary or nonrevolutionary nature of 17th-century science are necessarily of a purely terminological nature. Viewpoints for or against the truly revolutionary character of the Scientific Revolution may well represent a great deal more than arbitrary positions arbitrarily defended. Rather, they are at the heart of our historical understanding of the birth of early modern science. The search should be for good, historical criteria to help one decide whether it is more fruitful to speak of a break or rather of continuity in any given case. And the problem with such criteria is that it is so easy to make them perform a service already settled upon beforehand. The best criterion I know of to decide what would allow us to speak sensibly of one or another revolution in the history of ideas is a criterion applied by no other man than Dijksterhuis when, in *Val en worp,* he sums up Galileo's achievement in the domain of the problems of free fall and projectile motion. What Dijksterhuis does there is simply to compare the *status quaestionis* before and after Galileo had appeared on the scene.[340] The change then readily appears to him so incisive, so radical, that not even Dijksterhuis was able to resist calling the achievement revolutionary. To my mind, his result bears generalization: If the Scientific Revolution was not a revolution in the relative sense now discussed, we may safely conclude that no revolution has ever taken place throughout history.

⋘ THREE ⋙

THE NEW SCIENCE IN A WIDER SETTING

N O MORE THAN ONE FULL-FLEDGED, unbroken tradition has established itself in historians' efforts to define the essence of the Scientific Revolution: the Great Tradition analyzed in the previous chapter. Here a core set of issues was taken up by successive generations, its members agreeing at bottom—with all the modifications and extensions each of them found it necessary to apply—over what the whole event was about. Yet, in working on the core set they did not exhaust all there is to the Scientific Revolution. Few of them cared to devote more than isolated pages to the question of what may have *caused* the emergence of early modern science. But also, spread across the wider literature on the Scientific Revolution, one encounters a number of important features of the event that go beyond the primary concerns of the Great Tradition. Take, by way of one, by no means unusual, example Westfall's *Construction*. Seven of its eight chapters belong squarely to the core set. But chapter 6, "Organization of the Scientific Enterprise," stands apart from the principal argument in that the author takes up in quick succession the following topics: the place of the universities in the Scientific Revolution; societies and academies as foci of communication and stimulus for the new science; the rise of the scientific instrument; the experimental method; the changing attitude toward authority in science; and the Baconian idea that the new science not only yielded a better understanding of nature but could also be applied to exploit nature in the service of human needs.

This is quite a mixed bag of topics, which, in its turn, goes back to quite a mixed bag of literature. By and large, that literature stands outside the Great Tradition. Part of it stems from those contemporaries of Koyré *et alii* whose concerns did not fit in with the prevalent interest; for example, those historians of science who urged the crucial importance for the Scientific Revolution of Baconian ideals. But most authors in search of ways and means to put the new science in a wider setting have, in so doing, given rise to certain novel currents in the historiography of the Scientific Revolution that have developed over the past quarter-century or so. I discern three such main currents. One has been to demonstrate the contingent character of events and discoveries that together constitute the Scientific Revolution, while refusing to draw a neat dividing line between its winners and its losers; another, to consider science as the product of social forces rather than the embodiment of timeless truth; a third, to identify (largely by implication) the Scientific Revolution with events that took place in 17th-century England. For now we must remain content with sticking a label on these relatively novel research orientations. We shall see them at work throughout our treatment of the range of topics to be addressed in the present chapter. Thus prepared, the reader will find them discussed critically and in a generalized manner at its end.

What has been said above explains why a largely chronological account, such as presented in the previous chapter, would no longer make sense for our ongoing survey of the historiography of the Scientific Revolution. From here on the discussion moves from one topic to the next, adding to the list from Westfall's sixth chapter the following issues (most of which stem from more-recent literature): the cumulative nature of the new science and its new orientation in time; the subjection of 'feminine' nature; what has recently been called 'the experimental life'; the new values of science; patronage; the 'Rosicrucian' conception of early modern science; and the place of the new science in European history. We start with the question of the extent to which the new science of the 17th century may be characterized by a new method.

3.1. THE NEW SCIENCE AND ITS NEW METHOD

Was the new science of the 17th century the product of a new method?

In the old days of the 18th and 19th centuries, there was no doubt that the answer was an almost unqualified 'yes'. The history of answers given in more recent times to the question may well be summed up in stating that the rising number of qualifications subsequently added to the innocent, original 'yes' has rendered the topic engangled in perplexities to such a degree that no one has yet dared to go ahead and produce a comprehensive 'History of Method in the Scientific Revolution'. In the recognized absence of such a study, I can do no more here than outline briefly the principal threads that go into the knot in question and point at ways in which scholars have tried to disentangle them. Whether the knot is indeed of the Gordian variety only a future Alexander may be able to decide.

3.1.1. Pitfalls in the History of Scientific Method

Recall from the previous chapter that, before Mach's *Science of Mechanics*, Bacon and/or Descartes were taken in a self-evident manner for the principal figures in the 17th-century Revolution in Science. Either of these men was held to have laid down the fundamentally novel method which, henceforward, made it possible to do science in the right manner. 'Practical discoverers' like Galileo and Newton, in this view, had the great yet, at bottom, secondary merit of applying the method in practice, thus reaping the harvest sown by one or another of the two great contemporary methodologists.

The one remaining quarrel was about whether the method was primarily deductive (as Descartes', chiefly Continental, followers believed) or inductive (as upheld, particularly in the Anglo-Saxon world, by the Baconians). In either case the debaters felt assured that the key to the Revolution in Science had been a revolution in scientific method.

This is a belief that has not survived the shaping of the concept of the Scientific Revolution in the thirties of our century. (Only some specialized historians of scientific method tend to stick to the old conviction.)[1] Whether historians think that the Scientific Revolution signified a new world-view or just a spectacular amount of radically novel theories and/or observations, in any case the focus of their attention has shifted from formal method to actual discovery. In this respect Koyré's point against Crombie expresses well the position implicitly taken by most subsequent historians of the Scientific Revolution: "The place of methodology is not at the beginning of scientific development, but, we might say, in the middle of it."

What, then, was the place of methodology in the middle of the Scientific Revolution? A host of pitfalls faces anyone approaching this particular question.[2] For one thing, commentators find it hard to resist the temptation of ascribing to the protagonists of the Scientific Revolution the very method that they favor themselves. Mach, for example, was concerned above all to demonstrate how Galileo achieved his great successes by applying very neatly and carefully the empirical method advocated by Mach.

Also, conventional terminology as used in the 17th century was often highly misleading. Scientists might well adopt talk on induction or deduction while giving such terms meanings at substantial variance with what Bacon and Descartes had meant by them (also with what we nowadays mean by them). Many a scientist was quite happy to utter currently-in-vogue slogans about inductive or deductive reasoning, about combining reason with experience, and so on, whereas it is easy to see that the true problems begin where these slogans end: How exactly does one find a natural law through inductive generalization? At what point in a deductive line of reasoning does the empirical come in? In what precise manner ought experience to be combined with reason?

On occasion, too, new wine was sold in old bottles, in that under the cloak of the language of Aristotle's *Posterior Analytics,* altogether fresh scientific conceptions may in fact appear to lie hidden—Galileo provides one striking example.[3] Increasingly, historians have come to realize that the questions of to which contemporary methodologist 17th-century scientists paid allegiance, what they claimed their own method to be, and what they actually did when making a discovery are three distinct issues which need not at all coincide.

Hence, for a proper reconstruction to be made it is crucial to distinguish very carefully between what scientists *said* they did and what they did in reality. Burtt may well have been the first to observe that 17th-century scientists were not as a rule very adept at setting forth their own doings:

> One of the most curious and exasperating features of this whole magnificent movement [of the revolution in 17th-century science] is that none of its great representatives appears to have known with satisfying clarity just what he was doing or how he was doing it.[4]

Further complications arise by taking into account a distinction between method in *generating* knowledge and method in *validating* knowledge: the well-known 'context of discovery' versus 'context of justification' opposition. Unlike philosophers of science, who tend to rule the former out of their domain of interest altogether, historians must be concerned with both, and in examining matters of validation they must carefully avoid another trap set by the philosophy of science: the utilization of supposedly timeless standards by which to judge the validity of claims to scientific knowledge.

If matters of scientific method are treated with historians' aims in mind, it is hard indeed to keep these matters insulated from two overarching problems. One is the nature of causality—may not perhaps the birth of the new science be characterized in part as a transition from the predominance of final causes toward a search for efficient cause? Another question that is preliminary to the usage of one method or another is the extent to which a given scientist believes he can gain knowledge of nature at all. Is all of nature ultimately transparent to our understanding or not, and how certain can we be of what we believe we have come to understand of it?

Such are some of the pitfalls and complications that await the student of scientific method in the period of the Scientific Revolution. What guidelines may help avoid them? An elementary one must surely be that, even though historians of scientific method have no choice but to forgo their own preferences in this area and address the contemporary methodological scene, they cannot rely on Bacon and/or Descartes to have told us what, if any, novel method accompanied the Scientific Revolution. Neither can historians of scientific method rely on scientists' own words at the time for telling us. The one and only viable way that seems open is to take up the works of 17th-century scientists with a minimum of methodological preconceptions in mind and engage in a two-pronged search for each scientist's pronouncements on method in constant conjunction with his, and his contemporaries', actual scientific accomplishment. Although investigations along such lines have indeed been undertaken over the past decades for certain individual scientists,[5] no generalized treatment of method in 17th-century science allows us as yet to answer with sufficient confidence the question of whether the Scientific Revolution came with a new scientific method. Yet there is one essay, by Ernan McMullin, that comes closer to such an integrated treatment than any other discussion of the topic known to me. It was published in 1967 under the title "Empiricism and the Scientific Revolution." Our next section consists of a rendition of McMullin's principal results.

3.1.2. *From Demonstrative to Tentative Science*

McMullin's entire argument hinges on a distinction between 'conceptualist' and 'empiricist' conceptions of what constitutes scientific knowledge. In this dichotomy,

> the conceptualist supposes that a direct access to the essence or structure of natural objects is available. . . . The warrant for a scientific claim, in this view, is the self-evidence of the statement itself. . . . The science reached in this way will be certain and definitive.[6]

The conceptualist position, McMullin cautions, does not at all exclude the usage of empirical data; however, as Aristotle's example demonstrates in abundance, experience serves here to furnish the concepts which alone render science secure. Conversely, the empiricist position need not at all entail doing without concepts altogether, which, to the empiricist, serve as instruments. Only the extreme form of empiricism called positivism draws such a consequence, just as only the extreme form of conceptualism called rationalism forgoes experience entirely.[7] The key position of empiricism in the sense here defined is, rather, that

> the empiricist . . . believes that evidence for a scientific statement cannot be found in the concepts utilized in the statement, but only in the singular observations on which the statement rests.[8]

Whether the empiricist usage of empirical evidence is thought to be direct (through inductive generalization) or indirect (through hypothetico-deductive validation),

> the science one reaches by either of these ways will be tentative, approximate, progressive. . . . The empiricist claims that the world of sense is too opaque to be directly penetrated by human insight in the way the conceptualist assumes. . . . For the conceptualist, the principle (or the theory) is prior, even in the order of evidence: for the empiricist, the observation (the 'fact') is basic.
>
> It is a question, then, of the way in which the arrow of evidence points.[9]

It might seem that the stage has now been set by McMullin for arguing in a straightforward manner that the Scientific Revolution marks a one-way transition from a conceptualist to an empiricist stance. In essence, this is the very point he makes, yet he qualifies it in two, to him important, senses. First, the early stages of 17th-century science still displayed key features of conceptualism; second, the 18th century was to produce a somewhat surprising return to a conceptualist view of scientific knowledge. As a result, the second half of the 17th century appears as the high tide for empiricism—a view kept easier to defend by the exclusion of such outspoken, early empiricists as Mersenne and Pascal from treatment altogether. Nonetheless, the interest for us in McMullin's essay resides not so much in his thoughtful treatment of method in, successively, Aristotle, medieval scholastics, Copernicus, Galileo, Kepler, Descartes, Boyle, and Newton, however careful he is in linking these men's own pronunciations on method to their actual accomplishment.[10] Essential for us is that, spread across his essay, we find a number of illuminating definitions of the principal methodological shift that marks the Scientific Revolution, with Kepler serving as the chief transitional figure in this respect and Newton as the principal (though not quite trustworthy) witness. I shall not attempt to condense further McMullin's concise formulations, but render them as they stand:

> Scientific methodology before Galileo's time was defined largely in terms of a prior general theory of knowledge or a metaphysics. The procedures of test and proof one adopted in natural science were adopted not because of any proved successes on their part in this domain, but rather because it seemed to follow from the nature of man (or of knowing or of being) what procedures one should follow in such an inquiry. There is a definite movement away from 'epistemological' methodologies of this sort during the seventeenth century and a growing tendency to seek methods that will be adequate to the solution of limited definite problems in terms of some relatively pragmatic notion of adequacy.[11]

Further down, McMullin spells out in greater detail the nature of pragmatic adequacy, focusing on Newton's procedures:

> A new method of validation was in the making, in his own work and in that of many of his contemporaries. For what he and they were actually doing was to formulate hypotheses which were neither simple inductive generalizations nor deductions from the data, but explanatory models of a higher order; the next step was to 'verify' the model by means of successive experimental tests of predictions made with its aid.[12]

In short,

> Newton emphasized the observational origin of his specific mechanical laws . . . in a way Galileo had not; he emphasized the importance of a quantitative structuring of experience in a way Bacon had not. *This combination provided what amounted to a new method of scientific inquiry:* the formulation of hypotheses on the basis of empirical data sufficiently exactly stated to allow the hypotheses to be stringently tested. Thus to old types of 'proof' enumerated by Aristotle, deduction and induction, a third was added, that of hypothetico-deductive validation. It had been foreshadowed by many writers . . . , but *not until Newton were the data with which it could operate sufficiently exact to allow it to be used with*

any great success. It was this 'method' that ultimately detached natural science from its parent philosophy, by providing it with a new sort of empirical ground on which to stand. [My italics][13]

In a recent essay McMullin has elaborated further how this third type of proof (which, following Peirce, he now calls 'retroduction') grew out of a century of scientific trial and error.[14] Here he is concerned to show in particular how both a personal origin in Cartesian conceptualism (like Huygens') and a point of departure taken in Baconian empiricism (such as Newton's) could yield virtually the same result—the 'retroduction' advocated by both men in fact if not in unambiguous language. But we shall not follow McMullin further along this path.

What, in conclusion, I wish to call particular attention to is three key sentences I have italicized in the quotations from McMullin made above. Two of these recall familiar themes. One is the fertility of a two-current approach to the Scientific Revolution in that the success of the novel scientific method is shown here to have come from the fruitful marriage of conceptions derived from Galilean mechanics on the one hand and from the Baconian sciences on the other. The second italicized sentence helps to underline once more that not so much the formulation of new methods in the course of the Scientific Revolution marks a key difference with preceding modes of understanding nature as, rather, the application of a method that is novel principally for having entered, once and for good, the universe of quantitative precision. The final sentence indicates that, over and beyond a shift in method, we are witnessing here the operation of a process of even wider scope: science detaching itself from 'its parent philosophy'. The emancipation of science from philosophy is an extremely important aspect of the overall transition that we are studying in the present book. The next cluster of sections is devoted to a discussion of some of its principal ramifications.

3.2. THE NEW SCIENCE AND ITS NEW TIME FRAME

Great Knowledge enters in, and I don't know where it will ever end.—Chuang Tzu, 4th century B.C.[15]

3.2.1. *The Vanishing Role of Authority in Science*

Here is another simpleminded picture of the Scientific Revolution: Before Galileo, if one wished to acquire an understanding of nature, the best thing to do was to follow the sole authority in these matters, Aristotle. Then came Galileo, who once and for all taught us to think for ourselves, thus inaugurating early modern science and clearing the path toward its further improvement.

The question on which the present section centers is not whether such a picture is still tenable; it is not, because it is hopelessly naive and outdated.

Rather, the question is whether, after all the proper amendments and qualifications and refinements have been applied, the core of the picture still may be thought to hold in the sense of expressing some key characteristic of what the Scientific Revolution wrought.

The first amendment to be made concerns the question of whose authority came under attack in the 17th century. Whereas from the 13th century on Aristotle had been glorified as 'The Philosopher', "the master of those who know" (Dante),[16] and natural philosophy was largely cultivated in the format of commenting, Question by Question, upon his works, the Renaissance had witnessed a proliferation of ancient authorities. Atomist, neo-Platonist, and sceptical philosophies among many others were eagerly examined, and the freedom thus acquired to select one's own authority surely signified a certain weakening of the original posture. (There are even cases where it is hard to distinguish between taking a position because it derives from the authority adopted and selecting one's authority so as to gain legitimacy for a position of one's own independently arrived at.)[17] Nonetheless, in the 17th-century Quarrel of the Ancients and the Moderns, the upholders of the 'modern' point of view associated the authority worship of their opponents with the name of Aristotle more than with any other ancient thinker.

The '*locus classicus*' in the literature to turn to for the Quarrel in question is R. F. Jones' 1936 study *Ancients and Moderns: A Study of the Rise of the Scientific Movement in Seventeenth Century England*. In this book a large number of treatises are discussed, running from Bacon and Gilbert to the early years of the Royal Society, in which the two opposite points of view are adopted by scores of participants in the Quarrel. Jones sums up the issues as follows (outlined at this place from the point of view of the moderns):

> The revolt from the ancients arose from many causes. The old explanations had failed to satisfy the spirit of inquiry born of the Renaissance. The traditional philosophy seemed so far removed from reality that it evaporated into mere verbal mists, and also the discoveries of this and the preceding century had shown many traditional ideas to be erroneous. . . . Inspired by recent discoveries and sustained by an unwavering faith in the possibility of progress, new thinkers were eager to widen the bounds of knowledge. The realization that superstitious reverence for antiquity, embodying the disheartening belief that nothing new was to be discovered, was an effectual bar to the advancement of learning, incited them to vociferous iteration that there were no Pillars of Hercules with their *ne plus ultra* for modern thought, and to vigorous assaults upon the ancients.[18]

It was in this spirit that the Royal Society took as its motto *Nullius in verba* ('upon the words of no one'; an allusion to Horace's "*iurare in verba magistri*," 'to swear upon the words of the master').

To Jones the entire Quarrel appeared as one episode in "the history of the liberation of the mind."[19] More than that, "the clash between the two forces was really . . . between conservatism and liberalism, between progress and tradition. . . . The perennial clash between these two forces reveals itself in the rise of the new science, but it is also to be discovered in all the intellectual and social upheavals of modern times." [20]

Jones' book is almost exclusively confined to England. Yet the impression that the novel science of the 17th century may be characterized as a struggle for liberation from the authority of ancient thought can without difficulty be fortified with some celebrated examples on the Continent. Galileo's *Dialogo,* in particular, is famously filled with admonitions to poor Simplicio to give up his blind veneration of the books of Aristotle and his commentators and seek his inspiration rather in the Book of Nature. And Pascal, in taking the Descartes-inspired Aristotelian Father Noël to task in an exemplary fashion, delivered the punch line to the entire debate:

> On subjects in [the physical domain] we do not in the least rely on authorities—
> when we cite authors, we cite their demonstrations, not their names. [21]

In other words, we should not feel a blind reverence for past thinkers: We ought to weigh carefully the pros and cons of their arguments, but it will not do, in matters scientific, to forgo forming an independent judgment of our own.

The authenticity of a wealth of contemporary statements in the same vein is beyond dispute—this is how the proponents of the new science saw the issue. What may be disputed, though, is whether such statements do justice to the full situation at the time. At various places Jones himself talks of the "propaganda for the new science" these men engaged in. [22] But his siding with the victors in what he chooses to take for the perennial battle between the forces of progress and of conservatism makes him blind to the element of *distortion* usually to be encountered in propaganda battles. More recently, historians of science have become aware of the extent to which the proponents of the new science still had to gain a hearing for their unheard-of and potentially dangerous enterprise, utilizing the caricaturing of their opponents as one trusted weapon among others. In many respects, Simplicio is a caricature of Aristotelianism. As in all good caricatures, there is a striking resemblance with what is being caricatured, yet Aristotelians were not so dumb or so fully incapable of coming up with thoughts of their own as Galileo found it expedient in his campaign to picture them. On the other hand, it is also easy to overstate things here, and a picture of the Scientific Revolution which suggests major conceptual breakthroughs contributed to it by Aristotelian thought is a worse caricature than the one set up by Galileo, and lacks altogether its humorous qualities in the bargain.

Where is the proper balance to be struck here? No attempt has so far been made in the literature to produce a 'revised Jones', so some impressions are all there is to be put forward for now. One general point has recently been made by Peter Dear, who observed that it is not so much "any widespread and genuinely uncritical acceptance of Aristotelian texts . . . which prompted the seventeenth-century charges of slavish adherence to ancient authority" as, rather, the Aristotelians' "subordination of experiential statements to the structure of argument, without subjecting them to much investigation."[23] Other qualifications of the original idea can only be given by means of examples. On the one hand, here is a quotation from a first-rate Aristotelian, Alexander Ross, in a 1645 tract uncovered by Jones: "I follow the conduct of the most and wisest Philosophers, so that . . . I am not alone; and better it is to go astray with the best then with the worst, with company then alone."[24] To which confession Westfall appends the pertinent observation that "these words . . . suggest that Galileo was not dueling with a straw man."[25] On the other hand, it would also be wrong to suppose that 17th-century scientists managed to liberate themselves from the yoke of authority at one stroke. For example, it has been shown conclusively by Dijksterhuis and others to what, sometimes surprising, extent Galileo's thought remained governed to the end of his life by presuppositions unconsciously taken over from the man whose influence he believed to have rid himself of entirely.

Fully the same strictures apply to Descartes. In his *Discours de la méthode* he not only announces that he started his own career as an independent thinker by rejecting everything that philosophers had ever said before, but he goes on to claim that he went ahead to rethink the entire world from scratch. Here, too, it has been shown to what extent he often distorted ancient points of view, and how much more his own thought remained under the unconscious influence of what he ostensibly rejected. So far, nothing new. However, in the matter of authority Descartes strikes a tone not found with other protagonists of the Scientific Revolution. He attacks the authority of Aristotle, not so much because he wishes to combat in a general sense the veneration of authorities in science, but rather to set himself up as the new authority instead of Aristotle. It is this characteristic feature of Descartes' thought that makes it so hard to define his place in the Scientific Revolution. Whereas he took part in, and to quite some extent stamped, the ongoing revolt against the reigning philosophies of nature, the overall frame in which his thought was cast was fundamentally at odds with the conception of the task of the scientist formed by his innovative contemporaries, such as Galileo and Pascal. Descartes saw himself as the man to pull off what previous thinkers had failed to achieve—to produce the definitively valid system of nature. Men like Galileo had a very different idea of how their own thought contributed to the ultimate completion of science. I shall proceed now to outline the major issues pertinent to this particular topic.

3.2.2. *The Reorientation of Science toward an Unknown Future*

Galileo compelled his contemporaries to acknowledge that scientific truth was not so much the possession of mankind as, rather, its remote aim. In the place of the complacent conviction that our knowledge of nature had already been comp-leted he put the exciting conception of truth as the product of time, of doubt, of progress, of an as yet indefinitely distant future. Instead of the totality of a system of nature, he offered nothing but impulses toward its formation, as well as an interminable variety of problems and riddles whose resolution he left to posterity.

—Leonardo Olschki, 1927 [26]

What about the completion of science? Does it belong to the past, as a treasure of knowledge once possessed but since lost? Or is it to be envisaged within the confines of our own lifetime? Or is it a matter of the future—a perhaps distant or even indefinite or never-to-be-attained future?

One remarkable thing about the Scientific Revolution is that traces of all these three conceptions of the place of science in time can be detected in the pronouncements of 17th-century scientists. People's notions about their own position in time are notoriously difficult to reconstruct; these belong to the unspoken, shared beliefs of a period rather than presenting a topic of lively contemporaneous debate. This is perhaps why, to my knowledge, no study has ever appeared of the time frame in which 17th-century scientists placed their own occupations.[27] What one does find in the literature are scattered remarks of sometimes great perspicacity, and I have placed the most perceptive obser-vation of those I came across at the head of the present section. At this place, I can do no more than structure the argument as I see it and cite a few more observations I encountered here and there in the literature.

The most commonly accepted point of view is that in the Middle Ages as well as in the Renaissance the primary business of the natural philosopher was recovery rather than discovery, or, rather, that even when a discovery was made it was presented as, and often felt to be nothing but, recovery. There is a differ-ence between the two periods, however. The reigning medieval view appears to have been that, in principle, Aristotle had left nothing unexplained. He had created a system in which all phenomena of nature had been assigned a fitting place of their own. Certain phenomena did not fit in particularly well—projec-tile motion is just one example—but the thing to do in such a case was to propose means for a better fit in accordance with Aristotelian principles. New phenomena were also occasionally observed, but again, since fundamentally science was already completed, what remained to be done was to fit these into the picture as well. Despite some real discoveries being made during the Middle Ages concerning such phenomena as light and the magnet, natural phi-losophy was at bottom a backward-looking business.

Although there is no fundamental break with this pattern to be discerned in the Renaissance, nonetheless a characteristic turn has often been observed. Here is how Frances Yates put forward these 'truisms', as she calls them, in stirring prose:

The great forward movements of the Renaissance all derive their vigour, their emotional impulse, from looking backwards. The cyclic view of time as a perpetual movement from pristine golden ages of purity and truth through successive brazen and iron ages still held sway and the search for truth was thus of necessity a search for the early, the ancient, the original gold from which the baser metals of the present and the immediate past were corrupt degenerations. . . . the past was always better than the present, and progress was revival, rebirth, renaissance of antiquity. The classical humanist recovered the literature and the monuments of classical antiquity with a sense of return to the pure gold of a civilisation better and higher than his own.[28]

Now compare such an attitude with the conception that pervades the work of protagonists of the Scientific Revolution and that comes to the surface in occasional remarks some of them made on the topic. For them, generally speaking, discovery replaced recovery, and the completion of science, if attainable at all, became a thing of the future rather than the past. To cite just one celebrated example, Galileo announces, right at the beginning of the Third Day of the *Discorsi,* that "the Author" is going to prove a number of theorems on free fall and projectile motion, as follows:

That this is so, as well as a great number of other things which are no less worthy of knowing, is going to be demonstrated by me, and what I find even more important to do, the path shall be cleared and made accessible for a very ample, very excellent science, of which these our labors will be the elements, and in the more hidden recesses of which farther-seeing minds than mine shall penetrate.[29]

In other words, subsequent advances to be made in the new science of kinematics are confidently left to a future generation, which can build on what its predecessors already accomplished.

In other programmatic pieces of contemporary writing (e.g., Bacon's *Novum organum,* Pascal's Letter to Noël, Newton's preface to the first edition of the *Principia*) a similar sentiment finds expression. I shall now qualify its general validity in three ways: by citing one obvious counterexample, by pointing at contemporary remnants of the older view, and by finding out a bit more about the time span in which the future completion of science was expected at the time to take place.

My first qualifier concerns, once again, Descartes. For Descartes the completion of science lay neither in a past of which he believed he had fully rid himself nor in that future to which his more modest contemporaries relegated the making of further new discoveries. Rather, in Descartes' view the task of completing science had been assigned to the present, that is, to his own generation, or, to be even more precise, to that very thinker to whom it was given to receive virtually all knowledge God permitted man to gain possession of—René Descartes himself. There is quite substantial evidence in his work to show that this is what his outlook came down to (with his attitude toward authority in science providing a close counterpart).[30] Once again Descartes' position in the ongoing process of the Scientific Revolution appears as truly exceptional.

Still, it would go too far to state that the new frame of mind that enabled 17th-century scientists to conceive of science as a forward-looking business was itself entirely unambiguous. Indeed, it would have been humanly impossible to cut off at one stroke the intimate link to the past which had nurtured all their predecessors and to face the unknown without any sense of support from the past whatsoever. The thing to do, then, was bravely to face the future and make independent discoveries of one's own, but subsequently to claim that these very same discoveries had already been known in the distant past, yet were hidden in the recesses of time, ensconced in undecipherable codes. Over the past decades historians of science have come to penetrate the surface of such seemingly odd claims. For example, Stevin's invocation of the "Age of the Sages," which to Dijksterhuis represented little more than an unexplicable aberration in an otherwise sane and down-to-earth mind, appears to be one example of a widely spread 17th-century *topos*.[31] Newton's ideas on how Pythagoras encoded the law of universal gravitation, which McGuire and Rattansi uncovered in his notebooks,[32] seem to make similar sense when interpreted as a relict of an earlier attitude toward the completion of science—no longer instrumental to the scientific work itself but a psychologically important reassurance for men facing an otherwise all-too-threatening, unknown future.

One other *topos,* to be encountered often in exchanges between 17th-century scientists, is a host of remarks like one Isaac Beeckman made in a letter of 1 October 1629 to Marin Mersenne, with Beeckman expressing in a flattering yet highly characteristic manner the high expectations he sets on Mersenne's forthcoming *Harmonie universelle* after having seen a preliminary outline:

> What you promise in those books on music is certainly great and splendid, indeed, truly philosophical. Once you have fittingly completed them you will surely have robbed all of us of all occasion to think any further about philosophical matters.[33]

The phrase 'nothing more will remain to be done in a given science' rings down the 17th century, and this suggests that for many a protagonist of the Scientific Revolution the completion of at least substantial portions of science was to be envisaged within a few generations at most.[34]

Which conclusion at once raises a wider question: How far, in the minds of 17th-century scientists, did the future extend? Was there sufficient future left for the completion of science? This question, in its turn, is intimately connected with contemporary expectations of the Day of Judgment and a Millennium possibly preceding it. There is a big jigsaw puzzle here, and no more than isolated pieces have yet been touched by historians of the Scientific Revolution.[35] (Connecting all this with the interest of men like Kepler and Newton in chronology is certainly germane to the topic, too.) Just one little piece of the puzzle is an intriguing side remark by Kepler, who, in his *Astronomia nova,* relegated to posterity the resolution of certain technical details regarding variations in the rate of precession, "if it has pleased God to allot to the human race enough time on this earth for learning these leftover things."[36]

Whatever meaning is ultimately to be assigned to utterances of 17th-century scientists on the issues treated here, it is clear that they betray a vision of the place of science in time fundamentally at variance with what had been the customary attitude only a few decades earlier. This vision had much to do with the newly arising view of science as a communal effort, with future generations to be taken as part of the community—in other words, with a conception of science as a *cumulative* enterprise.

3.2.3. *When Did Science Become Cumulative?*

When a scientist poses himself a new problem, he can attack it fortified by the pooled resources of all his predecessors.

—*Condorcet, c. 1782*[37]

Thomas Kuhn, in particular, has explained in illuminating detail one major structural difference between the sciences of nature, and disciplines in the humanities and the social sciences. In the latter, if theory formation is an important objective, debates over basic issues appear endemic; the one thing over which theorists are in agreement is that no common foundation for the discipline in question exists; a host of rival 'schools' tend to seek theoretical niches carefully insulated from one another; 'past masters' are not only venerated but also continue to be read for inspiration; extensive searches in the literature from previous decades must often be made to gather all the data and points of view one feels to be in need of; textbooks present either one pet viewpoint or an indiscriminate plurality of theoretical perspectives; in brief, anyone with sufficient daring may start from scratch without making himself *ipso facto* ridiculous. In the sciences, on the contrary, "when a scientist poses himself a new problem, he can attack it fortified by the pooled resources of all his predecessors"; there is a broad consensus on what the items are that constitute current problems in the field; 'great scientists' of the past are venerated but otherwise ignored, or at least considered irrelevant to the solution of current problems; fresh journal articles provide the bulk of one's literary needs; textbooks compete for little else but didactic clarity. Polar oppositions like these are surely ideal types rather than lifelike pictures; blurring edges, overlaps, and exceptions may be cited, and yet, few practitioners of the natural sciences on the one hand and of the humanities and social sciences on the other fail to acknowledge that, along lines like those just sketched, the former work in a broadly cumulative manner whereas the latter do not.[38]

In Kuhn's description, cumulativity is narrowly bound up with his theory of scientific revolutions. First, he insists, science is cumulative only in the context of a paradigm. That is to say, scientific revolutions are those very episodes in the history of a scientific discipline when cumulativity breaks down, and once the revolution is over cumulation of research findings starts all over again. Also, such cycles characterize exclusively those scholarly disciplines that have acquired (to adopt Kant's terminology for a change) 'the secure pace of science'—here we find again the very opposition between the sciences and other scholarly endeavors just outlined.[39]

There are other views. Few historians of science were so programmatically convinced of cumulativity as a feature uniquely characteristic of the sciences as George Sarton. For those historians who, like Sarton, stand in the positivist tradition science is an inherently cumulative enterprise. Cumulation, in this view, starts with science itself—with the Greeks, that is—and science has never lost this uniquely distinctive feature. It has also been suggested, however, that (in Henry Guerlac's words) "science since about 1600 has shown a steadily progressing and 'cumulative' character."[40] In this view science did not become cumulative until the onset of the Scientific Revolution, and Sarton's conception is—implicitly—unmasked as the projection backward upon the early Greeks of a feature not to be acquired by science until the 17th century.

An intuitive sense of how the Scientific Revolution might provide the proper date for when science acquired this novel feature of cumulativity may be gained from taking another look at Condorcet's fine summing-up (in a eulogy for Vaucanson) of what cumulativity means. Let us note in particular when he pronounced it, in an aside meant to do no more than rehearse the obvious (c. 1782). Could anyone imagine such a statement being made in, say, the middle of the 16th century by an Aristotelian or a neo-Platonist, be it Copernicus or Vesalius himself? Or, before that time, could such a statement truthfully have been made for any but the handful of 'classical mathematical sciences' as cultivated in ancient Greece?

The view that science became cumulative during, and as a result of, the Scientific Revolution is rarely set forth at any length in surveys of the birth of early modern science. But it is implicit in many. Particularly when we take a grand and somewhat superficial view of things, it may appear—choosing an obvious example—how smoothly Newton built upon the discoveries of Galileo and Kepler, uniting them in the synthesis of the *Principia* and thus displaying cumulation in science at work. Nor does such an overall view seem mistaken; something drastically new is surely happening here, on a scale never seen before. But again, the overall picture requires serious qualification.

In the first place, such finer-grained historical analyses of episodes in the Scientific Revolution as have multiplied over the past half-century tend to show

how uneven cumulative scientific progress over the 17th century really was, and how easily major findings could be ignored, misunderstood, stood on end or otherwise misapplied, or garbled by those who applied themselves to their further elaboration.[41] Once again the Scientific Revolution appears as transitional—in this case, between the previous overall absence of cumulativity in science and the developed product we take for granted in our own day. Still, I think, the key step—well worth calling a major break—was made in, and through, the Scientific Revolution.

One other signpost of how ambiguous the development of cumulativity in 17th-century science was may be gleaned from the principal programmatic statement on the topic made at the time. This statement is to be found, once again, in the *Discours de la méthode*. In Part 6 Descartes defines the idea clearly and distinctly and suggests in so doing that he offers his own work as one link in the chain thus forged with the future. A few pages further on, his true idea of his own place in the history of science is allowed to emerge when he effectively demolishes the picture of ongoing cumulation just called up: After all, he has already laid down for posterity the necessary first principles, and some filling in of remaining empirical details is all that remains because he has already performed the securely deduced groundwork.[42] For the third time in succession we find Descartes standing athwart a major novel characteristic of the Scientific Revolution, thus helping to throw that very novelty into relief for us. It has become time now to identify the underlying transition from which both the novel developments here discussed and Descartes' exceptional status in their regard can be seen to follow.

3.2.4. *From Natural Philosophy to Science*

In the 17th century, authority began to lose its power for scientists, but many among them still owed more to the ancient authorities than they cared to admit, with Descartes even seeking to set himself up as the new authority. In the 17th century, science began to orient itself toward the future rather than the past, with Descartes focusing rather on the present, and ambiguous relics of taking discovery for recovery continuing to satisfy a felt need. In the 17th century, science became a cumulative enterprise over a far broader domain than ever before, but this did not happen at one stroke, and Descartes rejected the notion right after phrasing it. The key transition that lurks behind such comparatively novel states of the scientific mind in the 17th century is, in my view, *the emancipation of science from natural philosophy* (in one particular sense of the term, which does not quite cover contemporary usage). That is to say, a transition took place some time around 1600 from natural philosophy to science in a new sense. A proliferation of systems aiming to encompass, at least in principle, the totality of nature and the world gave way to a much more limited conception of scientific knowledge.

Ideas on whether and, if so, when and how science and philosophy separated diverge very widely. Derek Gjertsen, for example, in his book *Science and Philosophy: Past and Present,* takes issue with suggestions, advanced by Ayer and Rorty, that the split occurred in the 19th century and proposes instead Socrates' appearance on the Greek scene as a more fitting date for the event.[43] In pursuing their respective arguments one easily finds that quite different ideas of what philosophy is about are at stake. In Gjertsen's case, for instance, the distinguishing characteristic appears to be that philosophy supposedly deals with man, whereas science deals with nature. In other conceptions of the relation between philosophy and science—those of Burtt and Koyré, which we have treated at length in the preceding chapter—the view lies implied that no separation ever took place, since scientific knowledge is necessarily bound up with philosophical notions in any case.

The latter view can easily be conceded without sacrificing the idea that a very important shift in their mutual relations occurred in the course of the Scientific Revolution, or, to put it more strongly, that the emancipation of science from an overarching entity called 'natural philosophy' is one defining characteristic of the Scientific Revolution. The key point is that, in earlier times, knowledge of nature was held to make sense only in the framework of a comprehensive view of the world and man and how these are connected. For example, in Aristotelian philosophy the fall of bodies was not a natural phenomenon to be examined in its own right but rather one exemplar in a huge, multileveled subdivision of categories of motion through which the humble, subordinate phenomenon itself was ultimately connected to the essence of Change and Becoming. As opposed to this, the new science, particularly as exemplified by its first outspoken pioneer, Galileo, allowed one to study the phenomenon without a prior knowledge of the totality of things.

The new science claimed to provide an understanding of reality that did not owe its warrant to whether or not it fitted in with an all-encompassing insight into the order of the world but only to the extent to which it satisfied inherent criteria of a methodical nature: quantitative precision and, above all, susceptibility to empirical checking on the level of detail. Recall here McMullin's argument as discussed in our sections on method. Before the Scientific Revolution, so he claimed (see section 3.1.2), procedures in science derived from "a prior general theory of knowledge or a metaphysics. . . . There is a definite movement away from 'epistemological' methodologies of this sort during the seventeenth century and a growing tendency to seek methods that will be adequate to the solution of limited definite problems in terms of some relatively pragmatic notion of adequacy."

No longer the totality of attainable knowledge but only seemingly random aggregates of partial insights gained in response to "limited definite problems"—is it to be wondered that not every contemporary thinker jumped upon the bandwagon? Descartes, with his customary perspicacity, marked the exact point where science took leave of the search for the totality of things in accusing Galileo's kinematics of being "built without foundation." How can one talk of the effects of gravity without a prior knowledge of what gravity *is?* And how can one know what gravity is without a prior knowledge of first principles?[44] To Descartes, the traditional ideal of aiming for the totality of knowledge was as alive as it had been for Aristotle, his principal quarrel with Aristotle and rival natural philosophers being that they had happened to totalize knowledge in the wrong manner, whereas (as has been stressed particularly in Jean-François Revel's *Histoire de la philosophie occidentale*) Descartes knew how to do the old job better. The revolt of most of his contemporaries with an interest in grasping the secrets of nature, however, was directed against the idea that the old job was worth doing anymore.[45]

The opposition here outlined has persisted through the centuries, taking on new guises as the confidence of scientists in the worth of their own activities grew. In fact, much of that confidence was gained in battles fought over this very issue. How to gain a more reliable knowledge of nature: through system-building or rather through the untiring construction of partial theories and concepts and experimental tests, with 'pragmatic adequacy' as the primary guideline? One significant 17th-century episode in the ongoing battle was between Pascal and adherents of Cartesianism; another, between Boyle and Hobbes. In a thought-provoking book published in 1985, *Leviathan and the Air-Pump,* to which I shall presently return in other contexts, Steven Shapin and Simon Schaffer have analyzed the latter contest from a variety of angles. The one that concerns us here is where we find Hobbes rejecting all claims that fail to derive from a prior natural philosophy. Here Hobbes constructs a completed system out of elements taken from *avant-garde* mechanical thought, with Boyle claiming the rights of experiment as a mode of attaining partial knowledge and laying down the accompanying rules in the process.[46]

From the 17th century on, science and philosophy went their largely separate ways, with philosophers continuing for a long time to produce total systems in which all things were to be assigned their fitting place. As scientists grew bolder over the centuries, some went so far as to inflate their own scientific findings and turn them into—allegedly—complete conceptions of the world and humanity, scientistic constructions whose history remains to be written.[47]

But this is no less true of the relative emancipation of science from philosophy, whose origins we have now assigned to the onset of the Scientific Revolution. This emancipation process, too, awaits its historian. To be sure, in the literature on the Scientific Revolution one finds countless allusions to the split, but hardly an integrated treatment along such lines as have been drawn in a preliminary fashion, and for one particular episode, by Shapin and Schaffer.

Nonetheless, there is one aspect of this entire movement of 17th-century thought that has in recent decades caught the attention of historians of science. Ever since the split occurred, much intellectual energy has been spent, by philosophers and scientists alike, on attempts at reconstructing whatever could be rescued from natural philosophy in light of the new scientific results.[48] After all, the split involved nothing less than giving up the ideal of the unity of knowledge in favor of a relentless process of specialization. Already in the 17th century itself there were those who, while recognizing that the new science represented the 'wave of the future', could not help regretting deeply the loss of unity that accompanied the apparently irresistible wave and therefore set out to recover unity at a higher level. In the sixties of the present century, efforts in such a vein began to elicit a great deal of historiographical attention, and it is to the debate surrounding these issues that we turn now.

3.3. THE NEW SCIENCE AND THE OLD MAGIC

A number of books dealing partly or wholly with the Scientific Revolution have come out under some variant of a 'From . . . to . . .' title. Examples are Hall's *From Galileo to Newton* and Crombie's *Augustine to Galileo.* Thus, when in 1982 Charles Webster's *From Paracelsus to Newton* appeared, this little book might, by the casual browser, have naturally been taken to belong to the same class of straightforward overviews of science during the period demarcated by the two names in the title. The name Paracelsus, to be sure, might cause an eyebrow to be raised somewhat, if its owner remembered how thoroughly this early 16th-century chemist had been banished by the Halls from early modern science as the irrational disseminator of what they called "esoteric chaff" (see section 2.4.3). The mild surprise of our hypothetical prospective buyer would be enhanced considerably when, on looking up the List of Contents, he found that the period between Paracelsus and Newton, after an introduction, was treated under the following three headings: Prophecy, Spiritual Magic, and Demonic Magic. And his bafflement as to what these subjects could possibly have to do with the history of *science* would not be resolved until his glance was drawn by the subtitle, which gives the secret away while exacerbating it at the same time: *Magic and the Making of Modern Science.*

Webster's book, although an interesting contribution to our understanding of the full world of ideas in which 16th- and 17th-century scientists lived, does not by itself mark a particularly dramatic break in the historiography of the Scientific Revolution. Rather, it adds one more wave to a comparatively recent current in this historiography, roughly to be labeled 'the Hermeticist approach to the Scientific Revolution'. This may conveniently be said to have started in 1964 with the publication of Frances Yates' *Giordano Bruno and the Hermetic Tradition.* Broadly defined, the new current came down to calling attention to mystical as well as magical patterns of thought in the Scientific Revolution, which before had either been largely ignored or been relegated to the dustbin of the irrational.

To be sure, magic and science had been connected before in another, more sympathetic sense than that of virtually total opposition. The pioneer here was the American historian of science Lynn Thorndike, whose monumental, eight-volume *History of Magic and Experimental Science* (1923–1958) already in its programmatic title drew a close analogy between the history of magical thought and practice and the experimental aspect of early modern science. However, Thorndike's work was more in the nature of sympathetically tracing and digesting a specific tradition in the history of ideas than of a definite, clear-cut historical thesis. It was not until such a thesis was framed and presented to the world in a plausible fashion and in rousing prose by Frances Yates that the work done previously by Thorndike as well as by a substantial number of other, mostly later investigators of Renaissance modes of thought (Kristeller, Garin, Rossi, Pagel, Walker, to name but a few) began to be seen in a different light. It appeared no longer as the filling in of details of marginal interest to the mainstream of scientific thought in the 16th and 17th centuries but rather as the definition of elements which somehow seemed constitutive of the very core of the scientific renewal of the early modern period.

In its original guise, as set forth principally in her book on Bruno, Yates' thesis is chiefly concerned with *explaining* the Scientific Revolution by connecting the event with the magical philosophy contained in what she called the Hermetic core of 16th-century neo-Platonism. (This explanatory side to the Hermeticist historiography of science is discussed in section 4.4.4). But in the late sixties and in the seventies the new approach increasingly spilled over into the history of 17th-century science and began to give a quite distinct, new coloring to our conception of what early modern science itself stood for. This happened in Yates' own work in the first place, as her scholarly interests progressed more and more from the 16th to the early 17th century. It also became a prominent mark of the work of a number of other historians of science (examples are Rossi, Debus, and Webster), not all of whom were prepared to share to the full the lengths to which Yates, in her enthusiastic search for radical reinterpretations, never hesitated to go. Finally, it opened the floodgates for an even more drastically altered conception which made the Scientific Revolution, not so much the beneficial triumph of rational thought about nature, but rather the agent chiefly responsible for the destructive handling of nature, whose consequences we find ourselves facing in the late 20th century. We shall start our account of this great overhaul in the historiography of the Scientific Revolution with an outline of Yates' pertinent ideas.

3.3.1. *The 'Rosicrucian' Conception of Early Modern Science*

One predominant, very attractive, though also rather tricky feature of Yates' mode of history writing was her habit of linking, in the framework of one overarching conception, elements from the history of science, of ideas in general, of art, of religion, and of political events as so many contributing pieces to one grand jigsaw puzzle. In her treatment, such historical puzzles consisted of the attempted reconstruction of 'lost causes', that is, of fleeting historical events which, although watched with intense expectation at the time, ended in catastrophe and were subsequently wiped from our collective memory by the victors—free as these now were to rewrite their propaganda into official history. Whereas the careful attention to such neglected episodes in history is typically Yates' own, her smooth application of an immense erudition fed by the most variegated historical disciplines was greatly enhanced by her association, from the late thirties onward, with the Warburg Institute in London.

Frances Amelia Yates lived from 1899 to 1981. Although a productive historian from the thirties onward, she did not acquire world fame as a perhaps overspeculative but certainly most creative and imaginative historian of the Renaissance world of ideas and events until the late sixties.[49] With hindsight one can note that many things she said had already been said before by herself as well as by a variety of more specialized scholars, none of whom she ever failed to acknowledge in a most generous way. Yet there is something uniquely suggestive about her peculiar mode of expression that greatly helped to turn her work rather than anybody else's into the catalyst for the most drastic upheaval in our conception of the birth of early modern science to take place since the days of Burtt, Dijksterhuis, and Koyré.

Not that the inner core of this conception itself, centering as it did and does on such notions as mathematization, experiment, and mechanization and on how these hang together, was significantly altered by her—the debate over the relative importance of these strands of the Scientific Revolution has gone on largely independently of the 'Hermetic' issue, and no 'Hermeticist' historian has tried to intervene in this particular dispute. Rather, our entire view of the world of ideas in which the protagonists of the Scientific Revolution lived has undergone a quite radical overhaul, and the debate has principally been over what this means for our view of the Scientific Revolution as a historical phenomenon.

The Rosicrucian Enlightenment. In 1964, in her book on Bruno, Yates took the year 1614 as the watershed in the history of Hermetic thought. In that year, Isaac Casaubon had proved that, rather than dating from Mosaic times and thus foreshadowing important Christian truths, the Hermetic corpus was in reality a syncretist product of the 2nd century AD with little right to claim particular reverence. Hence, so Yates thought, the rapid waning, after 1614, of the magical view of the world embodied in the adherence of various influential Renaissance philosophers to the Hermetic corpus of writing.[50] However, in her later book *The Rosicrucian Enlightenment* (1972), Yates found that, mediated through John Dee (whom she increasingly came to regard as *the* prototype of the Renaissance Magus), and proclaimed to the world through the Rosicrucian manifestoes of 1614–1615, originally Hermetic thought patterns had been preserved and continued to exert considerable influence on the emerging new science. This 'influencing' was a typically ambiguous affair. In Yates' account, the Rosicrucian movement proclaimed the advent of new knowledge all over Europe and thus prepared the European mind for the Scientific Revolution, while at the same time standing in opposition to central lines of approach of the newly emerging science. Here is how Yates expressed her intimations about the several ways in which the Rosicrucian movement of 'Magia, Cabala, and Alchemy' might have tinged the new science:

> The Rosicrucian movement is aware that large new revelations of knowledge are at hand, that man is about to arrive at another stage of advance, far beyond that already achieved. This sense of standing on tiptoe in expectation of new knowledge is most characteristic of the Rosicrucian outlook. And the Rosicrucians, who know that they hold in their hands potentialities for great advance, are concerned to integrate these into a religious philosophy. Hence the Rosicrucian alchemy expresses both the scientific outlook, penetrating into new worlds of discovery, and also an attitude of religious expectation, of penetrating into new fields of religious experience. . . . It was the effort to avoid doctrinal differences, to turn from them to exploration of nature in a religious spirit, which constituted the atmosphere in which science could advance. . . .
>
> Paracelsist physicians like Fludd, Maier, Croll, represent the thought of the movement. But there is in Dee's *Monas* and in Maier's alchemical and cabalist formulations a further aspect which it is difficult to seize and which may represent an approach to nature in which alchemical and cabalist formulations have combined with mathematics to form something new. It may have been this germ in Rosicrucian thought which caused the bearers of some of the greatest names in the history of the scientific revolution to hover around it.[51]

Among these great names, she mentions, and occasionally discusses in some detail, Kepler, Descartes, Bacon, the various groups that were to merge into the Royal Society, and Isaac Newton. In each case she is concerned to trace specific associations, whether religious, political, or in the domain of alchemy, of these scientists with the Rosicrucian movement. Examples are Kepler's immersion in the Hermetic atmosphere of the court of Rudolf II in Prague, Descartes' anxious and demonstrative refusal to traffic with the 'invisible' Rosicrucians in Paris, and Newton's alchemical efforts. All this and much more, set in the overall, largely political framework of the historical recovery of a 'Rosicrucian' culture centered in Bohemia and the German Palatinate but irretrievably destroyed in the early struggles of the Thirty Years' War, certainly makes for attractive reading. But what does it in reality have to do with the constitutive features of early modern science?

Extreme positions. This, of course, is the question that has continued to be asked ever since.

As increasing doubts emerged in the wider society of the late sixties and seventies about the beneficial effects of science, historians of science, too, began to look from a somewhat different viewpoint at the 17th-century roots of present-day science and, in so doing, seized the 'Hermeticist' issue as a convenient focal point for their debates.

Some historians went to amazing lengths in their reinterpretations. In such a book as Webster's *From Paracelsus to Newton,* for example, the 'magical' ideas of his protagonists are focused upon to the virtual exclusion of their scientific pursuits, with men like Galileo not being mentioned even once.

On the other extreme we find the Halls, who, in the fifties and early sixties, had indeed paid considerable attention to the esoteric undercurrent of 16th-century science, but only as something early modern science had to free itself from in order to establish its own identity. They persisted fully in their original attitude; continued to make a sharp distinction between rational science and irrational superstitions; and, in later accounts of the Scientific Revolution, left only so much room for 'irrational' elements in the thinking of its protagonists as ought to be reserved for remnants of outdated modes of thought. Rupert Hall, in 1983, made the point quite clearly:

> If one is interested in creativity one must, to a large extent, follow the victorious and not the defeated. Atavism must be accepted, but not supposed to be more interesting and significant than creativity leading to the abandonment of traditional ideas.[52]

A more flexible, middle-of-the-road attitude was displayed by Westfall. Whereas, in accounts of the Scientific Revolution from the fifties and sixties, the introduction of the mechanical philosophy had usually been set off against Aristotelian conceptions, Westfall, in his *Construction* of 1971, rather interpreted the mechanicist world-view as the overdrastic excision of a panpsychism characteristic of Renaissance Naturalism with its tendency to project the human soul upon the cosmos. In this way Hermetic thought was indeed made to enter into the story of the Scientific Revolution, if chiefly on the negative side. A more directly fertile role in the ongoing process of the Scientific Revolution was attributed to it by Westfall through his study of Newton's alchemical work. And here Westfall (in conjunction with two other students of Newton's alchemy, Dobbs and Figala) achieved a feat that considerably transcends the mutual sloganizing characteristic of substantial portions of the debate over the significance of esoteric currents of thought during the Scientific Revolution. For it must in fairness be noted that Yates' ever more strongly displayed habit of approaching the past by primarily collecting suggestive pieces of an intriguing jigsaw puzzle rather than by constructing a clinching historical argument has contributed considerably to the somewhat sterile nature of many exchanges about topics she brought to the forefront of historical attention.

It is on the more productive reinterpretations that I shall focus in the following sections, before rounding off the entire discussion by coming back to the deepest issue raised by Yates: whether or not the consideration of Hermetic thought patterns in the 17th century signifies anything of profound importance about the nature of science itself.

3.3.2. *Rosicrucians, Chemists, and Alchemists in 17th-century Science*

There is something elusive about the Hermeticist reinterpretation of the Scientific Revolution. Unlike, for example, the Duhem thesis, it cannot easily be summed up in one clear-cut statement like his 'The Revolution in Science really started in the 14th century'. Perhaps the best way to sum it up is as follows: The reinterpretation came down to an ongoing demonstration that esoteric notions not nowadays considered scientific persisted until a far later date into the 17th century than had previously been taken for granted. Thus the historian of chemistry Allen Debus, in a large number of books and articles, called particular attention to the huge interest evoked in the 17th century by debates in chemistry, in particular involving Paracelsism and the critique of Paracelsus by van Helmont—subjects written about before but only now seeming to have found the right, Hermetic or Rosicrucian, framework, in which they began to make proper sense. In a similar vein, Charles Webster was concerned to show that such issues as the dating of the Millennium and the Day of Judgment, the portentous meaning of comets, the supposition of vital principles in nature and their relation to natural magic, belief in witchcraft, and the like continued to occupy men like Kepler, Bacon, Boyle, and Newton in a spirit which Webster, too, traced back to Paracelsus' writings. Not by chance, neither in *The Rosicrucian Enlightenment* nor in Webster's *Paracelsus to Newton* is any pertinent mention made of such men as Galileo and Huygens, who were as far removed from esoteric modes of thought as can be imagined. However, the issue of how to reconcile the 'no-nonsense' attitude displayed by such men with the new, more encompassing picture of the Scientific Revolution gradually arising from contributions like those just mentioned was not likely to be tackled by contributors to the dispute.

What tended to remain similarly elusive was the issue—prominent in a number of debates—of whether the split experienced by historians living in the 20th century between 'hard-core science' and esoteric interests similarly existed in the minds of the protagonists of the Scientific Revolution themselves. Or, asked in a more concrete fashion: What, if any, relationship was there between the hard-core scientific and the esoteric thought of one and the same thinker? It is clear that, if such factual links can be demonstrated conclusively, the significance of the Hermetic contribution to the debate on the nature of the Scientific Revolution is considerably enhanced. If not, we still gain a clearer insight into the full, as distinct from the merely scientific, world-view of its principal protagonists, whereas if yes, the making of early modern science itself is obviously at stake.

But here pre-Newtonian chemistry, taken as a budding discipline in its own right, does not seem to have particularly much to offer. One may surely point in a general way to "a new observational approach to nature" which was stimulated by "Renaissance magical science," as Debus affirms.[53] But this only recalls all the earlier disputes over whether, indeed, the new thing about early modern science is to be sought in its 'observational approach', thus leaving out both the mathematization process and the manipulation of nature by experiment. To my knowledge, there is one case only, though indeed the most baffling of all, where a tangible link between 'esoteric' and 'hard-core' thinking has been suggested and underpinned with a wealth of factual evidence. This is the riddle of Isaac Newton's alchemy. It was one thing for Frances Yates to show that the chief contribution made by the Rosicrucians to the Hermetic tradition was a renewed interest in alchemy. She linked this in a vaguely suggestive way to the further history of scientific thought in the 17th century, and one pioneer student of Newton's alchemical papers, Betty Jo Teeter Dobbs, quickly pointed out the superficiality of her supposed connections. It was quite something else to propose, as was done by Dobbs and Westfall in particular, that the concept of force which Isaac Newton needed in order to overcome the fundamental defects he perceived in the mechanical philosophy may very well have derived from his alchemical pursuits.[54] Here, so Dobbs as well as Westfall contended, was a probable source of inspiration for Newton's conception of the world as filled with forces, both attractive and repulsive, operating over varying distances, and acting upon different bodies. What, for Newton, made such a conception of things distinct from Renaissance operations with occult forces, however, is that these forces could, in principle, be measured, and that a scientific treatment of them was possible only to the extent that they had been measured in fact. But the sole forces which Newton ever succeeded in actually quantifying were those which make up the fundamental topic of the *Principia*. In this vision of Newton's achievement, the *Principia* appears at one and the same time as the point of culmination of the Scientific Revolution, a view commonly held by historians, and as just one fragment of Newton's much more widely cast, yet fundamentally uncompleted, life's work.

The above is not to imply that such a drastic revision of the received picture of Newton necessarily had to await the advent of the Hermetic reinterpretation of the Scientific Revolution. Nonetheless, such a revised picture makes considerably more sense in the new, wider perspective provided by the Hermeticist current in the historiography of early modern science than it does in the 'regrettable remnants' tradition of old.

3.3.3. *The Elusive Core of the Debate*

The Hermeticist debate has been with us now for a quarter of a century. Much of the original emotion brought to it has meanwhile been spent, yet many of the issues still seem unresolved. The debate largely precipitated in pertinent collections of essays, with often polemical or programmatic forewords, and in essay reviews devoted to such collections.[55] The debate has justly been called

> an intellectual minefield, littered with dangerous explosives like 'the Yates thesis', 'rational' and 'irrational', and even . . . the words 'magic', 'the occult' and 'science' themselves.[56]

Deep issues about what science ultimately is and what it has, or has not, meant for the liberation of humanity from superstition are obviously at stake but have rarely been addressed in a completely straightforward manner. Two fundamentally different views of science seem to clash. For some contributors to the debate, the core tenet to which they wish to cling is that early modern science has revealed for the first time the superstitious nature of many of humanity's traditional conceptions of the natural world. Thanks to early modern science and its subsequent elaboration into the highly intricate structure of science in its present-day state, the fear of nature that characterizes premodern societies could give way to the quiet certainty that we know, and can predict, nature's operations. If only every citizen could be educated to an understanding of what science really is about, all reason for fear of nature's workings would disappear.

In essence, this is the Enlightenment view of science—modified, to be sure, among the more sophisticated and professional historians of science by a 20th-century awareness that scientific progress has its drawbacks; yet fundamentally the conception survives unscathed. It implies that yes, there is something perennial to science in the guise it has taken ever since the Scientific Revolution: its rationality and its potential as a force of liberation. For many historians of science, a view such as this has been their primary driving force in turning to the field in the first place, rather than sticking to research in the particular branch of science in which they had originally been educated. It is not surprising that many of these historians tended to feel challenged, if not downright threatened, when a new historiographical current suddenly appeared that evidently negated the basic tenets of their conception of science while remaining rather elusive itself about whatever alternative conception might lurk at the background of this new approach to the origins of early modern science. Overall it seems true to say that historians who explored the Hermeticist approach have in general confined themselves to upbraiding their opponents for taking a fundamentally unhistorical stance in sticking to a supposedly perennial, timeless characteristic of science. True, it has sometimes been acknowledged that the new approach was quite as one-sided as that of the opponents to the Hermetic reinterpretation. Frances Yates made the point in a charming and characteristically open-minded way:

> I would thus urge that the history of science in this period, instead of being read solely forwards for its premonitions of what was to come, should also be read backwards, seeking its connections with what had gone before. A history of science may emerge from such efforts which will be exaggerated and partly wrong. But then the history of science from the solely forward-looking point of view has also been exaggerated and partly wrong, misinterpreting the old thinkers by picking out from the context of their thought as a whole only what seems to point in the direction of modern developments. Only in the perhaps fairly distant future will a proper balance be established in which the two types of inquiry, both of which are essential, will each contribute their quota to a new assessment.[57]

This might seem, and in many respects is, a commendable, live-and-let-live compromise on which both parties to the debate might be happy to settle, yet it also begs the fundamental question of whether 'a history of science may emerge from [Hermeticist] efforts' *at all*. I shall now cut two paths through the jungle created by the implications of this fundamental question. The first is concerned with the conception of science as a force of liberation; the second with science as the embodiment of rationality.

3.3.4. *The Scientific Revolution and the Disenchantment of the World*

In a general sense, it seems obvious to regard the Scientific Revolution as the direct precursor to the European Enlightenment—that is, to ascribe to the rise of early modern science the advent of a new, joyous, and pervasive feeling that the natural world has finally become fully transparent to humans and that the superstitions of bygone ages may now for the first time be put to rest.

However, with good reason two qualifiers have been built into the previous sentence. The connection in question does seem obvious to historians of science, but in no other than a *general* sense indeed. They have bothered only in the rarest of cases to find out in detail what the precise effects of early modern science may have been upon magical practices, upon mystical thought, upon contemporary and later accusations of sorcery, upon the great European witch-craze, and so on and so forth. Did early modern science eradicate these things at one stroke? or not at all? or only in a slow process? or just some of them, but not others? Did it perhaps drive them underground? or leave them untouched? or reach them only at the level of an intellectual or social elite? In short, is there any firmly established, rather than generally plausible, connection between the Scientific Revolution and what the German sociologist Max Weber called, in an unforgettable phrase, "the disenchantment of the world" (*die Entzauberung der Welt*)?[58]

My other qualifier was that "it seems obvious" to state such a connection.

For many scholars who have indeed gone ahead to investigate detailed aspects of these historical processes that together made up the disenchantment of the European world, the connection with the Scientific Revolution seems hardly to exist.

Here Max Weber himself provides a very interesting case in point. There are indications in his work (to which I come back in another connection; see section 3.6.3) that he was aware to what considerable extent the disenchantment of the European world took place under the aegis of early modern science. For example, here is the definition he gave in "Wissenschaft als Beruf" (1919; Science as a Vocation):

> Increasing intellectualization and rationalization does not therefore mean an increase in overall knowledge of the conditions of life around one. But it means something else: one knows—or believes one knows—that, if only one wants to, one might pick up that knowledge; in other words, one knows that, as a matter of principle, no mysterious, incalculable powers and forces exist that come into play here. Rather, in principle one can dominate all things by computing them. This is what the disenchantment of the world signifies.[59]

Weber goes on to indicate that in Western civilization the process of disenchantment has taken thousands of years, adding in one subordinate clause that "science [i.e., both the natural sciences and other scholarly endeavors] belongs to [the process] too, both as a member and as a motor."[60]

However, at this point he drops the matter, and if one surveys his life's work one finds that Weber investigated very extensively such roots of the process as he found in the medieval city, in early capitalism, in the economic ethos of Calvinism, in the secularization of law, etc., but scarcely in the advent of early modern science. Whatever the causes of this neglect, its negative impact upon entire areas of scholarship has been drastic. Suppose that Weber, with his formidable powers of imaginative generalization built upon vast amounts of detailed study in the most diverse fields of scholarship, had indeed sought for roots of the western European rationalization process in the emergence of early modern science. Might not then the historiography of science, from its inception onward, have found a most fertile source of support in a wider research program, embodied in the issue of the how and why of western Europe's emancipation from the ways of traditional society? And would the entire process of the disenchantment of the world—what exactly it consisted of; what its limits were; what strata of the population at large were affected by it, and when, and how deeply—not have been handled right from the start with the fortunes and the results of that great agent of disenchantment, early modern science, permanently in view? Present-day research on the 'history of mentalities' is still often marked by the virtual absence of the particular viewpoint that could be introduced by looking at what, precisely, early modern science did to sorcery, to the belief in witches, to magic, to astrology, to everything, in short, that nowadays goes under the name of 'the occult'.

It is true that a variety of pitfalls awaits the student of this particular domain of research. After all, a great deal of patience is needed to acquire a sympathetic (though not necessarily uncritical) understanding of what it must have felt like (from within, as it were) to believe in sorcery or in astrology, or to engage in magical practices. Too easily, if the point of view of modern science is adopted at all, it is done simply by *identifying* with modern science and proclaiming magical thought and practice outdated from the moment when early modern science arose.[61] Also, can science and the occult be so neatly dichotomized as has often been done in the course of the Hermeticist debate? If the question is intended to mean whether any definition can be given that for all times fixes the demarcation between the two, the answer for the self-respecting historian can hardly be other than 'no'.[62] But if we interpret the question as meaning that, from the very beginnings of science onward, there has been a distinction between science and the occult which, however, has continued to vary over time, then sufficient room has perhaps been created for inquiring into the historical relations between science and the occult without either *a priori* freezing the distinction or obliterating it entirely. I think that the general question of the contribution made by the Scientific Revolution to the disenchantment of the world would provide an excellent umbrella for such much-needed research.[63]

3.3.5. *The Debate over the Rationality of Early Modern Science*

In 1974, on the island of Capri, a conference was held, with the papers being published one year later under the title *Reason, Experiment, and Mysticism in the Scientific Revolution*. On this occasion the issue of rationality as situated at the center of the Hermeticist debate was taken up with particular cogency by the Italian philosopher and historian of science Paolo Rossi and by his commentator Rupert Hall.

Hall put the stakes particularly high when affirming that

> Rossi raises the issue quite simply—and he is right to do so: if the history of science is concerned with rational discourse between men, then the study of alternative models of discourse is certainly of auxiliary interest only; if on the other hand it is *not* (perhaps because of some special link between the 'pseudo-sciences' and the deepest levels of the human psyche, for example), then not only has the history of science as understood for the last three hundred years been a colossal fraud, but so has science itself.[64]

This seems rather strong stuff from a historian who, seventeen years earlier, had gone on record with the assertion that "I dislike dichotomies; of two propositions, so often neither *a* nor *b* by itself can be wholly true."[65]

The significance of the contribution Rossi made to the conference was enhanced not a little by the fact that he had himself contributed to, and in fact anticipated, the Hermeticist flood with a book on Francis Bacon in which he had aimed at uncovering the background in Renaissance magic of many of Bacon's key ideas.[66] Small wonder, then, that those who had never liked the Hermeticist fashion at all exulted when Rossi used the occasion to dissociate himself from what he regarded as the excesses to which the Hermeticist reinterpretation of the Scientific Revolution appeared to be leading. For himself Rossi chose the middle position of demanding liberty for research that indeed implies

> a recognition of the troubled waters at the origin of modern science, an awareness that the birth of scientific learning is not quite as aseptic as the men of the Enlightenment and the positivists naively assumed.[67]

But such recognition, Rossi maintained,

> does not imply either a denial of the existence of scientific knowledge, or a surrender to primitivism and the cult of magic.[68]

It was one thing to acknowledge that

> the study of the interconnections between hermeticism and modern science have [*sic*] greatly enlarged our historical horizon. [However,] the recognition of the 'hidden presences' within the hermetic tradition of modern science does not entitle us to *reduce the latter to the former,* and to forget that in the case of the history of science—at least from the age of Galileo and quite apart from what was happening in the world of magic—it is justifiable to speak of theories that are more or less rigorous, have greater or lesser explanatory and/or predictive power, and are verifiable to a greater or lesser degree.[69]

The position, in short, from which Rossi wished to dissociate himself is the view of science as just one among a variety of possible belief systems, each with its own standards of rationality or lack thereof.

Whom, then, did Rossi count among the upholders of such a conception of science as, if not downright irrational, at least quite arbitrary and devoid of proper standards? In his paper, Rossi identifies—and combats—three distinct strands of irrationality about science, located by him in contemporary philosophy of science (Kuhn, Feyerabend), in countercultural currents of the late sixties and seventies (focused on Roszak), and among Hermeticist historians of science, with Yates (together with Rattansi) being singled out in particular. Now this is quite a mixed bag, treated by Rossi as a far more homogeneous group than these distinct people would seem to deserve. Rossi's paper has, among historians of science, been widely acclaimed as the recognition, made by a henceforth respectable contributor to the Hermeticist fashion, that 'Yates had gone too far'.[70] Therefore it seems proper to take up the issue and find out whether Yates' conception of science ought indeed to be categorized under the fashionable antiscientism Rossi wished to associate with her, or whether perhaps something fundamental about the nature of early modern science itself is being swept under the carpet here.

The meaning of science for Frances Yates. Frances Yates did never, to my knowledge, set forth in clear and unambiguous language what, to her, science meant. Yet in the final pages of both *Giordano Bruno and the Hermetic Tradition* and *The Rosicrucian Enlightenment* some indications can be found of her pertinent views. Perhaps the most revealing in this respect is a rhetorical question at the end of her earlier book:

> Is not all science a gnosis, an insight into the nature of the All, which proceeds by successive revelations?[71]

And on the two final pages of the later book we find:

> The most striking aspect of the Rosicrucian movement is . . . its insistence on a coming Enlightenment. The world, nearing its end, is to receive a new illumination in which the advances in knowledge made in the preceding age of the Renaissance will be immensely expanded. New discoveries are at hand, a new age is dawning. And this illumination shines inward as well as outward; it is an inward spiritual illumination revealing to man new possibilities in himself, teaching him to understand his own dignity and worth and the part he is called upon to play in the divine scheme.
>
> . . . *Rosicrucian thinkers were aware of the dangers of the new science, of its diabolical as well as its angelical possibilities, and they saw that its arrival should be accompanied by a general reformation of the whole wide world.* [My italics][72]

We find expressed here the idea that the Scientific Revolution meant not only an immense increase in man's knowledge and powers but that something essential has got lost in the process as well. And this 'something' has to do with an insight into the soul of man; into its intricate layers on both the conscious and the subconscious levels; into its capacities for good and evil; into the secret of its creative powers revealed, but at the same time obscured or downright ignored, by the advent of early modern science itself.[73]

The awareness that through the Scientific Revolution, beside the impressive gains the new science went on to make, an essence on another level was being lost at the same time had been expressed previously in E.A. Burtt's *Metaphysical Foundations*. It will be recalled from our account in section 2.3.4 that this particular side of Burtt's message—the notion that the immense progress of the mathematization of nature in the 17th century had brought with it a philosophical loss at another level, that of the dignity and autonomy of the human mind—has been rather neglected by historians of science. However,

Frances Yates caught it; it is not at all by chance that, precisely on the final pages of her book on Bruno, she invokes Burtt and his keen perception of the problem of the mathematical reductionism that went with the advent of early modern science. And it is at this very point that Yates goes on to situate the deepest connection between science and Hermetic thought:

> The basic difference between the attitude of the magician to the world and the attitude of the scientist to the world is that the former wants to draw the world into himself, whilst the scientist does just the opposite, he externalises and im-personalises the world by a movement of will in an entirely opposite direction to that described in the Hermetic writings, the whole emphasis of which is pre-cisely on the reflection of the world in the *mens* [i.e., mind]. . . . Thus, from the point of view of the history of the problem of mind and of why it has become such a problem through the neglect of it at the beginning of the modern period, 'Hermes Trismegistus' and his history is important. . . . In the company of 'Hermes Trismegistus' one treads the borderlands between magic and religion, magic and science, magic and art or poetry or music. It was in those elusive realms that the man of the Renaissance dwelt, and the seventeenth century lost some clue to the personality of that *magnum miraculum*.[74]

What, in short, is suggested here, is that the persistence of Hermetic patterns of thought throughout much of the 17th-century adventure in science betrays an acute awareness, among many though not all of the pioneers of the Scientific Revolution, that their new science, however irresistible in its intellectual sweep, caused an attendant loss of insight into the endlessly complex makeup of the human personality—not without consequence for man's future handling of nature. Throughout the Renaissance such an insight had been expressed in largely symbolic language, much of which persisted, in a new setting and in adapted forms, during the 17th century. Calling such Hermetic remnants mere atavisms is to ignore the authenticity of what it was that these men sought to express, in a language, to be sure, that has become rather foreign to us. Such language is the common mode of expression of 'wisdom literature', that great strand running through the world's philosophies and religions that aims at put-ting into words man's stance in the cosmos as well as his own capacities to come to terms with that stance—all of which is preliminary, in this view, to viable efforts of man to arrive at a full understanding of his natural sur-roundings.

Throughout the history of western European culture a dual attitude toward science can be discerned: the enthusiastic embrace of science as the embodi-ment of our triumph over nature, accompanied by bitter denunciations of sci-ence for its dehumanizing reductionism. Later high points of such persistent countercurrents are to be found in, for example, Goethe's opposition to experi-mental science as a disfigurement of the reality of nature and in the debate over the promise and threat of science in our own day. We can now see that the onset of this dual attitude may be traced back to the very inception of the Scientific Revolution itself, and to its enduring Hermetic undercurrent.

3.4. THE NEW SCIENCE AND THE CREATION OF 'ARTIFICIALLY PRODUCED NATURE'

One recurring theme in discussions by 17th-century scientists is the relation between Art and Nature. These discussions have barely been charted by historians so far,[75] yet so much is clear that the Aristotelian position of artificial operations being inherently incapable of imitating (let alone surpassing) the phenomena of nature ceased to govern their views on the matter. Whatever conception took over in the end, a significant change was taking place here, which is pertinent in one way or another to a number of topics taken up by historians as constitutive of the Scientific Revolution—topics that come up for treatment in the present cluster of sections.

By and large, natural philosophies prior to the Scientific Revolution were concerned with such phenomena of nature as present themselves immediately to the senses (e.g., trees burning, stars shining, stones being thrown). In contrast, the Scientific Revolution may be characterized by the *expansion* into, as well as the *exploration* of, a new phenomenal domain previously hidden to the senses (e.g., infusoria, Jupiter's satellites, an apparently void space on top of a mercury column in a glass tube deliberately erected in a dish of the same fluid). How to characterize this notable shift? The French engineer and social philosopher Georges Sorel suggested in 1905 distinguishing between 'natural nature' as the sole object of investigation prior to the advent of early modern science and an entity he proposed to call 'artificial nature'.[76] In this novel mode nature is compelled to display features quite beyond its repertoire if left to itself. Natural phenomena, rather than being produced spontaneously, are forcibly brought forth in a man-made environment and/or are being made perceptible to human sense by means of man-made gadgets. Nature, in short, is being subjected to artificial treatment.

While Sorel's terminology carries an undue suggestion that there are two distinct natures rather than one single nature apprehended in more or less radically different modes, it does give occasion to categorize some tough problems that occupied many of the best minds in 17th-century science. The 17th-century turn toward what we shall from here on call 'artificially produced nature' did not fail to raise profound questions concerning the status (in terms of the type of knowledge yielded) of the scientific instrument and *a fortiori* of the scientific experiment the instrument helped to make possible. Ultimately, such questions appear to come down to the root problem of what relationship 'artificially produced nature', explored for the first time in the course of the Scientific Revolution, bears to 'natural nature'. It is against the background of this underlying issue that, in the sections that follow, we shall pursue the historiography of the role of experiment in early modern science and of the advent of the scientific instrument.

The subjection nature underwent in experiment and by means of the scientific instrument brought with it, as an obvious counterpart, the idea that man can put nature to use. 'Artificially produced nature' lends itself to employment in the service of human needs: Man (masculine) need not confine himself to the contemplation of nature, but he may go ahead and turn himself into her 'master' (as the contemporary phrase ran). What historians have had to say about the implications of this novel idea and of its subsequent realization is equally to be taken up in the present context.

3.4.1. *The Nature of Early Modern Experiment*

The issue that precedes all others in the area just defined is the role experience fulfills in science. In the 18th and 19th centuries it was quite customary to contrast the 'empiricism' of the new science with the sterile *a priori* reasoning taken to be characteristic of previous philosophies of nature, in particular, of Aristotle's. Both Whewell and Mach, for example, however distinct in their approaches to history otherwise, saw the issue in this light. But around the turn of the 20th century the terms in which the matter was put changed considerably. It was discovered how deeply the empirical fact had gone into the making, not only of early modern science, but also of its apparent antithesis, Aristotelianism. The crucial difference that did exist between the two in this respect could only be pinpointed by defining the key characteristic of experimental science in a narrower sense. In line with Kant's distinction between unguided observation and the active interrogation of nature (see above, section 2.2.1), both Pierre Duhem and his contemporary Paul Tannery demonstrated the difference between the Aristotelian conception of nature, with its commonsense reliance on daily experience, and the deliberate subjection of nature to searching questioning called 'experiment' that became one hallmark of the new science of the 17th century.[77]

This distinction has remained basic to subsequent historiography. Usually, when early modern science is being characterized as 'empirical', this is done with the implication in mind that active interrogation is meant here rather than 'naive', more or less unguided observation. Thus, the idea that experiment as distinct from 'raw' experience provides one principally novel feature of the Scientific Revolution is, by itself, hardly controversial anymore. What has remained thoroughly controversial, though, is what experiment did and what it did not contribute to the inception and the subsequent fate of the Scientific Revolution.

Two views of the role of experiment in the making of early modern science. So far we have encountered experiment chiefly in the framework of the 'mathematical' conception of the Scientific Revolution, with both Koyré and Dijksterhuis denying any independently constitutive role to experiment at all. In their conception, the role of experiment was almost entirely confined to bridging the gap between the abstract, idealized domain of mathematical theory and the empirical domain that is, or can be made, manifest to the senses. Not until mathematical theorizing on a given topic had been developed sufficiently to yield empirically verifiable predictions was experiment called in to settle whether or not the abstract theory dealt indeed with the real world. Thus, the job of experiment was no other than *a posteriori* checking, and this included thought-experiments, too, in those many cases when proper instruments to carry them out in reality could not be furnished.

Consequently, experiment was almost fully denied a heuristic purpose by those historians who, in the twenties through sixties, created the concept of the Scientific Revolution and sought the key to it in the mathematization of nature. However, contemporary historians of science with a background in chemistry rather than in mathematics or mathematical physics felt ill at ease with confining experiment to the role of *a posteriori* verification. Among these pioneer historians should be mentioned in particular Pagel and Hooykaas. These men put forward, and worked out to some extent, the idea that the Scientific Revolution is principally to be characterized as the transition from an organic to a mechanical conception of the world—the word 'mechanical' here meaning 'machinelike'. In the outlook adopted by the ancients, nature is an organism; art is incapable of imitating, much less improving upon, natural effects—to dissect nature by means of experiment would teach us little because the property to be uncovered would change beyond recognition under the dissector's knife. As opposed to this, 17th-century scientists considered nature primarily as a machine; witness the ubiquity at the time of the 'clockwork' metaphor for nature. The organic view of things meant that nature was taken as alive and animated rather than conceived of as machinelike; as directed toward higher purposes rather than ruled by the chance motions of material particles; as generated rather than fabricated. In such an overall conception of what characterized the Scientific Revolution, carrying out an experiment may indeed enlighten us about the true nature of things just as taking a machine apart helps us understand how it operates. More than that: The novel habit of doing experiments should be seen as the primary expression of one essential change wrought by the Scientific Revolution.[78]

In summarily redefining the Scientific Revolution along such lines the 'chemists' elaborated their alternative position in two distinct directions. Pagel and his disciple Debus posited a late 16th-century 'chemical revolution' with roots in Paracelsus' magical thought and practice. This line of thought eventually found itself incorporated in the Hermeticist current discussed in previous sections (3.3.2, in particular). Hooykaas rather stressed links between the rise of the mechanical, as opposed to the organic, conception of the world and a new appreciation of manual labor arising in the Renaissance. Not until man had learned to 'think with his hands' and to gain "from his own experience" an intimate knowledge "of the whimsical tricks of nature"[79] did early modern science become possible. In this manner he uncovered a nexus of explanations for the birth of early modern science that will occupy us further in Part II (sections 5.1.1 and 5.2.8).

Let us return now to the implications such alternative conceptions of the Scientific Revolution held for the role to be attributed in the event to experiment. The man to take up both the mathematicians' and chemists' apparently contrasting views of things and to attempt a reconciliation between them by inserting them in one coherent picture of the Scientific Revolution was Thomas Kuhn.

'Mathematical' and 'Baconian' experiments. We are already aware of that picture—it is none but the two-current conception of the Scientific Revolution proposed by Kuhn in 1972 (see section 2.4.4). Here Kuhn showed (among many other things) that the terms in which the role of experiment in the Scientific Revolution—independently formative or subordinate—was being debated tended to doom the debate to sterility. Rather, two types of experiment must be distinguished, which belonged to two relatively independent clusters of scientific disciplines, namely, the classical and the Baconian sciences, respectively. Whereas the subordinate status of experiment in the tradition of the classical sciences could hardly be denied, the other current that made up the Scientific Revolution, the Baconian one, was marked by the rise of a novel type of experiment, which was heuristic rather than directed toward confirmation. One striking difference between the two experimental traditions, so Kuhn observed, was revealed in the manner in which experiments were reported at the time: casually and 'idealized' in the case of the 'mathematicians' (Galileo and Pascal, for example); carefully and circumstantially in the case of the 'Baconians' (Boyle and others who gathered around the Royal Society).

These few passing remarks by Kuhn have remained thoroughly underexploited.[80] To my knowledge, they still constitute the one and only attempt ever made to analyze the issue of the role of experiment in the Scientific Revolution in another framework than that of the either/or positions we have encountered so far. We have found, as one major conclusion of the previous chapter, that here lay the principal point of attack against the Great Tradition embodied by Koyré and others. Equally, in the domain of the role of experiment lay the principal stumbling block in the way of Westfall's two-current account, where experiment more or less fell off the table in the turmoil between the mathematical and the 'mechanicist' currents. Nor has the issue been resolved to satisfaction in the most daring effort that has since been devoted to the question of the nature of experiment in 17th-century science: Shapin and Schaffer's book of 1985, *Leviathan and the Air-Pump*.

Experiment and the reality of nature.   In their book, Shapin and Schaffer take up the issue by way of an analysis of how the rules of the experimental method, far from being self-evident once codified, were laid down by Boyle and other British experimentalists in a contest with Hobbes that involved much more than purely scientific matters, the achievement and maintenance of social order being the underlying bone of contention between them. The authors' point of departure is the idea—adopted from recent social-constructivist theorizing in the 'science studies' tradition—that the determination of what may count as a scientific fact derives not so much from the detection by the researcher of a natural given as, rather, from the outcome of a process of negotiation between scientists. The appeal to nature is a social *a posteriori* act, meant to seal the point of view of the victors after they have won the battle, but the battle itself is fought over quite different matters, in which all kinds of extrascientific interests, arguments, etc., play the key role. Thus when, in his dispute with Hobbes over the status of his results with the air pump, Boyle claimed that his experiments were public, accessible to all, open for replication by everyone, exempt from the search for causes, and so on, he was doing nothing but laying down rules for the experimental method in a manner that suited the Restoration order in which the new experimental science was meant, by the ideological leaders of the Royal Society, to function. All such attributes of experimental science as were claimed by Boyle served to obfuscate the fact that he had failed to make his results truly conclusive, as Hobbes relentlessly and untiringly went on and on to point out.

One wonders what Shapin and Schaffer's argument—set forth here only in some bare essentials—has to do with the Scientific Revolution at large. On the one hand, the very expression is absent from the book, which is defined by the authors as "a study . . . of conflicting strategies for generating natural knowledge in mid-seventeenth-century England."[81] On the other, Boyle is without more ado taken for the founder of early modern science.[82] Thus, while dealing—in a virtuoso manner, to be sure—with just one episode in the history of science, the book truly carries messages of far wider scope, and this accounts at least in part for the predominant position it has acquired since its appearance in historians' thinking about the Scientific Revolution—it looms over oral and written discussions almost as much as Koyré's work did in the forties and fifties and Yates' in the sixties and seventies.

Still, there is something deeply unfortunate about the historiographical situation the book has thus helped to create. In it, one finds an argument made without really being made. To some extent, this cannot be otherwise in history writing. For example, Koyré talked about Galileo, and meant the Scientific

Revolution. (True, originally he *defined* the Scientific Revolution as the novel things done by Galileo, but his concept soon expanded way beyond that particular boundary.) Inevitably, generalized conceptions of an event like the Scientific Revolution are built upon partial studies which may always be contested for lack of representativeness or of fulness of coverage. But this is not quite the same thing as making a case without making it. There is a subtle balance to be respected here, which is not always maintained. In my view it has become increasingly fashionable among historians of science to rest a case for a large claim upon empirical research of far smaller scope, while filling the gap with rhetoric or ideological talk or silence. In the case of *Leviathan and the Air-Pump* all three stopgaps are applied with equal brilliance, which is what lends so much interest to the study beyond its formal topic. Nonetheless, it is hard to see what the account means for science on the Continent, where no Restoration social order had to be attacked or defended, or for experiments in the mathematical, as opposed to the Baconian, tradition, where, as Kuhn indicated, significantly different rules obtained, and where experimental research was not equated with an 'acausal' mode of doing science in any case.[83]

To the extent that Shapin and Schaffer have shown that no truly conclusive experiment can ever be made because one can always construct a rival theoretical framework in which the outcome may also be fitted, they certainly have a point—though hardly an original one, as they themselves acknowledge.[84] But a larger topic appears to loom behind this. What, at bottom, is at issue here is the question of the element of artificiality in experimentally revealed nature. Why, for example, did Galileo make it appear, in his published writings, as if his experimentally attained outcomes coincided exactly with the ones predicted by theory? Not because he did not know better, or because he had not bothered to carry them out (as we know now), but because he realized that, to an audience entirely uneducated about the gap between 'natural' and 'artificially produced' nature, any outcome that reflected the gap at all was most likely to count as a *rejection*, rather than as a *confirmation*, of the theory.

Every interested thinker in the 17th century had to grapple with the distinction between 'natural' and 'artificially produced' nature and the root issue of how the one was related to the other. Shapin and Schaffer have marvelously shown how Hobbes exploited this feature of the relative artificiality of scientific experiment for his own ends with great polemical skill.[85] What they have neglected to point out is how all this forms just one episode in an ongoing 17th-century search to settle the pertinent questions that started already with Galileo and found other high points in Pascal's and Huygens' strictures against comparable objections made from a mixed Aristotelian/Cartesian point of view against the relevance of the artificial experiment for our understanding of nature.[86] A much fuller study of this 17th-century debate remains to be written. We may indeed require it not to side *a priori* with the victors: the 17th-century pioneers who laid down those rules of the experimental method still by and large held to be valid today. But it may also be asked not to decide *a priori* that the constraints set by nature were quite irrelevant to the issue. Once in possession of such a study, we would understand a great deal better what role 'mathematical' as well as 'Baconian' experiments fulfilled in the full west European movement that we call the Scientific Revolution.

3.4.2. *The Rise of the Scientific Instrument*

The key idea that underlies the discovery, early in the 17th century, of the domain of 'artificially produced nature' is that nature may display many other observable phenomena than the senses immediately reveal to us. By and large, the discovery of this fundamental fact may be attributed to the Scientific Revolution. As always, there are borderline cases. As usual, these tend to highlight rather than call into doubt the fact of the revolution. One revealing example of such a borderline case is presented by the harmonics, the existence of which was unknown to Greek musical theory with the one exception of Aristotle's observation that, in some undefined way, any given musical note contains its higher octave. On the one hand, a substantial number of harmonics can be heard without the help of any special device. As their discoverer, Mersenne, insisted in 1636/37, only silence and concentration are required for observing that any tone produced by a voice or a musical instrument actually consists of a fundamental tone blended with a number (for Mersenne: five) of higher-pitched but much weaker partial tones. On the other hand, but for the *a priori* expectation that there is more to nature than is revealed by daily experience, there would have been no reason to try and listen in the first place; during fully eighteen hundred years between Aristotle and Mersenne no one had bothered to find out whether there was something there.[87] One way to grasp the novelty of the Scientific Revolution is to notice this quite sudden appearance of the suspicion that there is another set of natural phenomena ready to be discovered beyond the realities of daily observation.

A variety of roles for the scientific instrument. The principal mediator between humans and their discovery and subsequent mapping of the boundless, novel world of artificially produced nature has, of course, been the scientific instrument. Just as it would be an overstatement to say that nothing in this novel domain could be detected but through instruments (*vide* the harmonics), so have scientific instruments served more purposes than the exploration of the domain of artificially produced nature. Instruments prior to the 17th century served Tycho Brahe and contemporaries as an aid to the naked-eye observation of the heavens, whereas others had been designed for purposes of representation (astrolabes, planetaria) or as computing aids. Yet by and large it is true to say that the early 17th-century discovery of artificially produced nature and the early 17th-century outburst in the invention and subsequent development of such scientific instruments as the telescope, the microscope, the barometer, and the air pump are very closely connected. To be sure, the two first-mentioned instruments expand our knowledge of natural nature rather than create a domain in artificially produced nature. But the use of both ultimately propelled human civilization away from the world that is given by our senses only.

Although there is a great deal of literature on the history of specific scientific instruments, only a very small portion is devoted to the discussion of the place of the instrument as such in the history of science in general, or in the Scientific Revolution, in particular. For a few general points, one must still turn to Maurice Daumas' standard work of 1953, translated from French into English under the title *Scientific Instruments of the Seventeenth and Eighteenth Centuries.* Here the Scientific Revolution is taken as the agent of a sudden break in the history of the scientific instrument. Albert Van Helden, in a pioneering paper, has made the case that, however novel and sudden its appearance, only in the course of the 17th century did the scientific instrument gradually acquire its prominent position in science we nowadays take for granted.

Scientific instruments and the rise of early modern science.   In Daumas' view, the Scientific Revolution caused a major break in the history of the construction of scientific instruments. Up to the end of the sixteenth century, what scientific equipment there was resulted from "the empirical exploitation of traditional knowledge," which had been accumulating in a small number of specialized workshops.[88] "The ingenuity and skill of the craftsmen was devoted to perfecting instruments whose underlying principles had in some cases been known for a millennium." The slow rhythm of gradual improvement was abruptly broken by the new science, resulting in "an expansion no longer based on evolution, but on invention." The breakthrough in instrument-making and the breakthrough in science conditioned one another to some extent. On the one hand, at the end of the 16th century "technical progress had attained a level which permitted the solution of the new manufacturing problems by recent discoveries." On the other, the new science created the demand necessary "to stimulate makers to solve these problems." As one result, workshops rapidly specialized, moving with increasing freedom inside the framework of the traditional craft guilds. However, throughout the century they had to compete with quite a few among the great 17th-century scientists themselves: Galileo and Huygens and Newton and many others were (or at least thought themselves to be) as gifted craftsmen as any to be found in the regular trade.[89]

Investigating further the interaction between instrument and science, Van Helden has argued that the "integrated status [of the scientific instrument] in mathematical and experimental science . . . did not spring forth fully formed around 1600; it grew in the course of the seventeenth century."[90] For example, to seek maximal accuracy as an aim in its own right did not begin to appeal to either scientists or instrument-makers until deeply into the century. Originally a foreign intruder into the realm of theoretical reflection, the instrument needed time for the creation of a research context around it (this applies to the areas of telescopic astronomy and of microscopical research, in particular); for having its results and its readings accepted as a matter of course; in short, for turning into an indispensable tool of science as we know it. By the end of the Scientific Revolution, all this had been achieved.

We note that, in the view of Daumas and Van Helden alike, both experimental science and its principal tool, the scientific instrument, made their entrance on a more than incidental scale around 1600, thus adding one more powerful reason to many others for making the Scientific Revolution start at that time rather than with Copernicus in 1543. For the rest, we observe that many novel features indicated by these two authors are still up for in-depth scrutiny of a very exciting kind. Attempts to investigate, from the viewpoint of economic history, the demand-and-supply structure of the instrument trade and the many, radical alterations it underwent in the course of the 17th century have scarcely been made so far.[91] Nor do we possess more than incidental inquiries into that markedly novel habit of 17th-century theorists: to work with their own hands and to produce and even sell, on occasion, some of their own instruments.[92] Nor has a quite crucial issue mentioned by both Daumas and Van Helden been systematically explored so far. This is the extent to which scientific research during the Scientific Revolution was spurred on or, conversely, was nipped in the bud by the 'technological front line'—by the limits, that is, set by what was technically feasible at any given moment in time. In short, even though in the literature many good, pertinent questions have been asked more or less in passing, the part played in the Scientific Revolution by the scientific instrument remains remarkably underexplored.

3.4.3. Science Applied: Idea and Reality

Most accounts of the Scientific Revolution reserve a place for Francis Bacon's stirring battle cries about the dual purpose of the study of nature. In Bacon's view, the traditional objective of intellectual mastery was no more than preliminary to the ultimate goal of the study of nature. The novel, radically improved understanding of nature's laws Bacon advocated was to yield, in addition, man's effective mastery of nature, the tapping of nature's resources for the benefit of man, 'to the effecting of all things possible'. Quite similar sentiments were voiced slightly later in Book 6 of the *Discours de la méthode,* where Descartes confidently announced our capacity "to turn ourselves into the masters and possessors of nature."[93]

Clearly, the new ideal expressed so eloquently by Bacon and Descartes and echoed by countless others made sense only in the context of the discovery of 'artificially produced nature'. Aristotelian natural philosophy, in particular, was quite satisfied with interpreting nature as a given that cannot be changed by man. Its quite conscious limitation to the domain of readily observed facts precluded any idea of nature as an entity that can be exploited as well as understood.

The idea of the applicability of science is undoubtedly one of the great novelties produced by the Scientific Revolution. (This notwithstanding recent findings that, in Bacon's own work, the new ideal made its appearance in a context of alchemy and natural magic and should in fact be seen as a transformed descendant of the activist, magical tradition.)[94] Baconian phrases about the uses of science for the benefit of mankind ring down the 17th century. But what actual state of affairs were these phrases meant to express?

Science-based technology now and then.   The close intertwinement of science and technology is among the most predominant features of modern life. This is true even though quite opposite opinions may be heard on how the two are actually related. At one extreme, science is held to be little more than generalized technology; at the other, technology is taken for not much more than applied science. A more realistic assessment is surely situated at some point of the broad spectrum that extends between these two extremes. However, what all possible points of view in this spectrum of opinions have in common is that they presuppose a very close link between science and technology. For our own time this is only natural: A largely science-based technology determines so much of our everyday lives. The temptation is quite strong to project this close intertwinement back onto the past. But if one considers the relation between science and technology as it must have impressed the contemporary observer before, say, 1800, is it then true to say that an equally science-based technology is what met the eye?

Quite a few historians of science have taken it more or less for granted that this is indeed the case, in that science-based technology must be regarded as a product of the 17th century. They rest their case to a considerable extent upon the overall environment of the Royal Society, with its largely utilitarian atmosphere, or upon the Society's immediate predecessors among the Puritans, in particular.[95]

The application of 17th-century science: Words and deeds.   There is no doubt at all that 17th-century scientists were prone to *justify* their activity by pointing at its practical benefits. Koyré—so eager to vindicate the Platonic purity of the scientific endeavor—once somewhat cynically suggested that the ongoing invocation by 17th-century scientists of the use to which an applied science may be put is wholly to be attributed to an early-felt need "to 'sell' their science to wealthy and ignorant patrons."[96] The sociologist R.K. Merton raised the same issue to a higher level of generality by pointing out that the legitimacy science has come to enjoy as a matter of course in the society of our own day had to be attained before it could be enjoyed, and that the legitimation 17th-century scientists sought for their unprecedentedly novel enterprise was naturally found by them in the manifold uses to which the new science could be put. Here is a list of such uses as assembled by Merton from an examination of reasons advanced for doing science in 17th-century England:

> To the implicit and sometimes explicit question raised by the doubting Thomases of the time—why practice science and why support it?—natural philosophers, clergymen, merchants, mine-owners, soldiers and civilian officials developed an impressive inventory of the diverse 'utilities' of science:
> —the religious utility of exhibiting the wisdom of the divine handiwork;
> —the economic and technological utility of enabling mines to be workable at increasing depths;
> —the economic and technological utility of helping mariners to sail safely to ever more far off places, in quest of adventure and trade;
> —the military utility of providing for ever more efficient and inexpensive ways of killing the enemy;
> —the self-development utility of providing a form of mental discipline . . . ; and
> —the nationalistic utility of enlarging and deepening the collective self-esteem of Englishmen as they advanced their claims to priority of discovery and invention.[97]

This remarkable list reminds us in a very effective manner that beyond our present-day tendency to dichotomize between 'science for its own sake' and 'science for money and fame' the 17th century recognized other valid objectives for science. It is, however, evident that only three out of Merton's six items are suitable for supporting an argument in favor of the emergence, in the 17th century, of a science-based technology. All three have found their rebuttal in the literature. The drift of the argument has been in each case that, even though contemporary claims ran very high, actual performance fell far short. Thus, it is certainly true that during almost the entire century science was—quite rightly—held to possess the key to the resolution of the problem of the determination of geographical longitude at sea. However, it is also well known that the solution continued to elude scientists. It was finally attained by an 18th-century artisan (Harrison) who had absorbed the relevant scientific background without himself being among its practitioners. Similarly, at least from Guericke's experiments onward it was realized that atmospheric pressure and the void could be used to harness natural forces of which even the existence had escaped earlier philosophers of nature. However, the practical execution of what was in theory well understood had, once more, to await indispensable inventions by an 18th-century artisan (Newcomen) who had absorbed the relevant scientific background without himself being among its practitioners.[98] Finally, the case for the improvement by science of the actual performance of contemporary shotguns has been invalidated by Hall along similar lines.[99]

In each case, the gap between a solution that worked in theory and a solution that worked in practice proved to be far greater than could be overcome at the time. The general conclusion that appears to follow was drawn by J.D. Bernal in 1954. Granting that a "most characteristic unifying principle of the new science was its concern with the major technical problems of the day,"[100] he nonetheless pointed at the nature of the as yet unbridgeable gap: "Working in wood or in roughly cast metal it was impossible to make use of the refinements which the new mathematics and dynamics could provide."[101]

Even though for mostly ideological reasons Bernal himself went on to qualify his point by arguing that navigation was the one, exceptional domain where science indeed proved useful already at the time of the Scientific Revolution, his general conclusion has been upheld since by others. It was confirmed and amended in an article "Robert Hooke, Mechanical Technology, and Scientific Investigation" published in 1983 by Westfall. His fundamental point is that in the 17th century science did not as yet provide sufficient basis for erecting upon it a scientifically infused technology: To talk about the possibility of a science-based technology was one thing; to achieve it, quite another. Westfall derived empirical support for his thesis from the consideration that, if any science-based technology is to be encountered in the period at all, Robert Hooke's work must be where to find it. But here, so he shows, pre-Newtonian dynamical insights, in particular, stood in the way of the successful solution of such practical problems as Hooke had posed himself and believed he could solve with science (e.g., how to make a lamp burn steadily; or the optimal way to trim sails). Precisely because Hooke was such a marvelously ingenious inventor of practical devices, and precisely because his scientific knowledge was as great as that of any of his contemporaries, barring Newton, he should have been the candidate *par excellence* to realize at least samples of a science-based technology if such a thing was to be had at the time at all. From their absence Westfall tentatively infers the conclusion indicated above on other grounds: that a later age than that of the Scientific Revolution was to make good the promise its protagonists had so keenly advanced.

If one accepts the conclusion, it remains to be asked why this was so— why the new science was unable to overcome the distance still separating it from contemporary art-and-craft technology at one stroke. Two answers have been pointed at above. One answer resides in the gradually dawning awareness that between a theoretical and a truly practicable solution to a practical problem posed by science a huge gap yawns that can be bridged neither by the scientist nor by the traditional craftsman, but only by the kind of skilled ingenuity a new kind of engineer (the Newcomen/Harrison type) had to learn to bring to the matter. The other is that pre-Newtonian dynamics was in too confused a state to render its application in practice possible. Additional reasons will scarcely fail to be advanced once historians take up the issue of the absence of a science-based technology in the 17th century at greater length. Or, to rephrase in a more positive vein, we need an overall, in-depth investigation into that striking historical process of the 17th century of how scientists and artisans, both separately and in the course of their increasing interaction, began to discover, to explore, and to attempt to narrow the gap between their respective enterprises so as to fulfill the Baconian promise.

In spite of all that has been said so far, one borderline area where science and technology fruitfully combined in the 17th century is recognized in Westfall's article. This is the domain of the scientific instrument. However significant, it must be clearly distinguished from the Baconian claim:

> The pendulum clock was perhaps the epitome of a technology to which we cannot deny the adjective scientific. It was not, however, a technology that answered the expectations of Baconian utilitarianism. Turned inward upon the scientific enterprise itself, it promoted the investigation of nature but did nothing to augment the comfort and convenience of life.[102]

A caveat to round off the conclusion. The conclusions drawn above are not at all meant to imply that there was no 17th-century technology, or that 17th-century science, its Baconian ambitions notwithstanding, was not applicable at all. In the 17th century, technology was surely alive and kicking,[103] only, just as had been true for the entire previous history of 'man the maker', technology still benefited little from scientific findings (the scientific instrument providing the sole exception to this rule). Not until the middle of the 19th century, when chemical dyestuffs were mass-produced for the huge quantities of textile spawned by the machinery of the Industrial Revolution, did a fully developed science-based technology come into its own.[104]

Francis Bacon had possessed the right intuition: The new science could be applied. Only, more than a century and a half had to pass before even the seemingly most promising 17th-century claims—for the harnessing of natural forces and for the determination of longitude—could at long last be attained. Not until the advent of the Industrial Revolution did the Baconian ideal of applied science find its first, tentative realization.

Therefore none of what has been said in the present section should be read as a disparagement of the Baconian idea as such. One glance out of the window is sufficient to persuade us in case we have forgotten that the application of science has altered, and continues to alter, our daily lives beyond recognition, not exclusively for the good, as Bacon had optimistically believed, but not exclusively for the worse either. By expressing a vital new attribute of this new science born in the 17th century, the Baconian message added a new dimension indeed to whatever had previously gone under the name of science.

3.4.4. *The Subjection of Feminine Nature*

Our next topic is that darker side, not foreseen by Francis Bacon, to the exploitation of nature he heralded. We are to deal with a thesis on the subjection and subsequent destruction of nature as one key effect of the Scientific Revolution. In order to locate the subject historiographically, let us recall two strands of thought from previous sections. One is the conception of the emergence of early modern science as characterized principally by a transition from an organic to a mechanical, in the sense of machinelike, idea of nature (section 3.4.1). The other is the tendency, which came up in the sixties and seventies and centered around the Hermeticist issue, to ascribe an independent value to 16th/17th-century modes of thought and feeling that were more or less vanquished by the victorious new science (section 3.3). These two lines of thought were not only combined but also extended from a feminist point of view by Carolyn Merchant in her 1980 book, *The Death of Nature: Women, Ecology, and the Scientific Revolution.*

The central thesis of Merchant's book may be summed up in two coupled assertions. One: The Scientific Revolution signified the death of nature. Two: The rise to prominence of an image of nature as a passive woman to be subdued rather than as a nurturing mother to be revered removed one crucial constraint that, before the 17th century, had stood in the way of the merciless exploitation of nature which has inexorably led to our present-day environmental problems.

One limitation of Merchant's study that she herself recognizes is that the content of 17th-century science is largely left out of her account. It is "the values associated with the images of women and nature as they relate to the formation of our modern world and their implications for our lives today" that she wishes to examine.[105] This is done chiefly by making a general statement on the first one or two pages of each chapter, followed by an enumeration of contemporary ideas, literary images, and activities gleaned largely from the intellectual and social history of England. Thus, rather than with a sustained, historical argument we deal here with a cluster of subtheses loosely linking a huge number of diverse facts from such varied domains as thought about women, images of nature, machine technology, philosophizing on how matter and spirit are related, and deforestation and pollution in late 17th-century England.

Merchant's central claim takes its point of departure in the assertion that nature has always been metaphorically compared to a woman. She maintains that this image has right from the start in both Greek and Christian thought carried connotations of nurturing as well as of dominion. For a very long time, though, the former connotation was experienced in a far more lively and 'natural' manner than the latter, with the prevalence of the 'nurturing' metaphor acting as a powerful constraint against possible ideas, or acts, of exploitation. Such activities as mining (which, in the metaphor of nature as a nurturing mother, comes down to digging in her flesh, and thus was an ethically most dubious activity) lacked a sanction and thus could not become a model of how humanity should treat nature. However, the onset of commercial capitalism required the products of such 'counternatural' activity to an increasing degree.

It thus put a premium on the removal of the ethical inhibitions provided by the prevalent image of nature:

> Because the needs and purposes of society as a whole were changing with the commercial revolution, the values associated with the organic view of nature were no longer applicable; hence the plausibility of the conceptual framework itself was slowly, but continuously, being threatened.[106]

How, therefore, to sanction the desired exploitation of nature? The Scientific Revolution, in making the intellectual transition from an organic to a mechanical picture of the world, provided the required sanction. It found its expression in a metaphor of nature that had always been there but only now managed to prevail: the metaphor of nature as a woman to be subdued.

Merchant's crown witness both for the upgrading of the 'dominion' image and for its close association with a new, activist approach to nature is Francis Bacon. The language in which he advocated his experimental method invokes, time and again, the image of the subjection of feminine nature (e.g., from Bacon's *The Masculine Birth of Time,* "I am come in very truth leading to you nature with all her children to bind her to your service and make her your slave"; and elsewhere, "you have but to follow and as it were hound nature in her wanderings, and you will be able, when you like, to lead and drive her afterwards to the same place again").[107] Together with Bacon's frequently employed imagery of nature being put on trial, his talk of nature as feminine "strongly suggests the interrogations of the witch trials and the mechanical devices used to torture witches."[108] In this manner such an activity as mining, rather than trespassing the limits set to human conduct vis-à-vis nature, could become, for Bacon, its very model. Thus Merchant reaches the conclusion that

> here, in bold sexual imagery, is the key feature of the modern experimental method. . . . The constraints against penetration in Natura's lament over her torn garments of modesty have been turned into sanctions in language that legitimates the exploitation and 'rape' of nature for human good. . . .
>
> The interrogation of witches as symbol for the interrogation of nature, the courtroom as model for its inquisition, and torture through mechanical devices as a tool for the subjugation of disorder were fundamental to the scientific method as power.[109]

From here on, Merchant is chiefly concerned to show how the mechanical philosophy, in replacing the living forces of organic nature by the minute particles in inertial motion that make up Descartes' *res extensa,* caused the "death of nature"—a phrase she nowhere defines, but which may be taken to mean the reduction of nature to the passive object of an inanimate physics, with her relentless destruction as the inevitable consequence. She ends her book by discussing a number of attempts, undertaken in the second half of the 17th century, to rescue at least parts of the previous, organic mode of thought, with the philosophy of Anne Conway as one example and of Leibniz (in this respect, so she maintains, largely derivative) as another. Efforts to fuse mechanical with organic modes of thought, so she suggests by way of a conclusion, may be precisely what humanity needs in order to avert the ultimate environmental catastrophe and to survive.[110]

We have come a long way now in discussing historians' explorations of the 17th-century discovery of the domain of 'artificially produced nature', from its being described as the one and only possible, in retrospect obvious, road to truth about our world, to being decried as the principal carrier—sanctioned by the image of the enslavement of feminine nature—of the death and destruction of 'natural nature'. At one end of the spectrum stands the time-honored image of science as the embodiment of timeless truth; at the other, a much more recent conception of science as a corpus of mostly destructive notions reflecting chiefly alterations in social values.

A threefold issue lies implied in this historiographical progression. Should the birth of early modern science be seen as a product of social forces? If so, to what extent? And if so, in what manner? It has become time for us to take up these questions in explicit fashion and to find out how far attempts to resolve them have deepened our understanding of the Scientific Revolution.

3.5. THE NEW SCIENCE IN ITS SOCIAL SETTING

To a considerable extent the history of the Scientific Revolution is a history of ideas, and the history of ideas is to a considerable extent the history of ideas of individuals. Historians' attempts at finding general patterns in individual ideas of the past is what has almost exclusively occupied us so far. The development of ideas over time displays a certain logic, or inner autonomy, of its own. But this autonomy is not absolute. Ideas develop (or do not) in a wider setting which turns their autonomy into a relative one. Just how far this autonomy extends—in other words, to what extent external circumstances impinge on it—is not a matter that can be decided *a priori* through abstract discourse. Generally speaking, scientists, who have personal experience of the exciting 'problem—discovery—new problem—new discovery' sequence, strongly tend to take such sequences as all there is to scientific advance. Those who by vocation or ideology tend rather to study the social history of ideas have a general preference to make the development of ideas partly or wholly dependent on outer circumstances, which may vary from matters of funding, institutional setting, and the like, to the fully relativist position that the establishment of 'scientific facts' is little more than the outcome of 'negotiations' between scientists which reflect matters of authority, power, and status much more than the constraints of nature.

The ensuing debate over the 'internal' versus the 'external' approach to the history of science has gone on for many decades and seems at least in theory resolved nowadays on the general line that somehow we need both.[111] I fully agree with that happy outcome but would like to sharpen it somewhat.

To begin with, it seems to me that the sense of the autonomy of ideas as experienced by working scientists should be taken with the utmost seriousness. If we do not, such a sequence of discoveries as analyzed in detail in, for example, Joella Yoder's book of 1988, *Unrolling Time*, would be quite enigmatic—here we see Christiaan Huygens moving from a limited, and inherited, problem to a first discovery, which opens up a new, unexpected, wider problem, which he resolves through his next discovery, which in its turn. . . . There is no social history that could even begin to reduce such a sequence to whatever else might be picked up from Huygens' social environment.

This is not to say, however, that the closely knit sequence of ideas which carried Huygens from the problem of determining the gravitational constant to the discovery of the theory of evolutes took place in full insulation from his intellectual, political, and socioeconomic surroundings. First, we ought not to take it for granted that such sequences are the alpha and omega of all sciences, let alone of all sciences at all times. 'Big Science' sets constraints in terms of, for example, funding, collaboration, and publication different from those of 'little science'. Also, Thomas Kuhn, in particular, has repeatedly stressed that a high degree of internal autonomy can be found in those sciences which, in his terminology, have reached paradigmatic status. The shared paradigm, he has urged, lends itself particularly well to such problem—discovery—problem sequences. But the existence of a paradigm presupposes a community of scientists that adheres to it. And thus we may ask—sticking for the moment to the Huygens example—how he picked up the problem of finding the gravitational constant in the first place. Evidently, seeking an answer to such a question leads us into issues of the communication and the institutionalization of science, which belong squarely to the domain of the social history of scientific ideas.

One may also ask what it was in the wider society that enabled science to become the lifelong vocation of this son, grandson, and brother of diplomats, who was so keen on being an appreciated member of polite society, and who did not need anybody's salary for a living. Such a question leads to inquiries into the formation of the role of the scientist as a socially recognized vocation in its own right, as well as to the question of how the pioneers of the Scientific Revolution got paid for their efforts.

Again, one may notice the many priority quarrels Huygens got embroiled in as a result of his discoveries, and then ask questions about available means to establish priority—which again is a largely institutional matter—or, in a deeper vein, what underlying values made the issue of priority come up in the first place as an outstanding feature of 17th-century science.

Such topics as these—matters of institutionalization, of values, of funding, and the like—have found quite some attention in the literature, much of it rather recent, some of it much less so. Still, we are a long way from possessing a full-fledged 'Social History of the Scientific Revolution'. In what follows I shall treat a limited number of topics as fragments of such a still-to-be-written social history. I concentrate on those accounts which, in my opinion, have managed to combine in a fruitful manner the history of relatively autonomous scientific ideas with their wider social setting. For my view on the matter is that, in general, the only type of 'external' history of science likely to contribute to our understanding of the Scientific Revolution is that which takes its origin in questions derived from the contemporary development of scientific ideas.

Thus, I cannot go along with the suggestion, occasionally put forward as an easy way out of the dispute, that the 'internal versus external' debate has revolved around a pseudo-issue. The distinction itself is real enough. Saying so does not, however, entail a value judgment. There is no reason to regard the internal history of scientific ideas as less or, for that matter, as more valuable than the external approach. Yet the relationship between the two is not quite symmetrical, either. Internal historiography of 17th-century science can, if need be, stand largely on its own; the external approach cannot. The latter can make fruitful contributions to our understanding of the past of science to the extent that it takes its leading questions from the internal history and pursues the wider setting of science from such points of view as are yielded by these questions. If science is *a priori* regarded as little more than an epiphenomenon of wider social processes, we lose sight of precisely what makes science such a unique social phenomenon in the first place.

3.5.1. *The New Values of Science*

The American sociologist Robert K. Merton (born 1910) is the chief pioneer of the sociology of science. The subject of his 1935 doctoral thesis, *Science, Technology and Society in Seventeenth-Century England,* has greatly contributed to the strongly historical underpinnings of the many studies he has devoted since to sociological aspects of the scientific enterprise.

The theme that runs through Merton's later studies is the institutional setting of the norms and values guiding scientists in their conduct vis-à-vis their peers and the scientific laity. Although what he has to say on the subject is meant to apply to science as we know it, there lies implied in his work the idea that the scientific ethos underlying these values is a characteristic of both modern and early modern science. In other words, implicit in Merton's work is the idea that the ethos of science with which we are familiar nowadays was in its essential features established in the course of the Scientific Revolution.[112] It is this implicit notion that makes Merton's work on the scientific ethos (for present purposes to be distinguished from the famous 'Merton thesis', which goes back to his doctoral dissertation and which will be discussed at length in Part

II) of importance for our present overview of historical interpretations of the wider social and intellectual ramifications of the Scientific Revolution.

Merton's four values.   In two articles that have become classics in the sociology of science, Merton distinguished four particular values that regulate the professional conduct of scientists. In the edition of 1973 that I have used, these articles, which date from 1938 and 1942, are entitled "Science and the Social Order" and "The Normative Structure of Science," respectively.[113] They heavily reflect the original time of writing. The politicization of science that was much in evidence at that time in both Nazi Germany and the Soviet Union led Merton to reflect upon the values that were under systematic assault in totalitarian societies and to reaffirm their broadly democratic nature. Before the totalitarian onslaught upon science, Merton wrote, it had almost seemed to scientists as if the integrity of their enterprise went unquestioned. But these recent attacks brought science back to a situation it had known before:

> Three centuries ago, when the institution of science could claim little independent warrant for social support, natural philosophers were likewise led to justify science as a means to the culturally validated ends of economic utility and the glorification of God. The pursuit of science was then no self-evident value. With the unending flow of achievement, however, the instrumental was transformed into the terminal, the means into the end. Thus fortified, the scientist came to regard himself as independent of society and to consider science as a self-validating enterprise which was in society but not of it. A frontal assault on the autonomy of science was required to convert this sanguine isolationism into realistic participation in the revolutionary conflict of cultures. The joining of the issue has led to a clarification and reaffirmation of the ethos of modern science.[114]

The scientific ethos is defined by Merton as "that affectively toned complex of values and norms which is held to be binding on the man of science."[115] Although not previously codified, it can be inferred from the observation of scientists' actual conduct. In so doing, observable regularities are shorn of what is personal and accidental about them and are generalized into what appear to be binding values which regulate conduct. These values are functional to the goals and the methods of science in that they help achieve the objective validation of empirical evidence, which is the methical hallmark of modern science.

> The mores of science possess a methodologic rationale but they are binding, not only because they are procedurally efficient, but because they are believed right and good. They are moral as well as technical prescriptions.
>
> Four sets of institutional imperatives—universalism, communism, disinterestedness, organized skepticism—are taken to comprise the ethos of modern science.[116]

By 'universalism' Merton meant that the criteria for deciding whether a scientific discovery is empirically valid are objective; that is, they are applied without consideration of the personality, creed, race, nationality, or social status of the scientist who made the discovery. This is not to say that such objectivity is invariably upheld in practice. The point is rather that, when violations of the rule take place, they are felt by the great majority of scientists to be precisely that—violations of a rule. The norm of universalism entails that careers in science ought to be open to talents irrespective of their owners' position in society. The plebeian is as welcome as the aristocrat, provided only he has something substantial to contribute to the ongoing scientific enterprise. (Remarkably, Merton overlooked here one serious exception which for centuries was not felt to be an exception to the rule of universalism at all, in that the rule was self-evidently taken to apply to a universe made up of men only.)

The second value was called by Merton, in a singularly unfortunate choice of terms, 'communism', used in the "sense of common ownership of goods." [117] The point is that the ethos of science requires a scientific discovery to be made public, and that, once it has entered the public domain, it has become the common property of the scientific community as a whole. The only claim the scientist can legitimately lay upon property rights for his discovery is the recognition and esteem extended by his peers in roughly approximate proportion to its intrinsic importance. It follows from this that the question of who was the first to make a certain discovery becomes an issue of the highest importance and thus leads to "the concern with scientific priority" and the ensuing "controversies . . . which punctuate the history of modern science [and are thus] generated by the institutional accent on originality." [118] Another consequence is that a scientist is supposed to communicate his findings to his peers. Finally, 'communism' involves an awareness that the scientist is indebted to the collectivity of preceding scientists and to their accumulation of validated evidence upon which his own contribution is built: "the realization that scientific advance involves the collaboration of past and present generations." [119]

The third value, disinterestedness, means that the scientist is supposed not to have a personal stake in scientific discovery. His personal interest in his discovery is subordinate to the objective establishment of the discovery by the rigorous scrutiny of peers. The best vindication of this particular value, Merton claims, is "the virtual absence of fraud in the annals of science." [120]

Finally, the scientific ethos is one of 'organized skepticism', that is to say, a methodical determination to bring to the investigation of phenomena an unchanging measure of objectivity irrespective of whether such phenomena hold for the society at large a sacred, or at least emotionally charged, character. "The scientific investigator does not preserve the cleavage between the sacred and the profane, between that which requires uncritical respect and that which can be objectively analyzed." [121] This is what makes science so unique a social phenomenon: "Most institutions demand unqualified faith; but the institution of science makes skepticism a virtue." [122] It is this feature in particular that, together with the norm of universalism, lays science open to the sort of intrusions which, at the time of writing, had come to threaten science on the part of totalitarian powers. These, Merton affirmed, cannot by their nature abide with the relative social autonomy of science that is embodied in these four values, which find a more natural (though surely not an exclusive) place in democratic society.

The four values and the Scientific Revolution. The two articles in which Merton established these four values were quite brief, and written as they were to reaffirm the relative social autonomy of science in the face of the totalitarian threat, the examples that illuminate the overall argument are largely taken from contemporary science. There is nonetheless an important connection with the Scientific Revolution. What underlies the entire analysis is a conception of the 17th century as the very period when scientists, through their actual conduct in publishing their work, in criticizing that of others, and in engaging in battle with one another over priority, established in practice the rules which were later internalized into the values henceforth upheld by the community of scientists. For example, there had been isolated priority quarrels in the 16th century (the cases of Cardano versus Tartaglia and of Tycho versus Ursus come readily to mind). But the virtual ubiquity of the phenomenon is a feature of the 17th century. Altogether, the Scientific Revolution may well be characterized as a laboratory for contemporary scientists in laying down the rules on how to handle disputes, both between themselves and vis-à-vis society at large.

Not a great deal of historical work has been done in either confirming or challenging the validity of Merton's four values as observable in scientists' actual conduct. Nor do we possess as yet a full-scale study of how, all across the various European centers of the Scientific Revolution, scientists went about laying down rules of conduct. The chief exception here is Shapin and Schaffer's study *Leviathan and the Air-Pump.* [123] Among the manifold objectives of their book is an attempt to show how Robert Boyle, in particular, defined the *mores* that, in his view, went with the cultivation of experimental science. There is one important respect in which Shapin and Schaffer seem to diverge from Merton's rules. One of their principal arguments is aimed at showing that what Merton called 'communism' was in practice subject to serious limitations: access to experiments, at least in the case of the Royal Society, being confined largely to those who were willing to acknowledge the experimental approach as a valid mode of discovery. More generally speaking, to those who, like Thomas Hobbes, rejected the very foundations of the new science, the values in process of being established appeared as arbitrary devices meant to foreclose a fundamental critique of the scientific enterprise. Such an interpretation does not invalidate Merton's four values so much as demonstrate the historical contingency of their establishment in the course of the Scientific Revolution.

One other amendment to the notion that already in the 17th century scientists came to endorse and exemplify the value of community-shared, accessible, published knowledge must in addition be put forward. This concerns the publication habits of many protagonists of the Scientific Revolution. Somehow we tend to take it for granted that the pattern we associate with science publishing nowadays—rush into print as soon as you believe you have made a discovery, or even before you have—applies equally well to the 17th century. We know of a great many individual exceptions to this rule, yet we tend to go on seeking individual explanations for every individual exception, instead of calling into question whether there was such a rule in the first place. Oddly enough, the historian of science in the 20th century knows so very much more about the Scientific Revolution than even its own protagonists possibly could precisely because they themselves left so many pathbreaking treatises unpublished that failed to see the light of day until our own century. Only very few scientists at the time made a consistent habit of publishing their findings as soon as they thought they had made a sufficiently good case: Kepler and Mersenne are two examples that come to mind. But cases of decades-long delay (Galileo, Huygens, Newton) or of failure to publish parts (Descartes, Pascal, Huygens, Newton) or the whole of one's findings (Beeckman, Harriot) are just too numerous to avoid the suspicion that, in this domain, the new scientific movement lived by somewhat other *mores* than those we have got accustomed to.[124]

Eisenstein has perceptively called attention to how often intermediaries intervened in getting major scientific work published, for example, Rheticus for *De revolutionibus,* Diodati for the *Discorsi,* Halley for the *Principia.*[125] This adds relief to the riddle but does not resolve it. In taking up the issue of how, in the course of the Scientific Revolution, scientists worked out habits and rules of conduct of their own, historians of science are still facing a huge task lying before them.

3.5.2. Societies and Universities

Projecting backward our present-day association of scientific discovery with the institutional setting of the university, it seems natural to assume that the universities provided the locus for the new science of the 17th century. The classic book that exploded this convenient assumption was written in 1913. It was the doctoral dissertation of an Austrian mathematician who had recently come to the United States, Martha Ornstein. She was born in 1878 and died in 1915 in a motor accident. The book, entitled *The Rôle of Scientific Societies in the Seventeenth Century,* was not published until 1928. Its final conclusion is that

> the organized support which science needed in order to penetrate into the thought and lives of people was not obtained from universities, but was derived

from those forms of corporate activity which it had created for itself, the scientific societies.[126]

For Ornstein (under the indirect influence of Mach) the key to what she was the first to call "the scientific revolution of 1600–1650" was to be found in experiment.[127] The first half of the century, she affirmed, constituted the formative period of the new science, which

> seems more like a 'mutation' than a normal, gradual evolution from previous times. It accomplished through the work of a few men a revolution in the established habits of thought and inquiry, compared to which most revolutions registered in history seem insignificant.[128]

Unlike the scholastic and humanistic predecessors of the new science, its novel experimental bent naturally lent itself to "the popularization and hence the democratization of knowledge."[129] All over Europe, kings and aristocrats took an interest in the new experimental science and provided the instruments and laboratories by means of which experiments could in fact be carried out. Ornstein provides a number of examples of amateur interest in the new science and then goes on to make the crucial transition to the origin of the scientific societies:

> Enthusiasm for experimentation and the widespread interest it aroused apparently led those devoted to science to enter into more or less formal affiliations. The rich and noble amateur devoted some of his wealth to gathering about him men who would jointly experiment and benefit by this collaboration. The professional scientist would become the center of people who joined him for instruction and whom he needed for assistance. Or at times without any such external stimulus the experimenters would band together. . . .
>
> It cannot be sufficiently emphasized that it was the experimental feature of science which called forth the societies. The mathematician could have worked out his problems in seclusion; the experimenter needed the laboratory and this in turn could not be supplied under usual circumstances by an individual, but only by a society.[130]

The net result of this was that such "unions of scientific workers . . . became the dominating feature of scientific work in the second half of the century."[131] The great scientists who followed upon the generation of pioneers, although not strictly dependent on the new societies, were nonetheless associated with them closely enough for their work to be considered as fitting in with the institutional framework provided by these societies.

The remainder, and by far the largest part, of the dissertation consists of a description of the origins and activities of what Ornstein took to be the four principal societies: the Italian Accademia del Cimento as a successor to the Accademia dei Lincei, the Royal Society of London, the French Académie Royale des Sciences, and finally the Berlin Akademie. Of particular importance in this connection is the rise of the scientific journal as a product of these societies, taking the place of the networks of correspondence that informally had developed around Mersenne in the first half, and Oldenburg in the beginning of the second half, of the 17th century.

In a final chapter before the conclusion is drawn, Ornstein asks what the universities contributed to the advancement of science during the 17th century. The verdict 'virtually nothing' is reached through an overview of statutes of various universities scattered all over Europe; of contemporary complaints leveled against their educational methods as well as of proposals to reform them; and of biographical sketches revealing how little the major individual scientists of the period owed to the universities during, and in particular after, their formative periods. Ornstein lists twelve reforms which should have been carried out by the universities so as to become viable bearers of the new science and finds that such reforms eventually occurred in the middle and late 18th, not 17th, century. During the earlier period the scientific societies, not the universities, were the institutions "to foster the cause of experimental science." [132]

In view of the outdatedness of the overall picture of the Scientific Revolution that structures her study, it is quite remarkable that in essence Ornstein's conception of the institutional framework of 17th-century science still stands—it continues to be upheld in current histories of the Scientific Revolution. The principal thing to have been added since is a great deal of research on individual societies and what they did for the new science, with the chief focus of attention being the Royal Society. The general picture drawn by Ornstein has not been fundamentally affected by such research. This, however, is no longer quite true of the negative counterpart to Ornstein's picture: her view of the universities as fortresses of conservative Aristotelianism run by religious interests and inherently incapable of providing the new scientific movement with the support it needed.

The contribution of the universities: Recent reassessments. In recent years the accepted, quite bleak view of the role of the universities in the Scientific Revolution has been challenged by a number of authors. One outspoken defense of their importance for the Scientific Revolution is to be found in a book published in 1984 by Mordechai Feingold under the title *The Mathematicians' Apprenticeship: Science, Universities and Society in England, 1560–1640*. In the introductory chapter the book is launched as a general attack on the view that the universities stood outside the movement that brought forth the Scientific Revolution. In effect, however, it appears to establish the narrower claim that the universities of Oxford and Cambridge provided a sufficiently up-to-date education in the mathematical sciences to serve as a general preparation for the great upsurge in science that England witnessed in the second half of the 17th century. Rather similar considerations have been advanced by van Berkel for the Dutch universities (calling attention to such exercises in compromise as the *philosophia novantiqua*)[133] and by Heilbron for the contribution, both inside and outside the university system, made to 17th-century science by the Jesuits.[134]

Studies such as these certainly serve a most useful purpose in showing that scholasticism, as it continued to reign at the universities all across Europe in the late 16th and in the 17th centuries, was not quite so crudely dogmatic and inflexible as contemporary propagandists of the Scientific Revolution found it expedient to picture. But when one wishes to assess what this entails for our picture of the Scientific Revolution, it is not always clear with studies in this vein how far the evidence adduced may be taken to stretch. After all, it makes some difference whether one sets out to argue, for example, that the universities *stood in the way* of the Scientific Revolution or even *fought* the Scientific Revolution, or rather were *irrelevant* to the Scientific Revolution, failed to *assimilate* the Scientific Revolution, failed to *disseminate* results of the Scientific Revolution, or failed to *originate* the Scientific Revolution. Too often the stronger case ('the universities contributed in a major way to the Scientific Revolution') is argued by presenting evidence that sustains a far weaker case only ('certain universities were capable of absorbing more of the products of the new science than we had been led to believe so far').[135] Also, one strong point of Ornstein's study continues to be that (however limited her source material was otherwise) she addressed the state of the universities all across Europe, rather than in one country only. It is very well possible that Oxford and Cambridge provided a much more advanced education in science than was generally true of universities on the Continent, so that altogether Feingold may have established at most that the English university system was better equipped than had been realized before for absorbing at least some fresh developments in the new science. This is an important conclusion, yet it yields quite a different claim from the one advanced in his introductory chapter, which suggests that the universities were to some extent involved in active contributions to the Scientific Revolution. It is certainly true that virtually all protagonists of the Scientific Revolution were rooted in the science of their own time, which was largely that of the university curriculum. But it is no less true that what makes these men stand out is precisely the degree to which they managed to *transcend* these origins.

The ongoing collection of fresh evidence on the matter is unlikely to do more than to grant the universities of the 17th century somewhat greater flexibility than they have previously been credited with.[136] Nor could it be otherwise. After all, professors' assigning to students some advanced books among a great many traditional ones or cautiously exploring the strict limits of tolerance set by the university and religious authorities for 'modernizing' tendencies in the curriculum reflect practices on a plane altogether different from the relentless search for a new truth that provided the powerful, ultimate drive for the unique movement embodied in the Scientific Revolution.

3.5.3. *Patronage*

A more promising venue toward increasing our understanding of the social ramifications of the Scientific Revolution comes as a by-product of Feingold's original research. The final chapter of his book is devoted to uncovering a comprehensive system of patronage that deeply affected the direction of careers toward (or, on occasion, away from) the pursuit of science.

In historical discussions of the Scientific Revolution, accounts of the financial means of scientists in the 17th century have rarely been raised to a higher level than the anecdotal and the biographical. Thus, we have known all along that certain individual protagonists of the Scientific Revolution were in a position to pursue their intellectual concerns while living on independent means (e.g., Descartes, Huygens, Boyle) or through close association with a court (Kepler, Galileo, Stevin, countless others) or with the Church and the convent (Mersenne, Gassendi, again many others). As yet, no systematic survey has been published of what might somewhat roughly be called the funding of the Scientific Revolution. This is at present being examined by Westfall in the framework of a prosopographical study of the social history of the Scientific Revolution.[137] Here his subquestion of how 630 scientists between Copernicus and Newton made a living is examined in close connection with a wide area of study first trodden, beside Feingold, by Westfall himself: the 17th-century scientist as a client (along with artists and other famous men) of a wealthy patron. Patronage was a pervasive social reality of the 17th century, and it seems quite likely that the concepts of patronage and clientship go a long way toward providing the proper framework for a unitary account of this vital aspect of the Scientific Revolution. Feingold, in the chapter mentioned above, and Westfall, in some preliminary studies of how Galileo's relations with the Medici court affected his scientific work, have begun to show the way.[138]

Questions that come up in this regard are: What motivated a patron to support a scientist? What obligations was the latter expected to meet so as to keep his patron happy? How did the inevitable competition among clients affect scientists' research topics and findings? How secure was the client's position when the patron died? And so on, and so forth. Recently, Roger Hahn has opened up even wider vistas by suggesting that the court of Louis XIV achieved a wholesale transformation of traditional forms of patronage, engaging in it on a scale hitherto undreamt of and in effect ushering in the notion of "planned research often employing a collection of salaried scientists engaged in mission-oriented projects."[139]

In sum, the program of unraveling the clientele systems behind the Scientific Revolution, although still rather in its infancy, stands a good chance of providing the overall viewpoint from which a full-fledged *Social History of the Scientific Revolution* may in future be written.

3.6. THE NEW SCIENCE IN EUROPEAN HISTORY

We widen the perspective further. Not only were there, in the 17th century, social forms within which the novel science grew as well as those which it created for itself (topics we have treated in the previous cluster of sections), but, obviously, the Scientific Revolution itself took place against a background of further history. Take, first, the 17th century itself. At least three 'crises' have been assigned by historians to that sorry century, and the Scientific Revolution has been linked to each of them. But the Scientific Revolution may also be considered against an even wider background than that of the century in which it occurred. From a Marxist viewpoint the Scientific Revolution has been studied as but one manifestation of the dissolution of feudalism and its replacement by commercial capitalism—a seminal event in western Europe that equally manifested itself in such cognate landmarks as the Renaissance and the Reformation. Finally, one may take the entire course of Western civilization as a frame of reference and attempt to assign the Scientific Revolution its fitting place inside that frame. Even though I have argued, in my introductory chapter, that this is the very task still awaiting us, a few remarks are nonetheless apposite, not only on the rare efforts that have been undertaken so far but also on a remarkable case of just missing the occasion to take up, in all its countless ramifications, 'the place of the Scientific Revolution in the history of Western civilization'. These, then, are the subjects that come up in the present cluster of sections.

3.6.1. *The Scientific Revolution and the Crises of the 17th Century*

Not so long ago 'general' historians found the 17th century to be a comparatively dull one. Evidently, sufficient events took place all over Europe to fill huge tomes, yet the century seemed somewhat aimlessly adrift between two episodes which lent themselves far better to unified treatment: the wars of religion and the creation of the 'new monarchies' in the century that preceded, and the Enlightenment, culminating in the French Revolution which came to crown its successor. In short, the 17th century seemed to lack a character of its own (which need not astonish us, given how artificial the 'century' is as a historical category). No obvious label came readily to hand,[140] and if one peruses, for example, G. N. Clark's study of 1929, *The Seventeenth Century,* one finds that the author, while treating manifold aspects of the period, and while doing his best to connect these with one another, failed to find an organizing principle enabling him to hold the century together.

Since then, a great deal has changed. At least three major crises have been ascribed to Europe in the 17th century, and—this is the important thing for us—the new science has been claimed to have played a part in all of them.

Rabb and the crisis of authority in 17th-century Europe. Stimulated both by novel currents in history writing and by the example of a unified treatment set by historians of the Scientific Revolution,[141] 'general' historians in the fifties and sixties began to identify, and to debate the precise nature of, a 'crisis of the seventeenth century'. In his book of 1975, *The Struggle for Stability in Early Modern Europe,* the American historian Theodore K. Rabb summed up the debate and went on to outline an interpretation of his own which, exceptionally, included the Scientific Revolution as a major participant in both the crisis and its resolution.

Rabb identified the crisis as one of *authority.* The stormy events that clustered in the first decades after 1500—the rapid growth of centralized government, the establishment of international diplomacy, the Reformation, "a soaring ascent in population, trade, and eventually prices,"[142] the trying out of pragmatist and sceptical ideas, the breakdown of previously self-contained villages and regions—had cut Europeans loose from what had seemed forever secure and stable. "No succession of events so disruptive of safe and comfortable suppositions had occurred for hundreds of years. As expectations lost their cogency, an atmosphere of groping and unease descended."[143] Thus was ushered in a period of unrest, of searching, and of endless fighting leading to wanton destruction. This period was not to come to an end (at least in people's perceptions) until the third quarter of the 17th century. From then onward, stability reigned once more, not to be disturbed in any comparable degree until the advent of the French Revolution.

The most striking feature of this thesis is that Rabb seeks, and often finds, evidence for it in a variety of domains of human endeavor at the time: domestic politics, international relations, economic and demographic trends (where, because of much regional diversity, he finds the evidence to be least conclusive), but also literature, art (which he employs in a truly virtuoso manner), and science.[144] One lengthy quotation may give the reader a sense of the scope of Rabb's argument about "the discontinuity that has been our central concern":

> The transition from the age of Reformation to the age of Enlightenment . . . , and the distinctiveness and homogeneity of the period from the death of Calvin to the birth of Voltaire should be apparent. To return to the contrasts posed at the beginning of this essay, we can appreciate that, on one side of the divide, Rubens, Milton, Charles I, Condé in the 1640's, Wallenstein, Galileo, Descartes, Hobbes, Gustavus Adolphus, Paul V, and a witch-ridden society were all engaged, in a setting where all standards and institutions were open to question and total overhaul, in correspondingly vast ambitions—to overwhelm the senses, to change the form of politics, to find a new order, and to control or evade unresolvable doubts. On the other side of the divide stand those who either established, or worked within, a structure that was no longer bitterly disputed—for whom the uncertainties had essentially been settled, possibilities had been limited, and re-

laxation had become more attainable: Claude, Dryden, Charles II, Condé in the 1680's, Eugène, Newton, Locke, Charles XII, Innocent XI, and a witch-free (or at least relatively impervious) society. The major division that occurred approximately in the third quarter of the seventeenth century between disequilibrium and equilibrium may have been the result of massive delusion, but it was real nevertheless. And it is from the phenomenon of resolution that we can deduce Europe's passage through a 'crisis': a 'crisis' so general that it affected all forms of human activity, from diplomacy to drama, and every country, from Russia to Portugal; a 'crisis' primarily in perception, although it had its material components, and a 'crisis' of authority, if a single label seems appropriate to such disparate manifestations.[145]

Rabb ends his account by arguing that the return to stability starting in the 1660s and 1670s came primarily from a sense of exhaustion after a century and a half of bitter strife, both induced and exacerbated by the unprecedented destructiveness of the Thirty Years' War. The 1648 Peace of Westphalia, rather than temporarily settling one or another local dispute as had been customary for so long, was consciously set up as a package deal meant to put an end for good to forms of disintegration and destruction that threatened to rend all of Europe apart. From then on, a sense of finally returning stability began to set in, and later convulsions whose counterparts in preceding decades would assuredly have ushered in new rounds of fighting now dampened out more or less by themselves, as a kind of aftertremor.

What role, in all this, is attributed by Rabb to science? A dual one. In the stage of crisis, science was the domain where the insecurities of the age manifested themselves most acutely. Not only was the study of man and nature taken up as an escape from the turmoil of the times,[146] but, more importantly, in science the contemporary quest for novel forms of authority and certainty found its most outspoken expression.[147] Overturning comprehensive systems of thought that previously had gone unquestioned, the protagonists of the Scientific Revolution contributed even more to the confusion by the sheer plurality of new approaches and viewpoints they offered instead.

However, and this is the other role Rabb attributes to the new science, having exacerbated the reigning confusion and disruption it went on to provide much of the medicine needed. Even before Newton rounded off the Scientific Revolution, the essential unity common to the new, competing approaches had become apparent. It was seized at once by the "elite culture. . . . For here was the principle of radical doubt being put to marvelous advantage, transformed from a curse into a blessing. The old, shaky reliance on tradition as the root of knowledge had been replaced with an emphasis on abstract reasoning and sense experience."[148] A new source of security could henceforward be tapped, and this explains in large measure a phenomenon otherwise hard to fathom. After all, Rabb observes, "the quick and decisive triumph of this handful of

scientists is one of the most amazing episodes in European history." [149] Undoubtedly the intellectual genius of Newton accounts in part for the ready acceptance of Newtonianism by nonscientific authorities, yet the new stability
they sought for other reasons found its intellectual counterpart in both the accomplishment and the further promise held out by the new science and thus
helped make possible its smooth espousal. "What the age wanted to hear was
that the world was harmonious and sensible," [150] and this the new science could
tell its contemporaries to its own and almost everybody else's satisfaction. In
this manner, "the triumph of science was as much a symptom as a cause of the
wave of settlement of the late seventeenth century." [151] The previous stability
science had helped so much to disrupt was now restored with the help of a
science meanwhile stabilized from within and called in from without when
needed most.

It does not seem hard to criticize such a daring, widely cast argument in
detail. For example, it follows from Rabb's ideas about the ready acceptance
of the new science that the patricians of the Dutch Republic, where a bitter
struggle over the Oranges' stadholderate had been rampant for many decades,
would have embraced the new science as eagerly as elites in other countries
did, whereas in historical reality there indifference was almost palpable. [152]

Still, objections along such lines, even if multiplied, do not exhaust the
essence of Rabb's argument. His effort may be called a study in the secular
shift of sensibilities. Its value depends critically on whether or not the resulting
picture of the period succeeds in capturing the reader's imagination and in
persuading him or her that previously unconnected facts from widely separated
domains now begin to throw light upon one another. Obviously, this is not the
place to deal further with its merits as an evocation of a period. Yet I do think
that Rabb's account of the role of the Scientific Revolution in the stabilization
of late 17th-century, war-ridden Europe is of considerable merit in that it em-
braces in a comprisive manner two closely related topics to be encountered
elsewhere in the literature but handled there in a narrower fashion. On the
one hand, Rabb consciously expanded Paul Hazard's 1935 thesis of the
'crisis of the European mind', located by Hazard in the period 1680–1715,
a thesis confined to the domain of ideas. On the other hand, Rabb appears
to prefigure a line of argument that has lately become much *en vogue* owing
chiefly to Shapin and Schaffer's many-sided book, *Leviathan and the Air-
Pump*. Here, in an argument confined solely to England, the convulsions of
English politics before and after the Restoration are connected to contemporary developments in science. We take up their respective ideas in reverse
order. [153]

Shapin and Schaffer on the problem of order in late 17th-century England. In
their book, Shapin and Schaffer are concerned to demonstrate an analogy between Boyle's and Hobbes' respective approaches to the understanding of na-

ture and the need felt by the authorities of the Restoration settlement to keep society in good order. The analogy runs as follows.

On the side of science, they make a case for showing that the experimental approach advertised by Boyle and the demonstrative approach defended by Hobbes entailed much more than the issue of how to gain the best understanding of nature. Rather, distinct 'forms of life' were at stake. Boyle claimed that the community of experimental scientists provided a social space inside which dispute of a nondisruptive type was possible. In the end, he believed, the outcome of an experiment could always settle the dispute; once an objective fact had been ascertained by means of experiment the dispute would, as it were, resolve itself. Hence Boyle's incessant efforts to persuade his audience by means of suggestive rhetoric that the verdict rendered by experiment was the word of Nature herself.[154] Shapin and Schaffer then go to great lengths to make a point that had been made by others (by Ben-David, in particular; see section 5.3) in a different context. This is that the founding fathers of the Royal Society recommended their approach to nature as one which, due to its lack of ideological contention and its offer of a neutral meeting ground where disputes could be resolved peacefully, would contribute to peace in the community at large. In the words of Shapin and Schaffer:

Boyle aimed to achieve peace and to terminate scandal in natural philosophy by securing a space within which a specified kind of dissent was manageable and safe. In the experimental form of life it was legitimate for philosophers to disagree about the cause of natural effects: causal knowledge was removed from the domain of the certain or even the morally certain.

. . . Dispute *within* the experimental space was possible, even necessary. While the variety of professions and beliefs of the experimental philosophers was a virtue, that variety *outside* this space had been disastrous. Inside this boundary, Sprat [the first 'biographer' of the Royal Society, 1667] claimed, the *exact* equivalent of a civil war could be staged, as in a theatre, with no harmful result. If subjects did not know how safely to dispute, they should go and watch the experimenters: 'There we behold an unusual sight to the *English Nation,* that men of disagreeing parties, and ways of life, have forgotten to hate, and have met in the unanimous advancement of the same *Works.*' [155]

Hobbes' philosophy embodied quite a different 'form of life':

For Hobbes civil war flowed from any programme which failed to ensure absolute compulsion. What was a judicious and liberal bracketing strategy to the Greshamites [experimentalists] was, to Hobbes, a wedge opening the door which looked out on the war of each against each. Any working solution to the problem of knowledge was a solution to the problem of order. That solution had to be absolute.[156]

Only a natural philosophy founded upon absolutely secure foundations, such as the mechanicist natural philosophy Hobbes himself put forward, could ensure social order. "Causal inquiry *could* be concluded; it did not breed dissent but could provide the surest remedy for dissent." [157] One tenet to be derived from Hobbes' indubitably secure first principles was the impossibility of the void; hence, Shapin and Schaffer could render his position in his dispute with Boyle over the air pump in the following, marvelously concise line: "In Hobbes' view the elimination of vacuum was a contribution to the avoidance of civil war." [158]

So far, so good, or rather, so far taken for granted from the authors' surely imaginative, not to say idiosyncratic, reading of their sources. Let us now look at the other side, that of the state. In their 'sociopolitical' chapter "Natural Philosophy and the Restoration: Interests in Dispute" the authors claim to examine how these rival 'forms of life', with their rival solutions to the problem of order inside the community of thinkers about nature, impinged upon the state at large. Despite some protestations to the contrary,[159] their aim here is to explain the victory of the experimentalist program over Hobbes' by showing that it fitted in far better with the hope of Charles II's administration to put an end for good to the social strife that had torn England apart for decades.

However, to make such a point two lines of argument should in my opinion have been pursued that are scarcely to be encountered in the book. In the first place, the analogy between the solutions to the problem of order offered by Boyle and Hobbes, on the one hand, and the search for such a solution by the state, on the other, should have been turned from an interesting coincidence into a substantial link. The authors could only have accomplished this by moving, for the time being, from the history of ideas to true social history, setting forth at least in brief outline the sociopolitical history of Restoration England. Instead, we get a largely achronological jumble of bits and pieces held together by rhetoric. For example, the authors state that "the rules of the experimental community offered this solution to the fundamental political problem of liberty and coercion," [160] yet no evidence is presented showing that, or how, the offer was in fact accepted. The other line of argument that is lacking is the notion that the state was in such dire need of order. Instead, one finds some efforts to portray Charles II's rather lax and remarkably unrevengeful regime as little short of totalitarian in its enforcement of silence among its subjects.[161] Added to this is Hobbesian rhetoric on "the crisis of the Restoration settlement [which] made proposals for a means of guaranteeing assent extremely urgent. . . . The experience of the War and the Republic showed that *disputed knowledge produced social strife*" [my italics].[162]

Overdone and underargued as all this appears, however, much of the argument could be substantially strengthened in my opinion if considered in the larger framework of Rabb's thesis on the struggle for stability in the 17th century.[163] Seen in that light, Hobbes appears as one of those frantic seekers for absolute certainty from before the 'divide' of the sixties/seventies, whereas Boyle *cum suis* comes to represent the willingness to settle for the middle ground of mitigated, tentative, 'demetaphysicalized' certainty that helped usher in the eventual solution to the 17th-century crisis of stability. In such a generalized, European context Shapin and Schaffer's claims begin to make much more sense than they do when standing on their own feet. With the essential difference that, in Rabb's argument, the speedy acceptance of the new science by the political and cultural elites of Europe appeared as a consequence as much of its intellectual superiority as of its capacity to gain general assent, whereas Shapin and Schaffer, in their closing lines as well as everywhere else, picture the issue in typical either/or fashion: "it is ourselves and not reality that is responsible for what we know. Knowledge, as much as the state, is the product of human actions. Hobbes was right." [164] In the conclusion to the present chapter (section 3.7) we come back to the general issues that are involved here.

Paul Hazard's 'crisis of the European mind'. In 1935 Paul Hazard (1878–1944), a historian of French literature, published a book entitled *La crise de la conscience Européenne (1680–1715)* (translated in 1963 as *The European Mind, 1680–1715*). Driven by the conviction that in the end it is ideas that change the world, Hazard advanced the challenging thesis that all basic ideas of the Enlightenment era up to and including the French Revolution were already given shape and content during the period indicated in his title. Around the middle of the 17th century, he argues, Europe appeared poised for a moment on the serene, balanced, stately style of classicism both in thought and in manners. But the eternally restless European mind soon embarked on another of its searching voyages and, in a forward movement worth calling a second Renaissance, inaugurated an era during which Reason and Sentiment were to reign supreme.

Hazard's thesis, according to present-day historians of ideas, has largely withstood the test of time. His enthusiasm, his wonderful erudition, his gift for synthesis, the focus on all of Europe rather than on France alone, the powerful yet elegant style of the book—all contribute to the air of general persuasiveness it exudes. For us the important thing is the role the new science is accorded in the account. Four of its aspects make their appearance.

The most important of these is how, during the period of 'crisis', Descartes' method (as distinct from his results) is turned against everything once held venerable and impervious to attack. The floodgates of purely rational analysis have been opened for good; neither religion, political authority, the idea of divine law, nor the belief in miracles, in witches, etc., can be kept outside its realm. Next, Fontenelle's elegant discourses are taken as the prime example of how the new science could be popularized and spread beyond the domain of the experts. Newton's *Principia* is treated as the crowning example of how two important new approaches to nature—the geometrical and the experimental—could be fused. Finally, Locke's philosophy is taken for the epitome of the entire crisis of the European mind, in that he eliminates metaphysics—that source of perennial contention—from what humanity may securely know. This move is made possible by Locke's quiet conviction that after such drastic surgery upon traditional modes of thought, something tangible and reasonably secure remains: science.

Despite such prominence accorded to science, when in his conclusion Hazard asks about the entire movement of liberating thought he has been describing: "Who nourished this critical mode of thought? From where did it take its force and its boldness? And where did it ultimately come from?" [165] his unhesitating response is: from the Renaissance. Remarkably, he supports this contention solely with reference to the resemblance his protagonists of the 'crisis' felt themselves with that other Renaissance that had preceded them. Science does not turn up in his final considerations, but does it not follow from the treatment accorded to science in the body of the book that here a substantial part of the driving power for Hazard's proto-Enlightenment is to be found?

In a previous section (3.3.4) I enumerated a number of questions to be asked in connection with the issue, so far largely unexplored, of precisely how the Scientific Revolution helped to give rise to Enlightenment modes of thought. Hazard published his book five years before the concept of the Scientific Revolution was coined, so the relatively subordinate place he ultimately allotted to the new science is quite understandable. Nonetheless, Hazard's masterwork still offers an excellent framework in which to pursue the more pointed questions his own work suggests.

3.6.2. The Scientific Revolution and the Dissolution of Feudalism

The Scientific Revolution has been treated as one manifestation of the dissolution of feudalism and of the rise of capitalism by the British physicist of Irish descent J. D. Bernal (1901–1971). He was a staunch supporter of the Soviet Union from the thirties till the last day of his life, and his four-volume effort *Science in History* (first published in 1954, one year after he received the Stalin Peace Prize) bears the mark of his personal brand of orthodox Marxism throughout.[166]

Marxist thought of varying degrees of subtlety has pervaded a number of analyses of the Scientific Revolution that are to occupy us, so there is good reason to introduce Bernal's conception of the emergence of early modern science by means of a note on Marxist historiography of science in general. Today, after the dissolution of the communist world, such an effort may well appear rather futile. In 1987, when I originally wrote the following note, it was assuredly not. Marxism as a theoretical position has wrongly been declared dead before, and even though I grant that it may well at present be undergoing its final spasms, it seems to me that a brief, general discussion may still retain some merit in view of a body of literature that is to be with us for as long as libraries exist. Also, however unduly Marxist-inspired accounts were exalted by many during the heyday of the doctrine, some truly valuable contributions would be lost if allowed to fade out of sight together with its present, overall decay.

A note on Marxist historiography of science.   Marxism is not the establishment of occasional links between human thought and human economic activity.[167] Rather, Marxism, at least in its original meaning, is the *a priori* conviction that human thought is ultimately determined by human economic activity. Thus, in order to argue for the plausibility of the Marxist interpretation of history, it is not enough to point at specific cases in history when a given human thought, or a comprehensive set of thoughts, was demonstrably influenced, or even determined, by a given economic activity. Rather, one has to show that this is always true, in other words, that human thinking has no autonomy of its own but is guided, 'in the final analysis', by the material substructure of society.

Leaving aside the thorny question of how this could possibly be demonstrated other than by giving examples which will never establish the general rule, we may note that the converse of the statement in the previous paragraph is also true. The assertion that a certain human thought was indeed, in a given historical setting, determined by a certain economic activity is by itself insufficient to qualify the person who makes the assertion as a Marxist.[168] To state that there are certain links between, for example, scientific thought and economic activity is a specifically Marxist statement only if two conditions are fulfilled. First, the Marxist supposes that such links are always there. Second, the nature of the connection is taken to be one of 'determination'; that is, the way human thought changes over time is 'determined', rather than loosely 'influenced', by economic activity.

This connection runs as follows. The truly determining agent in history is the forces of production in their gradual change over time. These forces (tools, machines, etc.) determine the modes of production (that is, labor relations as these reflect the class division of society; who gets what of the produced commodities, and so on). The modes of production determine, in their turn, human thought (political, legal, social, etc.). In this sense all human thought is ideological; that is to say, it does not reflect any objective state of reality but rather the class interests of those who think these thoughts. The one and only exception in world history is proletarian thought, which reflects no class interest because the interests of the proletariat coincide with the true interests of all humanity and therefore enable proletarian thought (and, crucially, the thought of its self-appointed representatives) to reflect reality in a straightforward manner.[169]

For those who take such an account of change in world history for granted, or as proven, there is no longer any need to point out, in any specific case of an alleged connection between thought and economic activity, what precisely the nature of the connection is, since this has been shown once and for all in Marx's classic works. Thus the true difference between Marxist and non-Marxist analyses of the relation between thought and structural elements of society is that, for the non-Marxist, the validity of the connection depends entirely on the quality of the argument that is made to establish the connection in any given case. He or she cannot take it for granted that (to take just one arbitrary example) a reduction of Kepler's area law to changes in the forces and concomitant modes of production in early 17th-century Austria follows automatically from a general point made in 1859 in two pages of Marx's preface to his *Zur Kritik der politischen Oekonomie,* the only place Marx ever formulated as one consistent whole, rather than in scattered bits and pieces, his world-encompassing theory.[170]

One good reason among many for not taking for granted that human thought is always and everywhere dependent on changes in the forces and, hence, in the modes of production is that, in Marx's own schema of things, the nature of the connection is not made very clear. How precisely does the 'determination' work? Is it a matter of motives in the sense that people think certain thoughts and not others because these thoughts further their economic interests? Or is it rather a matter of institutional constraints, in the sense that the changes in the mode of production drive our thoughts in a certain direction without our even being aware of it? Questions such as these can be, and have been in fact, multiplied at length.[171] Among these there is one which lends itself particularly to being asked by historians of science.

The question has to do with what may be called the 'prime mover' of the entire system: the changes in the forces of production which, by way of a 'substructure' (*Unterbau*), are supposed to carry with them the entire, two-storied superstructure (*Ueberbau*). Not only can it be asked in a general sense where this most fundamental of all changes comes from (a question that usually receives the response that changes in the forces of production result from their 'inner development'),[172] but, more specifically, the historian of science and technology is in a good position to observe that changes in the technological level reached by a given society are what truly lie at the bottom of all social and intellectual change in the Marxian scheme of things. In other words, technology is not part of the superstructure but rather of the substructure. And because the level of the forces of production has been raised to a large extent by an increasingly science-based technology ever since the middle of the 19th century at the latest, this means that at least from Marx's own time of writing onward science has been part, not only of the superstructure, but simultaneously of the substructure as well.

This curious paradox can be resolved in two directions. One may conclude that at this very point—the dual nature of science—Marx's schema of human history breaks down. Alternatively, one may declare science, from the Scientific Revolution onward, exempt from the domain of 'ideological' thought and treat it instead as (1) objectively true and (2) a motor, rather than a 'moved', of societal advance. The man to take this latter route—selling it as orthodox Marxism to boot—was one of the first in our time to call attention to the power of science to alter society: J. Desmond Bernal. He did so in a variety of works, among which a 1325-page, very influential study that he effectively entitled *Science in History* will now hold our attention.

Bernal and the simultaneous birth of capitalism and early modern science. Bernal's enduring claim to fame is that from the thirties onward, when few scientists were yet aware of their own seminal role in the ongoing transformation of modern society, or politicians of the power of science, he consistently pointed out the newly acquired function of science as a force of production, or, in non-Marxist vocabulary, the capacity of science to be applied and of its applications to change deeply the human environment in the widest sense. He urged the point in his book of 1939, *The Social Function of Science.* Here he expressed his root conviction that science, although brought forth in its modern guise by capitalism in its early, progressive stages, could only unfold its full potential in socialist society. The further elaboration of his viewpoint was what made turn Bernal to history. As he writes in his resulting book, *Science in History:*

> It is the need to achieve this transformation [of present-day society under the influence of science] in the best way, and to secure the intelligent utilization of science at every stage in it, that is the strongest reason for the study of the relations of science and society in the past, for only through this study can it be adequately understood.[173]

Undaunted by the grandiosity of his self-set task, Bernal took up the study of how science and society interacted over the entire past course of science up to the present time (not omitting the social sciences, either), and completed it in six years. For present purposes, I focus on what he had to say on the Scientific Revolution, which occupies a key position in the story.

Bernal's account of the intellectual content of the Scientific Revolution is conventional enough: In what he has to say on the accomplishments of its leading men no insights are to be gained that had been overlooked by such previous historians as Burtt and Koyré and Butterfield and many others. This may seem odd for such an iconoclastic work as Bernal meant *Science in History* to be.[174] But it is not so odd, and explaining why not carries us straight to the heart of the peculiarities of Bernal's study. For one, it has often been observed that, to a large extent, Marxist historians are in the habit of adopting factual accounts wholesale from their 'bourgeois' counterparts, while supplementing them with their economic interpretations. Rare is the case of a Marxist historian who presents a fresh look at the factual material, and Bernal's account is no exception to this rule.

Also, more importantly, for Bernal there was nothing 'ideological' about science—at least, about scientific findings still by and large held valid today. Thus, we find no trace of attempts to reduce, for example, Kepler's area law to any forces or modes of production whatsoever; scientific thought as such is implicitly taken by Bernal to be fully autonomous. In fact, he was quite as 'idealist' a historian of scientific ideas as anyone one can imagine in the 'bourgeois camp' (also, inasmuch as he treated the advance of science as a triumphant march toward the present, a much more positivistically inclined one). Thus, when Bernal asks in his preface, "How in detail does a social transformation affect science?" the first specification he presents for this "central question" is: "What gave the science of ancient Athens, of Renaissance Florence, [etc.,] its particular drive and novelty?" [175] Not the *content* of science is socially determined in Bernal's view, but only its *drive and novelty*.

What, then, was the drive that gave the Scientific Revolution its particular novelty? It was the dissolution of feudalism and the birth of merchant capitalism. Or rather, early modern science and capitalism came about together in one great historical movement. The emergence of early modern science exemplifies a general truth: The "flourishing periods [of science] are found to coincide with economic activity and technical advance. The track science has followed—from Egypt to . . . England of the Industrial Revolution—is the same as that of commerce and industry." [176]

Feudalism is taken by Bernal to have been the dominant economic order in Europe from the fifth through the seventeenth century. Yet only from the eleventh to the fourteenth centuries do we find it in fully developed shape. In accordance with the Marxist schema of the successive transformations of society Bernal states that "the feudal system . . . contained the seeds of its own transformation." After all,

> however much the medieval system of thought might tend to be static the medieval economy could not stay still. . . .
>
> Greater trade and improved techniques of transport and manufacture drove relentlessly towards a commodity and money economy in place of one based on prescribed service. It was the technical aspect of this economic revolution that was to be the decisive factor in creating a new, progressive, experimental science to take the place of the static, rational science of the Middle Ages. It was to present the men of the Renaissance with situations and problems that the old knowledge was inadequate to deal with. [177]

As a result:

> The same period—1450–1690—that saw the development of capitalism as the leading method of production also witnessed that of experiment and calculation as the new method of natural science. The transformation was a complex one:

changes in techniques led to science, and science in turn was to lead to new and more rapid changes in technique. This combined technical, economic, and scientific revolution is a unique social phenomenon.[178]

The aggregate nature of the revolution thus defined allows Bernal to divide the Scientific Revolution into three successive stages, those of the Renaissance, of the Wars of Religion, and of the Restoration, respectively, which "are not three contrasting eras but three phases of a single process of transformation from the feudal to the capitalist economy."[179] Each stage is duly marked by Bernal according to events in the scientific as well as in the socio-economic-political domain. The first stage (1440–1540) is that of the Renaissance, the Voyages of Discovery, the Reformation, and the wars which made Spain the dominant world power. In science, it is a period of "description and criticism rather than constructive [with the exception of Copernicus] thought."[180] In the second stage (1540–1650) the new bourgeoisie becomes dominant in England and Holland, while "in science the period includes the first great triumphs of the new observational, experimental approach."[181] Finally, in the third stage (1650–1690) the bourgeois make their compromise with the reigning powers. During this period the new science establishes itself and is also turned into a force of production for the first time (exclusively, as yet, through the services rendered by science to navigation).

So far we have encountered no more than just the assertion of certain *parallels* between scientific and socio-economic-political development in late medieval and early modern Europe. At times Bernal tries to make the connection more specific, for example, when he seeks to link the accomplishment of a man of science like Gilbert to the experiences of contemporary craftsmen.[182] But on other occasions he blithely mentions states of affairs that make nonsense of his overall parallelism, for example, when he observes that "Italian intellectual pre-eminence was for some time to outlast political and economic decadence; for Italy, the first of the countries of western Europe to break away from the feudal tradition, remained the centre of European culture long after she had lost her political and economic importance."[183] It is true that the author has given himself sufficient leeway for such an exception to his rule of combined transformation by asserting that "naturally the process of transformation was slow and uneven,"[184] but this only amounts to saying that even when events do not conform to his schema they still do. Bernal's rhetorical strength throughout was to handle with a straight face facts wholly refractory to his overall claim as if they fitted in perfectly.

As so often in Marxist thought, we need only peel off a surface layer of strident expressions of orthodoxy—helpfully reinforced by the author with ugly Stalinist invective—to find quite other forces at work underneath.[185] An illuminating glimpse is provided by a passage that aims at summing up the entire episode of the Scientific Revolution: "The birth of science follows closely after that of capitalism. The same *spirit* that broke the fixed forms of feudalism and the Church . . ." [my italics].[186] Apparently, not even for the most materialist of scientists is it that easy to shed the suspicion that there may be something spiritual about human thought and action after all.

The question that, by way of a hopefully balanced conclusion to Bernal's treatment of the Scientific Revolution, must be answered is therefore a dual one: To what extent has Bernal's conscious Marxism, and to what extent has his unconscious idealism, contributed to our enduring insight into the event?

The Scientific Revolution in history.   Let us look first at Bernal's ideas about science turning into a factor of production during the final stage of the Scientific Revolution. Although ideologically almost predestined to lump science and technology together in taking, as other Marxist historians had done,[187] the Scientific Revolution for the time of birth of a science-based technology, Bernal's personal experiences with the complex relationship between science and technology helped to make his treatment of the topic in its historical dimension quite sophisticated. As noted in section 3.4.3, he was among the first to call attention to the wide gap between the utilitarian promise and the actual achievement of science in the 17th century, even though, in order to make the social function of science start with the Scientific Revolution, he insisted on an otherwise hard-to-explain exception for the issue of the determination of geographical longitude. Nonetheless, the passages devoted in *Science in History* to the topic are among the most nuanced and the best-informed in his entire treatment of the Scientific Revolution.[188]

For Bernal, the impact of science upon society over time was the most important theme of his study.[189] And indeed, his treatment of how, conversely, society impinged upon science displays an altogether lower degree of sophistication. On the negative side we note four serious defects to the parallelism he proclaimed throughout between the transformation of science and the transformation of the feudal order.

The first is that obvious exceptions to his general rule are either ignored entirely or cheerfully handled as if providing further evidence for it.

The second is that instead of what he really needed, and occasionally promised as well—solid connections arising out of an integrated treatment—Bernal rarely offers more than declaratory juxtapositions.

The third is that he makes these juxtapositions look conclusive through sheer reiteration. Announced in the general introduction, they are repeated in greater detail in sections that introduce the successive parts of the book, are repeated once more as the story unfolds, finally to be reiterated almost *verbatim* in concluding sections to the part in question and, in the end, to the entire work. What, in following such a procedure, Bernal may be blamed for is *not* that in his treatment a huge gap remains yawning between scientific and social development, for there is almost no historical problem harder to tackle than this, but rather that his dogmatic self-assurance prevented him from facing the gap squarely.

The fourth defect is the crudeness of the analytical tools with which Bernal approached his subject. Such sloganized, reified concepts as Feudalism, Capitalism, and Renaissance take the place of serious social-historical analysis in which raw facts of history are bound together by means of tools proving their worth in their very capacity to make sense of those facts. Bernal, in the only passage in his book that betrays some humility before his subject, acknowledged himself to be unfamiliar with the modes of analysis employed by the historian.[190] It would surely be unfair to reproach a working scientist for not being an *uomo universale*. Still, we must conclude that the ready-made schema he applied instead is quite inadequate to do the job.

Coming now to the positive side of Bernal's achievement in placing the Scientific Revolution in the context of late medieval and early modern European history, we begin by pointing at his extraordinary vision in sensing that there is a job to be done here in the first place. What Bernal accomplished was to call attention, in the inadequate language of Marxist social analysis to be sure, to the contribution of the Scientific Revolution to the making of the modern world. What kind of society was able to bring forth the unique phenomenon of early modern science and allow itself to be transformed by it so swiftly and thoroughly? Not only did Bernal take this most important historical problem with the utmost seriousness, but also we find in his treatment of it a few isolated passages which—once elaborated further—hint at a much deeper level of understanding than could possibly be within reach of the schema he applied. One example is where Bernal calls attention to the role of the medieval city in the furtherance of new, scientific ideas;[191] another, his keen observation that

> culture in Europe at the end of the Middle Ages was hardly materially or even intellectually at a higher level than in the great empires of Asia. That it held greater promise could only be apparent from its relative lack of fixity and uniformity in social and economic forms.[192]

However vague such a remark is by itself, it does point toward a mode of historical analysis that allows the job of which Bernal realized the key importance to be attacked with arms forged in a far more sophisticated arsenal. Hints of the appropriateness of such an arsenal for the job at hand can be found in the work in historical sociology of a scholar Bernal chose to ignore entirely: Max Weber. It is to the promise held in Weber's work that we turn now.

3.6.3. 'The Place of the Scientific Revolution in the History of Western Civilization'

When, in the preface to his unfinished series of papers "The economic ethos of the world's religions," the German social scientist Max Weber (1864–1920) enumerated a set of features by which the West distinguished itself from all other civilizations in the history of the world, he started his list with early modern science.[193] With this as with many other domains of human endeavor his point was *not* to argue that the West had brought forth an array of cultural products—science, doctrines of history and of law, polyphonic music, pointed arches employed in architecture, print from moveable type, universities, bureaucratic officialdom, political parties, states, and capitalist modes of money exchange—that were absent elsewhere. After all, in Asia, too, music made by various voices at a time, bureaucratic states, etc., were more or less ubiquitous. It is not in these overall phenomena that the uniqueness of the West is to be sought, Weber insisted. Rather, he was concerned to point out that

> precisely in the West, and only here, cultural phenomena appeared that—we like to think, at least—lay in a developmental direction of *universal* significance and validity.[194]

By 'universal', as we shall see, Weber meant 'rational' in a specific sense of the word: In the West these phenomena had acquired a rationality of their own. He presents this as an "inevitable and legitimate" way to put the problem, and the question to be asked about the uniqueness of the West thus defined is, simply: What concatenation of circumstances led to it?

This question, as formulated here by Weber in the very last year of his life, is among the two or three basic issues that animated his entire scholarly work—however far its varied elaborations may often seem removed from the original inspiration—from 1903 onward. Let us see what, in this connection, Weber had to say about the natural sciences.

Weber and the uniqueness of Western science. In the preface Weber acknowledges that, where the study of nature is concerned, many Eastern civilizations display wide empirical knowledge as well as deep philosophical thought of an often sublime kind. Still, what such science lacked everywhere but in Europe was a rational foundation in mathematics as discovered by the Greeks and in experiment as discovered in the Renaissance. Similarly, one finds a highly developed medicine in other civilizations as well, but a solid basis in science, that is, in "rational chemistry[,] is lacking in all civilized areas but the West."[195]

So it is in all other fields enumerated by Weber. Capitalism—the domain central to Weber's endeavor here—developed in the West in a manner to be distinguished fundamentally from the "adventurers' or merchants' type or from a capitalism oriented upon changes of gain from war, politics, or the administration."[196] What is typical of the uniquely Western brand of capitalism is, rather, "the emergence of productive, bourgeois capitalism with its rational organization of free labor." And the crucial question to be asked is, once again, what made it possible.

Essential to modern capitalism in the sense defined, Weber maintained, is the availability of an exact, rational calculation of costs and assets, without which the aim of long-term cost-effectiveness that marks modern capitalism cannot be attained. And here is where science comes in once again. I quote now in its entirety what Weber had to say on the relationship between the emergence of capitalism and of early modern science:

> The rationality [of modern capitalism] is nowadays essentially conditioned upon the *calculability* of its technically determining factors—of the foundations of exact calculation, that is. But this is to say, in truth: upon the peculiar character of occidental science and, in particular, of the natural sciences, which possess a mathematically and experimentally exact and rational foundation. Now the development of these sciences and of a technology based upon them received in its turn (and still receives) decisive impulses from the capitalist chances that are linked, by way of bonuses, to their being turned into economic values. The emergence of occidental science was not, to be sure, determined by these chances. Calculations were also made by the Indians, who discovered the system of numbering by position, but this system was put at the *service* of developing capitalism in the Occident, whereas in India it did not create modern accounting or auditing. The emergence of mathematics and mechanics was not conditioned upon capitalist interests, either. It is, however, the case that the *technical* application of scientific knowledge—decisive for the way of life of the masses—was conditioned upon economic bonuses which, just in the West, came with such application. But these bonuses arose from the peculiar nature of the Western *social* order. So the question to be asked is: from *what* components of this peculiar nature, since undoubtedly not all of these can have been equally weighty.[197]

So we have reached the issue of what was peculiar about the social order of the West which, over the detour of the need of modern capitalism for a stable, predictable administration governed by law, leads Weber back to "a specific type of 'rationalism' in Western culture."[198] To find out what this uniquely Western rationalism consisted of, so he goes on, is one of the objectives of the investigation that follows; another, to explain its uniqueness. At this point a remarkable twist occurs in the argument. From what he has said before one might expect an inquiry to follow in which, alongside a host of other uniquely Western products, the emergence of early modern science takes an important place of its own. In the present case, however, the uniqueness of Western capitalism takes over entirely, for Weber goes on to outline as his problem the mutual dependence upon one another of the economic relations of a given society and what he terms the 'economic ethos' of its predominant religion. Under this overall heading he put first his celebrated essay "The Protestant Ethic and the Spirit of Capitalism" and then a volumes-long, far less known overview of how Confucianism and Taoism, Hinduism and Buddhism, and Judaism (with counterparts on Islam and Christianity at large missing from the uncompleted study) impinged upon the way people came to orient their lives, and *vice versa*. The theme of how the formation of religious world-pictures put large portions of humanity on the way toward a 'disenchantment of the world' that reached its zenith in the West with its specific type of rationality receives ample treatment here, but the road to early modern science is never returned to.[199]

Nor is it in other studies by Weber that owe their ultimate significance to his fundamental problem of the uniqueness of Western development. In his equally unfinished *Wirtschaft und Gesellschaft* (Economy and Society) extensive chapters are devoted to the sociology of religion, of music, of law, of bureaucracy, of what Weber termed 'legal authority', of the medieval city, and so on, all of these with a view to pointing out their respective contributions to the peculiar development of the West. In the medieval city with its relative autonomy and its aura of liberty, in particular, Weber saw a most important ingredient of what set the West apart. But, as noted, science is scarcely ever mentioned again. In a life's work devoted to nothing so much as to an in-depth investigation into the unique ways of the civilization of the West, the names of Kepler and Newton do not occur; that of Galileo, just once. One offhand remark about the religious affiliation of Copernicus; one allusion to Jan Swammerdam as an exemplar of the 17th-century desire to cultivate science as a means to find God revealed; one reminder that Baconian utilitarianism was not what brought forth early modern science; one remark about the origin of experimental science in mining as well as in Renaissance art (in particular, music); one footnote suggesting a possible relationship between science and Puritanism in the 17th century: this is about all Weber has to say further on science and its place in Western civilization.[200]

There is a riddle here, and I am rather at a loss to solve it. Having upheld, in the passage quoted above, the relative independence of the rise of early modern science from its economic environment, Weber had put himself in an ideal position to analyze this seminal event, together with others he did investigate, and thus to assign to what we now call the Scientific Revolution its fitting place in the history of Western civilization.[201] The one thing I feel certain of here is that no lack of interest in, or knowledge of, the natural sciences prevented him from doing so; witness his most knowledgeable critique of Wilhelm Ostwald's 'energeticism' written in 1909.[202] I suspect that a key to the riddle may be found in an examination of the place science held in exchanges in the German Verein für Sozialpolitik (Society for Social Welfare), of which Weber, along with Sombart, Simmel, and many other prominent German economists, was a key member. I hope one day to be able to carry out such an investigation, put into the context of Weber's scholarly work as a whole. For now I wish to return to what Weber did accomplish and to seek to define what promising possibilities his intellectual legacy still holds for us. A good way to do so is offered by contrasting Weber's approach to Bernal's.

Weber and Bernal: Two reverse mirror images.   Weber possessed (through his own effort) what Bernal lacked: an infinitely sophisticated set of tools to analyze the society that brought forth early modern science. Unlike Bernal, Weber was not a dogmatist. He was far too careful to present causal relationships in either/or terms. He sought causal links in given situations defined with great precision and rigor rather than in a vague, overall philosophy of history. He possessed to an impressive degree the ability to combine an awareness of the infinite complexity of phenomena, which he expressed in the ongoing qualifications that make his sentences so involved and hard to follow, with a trenchant insight into what logic holds phenomena connected all the same. His chosen means for accomplishing this was what he called 'ideal-types'—concepts abstracted from history and reapplied to it so as to build up a historical sociology halfway between dehistoricized sociology, on the one hand, and more or less conceptless social history, on the other. Since Weber died, mainstream sociology has turned ahistorical to a remarkable degree, yet there seems to be an increasing awareness nowadays among sociologists that a return to Weber might restore to their craft a dimension it has sorely been missing. This not in the sense of a mindless repetition of Weber's specific points, some of which have inevitably become obsolete after almost a century, but rather in the sense of the overall approach he both advocated and demonstrated in a manner that continues to exude inspiration to almost everyone who makes its acquaintance.

Bernal, by contrast, possessed (again, much through his own effort) something Weber did not lack so much but somehow almost ignored: an acute awareness of the power of science. Bernal could never for one moment forget the importance of Science in History; Weber knew it, too, but went on to forget it. I believe that a future marriage between Bernal's fixed idea about science and Weber's overall approach to European history provides the most promising route yet available to that distant aim of assigning, on the grand scale, to the Scientific Revolution its fitting place in the course of Western civilization.

Pending such an effort to be undertaken in future, let us address what Herbert Butterfield, for one, accomplished in this area on a smaller scale.

Butterfield's chapter.   In Butterfield's book of 1949, *The Origins of Modern Science* (discussed in section 2.4.3) one chapter is devoted to "The Place of the Scientific Revolution in the History of Western Civilisation." This chapter, no more than fifteen pages long, was carefully designed to get across two notions in particular: the uniqueness of the Scientific Revolution as a world-historical event and the way in which early modern science, barely born, began at once to transform modes of thought and action that had remained more or less constant in human life over the ages. The tone of the chapter—at once thoughtful and resounding—was even more effective in bringing out these points than the, inevitably, quite sweepingly handled themes selected by the author to convey to the reader the overall idea.[203]

Some of these themes have meanwhile become familiar to us. For one, Butterfield elaborated Paul Hazard's thesis into a brief yet balanced discussion of how early modern science at once began to strengthen the gradual secularization in Europe already in process of trickling forth from other, independent sources. He also paid attention to the huge promise of the amelioration of daily life held by the new science, while cautioning against any pretense of knowing *a priori* how to disentangle the different forces, with science as a most potent one, that came together in the "general movement . . . which by the last quarter of the seventeenth century was palpably altering the face of the earth."[204]

The general conclusion drawn on this score is that, with the Scientific Revolution,

> not only was a new factor introduced into history at this time amongst other factors, but it proved to be so capable of growth, and so many-sided in its operations, that it consciously assumed a directing rôle from the very first, and, so to speak, began to take control of the other factors, just as Christianity in the middle ages had come to preside over everything else, percolating into every corner of life and thought.[205]

The most innovative, still unmatched pages of Butterfield's famous chapter are, however, those where he considers the Scientific Revolution as the landmark *par excellence* in a secular shift of Western civilization away from its traditional center. "Until . . . the sixteenth or the seventeenth century," so Butterfield observed, "such civilisation as existed in our whole portion of the globe had been centred for thousands of years in the Mediterranean, and during the Christian era had been composed largely out of Graeco-Roman and ancient Hebrew ingredients."[206] This by itself commonplace observation is taken as the starting point for two brief arguments. In one of these Butterfield dwells upon the ups and downs in the ongoing struggle between Europeans and Asians to gain domination over this particular center of civilization, the Mediterranean world, to which the northwest parts of Europe were as yet little more than appendages. The other is to show how the Scientific Revolution was part and parcel of a huge shift of the center of civilization away from the Mediterranean toward the Atlantic coast.[207] This shift, which in its origins was surely not caused by the Scientific Revolution, nevertheless gained a great deal of its further momentum through the agency of the new science created at that very time. Western civilization had taken a novel, unprecedented course, in a secular movement that could not as yet be perceived very clearly by its contemporaries. We in the West, however, living as we do in a 20th-century world still centered to a considerable degree around the Atlantic Ocean and endowed with a supremacy largely founded upon the new science and its further outgrowth, can

sense its decisive significance. Or, put in the rousing prose with which Butterfield ended his chapter:

> We know now that what was emerging towards the end of the seventeenth cen-
> tury was a civilisation exhilaratingly new perhaps, but strange as Nineveh and
> Babylon. That is why, since the rise of Christianity, there is no landmark in his-
> tory that is worthy to be compared with this.[208]

3.7. CONCLUSION: FROM AN "AURA OF SELF-EVIDENCE" TOWARD "MESSY CONTINGENCIES"

> Our goal is to break down the aura of self-evidence surrounding the experimental
> way of producing knowledge. . . . The 'von Moltkes' of the history of science
> [have preferred] idealizations and simplifications to messy contingencies.

> —S. Shapin and S. Schaffer, 1985[209]

The biggest feat accomplished by such pioneers of the Great Tradition as
Koyré and Dijksterhuis was to historicize the subject of the birth of early modern science previously in possession of the philosophers. Since their day much
in the historiography of the Scientific Revolution has changed. One way to
characterize in a single line a variety of complaints directed against them by a
successor generation is *that their effort at historization had not gone far
enough.*

Two sides to such a qualification must be distinguished. The pioneers
abolished the customary treatment of past scientific work as an aggregate of
mutually isolated statements and introduced instead the full body of a scientist's work, considered against the background of the state of knowledge of his
own time, as the proper context for study. This was called the 'contextual'
approach. Over the past thirty years or so, the search has been for what we are
now wont to call a 'new contextualism'; in the present chapter we have met
some products of this search. Here the idea is to consider the body of a scientist's work as an indissoluble part of its social, economic, and political context.
I have already advanced some thoughts on the 'external' historiography of the
Scientific Revolution in section 3.5, and I shall return to the topic presently.
First, however, I wish to delineate the other, related sense in which the historization of the Scientific Revolution wrought by the generation of the pioneers
may be considered incomplete.

An overall conception that lurks behind the work of Dijksterhuis and
Koyré—one apparently shared by most other members of their generation—
is that the fundamental accomplishment of the protagonists of the Scientific
Revolution was to remove the principal barriers that had for so long been standing in its way. With those barriers removed once and for all, though, science
could henceforward take its foreordained course, running under its own steam

and following nothing but its own internal logic. Deviations from the right path were bound to occur and could be studied with much profit; still, in the end scientific truth would prevail. The victorious protagonists of the new science could not but win, because they were closer to the truth about nature than any rivals, who for that very reason were bound to be losers.

As one result, historians of the Scientific Revolution for a long time failed to ask a particular question that 'general' historians find quite natural to ask in the case of, for example, political revolutions. The question is, simply: Once the revolution has got under way, what keeps it going? In political revolutions, a successful outcome is hardly guaranteed until the last moment. But in the case of the Scientific Revolution it had tacitly been assumed by its historians that its successful outcome was foreordained.

Aspects of this new question have been taken up, in a variety of ways, by members of a successor generation of historians of the Scientific Revolution. Much of their best work may be characterized as an ongoing effort *to turn apparent inevitabilities into contingencies.* This might seem to be the historian's ordinary business, yet for no type of historian is such a feat harder to accomplish than for the historian of science since, as I said, it seems so natural to assume that, once the path toward successful science has been cleared, nothing can halt its further advance.

For decades the search for ways and means to make the history of the Scientific Revolution a more contingent affair has been practice rather than program. It was turned into a program in 1985.

A manifesto for the contingent course of science.   Chapter 1 of Shapin and Schaffer's *Leviathan and the Air-Pump* presents itself as a manifesto for the contingent nature of the past course of science. It does so in a very peculiar manner. Inspired by Latour's and Woolgar's 'ethnographic' approach to present-day laboratory life, Shapin and Schaffer set out to accomplish their aim by adopting a 'stranger's perspective' upon the episode they wish to examine: Boyle's and Hobbes' dispute over the air pump in the sixties of the 17th century. They aim to understand what went on there by deliberately placing themselves at the sidelines, rather than sharing from the start the mode of thought of those who came out victorious: the experimentalists. Their chosen vehicle for taking such an outsiders' stance is the adoption of a fully relativist position about Nature, Reality, and Truth, such as has been worked out in the sociology of science by the school of 'social constructivism'.

Shapin and Schaffer achieve a wonderful coherence throughout their account by the happy coincidence that the loser's position they set out to describe with programmatic sympathy—Hobbes'—is social-constructivist *avant la lettre.* In the authors' rendition, Hobbes never missed an opportunity to ascribe his adversaries' ideas to their putative social effects (in most cases, civil war, no less), nor did he believe that disputes about natural phenomena could be settled other than through judication by an undivided, secular power, *"for want of a right reason constituted by nature"* [my italics].[210]

About the connection thus established by Shapin and Schaffer between the objective of a contingent mode of historiography and the social constructivism that must provide the proper vehicle, I wish to assert two things. In the first place, much of the singular appeal their book has exerted of late is, I think, due to this very combination. It holds out the liberation from the pressing bond of "the aura of self-evidence," while simultaneously offering a 'new contextualism' as the means to put the liberation to effect. A long-sought objective seems now to come within reach: to put past science in context in a systematic way.

My second point is that, contrary to what the authors suggest by implication, there is no necessary connection between objective and means at all. The very real freedom gained by examining "messy contengencies" need not be acquired by selling one's soul to relativism; other currencies are available in which to make the deal. The ideal of a more contingent mode of history writing—which I share—need not be bought at the expense of adopting a fully relativist position about nature and reality—which I detest. I do so because (1) no positive argument is ever given for the unlikely claim that nature is nothing but the name the victors assign *a posteriori* to the 'negotiated' outcome of their experiments; (2) although it is surely true that 'artificially produced nature' is man-made in the sense of 'called up through human operations', it hardly fails to manifest itself in reality in the most drastic manner (as one historian of science said, "If Shapin and Schaffer are right, I fail to comprehend how the explosion at Bhopal could have occurred");[211] (3) one principal reason for why science is the best way humanity has yet discovered to gain a grasp of nature is that experiment provides an inexhaustible source of feedback from reality and offers perennial caution against falling back into ideological thought; (4) the social-constructivist position, while heuristically fruitful up to a point, is typically put forward in none but either/or terms.[212]

From the point of view of historiography, my last point is the most important one. If compelled either to grant or to deny *a priori* that natural reality is *nothing but* a social construction, historians may lose sight of the much more fruitful question of the *extent to which,* in a given period, social conventions and circumstances impinged on the establishment of natural facts, or on how 17th-century scientists sought an understanding of their own doings while propagandizing their efforts and seeking to get them accepted by society at large. For example, we have seen in section 3.6.1 how Rabb very carefully put the matter in both/and terms; while acknowledging the intellectual superiority of the new science, he noted at the same time that its very ready acceptance by elites all over Europe constitutes an amazing fact of social history which cries out for explanation. The 'success' of the new science was partly inherent in its unprecedented grasp of nature, partly due to 'external' circumstances, and the art of the historian consists in nothing so much as in striking the proper balance in each particular instance.

The moral as I see it. From these considerations I should like to derive a twofold moral. The first is that much can be gained by putting science in a social context, provided that this is not taken as the sole road to salvation. However, I do not think that social constructivism, even if taken solely as a heuristic device, provides a particularly passable route toward that aim. In my view, the most promising questions concerning the new science of the 17th century considered as a social phenomenon come from merging history of science with social history broadly conceived, as the examples of Butterfield and Rabb and the promise held out by Weber's approach demonstrate.

My second moral is this. Let us go ahead and uncover contingencies wherever we can find them.[213] This need not be confined to developments 'external' to science; 'internal' developments, too, can be analyzed in a manner that may reveal a degree of historical contingency. However, let us not start our search by throwing out the child of the reality of nature together with the bathwater of self-evidence (keeping in mind perhaps Woody Allen's immortal line about a certain protagonist in one of his stories, who "hated reality but realized that it was still the only place to get a good steak").[214]

The fact is that we may adopt a 'stranger's perspective' by virtue of our being historians, and we are under no obligation to select or to justify such a perspective by invoking theories external to the very profession that is devoted to the writing of history, that is, to laying bare not only what happened in history but also what might have happened; what hidden possibilities can be brought to light that never materialized; what real forces prevented their occurrence.[215] My contention is that this is precisely what has been undertaken for some time in the historiography of the Scientific Revolution, too. Results—some still in their infancy—appear to converge toward a provisional conclusion that we may infer from much that has been said in the present chapter.

The Scientific Revolution—a radical break gradually worked out. Nothing in the view that the Scientific Revolution constituted a radical break with previous thought on nature is opposed to the idea that much time and effort were needed for its protagonists gradually to work out its full consequences, and that little in the way this happened was foreordained. This, so it seems to me, is one principal conclusion to emerge from many of our accounts in the preceding sections. In a perceptive yet slightly overstated formulation due to Kuhn:

> Like so many of the other novel attitudes displayed by the 'new philosophy',
> the significant effects of [the] new attitude towards measurement were scarcely
> manifested in the seventeenth century at all.[216]

Thus we have seen how the 17th century was portrayed as a transitional period during which protagonists of the Scientific Revolution came to the bare beginnings of an awareness of the nature of their own methods. They gradually emancipated these from the language of Aristotelian epistemology and worked out the first outlines of a tentative, rather than demonstrative, methodology for the new science. They believed they had liberated themselves from the authority of Aristotle at one stroke, but many remained under the unconscious influence of vital aspects of his thought to a much larger extent than appears from their pronouncements on the topic. A new orientation toward the future, with the ideal of discovery taking the place of recovery, appeared, accompanied by relicts from a bygone period. Cumulation of research findings began to take place, but was still very uneven. The loss of a sense of the unity of nature was acutely felt by many and manifested itself both in the construction of total systems based upon principles of the new science and in efforts to preserve or revive Hermetic patterns of thought. Protagonists of the new science, far from gaining acceptance at one stroke inside the framework of the reigning clientele systems, initiated a fierce propaganda battle against opponents who, in reality, were not quite so inflexible as it suited them to portray these. When acceptance came, it was aided not a little by an urgent desire felt all across Europe for peace and social order. In a process of gradual institutionalization adherents to the new science began to create values, and to work out attendant rules, meant to regulate their conduct both among themselves and toward those outside the domain of science. The nature of experiment, or, more precisely, the question of how experiment mediates between humans and nature, was gradually being sorted out in the course of the century. The scientific instrument as the principal aid to experiment, although introduced in the first decades of the century, needed several decades more to acquire a 'research context' around it. Even though almost at once the new science put itself under the aegis of its beneficial potential for man and society, achievement on this score failed as yet to match the original promise. Finally, and this is an important element of contingency adopted from our treatment in the previous chapter, the Scientific Revolution took shape originally in two distinct currents of thought, the mathematical and the mechanicist, whose mutual tensions had to be resolved in the Newtonian synthesis before a fully successful outcome could be said to have been attained.

In short, the overall picture of the Scientific Revolution that arises from these discussions marks the contingent nature of many of the ways in which the original insights of its pioneers were worked out and began to function in society at large. We find no straight path but rather a jumble of hesitant tryouts, mixed forms, and halfway positions. The jumble does display a pattern of overall progress to be sure, but only in the crooked and contingent mode that marks so much of human life anyway.

England and the Continent. Our summing-up suggests a somewhat greater homogeneity of results attained in the present chapter than is warranted by the vagaries of historiography over the past decades. Although nominally directed toward the 'Scientific Revolution' without exception, some of our accounts appeared to confine themselves in effect, if not always in intended scope, to events in England while ignoring the Continent more or less entirely. How crass such distortions may become at times can be illustrated from the example of J. R. Jacob and M. C. Jacob's attempted linkage of what they call the 'Whig Constitution' to late 17th-century science in England. Their claim to present a thesis about "the Anglican origins of modern science" was established by subjecting to treatment an entity especially created for the purpose: "the Scientific Revolution from Boyle to Newton."[217]

Causes of this rather novel historiographical habit need not be investigated here in depth. One readily thinks of the language barrier; of what Ian Hacking once called "the sceptred-isle version of the scientific revolution";[218] of how hard it is to make general statements about so widely flung an event as the Scientific Revolution undoubtedly was; of the historically contingent fact that England, if surely not the birthplace of the new science, equally assuredly was the country where the event culminated. Leaving diagnosis as well as remedy apart, I confine myself to noting that many facets of the Scientific Revolution in process of being investigated with much promise inside the framework of recent historiographical trends are not to attain their full fertility unless addressed to the event in its full geographical and temporal setting.

Having now rounded off our discussion of novel trends in recent historiography of the Scientific Revolution, with efforts at ferreting out elements of contingency being stressed in particular, I wish to end the present chapter by calling attention to what is surely the greatest contingency of all: the, on reflection, astounding fact that a Scientific Revolution ever did occur in the first place.

Einstein's famous letter and its meaning. Most historians of science have at one time or another come across a laconic response Einstein once gave to a query put before him by a certain Mr. Switzer:

> Dear Sir, Development of Western Science is based on two great achievements, the invention of the formal logical system (in Euclidean geometry) by the Greek philosophers, and the discovery of the possibility to find out causal relationships by systematic experiment (Renaissance). In my opinion one has not to be astonished that the Chinese sages have not made these steps. The astonishing thing is that these discoveries were made at all.
>
> Sincerely yours, Albert Einstein[219]

It is customary among historians of science, I believe, to take this spirited remark at a titillating reminder of some deep truth about the history of science that we tend to forget amid our everyday doings, but about which no more can, or need, be said. I am not aware of any printed discussion of the question that follows at once from Einstein's point: What, then, is it in human thought and action that makes the birth of early modern science such an astonishing event rather than one that was bound to occur sooner or later?

However, there exists at least one unpublished treatise which, in one of its chapters, takes as its central problem

> how, against all the usual societal odds, the historical 'miracle of science' could have come about at all. . . . In view of the societal forces against it, and in view of the possibilities afforded to the human mind to solve contradictions between objective insights and subjective desirabilities by rationalizing the contradictions out of apparent existence, the emergence of science as the recognized systematic pursuit of objective knowledge may indeed seem a miracle.[220]

This treatise, entitled "The Conquest of Bias in the Human Sciences" and still fully deserving to be published in my opinion, was written in 1976 by the Dutch social scientist R. Wentholt. He addresses the problem of why the 'human sciences' have not succeeded so far in what the natural sciences have attained since their emergence in modern shape in the 17th century: a pool of commonly shared, highly unified knowledge. Here is how Wentholt portrays the very same difference between the natural and the human sciences that we have earlier seen analyzed by Kuhn (section 3.2.3):

> The most outstanding difference between the sciences of nature and those of man is that the independent pursuit of knowledge, regardless of other considerations, has paid off handsomely in the natural sciences and not yet at all in the human sciences. In the search for the understanding of human actions nothing comparable to the discovery of laws controlling the phenomena of the natural world has been grasped. There is a wealth of empirical facts but there are no general agreed-on principles of ordering the facts for scientific usage; there are theories galore, but no one comprehensive scientific theory for human science. No basic regularities useful as major explanatory principles have been discovered, and there is no agreement that such principles can be discovered. There is a great deal of insight and knowledge, but as far as the cumulative effects of all these insights and knowledge is concerned, the situation is not basically different from before the rise of [early modern] science.[221]

Thus we are reminded that the overall state of the human sciences at present is structurally equal to the state of the natural sciences prior to the Scientific Revolution. A large part of Wentholt's subsequent argument is designed to demonstrate that the subject matter of "human science," which coincides with its practitioners, entails difficulties not to be met at all, or to a smaller extent, in natural science. Yet some of the difficulties are quite the same in either case, so altogether it is not by chance that it has taken so long for the natural sciences to attain the state the human sciences now envy them. Or rather, the common obstacles standing in their way were so powerful that it is rather a miracle that they were ever overcome in the case of the natural sciences.

Among these shared obstacles Wentholt analyzes in detail the proclivity of the human mind to distort the contemplation of reality through the agency of material interests, of immaterial, ego-centered interests, of forces of group pressure, and of the rewards that come from giving in to it, etc. He analyzes such mental biases, not as shorthand terms from everyday speech, but rather as concepts arising from an elaborate theory of human motivation nourished by thoroughly processed, pertinent literature in the psychology of the human mind and the sociology of group behavior. There is no need to discuss his ideas here further, for my point has already been made. This is that the 'miracle of the birth of early modern science' can be treated as any historical 'miracle' can: by applying all the pertinent conceptual apparatus we can muster, while acknowledging in the end that, in having reached a deeper plane of understanding, the miracle has become even more miraculous. Once we begin to understand what propensities of the human mind stood in the way of the by and large objective mode of acquiring knowledge that we call science, we become even more alert to the wisdom expressed in Einstein's words: "The astonishing thing is that these discoveries were made at all."

Which, inevitably, raises the question of how, if indeed at all, this miraculous event—the Scientific Revolution—can be *explained*. Scores of historians have set out to explain it; we are ready to give them a hearing and proceed to do so in Part II, which follows now.

Part Two

The Search for Causes of the Scientific Revolution

⊰ FOUR ⊱

THE EMERGENCE OF EARLY MODERN SCIENCE FROM PREVIOUS WESTERN THOUGHT ON NATURE

4.1. INTRODUCTION TO PART II

WHAT MAY COUNT AS AN 'EXPLANATION' of the Scientific Revolution? What does it mean to say that historical event P or historical state of affairs Q or any combination of such P's and Q's, 'caused' the birth of early modern science?

Seeking an answer to this question is perfectly superfluous so long as we confine our job to summing up in a proper order a selection of accounts of the Scientific Revolution intended by their authors to serve explanatory purposes. Once we come to a critical and comparative assessment of those purported explanations however, we can no longer avoid a confrontation with issues like: What is it that theory A actually has explained? To what extent may causal connection B be taken to do more than just couple a number of historical facts? Do we know now, after examining explanation C, how the Scientific Revolution actually came about? Or is that too much to ask from any given historical explanation? If so, what may we reasonably expect a historical explanation to accomplish? And should perhaps different types of historical causation be distinguished here?

In principle, two ways stand open to approach such issues. One is to start our overview of attempts at explanation of the Scientific Revolution with a brief theoretical treatise in which metahistorical debates are winnowed in order to find the best, or at least the most up-to-date, account of what, for philosophers of history, counts as a historical explanation. We could then use the result thus obtained as a yardstick by which to measure the various explanations successively under discussion. Current metahistorical debates being subject, however, to perennial disagreement over fundamentals and to fleeting fashions, with specific requirements from the history of science being left out of the ongoing debates entirely, a more inductive procedure is likely to yield results more cognate to our set of purported explanations.[1]

To begin with, the sheer panorama of explanations that follow is likely to elicit in the reader a more or less intuitive feeling for the sort of things an

239

explanation of the Scientific Revolution may reasonably be expected to accomplish. As we go along, I shall attempt to articulate such intuitions as the case under discussion may require. Finally, not all those authors who set out to explain the Scientific Revolution omitted to pause and ask themselves what the inherent scope and limitations of such an enterprise might be. We shall in due time come across the metatheoretical notions these admittedly rare historians developed in the course of devising their own explanations. Together with a few additional ideas to be taken up from the metahistorical literature, these three sources of metahistorical insight will stand us in good stead when it comes to making our final assessment.

Bringing order to a wide variety of explanations.　Given the wide spectrum they cover, any attempt to organize the multitude of explanations of the Scientific Revolution under suitable headings may be contested for a certain degree of arbitrariness. Most explanations stand entirely on their own, whereas a few were put forward in tandem with efforts to define the nature of the Scientific Revolution, making it on occasion hard to distinguish them effectively from those definitions. Some explanations take as their point of departure fresh results of ongoing research on the nature of the Scientific Revolution, whereas others rather ignore what earlier historians of science have had to say on the topic and start instead from unreflected notions of their own on what the Scientific Revolution was about, or from conceptions long ago discarded by responsible historians of science. Some explanations are purposely monocausal, in that one particular preceding event in history is assigned to bear the full responsibility for the emergence of early modern science, whereas others consist of a more cautious enumeration of a variety of circumstances which, in coming together, made the Scientific Revolution possible. Some explanations are in effect confined to one or another center of 17th-century science, whereas others aim to cover the event in its full geographical scope. Most explanations were meant to show in a straightforward manner how it is that early modern science came about in 17th-century western Europe, with some others seeking to explain its emergence there and then by examining what it was that prevented the birth of such a science elsewhere and/or in other periods. Some explanations appear to be chiefly or wholly limited to preceding elements in the history of thought on nature, whereas others cast about for causes of the Scientific Revolution in the domains of religion, politics, the economy, technology, the arts and crafts, or society at large.

　　To arrange all this in chronological order of appearance might make sense only if a lively debate had gone on among the propounders of these varying accounts—a debate in the sense of successively building on critically assessed results attained by predecessors. As pointed out in our introductory chapter, however, most purported explanations are striking for their 'monological' character. The debate does not in fact exist but dissolves in an array of separately conducted monologues (some of these, to be sure, being fitted out with their own, rather limited subdebates).

Given this peculiar historiographical situation, the explanatory efforts now facing us can best be put under three headings. We discuss in succession explanations aimed at showing that the emergence of early modern science was intimately bound up (1) with elements in scientific thought that developed in Western civilization from its very inception up to the 17th century or (2) with events and/or states of affairs to be discerned in the history of the West prior to the 17th century; finally, we discuss (3) efforts to throw light on the emergence of early modern science in the West by pointing at modes of thought and life in other (that is, in effect, Eastern) civilizations which prevented the birth of early modern science outside western Europe. In other words, we divide our explanations into 'internal' (this chapter) and 'external' ones, and subdivide the mostly 'external' explanations according to whether they take the 'Western' (ch. 5) or the 'Eastern' (ch. 6) approach. A few borderline cases and somewhat mixed bags inevitably remain, yet by following this route historiographical patterns come into view that would otherwise elude us.[2]

The topics of the present chapter. Hence, the present chapter is devoted to explanations of the origins of early modern science by means of the discussion of features of previous modes of thought on nature in Western civilization. Analyses that come under this heading took shape in either of two distinct formats. Some aimed at showing why the Scientific Revolution did *not* take place during an earlier stage of Western thought, whereas others sought to demonstrate that early modern science emerged as a more or less logical sequel to immediately preceding modes of thought on nature, that is to say, as little more than the obvious next stage in the ongoing development of scientific thought. The former type of analysis naturally dominates in accounts of Greek and, to a lesser extent, of medieval science. The latter type is characteristic of various attempts at explaining the Scientific Revolution as a sequel to the 15th-and 16th-century recovery of ancient learning—with the rise of Copernicanism, the rediscovery of Greek mathematical texts, the reform of Aristotelianism, the appearance of Hermetic patterns of thought, and the revival of scepticism, respectively, being put forward as possible causes. We start our survey with the question of why Greek science, in coming to an eventual standstill, failed to usher in the Scientific Revolution.

4.2. WHY DID THE SCIENTIFIC REVOLUTION NOT TAKE PLACE IN ANCIENT GREECE?

On the one hand, the Greeks in their grandiose zest for scientific knowledge made stunning advances in conquering and enhancing their understanding of the

world and of man, and, on the other, halted in view of many an insight lying
right on their own path, just as if they were running up against some invisible
glass wall, just as if some mysterious taboo forbade them the next step.

—Wolfgang Schadewaldt, 1957[3]

Why did ancient Greek civilization fail to bring forth the Scientific Revolution?

Put this way, the question seems grossly unfair. Greek science, starting
from scratch in many respects, accomplished so incredibly much in the course
of only a few generations that it seems quite unreasonable to require it, by
hindsight, to have accomplished even more. Why should the Greeks, on top of
an axiomatized geometry, of a sophisticated and highly successful predictive
mathematical astronomy, of a variety of theories of matter, and of a plethora
of fruitful ideas on a huge range of other scientific topics, also have made the
seemingly next step of producing more than incidentally such mathematical
laws of nature, experimentally confirmed, as have in fact culminated in the
Newtonian synthesis? On the other hand, the question, apparently so ungrate-
ful to what the Greeks actually did accomplish, nonetheless imposes itself for
the very reason that, in retrospect, they came so close. And it is precisely because they came so close that the issue of why the Scientific Revolution eluded
the Greeks can help clarify the problem that concerns us here of what it was
that made the Scientific Revolution possible. What, in other words, was absent
from Greek science that was indispensable for bringing forth early modern
science? We feel in our bones that, in many respects, the Greeks were 'on the
right track'; what precisely is it then that was nevertheless lacking?

Answers to this question have, without exception, come forward in the
format of chapters or parts thereof rather than in book-length treatments. The
books in which these discussions occur are devoted either to more or less the
entire history of science (Whewell, Dijksterhuis, Hooykaas), to one aspect of
the history of science as a whole (Ben-David), or to a specialized outline of
ancient science (Farrington, Clagett, Sambursky, Lloyd).[4] Discussions may be
categorized under two headings. One group of historians, while duly acknowledging the splendor of the Greek achievement overall, focused on the 'shortcomings' of Greek science: Many of the required ingredients were there, but
not their proper mix or their proper setting. Others, while acknowledging those
shortcomings to some extent, rather emphasized how close some exceptional
Greek thinkers nonetheless came to what we tend to regard as the 'right approach'. For the former group the problem was to determine what changes in
thought were needed to strike the right balance between the ingredients of
science and/or to provide the needed setting; for the latter, to find out what
subsequent changes in the intellectual and/or social environment of science
were effective in removing the "invisible glass wall" indicated by Schadewaldt
in the motto to the present section.

The first group consists of Whewell, Dijksterhuis, Hooykaas, and Sambur-sky; the second, of Farrington and Clagett (with a note by Koyré). We shall outline their respective views in the two sections that follow. Once having gained an overview of these, we proceed to discuss two other, more recent approaches: one, by Lloyd, to find out with greater rigor than applied before when exactly the decline of Greek science ought to be dated, and another, by Ben-David, to turn upside down one basic presupposition held in common by all other authors.

4.2.1. *Some Principal Shortcomings of Greek Science*

Whewell and the Greek failure to combine Ideas and Facts at the proper level. To William Whewell the problem presented itself in the following guise: How is it that the scientific enterprise undertaken by the Greeks, with their unique interest in the rational interpretation of nature, nevertheless culminated in the radically wrong natural philosophy of Aristotle? Not sur-prisingly, his answer derives in a straightforward manner from his own ideas on what is necessary for science to be cultivated in a productive, cumulative, and generally fruitful way. Recall from section 2.2.2 that, to achieve this, Whewell found indispensable the proper combination of Ideas and Facts, in the sense of a suitable Explication of Conceptions together with the Colliga-tion, by means of these Conceptions, of the Facts at hand. In his view this is precisely what the Greeks had mostly failed to accomplish. Whewell uses Herodotus' effort at explaining the seasonal overflowing of the Nile as an example to show how, as a rule, the Greeks tended to feel satisfied with explanations by means of ideas situated at a very high level of abstrac-tion and insufficiently explicated to make it even possible to combine them with the facts under investigation. The reverse side of this coin was that the Greeks only very rarely undertook a sufficiently patient investigation to find out all facts pertinent to the case at hand. As soon as the Greek 'specula-tors'

> had introduced into their philosophy any abstract and general conceptions, they proceeded to scrutinize these by the internal light of the mind alone, without any longer looking abroad into the world of sense. . . . They ought to have reformed and fixed their usual conceptions by Observations; they only analysed and ex-panded them by Reflection: they ought to have sought by trial, among the No-tions which passed through their minds, some one which admitted of exact appli-cation to Facts; they selected arbitrarily, and, consequently, erroneously, the Notions according to which Facts should be assembled and arranged: they ought to have collected clear Fundamental Ideas from the world of things by *inductive* acts of thought; they only derived results by *Deduction* from one or other of their familiar Conceptions.[5]

Here, then, lies the ultimate cause for why

> we cannot consider otherwise than as an utter failure, an endeavor to discover
> the causes of things, of which the most complete results are the Aristotelian
> physical treatises; and which, after reaching the point which these treatises mark,
> left the human mind to remain stationary, at any rate on all such subjects, for
> nearly two thousand years.[6]

Implicit in Whewell's diagnosis is the idea than an upper limit is set to what
efforts at understanding nature can accomplish in the absence of the proper
method; that this upper limit is situated fairly low; and that the path toward a
truly cumulative science is doomed to remain blocked until the proper way to
combine Ideas and Facts is discovered. In this view, the subsequent stagnation
was principally one of the mind, and finding the causes of the birth of early
modern science comes down to identifying the habits of thought that, throughout many centuries, tended to block the finding of the key that, in the 16th and
17th centuries, was to unlock the secret of what is the proper level at which to
combine the right Ideas with the pertinent Facts.

In this type of analysis, defining the nature of the Scientific Revolution
and explaining its origin almost seem to merge into one unified argument. The
essential feature of the Scientific Revolution was such-and-such; therefore the
root cause of the Scientific Revolution was the historical appearance of such-
and-such, and the failure of early modern science to manifest itself at any earlier date than it actually did is due to the absence of such-and-such at that
earlier date. Hence the pictures of science before and of science after the revolution become mirror images of one another. Apparently we are caught in a
fairly narrow circle. Whewell makes no effort either to widen it (for example,
by invoking explanatory factors outside the realm of scientific thought) or to
break out of it entirely. In creating such a circle, he set a pattern of explanation
that we are to encounter time and again.

In another sense, too, Whewell's account of the 'failure' of Greek science
set the tone for much subsequent discussion. It is striking that he took the
Revolution in Science as an event bound to happen in any case, whether earlier
or later, so that what separated Greek from early modern science was nothing
but an array of impediments successively to be removed. Although, as we shall
see, the question of why Greek science did not usher in the Scientific Revolution is not necessarily bound up with regarding the latter as the manifest destiny of the former, some of our historians will appear to have phrased the problem more or less that way, whereas at least one quite consciously did not. And
there is a third and final point made by Whewell that was to be taken up by
later historians. Even though Whewell's specific explanation of the Greek failure found as little favor as befell the overall conception of science from which
he derived it, a generalized version of his explanation has been adopted by
more than one later historian who set out to diagnose what was wrong with

Greek science. The verdict that Greek science suffered from an overdose of rash generalizations at the expense of a careful scrutiny, whether experimental or observational, of the relevant facts in their proper setting was also reached by Dijksterhuis and by Hooykaas.

Dijksterhuis and Hooykaas on Greek science. Part I of Dijksterhuis' *Mechanization* is devoted entirely to a thoughtful and overall admiring character sketch of Greek science. He does not attempt to reduce to one root cause a variety of shortcomings diagnosed by him along the way. Among these stand out the Platonic purity extolled by the geometers, which stood in the way of both a search for an applied mathematics and the treatment of variable magnitudes; the absence of a fruitful interaction between science and technology; and, finally, the defect that he stresses more than any other:

> the underestimation of the difficulties of the investigation of nature that characterizes Greek thinkers generally. No matter whether they adopted a more or a less empirical stance toward nature, *without a single exception* they overrated the power of unchecked, speculative thought in the natural sciences. They had no inkling of the tough, laborious effort—often getting lost in seemingly pointless detail—required for beginning to gain an understanding of nature. [My italics][7]

Dijksterhuis declined in so many words to disentangle what, in this list of shortcomings, was to be named 'cause' and what 'effect',[8] since the two seemed inextricably intertwined—a mental attitude that made him rather agnostic with respect to the explanation of the emergence of early modern science, too. To the extent that he did seek for causes, these were, naturally, counterparts to the shortcomings outlined above. The enrichment Greek mathematics was to acquire through Indian position-reckoning, Arab algebra, and Renaissance Italy's computing techniques; the fruitful interaction between science and technology that manifested itself in the course of the Renaissance; a gradually dawning awareness that the secrets of nature can only be revealed by combining ceaseless labor with daring speculation—all this had to come together to make early modern science possible; and Dijksterhuis felt certain that no enumeration of such elements could ever exhaust the list of possible causes.[9]

In this respect his countryman R. Hooykaas has been less agnostic. He found that, besides its many admirable traits, two fundamental shortcomings characterized Greek science. One of these concerns, once again, the lack of balance between reason and experience. Hooykaas gives the argument a twist of his own in characterizing such lack of balance as a case of intellectual arrogance. Far too easily were thought-constructions reflecting what humans at the time found reasonable used as constraints on nature itself: Nature *had* to be such-and-such, because otherwise it would run counter to *man's* view of what is rational. Thus the tropics were taken to be uninhabitable, for no other reason than that Aristotle's and Ptolemy's reason deemed the opposite inconceivable. This general posture, according to Hooykaas, is what made Greek science ultimately stagnant.[10] It follows from this that, as long as the inheritors of Greek science persisted in holding their reason to be superior to the lessons the facts of nature can teach us, no early modern science could arise. A more humble attitude toward what nature has to tell us was necessary, and the three partial causes Hooykaas went on to assign to the Scientific Revolution—Parisian Terminism, the Voyages of Discovery, and certain elements in the Reformation—were all events or developments in European history that fostered such an attitude. We shall encounter these in due course as we proceed, confining ourselves for now to recalling from section 3.4.1 the other principal shortcoming identified by Hooykaas in Greek science. This is that it reflected a conception of the world as an organism, whereas early modern experiment requires a mechanicist, in the sense of machinelike, world-view.

The root cause of the 'failure' of Greek science has been situated at this same point by the Israeli physicist and historian of ancient science S. Sambursky.

Sambursky and the organicist setting of Greek scientific thought.   Sambursky's book *The Physical World of the Greeks* was first published in Hebrew in 1954 and appeared in English in 1956.[11] Altogether the author is not very impressed with the overall accomplishment of Greek science, despite many striking similarities to what we, today, call science. In his view—forcefully set forth in his final chapter, which is entirely devoted to the problem at hand—the mathematization of science as achieved in the first decades of the 17th century, the experimental treatment of natural phenomena, the emancipation from philosophy, and the application of science for useful purposes, together constitute a coherent aggregate and share the property of *artificiality.* Presupposed in the early modern approach is the possibility of isolating pieces of nature without concern for the whole. This procedure implies a new conception of nature, peculiar to early modern science, "to include all phenomena whose existence is not ruled out by the laws governing the physical world."[12] Such an extended conception of nature, in which the artificial fits at least as well as the directly given, is incompatible with the world-view of the Greeks, which was organic rather than mechanical and hence precluded forever mathematization and experimentation in the area of terrestrial physics. As long as the cosmos is regarded as "a living organism, a body that can be understood and comprehended *in its entirety,*" and as long as there is "a profound awareness of the unity of man and the cosmos," what Bacon was to call "the dissection of nature" remains unthinkable.[13] Thus, Sambursky concludes, the crucial transition to comprehend is how this mythological attachment to a living cosmos could be transformed into the logical, modern attitude toward these things.

Having come thus far with his diagnosis, Sambursky seeks an illuminating yet all-too-rash way out in a two-sentence analysis, as follows. The agency he holds chiefly responsible for the transition in world-view is

> the influence of Christianity and the organized Church. By divorcing man and his vital interests from natural phenomena the Church helped to create the feeling that the cosmos was something alien and remote from man. It was this feeling that prepared men's minds for the next stage in which the investigator faced nature as its dissector and conqueror and thereby ushered in our own scientific era which still, after four centuries, retains its vigour undimmed.[14]

We are not told how the Church accomplished this fundamental change; we only learn that the process took place during the long period, including the Middle Ages and the Renaissance, when (so Sambursky took for granted) the heritage of Greek science remained immobilized in barren scholasticism.

4.2.2. Help Needed in Crossing the Threshold

We now pass on to those historians who believed that in the course of its development Greek science advanced so far that, in retrospect, the Scientific Revolution seems to lie just around the corner. They did not think, in particular, that there was anything inherent in Greek science to prevent the body of work of Archimedes and his Alexandrian contemporaries (by historians' consensus the scientists whose approach comes closest to the spirit of early modern science) from ushering in the Scientific Revolution.

Once the problem of Greek science is framed in such terms, the matter acquires particular urgency as soon as we realize that, in the Great Tradition, the printing of Archimedes' work during the Renaissance was considered to be a most powerful impulse toward the mathematization of nature held to be the primary characteristic of the ensuing Revolution in Science. Hence, inside the overall issue of the decline of Greek science there is a special problem, which we may call the 'Archimedean problem'. It may be phrased as follows: If the printing of Archimedes' works in the course of the 16th century was so instrumental to the emergence of early modern science, why, then, was the *original* impact of Archimedes' work, in the second century B.C., along with that of some contemporary and later Alexandrian scientists, apparently *not* sufficient for making the required transition? To my knowledge, only one historian cared to address the question in roughly such a guise. This was Alexandre Koyré. In his blunt, one-line answer it is in so many words affirmed of Archimedes *cum suis* that "by itself nothing is opposed to Copernicus and Galileo having succeeded them directly."[15]

This answer has remained unique in the literature. To the extent that other commentators have faced the Archimedean problem they have mostly done so in the context of the revival of ancient learning during the Renaissance, and we shall follow them in doing so (section 4.4.2). Meanwhile we proceed to an examination of what external factors were invoked by two specialized historians of Greek science—Farrington and Clagett—to explain why the very transition for which, *qua* content, Greek science seemed ready persistently eluded it.

Farrington and the fate of science in a slave society. Benjamin Farrington, an English historian of mildly Marxist persuasion, wrote several overviews of ancient science, in which the problem of its 'failure' is both cogently formulated and answered by means of reasoned argument. In the following account I take together considerations to be found in his books *Science in Antiquity* (1936) and *Greek Science* (1949).[16] Given that a new science appeared in the 16th century that was little but a continuation of the Greek heritage, so Farrington starts his argument, the problem arises of understanding "why did Greek science die if it had still such vitality that it was capable of a second birth?"[17] In his view stagnation had occurred relatively early, right after Archimedes' time, so that in the 1st century A.D., when Ptolemy and Galen flourished, the Greeks and Romans had already been "loitering on the threshold of the modern world for four hundred years" and, by that very fact, "had indeed demonstrated conclusively their inability to cross it."[18]

An odd paradox arises as soon as one realizes that such late 16th-century scientists as Stevin and Galileo, on encountering the very same threshold, were able to cross it at once. This paradox can only be resolved, Farrington asserts, by observing that, in ancient Greece, science failed to fulfill the social functions that have characterized early modern science ever since Francis Bacon proclaimed the ideal of the dominion over nature by man through the application of science. It was the fate of Greek science to function in a slave society; hence its becoming a pastime of a contemplative leisure class devoid of interest in advancing technology, let alone in bringing science into productive collaboration with arts and technologies that were little valued anyway.

Thus the fundamental brake upon the further progress of science in antiquity was slave labor. In a society in which all labor is done by slaves, and in which science is a purely contemplative occupation, no meaningful combination of theory and practice is possible. In contrast, two basic elements made for a fundamentally altered situation when, in the middle of the 16th century, the Greek heritage hit western Europe. One was the upheaval in technology that had in the meantime occurred in the Middle Ages. Citing Lynn White's pioneering studies in medieval technology, Farrington observes that medieval western Europe provides the first case in history of "a complex civilization which rested not on the backs of sweating slaves or coolies but primarily on non-human power."[19] This paved the way for Bacon's vision as well as for the reality of a science vitalized by its application to pressing, practical problems. The other novel element that contributed crucially to the changed atmosphere in which the heritage of Greek science was received in western Europe is the biblical world-view. This entailed a more positive appreciation of labor, of the arts, and of the possibility of the amelioration of man's future fate generally. Farrington underscores how deeply this attitude contrasted with the Greek mentality by quoting such passages from Bacon as:

> In things artificial Nature takes orders from Man. Without Man such things would never have been made. By the agency of Man a new aspect of things, a new universe, comes into view.[20]

He goes on to point out how Bacon derives this altogether new conception of the task and social function of science from God's exhortation to Adam to go forth and subject nature to his will.

Altogether, Farrington's argument consists of four successive steps. (1) Taking into account its content and methods only, Greek science from the 2nd century B.C. onward was ready for the Scientific Revolution. (2) The social environment of slavery, however, doomed ancient science to the ensuing stagnation that came from a lack of practical application and from the absence of fruitful interaction with contemporary technology. (3) When in western Europe, in the middle of the 16th century, Greek science was recovered, it found itself surrounded by a fresher atmosphere of free labor, where it was vitalized both by the achievements of medieval technology and by an optimistic, active world-view derived from the Bible. (4) In these new surroundings, "the seed" of ancient science could finally "raise a wholesome crop."[21]

Clagett and the leveling-off of Greek science.   Farrington's view that, intrinsically, Greek science had advanced far enough to allow a smooth transition to the next stage, the Scientific Revolution, has been shared by the American historian of Greek and medieval science Marshall Clagett. In his survey from 1955, *Greek Science in Antiquity,* he noted that the Greek scientific corpus served as the starting point, not only for major advances in science made during the Middle Ages, but also for

> the most fertile scientific activity in the sixteenth and seventeenth centuries. Thus Vesalius started from Galen and Benedetti and Galileo from Archimedes and Apollonius.[22]

After all, Greek science left much room for experimental activity, if we consider

> that a mathematical-experimental science existed in nascent form, at least, in optics, in statics, and in applied mechanics; that a mathematical-observational science was present in astronomy; and that an experimental science existed in zoology and physiology.[23]

Hence it seems obvious to ask "why it was that Greek science falls short of modern science."[24] Clagett's answer is that the techniques employed in these various sciences had not yet so expanded as to become the common property, felt to be indispensable, of all sciences. Before this stage could have been reached (which Clagett appears to regard as the inherently logical sequel), Greek science began, not so much to decline, but rather to level off. Causes of this leveling-off were the dominion of the Romans, the rise of Christianity (which absorbed many potentially scientific minds), and the spread of 'spiritual forces'. Among the last, Clagett mentions by way of one important example the Hermetic corpus, with its dependence on revelation rather than on the rational examination of nature and its "appeal to supernatural causation and to the continuous and direct influence of supernatural on natural phenomena."[25] It is ironic that, within ten years of the appearance of Clagett's book, the very same Hermetic corpus was to be heralded as the prime source of inspiration for those neo-Platonic Renaissance thinkers whose magical theories and activities were now said to have been instrumental in ushering in the Scientific Revolution (section 4.4.4).

4.2.3. The Problem of Decline

Looking back upon our survey of historians' thoughts on what, if anything, was lacking in and around Greek science as compared to the onset of the Scientific Revolution, we note a great variety of partly overlapping interpretations reached along many different lines of attack, and also one great dividing line: Was Greek science, at the end of its flourishing period, intrinsically ready for the Scientific Revolution, or was it not?

In the view of those who thought not, a fundamental reorientation of human thinking about nature was still required. Such a reorientation could be accomplished either through a transition from an organic to a machinelike picture of the world (Sambursky, Hooykaas) or through a reduction of abstract speculation in favor of a greater respect for empirical data. The latter type of reorientation could take place either through an alteration left unexplained (Whewell) or through a gradually dawning awareness amid an endless range of other, favorable changes (Dijksterhuis) or through three definite agents characteristic of developments in western Europe (Hooykaas).

Those who thought that, *qua* content, Greek science was indeed ready for a Scientific Revolution were of course aware that the sequel had nonetheless not manifested itself. They ascribed this to nothing at all (Koyré), to a leveling-off owing to changes in the political and spiritual climate (Clagett), or to the lack of a social function for science in a society based on slavery (Farrington).

Hence, what all conceptions save Koyré's have in common is that they posit chances for improvement in novel surroundings: The corpus of Greek science stood to benefit from a cultural transplantation, so to speak.

By the vagaries of subsequent history not one but two such transplantations have happened to Greek science. It is striking that, even though most of our historians of ancient science and the Scientific Revolution at least mentioned isolated examples of the expansion the Greek heritage underwent in Islamic civilization, they have been unanimous in forgoing the opportunity to check their overall conclusions about Greek science against its fate during the centuries when it was cultivated in Baghdad, Maragheh, Toledo, and many other centers of 8th- to 15th-century learning. When, in section 6.2, we come to a systematic discussion of why Islamic science (i.e., Greek science enriched by many novel elements) did not produce early modern science, we shall find that the converse is also true: None of the few experts of Islamic science to address the issue at all has taken note of the debate over Greek science that we are in process of surveying here.

This is the more to be regretted because intellectual cross-fertilization along such lines might help elucidate in particular the issue of the 'glass wall', or, in other words, the question of whether, indeed, there is only a very thin dividing line between ancient Greek and early modern science. Is it true to say that Stevin and Benedetti and Galileo went on at exactly the point where Archimedes *cum suis* had left off, as was more or less explicitly claimed by Koyré and Farrington and Clagett?

In the literature, the 'threshold', or the 'glass wall', has so far been posited rather than demonstrated in much detail. A systematic comparison between the corpus of Greek science and the state of west European science at the onset of the Scientific Revolution all across the board might help clarify matters here. In the present context I wish to do only two things: first and very briefly, to adduce some fresh empirical evidence in addition to what Farrington and Clagett, in particular, put forward in favor of the general idea of the 'glass wall' and, second, to suggest that there was a 'glass wall' in some sectors of science but not in others.

My piece of evidence concerns the science of music. The discovery of a neat correspondence between consonant intervals and ratios of the first few integers (with the octave being produced by strings with lengths in the ratio of 1:2; the fifth by 2:3; etc.) goes back to the Pythagoreans, and the resulting problem of why this is so was subject to much theorizing up to Euclid and Ptolemy and beyond. The idea that sound propagates by means of wavelike patterns goes back to the Stoics and provided one of two standard explanations of the phenomenon (the other being an emission theory of sound particles). Throughout antiquity it occurred to no one to apply the idea about sound to the problem of consonance. Without any preparation in any of the literature (including that of the best musical theorists of the Renaissance) Benedetti around 1563 advanced a theory in which the ancient wave analogy was linked up, in a quantitative fashion, with the explanation of consonance, thus ushering in a major transformation of the problem (which was subsequently put on a new, physical footing instead of the traditional, arithmetical one).[26] If one has searched in vain the works of Ptolemy and Boethius and many other ancient theorists for such a linking and then finds it casually expounded in a forty-line paragraph by Benedetti, one is hard put indeed not to begin to think in terms of a 'glass wall' separating Greek science from the onset of the Scientific Revolution.

The very nature of this piece of historical evidence also suggests its less-than-general validity. Could it not be that we may fruitfully invoke here once again Kuhn's distinction between the classical and the Baconian sciences, and come up with the compromise solution that a thin but as yet uncrossable dividing line separated Greek science in the Archimedean tradition from its seemingly logical sequel, whereas a far deeper chasm separated Greek science from almost everything that smacks of the artificial, the patiently empirical, and *a fortiori* the experimental?

One other way to untie somewhat further the knot at issue here is to delve a little more deeply below the surface of the notion of *decline*. Whether one describes what eventually happened to ancient science in terms of decay or stagnation or leveling-off, the fact of the decline of a once flourishing body of scientific pursuit is indisputable. All our commentators acknowledge it; each explains it in his own way. Thus, Greek science was bound to decline because of the lack of balance between reason and experience, because of its organicist setting, or because of its functioning in a slave society, or it happened to decline with the advent of the Romans and of Christianity, etc. Clearly, not all explanations fit the same period of decline, which, for example, is set quite early by Whewell and Koyré and Farrington and Clagett, in particular.

In short, virtually all our commentators assign a rough dating to the decline, and those datings are not the same in every case; also, all commentators without exception seek a special explanation for the phenomenon of decline. All post-Whewell discussions of Greek science treated so far were produced during the thirties through fifties of our century; in the early seventies two works appeared in which the problem of decline was set on a novel footing. The least revolutionary of these was by the British historian of Greek science, G.E.R. Lloyd, who made a patient effort to assign a less impressionist dating to the decline of ancient Greek science.

Lloyd and the dating of the decline of Greek science. Like every successive thinker on the topic of the decline of ancient science before him, Lloyd, too, started from scratch without entering into debate with his predecessors. In the final chapter, "The Decline of Ancient Science," of his 1973 book, *Greek Science after Aristotle,* he pointed at the small number of scientists involved throughout the entire period of ancient science. The relative nature of the problem of decline is thrown into relief once one allows oneself to become fully aware "that it was only in the fourth and third centuries B.C. that the scientists of whom [it] could be claimed [that they made an important original contribution to scientific thought] numbered more than a handful in any one generation." [27] Particularly in view of such small numbers Lloyd finds it mistaken to date decline prior to the appearance of two scientists of the stature of Ptolemy and Galen (both 1st century A.D.). For the ensuing period Lloyd carefully examines events in three distinct clusters of scientific pursuit: natural philosophy, including cosmology; mathematics and astronomy; and biology and medicine. Naturally he finds performance somewhat uneven (for example, the 6th century produced in the Byzantine Christian scholar Johannes Philoponos a quite original natural philosopher at a time when mathematics had ceased to flourish altogether). Yet an "overall picture" emerges that is not indeed one of an abrupt cessation of scientific activity after c. A.D. 200 but rather "one of declining originality, even though this decline is arrested, and even temporarily reversed, by exceptional individuals working in particular fields." [28]

Markers of such a pattern of overall decline are, in Lloyd's view, an increasing tendency toward preservation of an existent body of knowledge instead of its ongoing expansion and refreshment. This tendency is displayed in the production of commentaries and of outlines prepared for didactic purposes (guided by a general spirit of deference toward predecessors), as well as in a somewhat desperate "scepticism concerning the possibility of discovering the true causes of phenomena." [29] In short, after A.D. 200 much effort is being spent on preserving the results of scientific inquiry achieved in earlier times, but (with the sole exceptions of Diophantos in the 3rd, Proklos in the 5th, and Philoponos in the 6th century A.D.) the original spirit of fresh research has meanwhile been lost.

Lloyd goes on to point out that not until Greek science was recovered in 16th-century western Europe was the original spirit to be restored that had animated a small number of men from the pre-Socratics to Ptolemy and Galen onward. Evidently this leaves open our familiar question of why that spirit, rather than ushering in early modern science at once, had been allowed to evaporate after A.D. 200. Besides a thoughtful discussion of the extent to which Christian emphasis on revelation hampered the pursuit of knowledge based on observation and on reasoning, Lloyd offers no more than a few passing remarks on why, in his own words, "the conditions needed to insure the continuous growth of science did not exist, and were never created, in the ancient world." [30] The point is not so much that experimentation and the idea of the mathematization of physics were lacking, for the principles of both were known very well. Neither is it that scientific research was always carried out in the framework of a comprehensive philosophy, for it appears that Archimedes and the Alexandrian scientists had emancipated their researches from such wider pursuits. Nor is it that social support was lacking overall, for the Alexandrians benefited for many centuries from the patronage of successive Ptolemaic kings.

Yet it is in the last direction that the core defect must be sought. Paramount among the explanatory elements Lloyd briefly enumerates are the cultivation of science for the sake of pure knowledge rather than for practical application; the related absence of attempts to justify the examination of natural phenomena in terms of the material prosperity it might yield; and, finally, the absence of an "acknowledged place in ancient thought, or in ancient society, for science, or for the scientist, as such."[31] Altogether, to the small extent that Lloyd seeks to explain the lack of continuous growth displayed by Greek thought on nature, he ascribes it to "the weakness of the social and ideological basis of ancient science."[32]

Two years before Lloyd thus ended his account of Greek science after Aristotle, a book had appeared which, in one chapter, focused on the very same features of Greek science, but embedded here in a general theory of continuity and stagnation in scientific thought. That theory was proposed by the sociologist of science Joseph Ben-David in his 1971 book, *The Scientist's Role in Society*.

Turning the issue upside down: Ben-David's thesis. Joseph Ben-David was born in 1920 in Hungary as József Gross; he died in 1986 in Israel, where he had settled in 1941 (then Palestine).[33] His approach to the problem of decline in science differs fundamentally from all those we have discussed so far. From the sociologist's, as opposed to the modern scientist's, point of view, so he argues, it is not uninterrupted and by and large cumulative growth that constitutes the general rule for science. Rather, throughout the course of history we encounter a quite different pattern of scientific activity, which is characterized by slow, irregular, intermittent growth alternated with substantial periods of stagnation. The science that was born in western Europe in the 17th century provides the only case in history in which this rule has not been obeyed.

In other words, when we seek specific reasons for the decline of science in such-and-such a civilization at such-and-such a time, we are unconsciously projecting backward the notion that under normal circumstances science is bound to display continuous growth. In historical reality the uninterrupted growth of science to which we have become accustomed is rather the exception, and it is this exception that requires explanation. To the sociologist, Western science as we know it constitutes a major anomaly. When considered from the point of view of traditional (in the general sense of 'premodern') civilizations, their pattern of scientific growth, with the pursuit of the understanding of nature safely embedded in technological and medical activity and in the search for a satisfactory moral philosophy, represents a more balanced form of social and cultural development than the—in this perspective—truly pathological rate of growth characteristic of Western science since the Scientific Revolution.

Two questions now present themselves. One is: What general, as distinct from civilization-specific, cause explains the overall pattern? With the obvious counterpart to this question being: How did the pattern come to be broken in western Europe in the 17th century?

Ben-David's answer to the latter question (which amounts to a full-fledged theory on the origins of early modern science) comes up for discussion in section 5.3. For now we are concerned with the extent to which his answer to the former question helps illuminate our problem of how the Scientific Revolution eluded the Greeks.

The root cause that Ben-David, in *The Scientist's Role in Society*, assigned to this pattern of intermittent scientific flowering and decline is the absence, in traditional society, of a separate social role to be fulfilled by the scientist. This concept is crucial to Ben-David's argument throughout. By 'the social role of the scientist' he meant the recognition by society at large of the pursuit of knowledge about nature as a worthwhile activity (whether a vocation or a profession) in its own right. And he contends that in traditional society such social recognition was invariably lacking. The general pattern found in Eastern civilizations is that wherever the study of nature was cultivated, such activity either was subservient to practical pursuits felt to be useful or remained encapsulated in more comprehensive systems of thought directed toward finding out "what man's place is in the universe, what his destiny is, and how he should act to attain a state of perfection."[34]

To the extent that isolated individuals with a special talent for making sense of our natural environment came forward at all, either they were turned by posterity into magical heroes, or their interests were fitted into existing traditions of either a more pragmatic or a more overall philosophical nature. Thus the cultivation of astronomy, for example, remained a highly technical matter, supported by society for its practical usefulness in forecasting heavenly events.

For Ben-David the notion of 'usefulness' was not bound up with material interests exclusively, and he quite explicitly declined to go along with

> the accepted approach that the contents and structure of intellectual inquiry have to be rooted in political or economic concerns, as if the latter were more primary motivations than intellectual curiosity. The present study does not see any interests as more primary than others, although some interests are more widespread than others.[35]

Altogether, although there have always been exceptional thinkers to take an interest in our natural world, the independent pursuit of such an interest was, in traditional societies, invariably made subservient to a more comprehensive framework of thought and activity that could satisfy socially felt needs to heal, to predict, to build, or to give meaning to the world at large.

This, then, is the general pattern of scientific activity in traditional society throughout human history. And it holds for ancient Greece, too—but with a difference. Owing to the comparatively pluralistic nature of early Greek society in the centuries prior to its absorption into the empire created by Alexander the Great, so Ben-David went on, some remarkable deviations from the general pattern took place. These were treated by him in a chapter entitled "The Sociology of Greek Science."

Here he takes as his starting point the notion—frequently encountered by us in preceding pages—that early modern science, as it arose during the 16th/17th-century Scientific Revolution, must be taken as a straightforward extension of Greek science. If this is true, Ben-David observes, then the decisive transition and, hence, the origin of the social role of the scientist ought to be sought in the society of ancient Greece. But is it true that the Greeks regarded science as a separate activity in its own right? Or was science rather a marginal affair? What use was made of science? How was scientific knowledge transmitted? What pattern of scientific growth does ultimately emerge? These are the questions that guide the ensuing inquiry.

In order to answer his questions, Ben-David divides the history of Greek thought into three periods. The first, pre-Socratic stage fits in with the traditional pattern, despite a somewhat greater emphasis on mathematics and on natural as compared to other types of philosophy than is generally to be found in other traditional societies. The second stage was marked by the uncertainties created by the Persian wars. Now a need arose among the Greeks for a philosophy which, while being based upon insight into the structure of the cosmos, could point the way toward the right and just mode of living. Hence, although a considerable interest in the subject matter of natural science persisted, this interest remained derivative. Just as was true of, for example, Confucianism in China, the Academy and the Lyceum aimed for the codification of all knowledge in the framework of an essentially political and ethical philosophy. There remain of course quite substantial differences between Confucianism, on the one hand, and Platonism and Aristotelianism, on the other. The point, however, is that all "were variations within the same sociological role type."[36]

Still it is true that the Aristotelian framework in particular, with the importance of empirical research being stressed so much by its originator, allowed considerable (though not unlimited) freedom for autonomous scientific activity. This freedom, in being exploited by a successor generation, ushered in the third and, for a traditional society, unique stage of Greek science:

> There emerged intellectuals such as Aristarchus, Eratosthenes, Hipparchus, Euclides, Archimedes, and Apollonius whose work can be considered as specialized and professional science. This development did not have a parallel in other societies. It therefore poses the crucial question of the extent to which this development represented the first emergence of socially recognized scientific roles.[37]

From here on Ben-David is concerned to show that the reason for the different fates of science in ancient Greece and in 17th-century Europe must be sought at precisely this point in Greek history. Even if we grant the most 'modernizing' interpretation of the achievements of Greek scientists up to the second century B.C. (when decline set in), we find that no independent social role did, or could, emerge for the scientist. Aristotelian philosophy with its teleological explanatory model left insufficient room for the pursuit of a mathematical physics, in particular. Hence, only a complete divorce from such a philosophical framework could foster the further pursuit of science. This is precisely what happened. Yet when it happened, in the generation of the men just mentioned,

> this development, which may seem to be the beginning of the scientific role with a socially recognized purpose and dignity of its own, was, as a matter of fact, a sign of failure. The newly differentiated role was never given a dignity comparable to that of the moral philosopher. Independence from philosophy was a decline and not a rise in the status of the scientists. During the period when Plato and Aristotle tried to recast the moral-religious foundations of Greek society and of the intellectual foundations of Greek thought, science was drawn into the center of the intellectual concerns of society. . . . But starting in the third century, the few astronomers, mathematicians, natural historians, and geographers who worked mainly in Alexandria were completely isolated from any general intellectual or educational movement. [Thus] specialized science lost its moral importance.[38]

The loss of moral significance thus undergone by science swiftly proved fatal:

> The new autonomy did not, however, confer greater dignity on the scientists. On the contrary, it made evident the marginality of their concerns. As a result, the role did not develop any further and, starting from the second century BC, scientific activity declined.[39]

Here, then, we have in outline Ben-David's explanation of Schadewaldt's 'glass wall', or of Farrington's 'threshold'. The pattern displayed by ancient Greek science must be seen as a unique yet, in the end, typical case of scientific growth in traditional society. Greek science came closer to the Scientific Revolution than science in any other traditional society did, but the threshold was not crossed because Greek society failed to assign a distinct role to the scientist. At this point two questions arise, both taken up by Ben-David. How do we know that Greek science at this most promising moment in its career remained so marginal an affair? How can we explain, in its turn, why this was so?

Here is how Ben-David presents his evidence for the marginality of autonomous science in Greek society during the 2nd century B.C. He argues that, if indeed the newly won

> autonomy of specialized science [had] enhanced the dignity of the scientific role and the motivation to engage in research, one would have expected an acceleration of scientific creativity in the second century BC.[40]

It is at this point that the analysis becomes fully circular: Having set out to explain the discontinuous pattern of scientific growth in traditional society by an appeal to the absence of a social role for the scientist, the principal evidence for the absence of such a role in early Hellenistic society is now derived by Ben-David from the discontinuous nature of Greek science as apparent from its early decay. This logical flaw has important implications, yet it does not appear to be fatal to the argument at this particular point. This is so because the required evidence can be drawn from an independent consideration: the extremely tenuous social support on which the cultivation of science rested throughout the period. As Lloyd in particular has pointed out, even though we know very little of the patronage system set up by the Ptolemies in Alexandria, it is clear that patronage, in particular if virtually monopolized by one dynasty, is an inherently unstable and, in the long run, most unreliable source of support for activities, such as the cultivation of science, that lack an independent legitimation of their own.[41]

This lack of legitimation is what Ben-David adduces by way of an answer to his second question. For he goes on to explain the marginal role allotted to Greek scientists once their pursuit had succeeded in divesting itself of the cloak of a comprehensive philosophy, as follows. During the very period when the separation took place no attempt was made

> at the creation of an ideology claiming a dignity in its own right that would have been for specialized science equal to but independent from philosophy. There ought to have arisen some kind of ideology about the specificity and superiority of the scientific as compared to the philosophical way to knowledge. The rise of such ideologies marked the rise of modern science in the seventeenth century.[42]

So we can already foresee with confidence the drift of the argument by means of which, in his next chapter, Ben-David went on to explain the Scientific Revolution: Early modern science succeeded as its Alexandrian predecessor had not because it created an ideology for itself which, in being adopted by the wider society, granted it the legitimacy it needed in order to survive (see further section 5.3). Once again the historical explanation has taken the form of a mirror image. The same factor whose absence is invoked to explain the 'failure' of Greek science is made to account for the rise of early modern science some eighteen centuries later. And we may ask whether the argumentative circle thus constructed is vicious or not. We consider this question while taking stock of what eight students of Greek science have now taught us about possible causes of the Scientific Revolution.

The harvest yielded by our consideration of Greek science.   We start with Ben-David's thesis, which in spite of its overall neglect by historians of science strikes me as particularly illuminating in more than one respect. This is in part because through his dynamic, three-stage account he manages to give order and coherence to an otherwise somewhat arbitrary succession of advances in Greek scientific thought. But my principal reason is that Ben-David's insistence on the intermittent pattern of scientific growth in traditional society serves as an eye-opener of the very first rank. It brings out the uniqueness and the ultimate contingency of the Scientific Revolution, and it does so in a manner closely related to the type of society under discussion. We shall be able to pursue its eye-opening qualities further when we discuss answers to the question of why a Scientific Revolution eluded both the Islamic world and China. Here already we have seen how the problem of Greek science acquires a novel twist once Ben-David's perspective is applied to it. He treats Greek thinkers neither as our immediate forebears (the common posture of those scholars still tinged with an *Altertumswissenschaft* of humanist provenance) nor as members of savage tribes animated by mental habits altogether foreign to us (as has become one customary conception in more recent ancient historiography). Rather, his peculiar perspective on Greek science as traditional-with-a-difference allowed him to bring out with particular cogency the unique historical stance of Greek science: at once close to us in somehow foreshadowing early modern science and yet foreign to us in being enveloped in the ways of traditional society.

However, Ben-David's ground-breaking observation of the built-in lack of continuity of science in traditional society must be separated from the sole cause he assigned to the phenomenon—the absence of a social role for the scientist—because it leads so easily into a vicious logical circle. It is clear that in Ben-David's mind the emergence of early modern science and the emergence of a social role for the scientist had truly merged: The two phenomena are present together or are absent together. Here, however, it seems prudent to heed Dijksterhuis' wise warning that contemporaneous phenomena may well appear manifestly linked yet be hard to distinguish according to which should be called 'cause' and which 'effect'.

Once cut loose from a causal burden it cannot bear all by itself, Ben-David's identification of what was fatally lacking in Greek science takes its place besides features discussed in previous sections. The three explanatory venues outlined by our eight scholars taken conjointly—a degree of scientific autonomy to be both attained and legitimated so as to make the rise of a social role for the scientist possible; a technologically more vital environment; a positive valuation of manual labor and a humbler posture toward nature attained through the biblical world-view—already constitute almost the sum total of the 'external' explanations that will occupy us in due course (ch. 5).

Inevitably, all three are marked by our familiar circle, with the factor that was lacking in Greek science providing the mirror image to its counterpart being produced during the period leading up to the Scientific Revolution. By itself nothing need be wrong with this particular type of circular argument, provided that a plausible connection is constructed between the missing factor and its indispensability for early modern science. In this respect our explanations leave little to be desired. All stem from considerations that testify to a penetrating insight into certain characteristic features of Greek science as well as into requirements for its early modern counterpart. Having come this far, we turn to a later historical period, where our attention is centered upon the question of whether the particular portion of the full corpus of Greek science that was recovered in medieval western Europe was sufficient, in being elaborated from within, to usher in the Scientific Revolution.

4.3. MEDIEVAL SCIENCE AND THE SCIENTIFIC REVOLUTION

The Middle Ages, more than any other period in the history of science, have been subject to huge shifts of historiographical appraisal. Views on the extent to which medieval science did—or did not—help produce early modern science have shifted accordingly, and these shifts are to occupy us now.

We may conveniently divide a long period of historiographical pursuit into four stages, which partly overlap in time. The first stage (foreshadowed already in the Scientific Revolution) flourished until the first decade of the present century. Barring few exceptions medieval science was equated with what was customarily called 'barren scholasticism'. One historian representative for this viewpoint is William Whewell, who took pains to explain at some length in what manner the thought of the schoolmen impeded the required Revolution in Science. Pierre Duhem, in recovering the long-forgotten manuscripts of the Parisian Terminists, swung the pendulum to the opposite end by proclaiming the Revolution in Science to have started in the 14th century, taking shape as one fruitful outgrowth of scholastic thought. Hence, for Duhem, the job of explaining the revolution came down to seeking causes for this happy, 14th-century outgrowth. During the third stage, toned-down versions of the Duhem thesis appeared, with Hooykaas adopting a drastically modified version of Duhem's explanation. Maier started the fourth stage in identifying a small set of Aristotelian core tenets which, in persisting unabated throughout the Middle Ages, prevented 14th-century science from ushering in early modern science as the next step in scientific advance. This mode of analysis was taken one step further by the historian of medieval science Edward Grant, who returned in effect to the position of the first stage (on a far higher level of sophistication, to be sure) by pointing at a number of quite basic mental attitudes which inherently stood in the way of leading to early modern science in any straightforward manner.

Whewell against the Middle Ages. What little was still known about the scientific achievement of the Middle Ages was held in unanimous contempt from Galileo's time onward by those who adhered to the new science. Leibniz' scathing verdict "*barbarismus physicus*" neatly encapsulates the reigning sentiment.[43] In seeking to explain the agreed-upon barrenness of scholastic scientific thought, William Whewell invoked four specific features: its Indistinctness of Ideas, its Commentatorial Spirit, its Mysticism, and its Dogmatism. Whewell was careful enough not to condemn outright these characteristics as unreservedly bad for the fruitful cultivation of science (in particular, he acknowledged that 'mysticism' may assist the imagination of the scientific discoverer). But the four features together did nothing to alleviate, and in fact rather exacerbated, the fundamental shortcomings of Aristotelianism, which Whewell considered to be the root cause of the medieval failure.

In fact, so Whewell argued, the advent of Aristotelianism had nipped in the bud the ideas of Roger Bacon, who was the only thinker (together with Raimundus Lullus) from whom something better might have come during the period. Bacon's work succeeded in rousing Whewell's enthusiasm to the point where he interpreted his admonitions to investigate natural phenomena in an 'experimental' vein as an urgent program for a drastic reform of science. That reform, however, was delayed for three centuries, chiefly because of the intervening adoption by the Franciscan and the Dominican orders of the philosophy of Aristotle.[44]

Duhem and the decree of 1277. In the years between c. 1910 and 1916, so we recall from section 2.2.4, Pierre Duhem made a most vigorous effort to call into question the received view of medieval science and to replace it by the thesis, supported by a flood of wholly novel documentary evidence, that the Revolution in Science had occurred already in the 14th century. The three volumes of Duhem's *Études sur Léonard de Vinci* document, in their turn, how ongoing discoveries in manuscript treatises written by the Parisian Terminists and successively charted by Duhem led their proud discoverer to ever stronger claims about the merits of these long-forgotten thinkers. His claims culminated in the assertion that, in the process of systematically demolishing Aristotelian science, members of the Parisian Arts Faculty like Buridan and Oresme put a new science in its stead, which was left to sixteenth- and seventeenth-century thinkers to systematize and perfect.

In the *Études sur Léonard de Vinci* Duhem had only marginally touched the question of what had inspired their revolution. His short answer had been, in effect, 'Bishop Tempier's decree of 1277'. In another, more systematically arranged work undertaken almost simultaneously, this answer was both greatly expanded and, over long stretches, turned into its chief principle of organization. This series of books was entitled *Le système du monde* (The System of the World). If completed it would have contained, in accordance with its subtitle, 'A History of Cosmological Doctrines from Plato to Copernicus', with the term 'cosmology' to be taken in a very broad sense that almost matches 'science'.[45]

Let us first consider the grand picture as outlined by Duhem with his customary, enthusiastic eloquence in the fourth volume.[46] He begins his argument by invoking the extraordinary persuasiveness of Aristotle's system of natural philosophy. How, on making its acquaintance, could people fail to "believe that the human mind had attained the inexhaustible source from which it was henceforward permitted to quench its thirst; . . . that our intellect finally possessed the theory, so ardently desired, in which everything that exists finds its place and reveals its reason why"?[47] This is how Averroes and many other commentators felt about the Aristotelian legacy. If their sentiment had continued to prevail, the pursuit of science would forever have been confined to commenting on Aristotle's books, to elucidating obscure fine points, and to disputing remaining ambiguities.

However, two major forces intervened to liberate the human mind from such regrettable self-restraint and to "break the yoke of peripateticism."[48] One of these forces was what Duhem called now 'Experimental Science', then 'Positive Science', then 'Observational Science'. Called by whatever name, he invariably meant its admittedly sole representative available at the time, which was observational astronomy. Ptolemaean astronomy and Aristotelian cosmology contradicted one another on a number of points, and the growing realization that this was so guided thinkers toward following experience and their own common sense rather than Aristotle's prescriptions. However, what made this realization possible in the first place was an awareness of the heretical nature of many an Aristotelian tenet. This was made clear by the second force in play, which was called by Duhem, simply, 'Theology'. Very briefly said, the Aristotelian conception of the deity was held incompatible with the doctrine, shared by Jew, Christian, and Muslim alike, "that God was free in creating the World; that he governs it through an almighty providence; that he has made man free . . . ; that he has fitted him out with a personally immortal soul; and that, during a future life, he will reward or punish the deeds done by this soul in its present life."[49]

These two forces, then, allied against Aristotelian metaphysics and physics alike. Here is how the alliance worked:

> In condemning the heretical assertions of the peripatetic system, the Theology of the three monotheistic religions opened breaches in the solid bulwark of this system. In these breaches experimental Science found a passage which it widened up to the point that permitted its free expansion. This is why a historian is to comprehend imperfectly the upward flight taken in the Middle Ages by a Science liberated from Aristotelianism if he fails to recall the battering-ram-like blows with which Theology shook the prison walls.[50]

How and when were these blows administered? The crucial event took place on 7 March 1277, when Étienne Tempier, bishop of Paris, in obedience to an urgent request sent him in January by a former member of the Parisian Arts Faculty, now Pope John XXI, and after taking due counsel, issued a decree in which 219 specified 'errors' were formally declared anathema.[51] The key feature connecting many of these 'errors' was their shared tendency to subject God's omnipotence to limitations deriving from Aristotelian necessitarianism. Especially if considered from a background of Augustinian theological doctrine such a tendency was deemed inadmissible.[52]

Among the statements forbidden by Tempier, two strictures upon divine free will, in particular, appeared to Duhem to encapsulate the very essence of Aristotelianism. One is the notion that 'God would be unable to give the heavens a movement of translation because, thus moved, the heavens would leave a void behind'; the other, that 'the first Cause could not create several worlds'.[53] Here is how Duhem went on to magnify Tempier's injunction against these two heretical statements into a thesis of far wider import:

> Étienne Tempier and his council, in pronouncing these propositions anathema, declared that, in order to be subjected to the teaching of the Church and in order not to countervene God's omnipotence, one must reject peripatetic Physics. In so doing, they implicitly called for the creation of a new Physics—one that the reason of a Christian can accept. We shall see that, in the 14th century, the University of Paris put its best forces toward the construction of such a novel Physics, and that, by means of these attempts, it laid the foundations of modern Science.[54]

We know already that Duhem was less than fully consistent in assigning a date to the birth of 'modern science', which, in his treatment, varies from the day of Tempier's decree itself to the 14th-century work of Buridan and Oresme *cum suis* and beyond.[55] Altogether, it seems most in accordance with Duhem's intention to say that, for him, Tempier's decree *caused* the Revolution in Science in that his set of condemnations liberated science from its pagan, Aristotelian and Averroist constraints and thus opened the road to early modern science.

This happened—so Duhem cautioned in one revealing passage that scarcely matches his more radical pronouncements elsewhere—not at one stroke, but quite piecemeal,

> by means of a long succession of partial transformations, each of which claimed to do no more than just to retouch or to enlarge one piece or another of the building without changing anything in the whole. But when all these detailed modifications had been carried out, the human mind, taking into view the full result of this protracted labor, came to the surprised realization that nothing of the old palace remained and that a new palace now stood erected at its place.[56]

In the volumes of the *Système du monde* that follow, Duhem was concerned to demonstrate how this happened in rich historical detail. Taking up such topics as 'the infinitely small and large', 'place', 'motion and time', and so on, Duhem invariably split up his extensive citation-cum-analysis of pertinent manuscript passages into two parts, with the first part of his treatment being devoted to ideas prior to 1277, and the second part to discussions on the topic in question that took place after that fateful date. One example which Duhem himself found particularly representative was the issue of place. Aristotle's conception of location implied that every heavenly rotation required at its center the presence of an immobile body, whereas Ptolemy's planetary theory denied such a need. Tempier reinforced the Ptolemaean denial by condemning the statement that God could not possibly displace the entire universe. In thus working together, Positive Science and Theology prepared the downfall of Aristotelianism and, in its wake, the piecemeal, even unwitting, creation of 'modern science' by the learned doctors of the Faculty of Arts at Paris University.[57] These men had been destined to carry out the job called for by Tempier's decree, which may therefore be regarded as the principal cause of the Revolution in Science.

Hooykaas and the decree of 1277.   Naturally, the explanation of the Parisian Revolution through Tempier's decree stood or fell with the Duhem thesis as a whole. As the various weaknesses of the thesis were successively uncovered, and the study of medieval science was more and more cut loose from ongoing theorizing on the origins of early modern science (section 2.4.2), so was the matter of the decree. Its true significance for medieval science continues to be debated by specialized medievalists to the present day, yet the issue is no longer thought to carry particular significance for our understanding of the Scientific Revolution.

One scholar, however, has upheld that, in a manner rather different from how Duhem had construed the connection, there is indeed a significant causal link between the decree and the rise of early modern science. This scholar is R. Hooykaas. In his paper of 1954, "Science and Theology in the Middle Ages," he contributed his share to the consensus reemerging at the time that the downfall of Aristotelianism and the rise of the new science had been the work, not of the 14th, but of the 16th and 17th centuries. By itself, this consensus still left open the significance to be assigned to the performance of the Parisian Terminists, which had been remarkable in any case. May one follow Duhem at least so far as to allow that Buridan and Oresme *cum suis* made major advances toward the more radical revolution wrought by Copernicus and Galileo and Kepler? This is how Marshall Clagett, in particular, was to lay down the connection in his influential *The Science of Mechanics in the Middle Ages* of 1959, but without a causal analysis appended to the broad continuity conception upheld therein. Or should the Parisian achievement be considered in the very first place as a glaring demonstration of the hold the Aristotelian conception of things natural was able to exert even upon the most critical minds of the period? In the historiographical debate thus unloosed by the Duhem thesis Hooykaas took the former line. He followed Duhem, too, in acknowledging that Tempier's decree had been the principal agent to call forth the intellectual movement known as Parisian Terminism. By itself, so Hooykaas observed, there was nothing 'modern' about the decree, which, above all, was a milestone alongside the 13th-century *via antiqua*. Its philosophical context was the issue of which particular Greek conception was to be preferred, with Tempier attempting to stem the rising Aristotelian flood because of its heretical implications. Unwittingly and unintentionally the decree gave rise to the *via moderna:* the nominalist current inside the Aristotelian flood that had meanwhile prevailed despite Tempier's best efforts. Now the crucial ingredient Tempier's decree and the *via moderna* had in common was their shared emphasis upon the contingency of nature. For Hooykaas the story of the growth of science toward the Scientific Revolution is the story of the gradual liberation of science from rationalist presumption, from the arrogant idea that man may prescribe to nature her course rather than humbly submitting to what she has to tell us. Every single early contribution made in this direction must be regarded as a contribution to the Scientific Revolution, and this is how Tempier's decree, with its consequences unforeseen by its author, takes its place alongside other, equally contributing factors. In upholding God's absolute free will against both the extreme intellectualism of the Averroists and the moderate intellectualism of the Thomists, Parisian Terminism helped remove the constraints, not of theology on science, but rather of rationalist philosophy on science. In so doing this movement stimulated a healthy respect for the facts of nature over and above their rash rationalization in Greek thought. Hooykaas went on to unravel in an expert manner the diverse currents in medieval theology of which the manifold intricacies and interactions led to such an unexpected outcome (one principal force identified by him in this connection being the same Augustinianism that had only briefly been hinted at by Duhem).[58] We shall not follow Hooykaas along this path but rather go on to consider the consequences Anneliese Maier's differently shaded view of the Parisian achievement entailed for the particular contribution she made to our theme of the causes of the Scientific Revolution.

Maier and the limits of late medieval science. One constant element in Maier's conception of the significance of 14th-century scholasticism for the creation of early modern science, so we recall from sections 2.3.1 and 2.4.1, was that the 'revolution' then effected had remained incomplete. In the demolition of Aristotelianism that was indispensable for early modern science to emerge, 14th-century scholasticism represented a major, but not the decisive, stage. Whether that revolution had been marked by important adumbrations of scientific findings of the 17th century (as she upheld originally with much fervor) or whether the principal innovative contributions made by late scholasticism must be situated not so much in the area of science proper as in the fields of metaphysics and of scientific method (as she came to think later), in either case the matter of causation appeared in another light than that in which Duhem and his followers (and also Hooykaas) had seen it. The question to ask was, what had *prevented* the protagonists of the 14th-century revolution from completing their job, so to speak? What had held them back on the path they themselves had done so much to clear—the path toward a novel science? Maier sought the answer in certain constraints inherent in the Aristotelian doctrine, of which 14th-century scholastics had been able to liberate themselves to quite some extent, yet not fully so.

In her late essay "'Ergebnisse' der spätscholastischen Naturphilosophie" (The Achievements of Late Scholastic Natural Philosophy) she surveyed both the major results achieved by, and the principal limits set to, 14th-century scholastic thought. The intellectual liberty it conquered for itself was the liberty "to lay bare principles which make possible a direct knowledge and understanding of nature—a knowledge that is individual, empirical, and independent of all authority."[59] While seeking new orientations that might lead to a better understanding of nature, however, these pioneering thinkers continued to allow themselves to be constrained by certain fundamental tenets of Aristotle's natural philosophy. The two most important, and also the most damaging, of these were the following. One was the idea that in nature qualities have an independent, autonomous existence of their own; the other, the equally pernicious rule that lies at the heart of Aristotle's doctrine of motion and in which it is stated that every motion requires a specific motive force which is linked to the moving body and produces its motion without the intervention of anything else. The 'failure' to let go of these two ingrained habits of thought is what continued to prevent 14th-century natural philosophy from achieving more than incidental results pertaining to early modern science. As long as science remained locked up inside the framework of these constraints, the general reformation of the *content,* as opposed to the *method,* of science was inevitably to elude medieval thought.

Grant and the flexible integrity of Aristotelian doctrine. A further extension of Maier's explanation for why the impressive set of departures from Aristotelianism achieved by medieval science nevertheless "failed to produce genuine efforts to reconstruct, or replace, the Aristotelian world picture" is to be found in the concluding chapter to a booklet written in 1971 by the American student of medieval science Edward Grant, entitled *Physical Science in the Middle Ages.*[60] Having asked the question in the wording just quoted, he goes on to provide two answers. One is found in the "highly integrated structure" of Aristotelian thought:

> The rejection of certain crucial parts would have caused the collapse of much of the rest of it. As it happened, what was altered was frequently changed in accordance with Aristotelian principles. New changes and additions were often drawn to Aristotelian specifications and, although they were sometimes quite discor-

dant, were again made part of the system. Thus impetus theory replaced the
external contact of air with the push of an incorporeal force. . . .

These numerous additions and changes were disparate, and each was formulated in response to a separate problem and tradition. They remained unrelated and sometimes inharmonious parts of the Aristotelian system rather than
opening wedges.[61]

Thus, interestingly, the very same observation of the medieval alterations in
the Aristotelian system having been made one by one, in distinct settings and
separated from one another, which Duhem at one occasion took for the general
pattern of the 14th-century Revolution in Science, was now turned by Grant
into an explanation—much more plausible, in my view—of why the sum total
of these alterations never amounted to an all-out onslaught on the system at
large.

The resulting tenacity of the Aristotelian system of natural philosophy, so
Grant went on, was considerably reinforced by the *hypothetical* nature of much
medieval theorizing. The weakening confidence in Aristotle's physical explanations during the 14th century reflected a more basic sentiment: a sense of
scepticism regarding the possibility of explaining nature at all. 'Saving the
phenomena' became the predominant attitude; the thing to do was to think up
clever 'imaginations' about how things *might* be rather than to embark upon a
relentless investigation of reality. This, so Grant argued, is the principal reason
why 14th-century cosmological speculations failed to bring forth the Scientific
Revolution in the way Copernicus' astronomical reform was able to do two
centuries later. The key difference lies not so much in the nature of the arguments employed (in which Copernicus displays no superiority whatsoever) as
in the strongly realist thrust of Copernicus' claims. Copernicus believed that
the Earth truly rotates around its own axis and truly orbits the Sun. Hence, the
physical consequences of these views (to which he himself remained virtually
blind) could lead to the destruction of the Aristotelian thought-building in a
way medieval speculations never could.

The picture that emerges from Grant's account is that the physical realists
of the 13th century could not possibly bring forth the Scientific Revolution,
because they clung to Aristotle; that the 'sophisticated positivists'[62] of the
Paris Faculty of Arts and of Merton College at Oxford failed to produce early
modern science because of their very lack of confidence in the ability of the
human mind to penetrate nature; and that Copernicus succeeded because his
work made possible, for the first time,

a potent union of new ideas that would challenge the traditional physics and
cosmology . . . with the conviction, even if naive, that knowledge of physical
reality was fully attainable.[63]

And thus we come to Copernicanism as a key force in bringing about the Scientific Revolution.

4.4. THE EMERGENCE OF EARLY MODERN SCIENCE FROM RENAISSANCE THOUGHT

4.4.1. *Copernicanism*

In our survey of the Great Tradition as presented in Part I, we have encountered Copernicus as the quite commonly accepted, earliest protagonist of the Scientific Revolution. We are now to examine possible reasons for considering the doctrine Copernicus enunciated, not as the opening shot in the revolution, but rather as one principal force to call early modern science into being. In other words, we shall assemble here some considerations culled from the scholarly literature which converge toward the idea that the Scientific Revolution may profitably be taken to have started, not in 1543 with Copernicus' book, but rather in the decades around 1600 with Galileo and Kepler *cum suis,* a key *cause* of these men's performance being the manifold unintended consequences Copernicus' reform of mathematical astronomy increasingly appeared to entail.

What is to be gained by such an interpretation of what Copernicus wrought? Besides its inherent merits, to be set forth presently, the principal advantage is that it yields a smooth way out of the vexed question of Copernicus' 'modernity'. If Copernicus is taken as the first Scientific Revolutionary, he is quite easily turned into somewhat more of a modern figure than the facts of the case actually warrant.[64] After all, the arguments advanced in Book I of *De revolutionibus* in favor of the heliocentric hypothesis *taken as physically true* were clearly intended by their author to minimize damage to the traditional world-picture, and it is symptomatic that they drew upon ancient resources exclusively. Equally, the approach taken in Books II through VI, with their fictional machinery of epicycles and eccentrics, is none other than the one followed in the *Almagest* (with the exception of the elimination of Ptolemy's equant). All this has repeatedly been noted in the literature. One striking example is provided by Dijksterhuis. On the one hand, he readily recognizes that "barring the application of trigonometric methods of computation one finds nothing in [*De revolutionibus*] that might not just as well have been written in the second century AD by a successor of Ptolemy." On the other, he insists that "with Copernicus, a new era begins."[65] Now how can this be? Dijksterhuis grants that it has everything to do with the *consequences* of Copernicus' reform. But if so, why not pursue the reasoning down to its obvious conclusion and turn those who actually drew the consequences into the true inaugurators of the 'new era?' Is it not much more plausible to join Otto Neugebauer, the author of *The Exact Sciences in Antiquity,* in taking Copernicus for the last of the ancient astronomers, that is, for Ptolemy's truest and greatest heir, and in drawing the logical inference from this view?

There is no better way to convince oneself of the inner coherence of ancient and mediaeval astronomy than to place side by side the *Almagest,* al-Battani's [early 10th-century] *Opus astronomicum* and Copernicus's *De revolutionibus....* Chapter by chapter, theorem by theorem, table by table, these works run parallel. *With Tycho Brahe and Kepler the spell of tradition was broken.* [My italics][66]

In Neugebauer's approach (which differs subtly yet significantly from Dijksterhuis'), the history of mathematical astronomy is consistently being extended forward from antiquity rather than backward from our modern point of view. If one adopts it, the sense in which Copernicus may be regarded as 'modern despite himself' can be defined very neatly by handling, with Thomas S. Kuhn, Copernicus' proposal for a reform in planetary astronomy based upon the ancient idea of heliocentrism as a cause, rather than as the first act, of the emergence of early modern science. Let us find out how Kuhn did it.

Kuhn and the unintended consequences of Copernicus' hypothesis. In Kuhn's first book-length effort in the historiography of science, *The Copernican Revolution,* which appeared in 1957, Copernicus is presented as a highly proficient mathematical astronomer in the traditional sense whose very narrowmindedness outside his chosen domain blinded him to the destructive consequences his technical reform of astronomy entailed for the entire traditional world-view.

What considerations guided Copernicus toward his reform? Besides his rejection of the equant, a prime consideration as expressed in Book I of *De revolutionibus* was Copernicus' conviction that, over the centuries, Ptolemaic planetary theory had turned into a 'monster'. Kuhn makes a great deal of what this meant. It meant that the *Almagest* had allowed staggeringly accurate predictions which, however, as time went on appeared increasingly not to be quite accurate enough. Originally slight deviations piled up over the centuries, with mathematical astronomers confining themselves to correcting these by means of patchwork. Adding an epicycle here, changing a parameter there, moving an eccentric around for again another orbit, they inadvertently helped turn what had once been a harmonious representation of the universe into a 'monster'. Copernicus, so Kuhn insists, took it upon himself to restore the lost harmony. Casting about in antiquity for a fitting point of departure, he came up with the heliocentric hypothesis. Its apparent absurdity in view of the traditional worldview might have impeded a thinker less exclusively concerned with the technicalities of his domain of expertise—less equipped with 'blinders', as Kuhn has it. The need to meet objections in a more satisfactory manner than Copernicus himself had met them remained, however.

Now, if the response to the challenge implicit in Copernicus' theory had been kept confined to the domain of mathematical astronomy, to which Books II through VI of *De revolutionibus* themselves fully belong, then the heliocentric hypothesis might have remained locked forever inside the traditional 'saving the phenomena' posture adopted by mathematical astronomers through the centuries. But the realism of Book I had the potential of appealing to a small minority of scientists who found in the cosmological harmonies outlined there (of however technical a nature these were) sufficient reason to go ahead and turn the many implausibilities of the system into the fitting components of what ultimately took the form of a new, non-Aristotelian world-view. The stimulus that thus derived from the Copernican challenge was experienced in more than one domain of scientific inquiry. For one thing, it helped broaden mathematical astronomy into the cosmological search for a mathematically fitting picture of the world that could do more than just provide a useful computational model. Tycho and Kepler are, of course, the outstanding names in such late 16th-, early 17th-century efforts to reap (each in his own manner, to be sure) the harvest hidden in the Copernican heritage. However, so Kuhn observes, their efforts

> did not answer the *nonastronomical* evidence *against* [Copernicanism]. While they remained unanswered, each of those arguments, whether physical, or cosmological, or religious, testified to an immense disparity between the concepts of technical astronomy and those employed in other sciences and in philosophy. The more difficult it became to doubt the astronomical innovation, the more urgent was the need for adjustments in other fields of thought.[67]

It is rather odd that at no place in this book of his does Kuhn employ the term 'Scientific Revolution'. Nonetheless, if one takes the sum total of the 'adjustments' as these were being worked out at the time and considers them conjointly with the astronomical revolution that grew out of Copernicus' reform, one finds oneself in possession of a quite substantial portion of the emerging Scientific Revolution. In his final chapter, "The New Universe," Kuhn shows in some detail the manner in which many key ingredients of the new science were, as it were, triggered by the Copernican reform. For one, he shows how robbing the sphere of the stars of its motor function paved the way for ideas on the infinity of the universe and for filling such a universe all over with tiny corpuscles. Equally, Copernicus' depriving the center of the Earth of the status of point of orientation for all natural motion—a status it possessed in the Aristotelian conception of things—raised to a new level of urgency the problem of what forces moved the planets, thus leading to the rise of a celestial dynamics.

Confining himself as he does to the superlunary domain, Kuhn neglects to mention in this connection the well-known kinematic paradox emanating from Copernicanism that vertical free fall on a rotating Earth seems impossible. He equally says very little on the ensuing resolution of this paradox by Galileo's enunciation of his principle of inertia in the framework of a novel approach to terrestrial motion generally. Yet this vital contribution of Galileo's fits perfectly into the picture of the origins of the Scientific Revolution thus arising. This picture is summed up by Kuhn as follows:

The conception of a planetary earth was the first successful break with a constitutive element of the ancient world view. Though intended solely as an astronomical reform, it had destructive consequences which could be resolved only within a new fabric of thought. Copernicus himself did not supply that fabric; his own conception of the universe was closer to Aristotle's than to Newton's. But the new problems and suggestions that derived from his innovation are the most prominent landmarks in the development of the new universe which that innovation had itself called forth.[68]

It is in this sense that Copernicus' reform in mathematical astronomy may fruitfully be regarded as one decisive intellectual event which set in motion a preparatory movement in astronomical and physical thought, gradually expanding during the second half of the sixteenth century, which in its turn, during the decades around 1600, burst out in the unprecedented upheaval in science many are in the habit of calling the Scientific Revolution.

4.4.2. The Impact of Humanism

In thus discussing Copernicanism as one possible cause of the Scientific Revolution we have already plunged headlong into that most controversial of historical periods, the Renaissance. To the manifold complexities of the period recognized by historians of Renaissance thought, historians of the Scientific Revolution are apt to add some complications of their own making. Given that at least one commonly acknowledged feature of the Renaissance was a powerful drive to regain the wisdom of the ancient world, we seek an entrance into the many-layered complexities of the period through a brief survey of the ways by which western Europe came into possession of preserved remnants of ancient scientific learning.[69]

Generally speaking, the transmission of ancient scientific learning to western Europe occurred in three stages. The first stage was, as it were, a matter of direct inheritance. Various pieces of scholarship, preserved during the convulsions of the Roman Empire, trickled forth through the encyclopaedic works of such late Graeco-Roman scholars as Boethius and Macrobius and others, giving rise to the liberal arts curriculum in the early Middle Ages. The second stage followed during the High Middle Ages, in the 12th and 13th centuries, when a number of works, among which those by Aristotle proved paramount, were translated from Arabic (some directly from Greek) into Latin and thus became known in the West for the first time. This led to major developments in scholastic thought, whose significance for early modern science has been discussed in a previous section (4.3). A much fuller corpus of ancient texts became available in the 15th and 16th centuries through the efforts of the humanist movement. Seeking to vitalize what they felt to be a barbaric culture by means of a reorientation upon ancient letters and learning in their original, literal reading and in their full variety and scope, the humanists contributed their share to scientific development, too. This they did by putting western Europe in possession of the original Greek version of Aristotle as well as of Plato (hitherto chiefly known by the *Timaios*) and of texts from the Pythagorean, the atomist, the sceptical, the stoic, the neo-Platonic, and other traditions (some of these due to be unmasked in the course of time as less than fully authentic). Equally, works by Euclid, Archimedes, Ptolemy, and many other Greek authors in pure and applied mathematics either became known in Europe or appeared in nongarbled fashion for the first time.

The impact upon west European science was no doubt considerable. That is to say, under the inspiration of the humanist movement the thread was picked up again where the Greeks had left it. As the movement gained momentum, what had begun as an attempt at recovery of ancient scientific texts moved almost imperceptibly ahead to the more or less independent elaboration of their original contents. Thus, Copernicus' effort at 'restoration' of ancient planetary theory was an enterprise in many ways typical of Renaissance modes of thought. The humanist background to his work is unmistakable both in his adoption of the Ptolemaean legacy and in his seeking guidance from a tiny heliocentric tradition in Pythagoreanism.[70] At the same time, in the very process of 'restoring' the Greek heritage in astronomical science new insights were more or less inadvertently gained that can be seen in retrospect to have contributed greatly to the advent of early modern science. Which is not to say that the humanist movement can without reservation be taken as a 'cause' of that event.

In moving on now to a discussion of the extent to which, in the view of a number of historians of science, the Scientific Revolution can be explained by the humanist movement that preceded it, we confine ourselves for the time being to what humanism did for the Greek heritage in the exact sciences.[71] Thus, we reserve for treatment in subsequent sections the significance for the Scientific Revolution of certain philosophical currents that equally grew out of the restoration of Greek learning.

What humanism did for the emergence of early modern science. One concise and balanced discussion of the significance of humanism (as distinct from those Renaissance philosophies that were directly or indirectly founded upon humanists' efforts) for the emergence of early modern science is to be found in Dijksterhuis' *Mechanization*.[72] His position can be rendered by listing three principal points.

First, the idea that humanism furthered the required breakthrough in science through its outspoken hostility to the very mode of thought—Aristotelian scholasticism—that stood in the way of such a breakthrough rests on a number of mistaken assumptions. Thus, it is not true that scholastic science was a wholly barren affair. Nor is the possession of a common enemy enough to forge an alliance: In fact, the humanists had many more features in common with the scholastics they so loudly decried than they were willing to allow, such as "pride of caste, a one-sided intellectual orientation (philological in their own case, metaphysical with the scholastics), contempt of manual labor, and a lack of mathematical schooling." [73] Also, Aristotle's thought is marked throughout by an awareness (however defective, and however much obscured by many of his scholastic followers) of the epistemic importance of experience, an awareness lacking altogether in the "saint of the humanists," Plato. [74] Further, the overall orientation toward the past that marked humanist thought no less than that of the scholastics was not conducive to science taken as a search for discovery rather than for recovery.

Second, what humanist efforts did recover was quite a mixed blessing. Substantial chunks of the Greek legacy, such as nonmathematical physics, physical cosmology, meteorology, and alchemy, were most unsuitable for bringing science closer to the required breakthrough.

Third, the negative impression gained so far of the humanist contribution to the birth of early modern science must, however, be balanced against the positive side. Thus the revival of Plato, though harmful from the point of view of experience, was beneficial where the status of mathematics was concerned (even though this effect, in its turn, was offset in part by much of the numerology that also went by Plato's name). But the principal service rendered to science by humanism was the recovery of a number of texts in ancient mathematics and astronomy. It is remarkable how little this point was elaborated here by Dijksterhuis. He devotes just one sentence to the "salutary effect exerted upon thought" by "the works of Euclid, Archimedes, Apollonios, Pappos, Diophantos, and Ptolemy" and leaves any further demonstration of this overall "effect" to the treatment of individual protagonists of early modern science that follows in later chapters of the *Mechanization*. [75]

This omission of sorts is particularly noteworthy in view of the conception Dijksterhuis had developed a quarter-century earlier in *Val en worp*. There an acquaintance with the Archimedean mode of thought was almost the sole novel element invoked to explain what enabled Benedetti and, *a fortiori,* Galileo to transcend scholastic categories of thought on such topics as free fall and projectile motion. Equally, Koyré had invoked Archimedes as far and away the most important agent in the maturation of Galileo's thought: from his getting stuck in *De motu* to his consequent start from scratch to mathematize nature by means of principles obtained under the aegis of Archimedes' inspiration (sections 2.3.2–3). Indeed, whereas Dijksterhuis in his later years was willing to allow an almost infinite variety of causes of the birth of early modern science, Koyré clung throughout to scarcely more than just one such cause: Archimedes. Time and again Koyré sought to demonstrate that neither the concept of impetus nor a recourse to experience gained from arts-and-crafts practice was sufficient to overcome the obstacles standing in the way of a new dynamics—the key to the secret of the mathematization of nature lying firmly in the hands of those who knew how to extend Archimedes' thought.

There is a certain tinge of the *a priori* about such a conception of the significance of Archimedes' works for the birth of early modern science. However, the documentation on the work of the mathematical humanists largely presupposed in Koyré's conception as well as in Dijksterhuis' early view was supplied in large measure in a study by Paul Lawrence Rose which appeared in 1975. It is to the thesis modestly put forward and exhaustively documented in his *The Italian Renaissance of Mathematics* that we turn now.[76]

The recovery of the Greek legacy in mathematics and mathematical physics. Put in the briefest of terms, Rose's thesis is, simply, "that in the renaissance of mathematics is to be found the indispensable prelude to the scientific revolution."[77]s[77] Taken by itself, the assertion sounds rather platitudinous. With the sole exception of Duhem, for whom mathematization was just a second-order supplement to the true Revolution in Science (section 2.2.4), no one has ever doubted that works deemed by historians' consensus to constitute together a substantial portion of the Scientific Revolution display a level of mathematical expertise at least equal to that attained by the Greeks, and are therefore inconceivable but for the prior appropriation of the Greek legacy. The question to ask is, therefore, what it was that enabled west Europeans to *transcend* the Greek legacy (embodying as it did enrichments meanwhile acquired in the Islamic world). Now one strength of Rose's book resides in the manner in which he takes up precisely this question. In the first place, he takes 'mathematics' in the broad, contemporary sense of the term, thus including both astronomy and the Archimedean tradition in mathematical physics.[78] Second, he builds up a position from which the point where restoration of the Greek legacy subtly shifted into its further elaboration and transformation can be determined with considerable accuracy by means of a very detailed account of the entire humanist movement in mathematics. He does so by dwelling at length upon such topics (confined to what happened in Italy, to be sure) as the successive urban centers of pertinent learning; the libraries where Greek manuscripts were stored; their lending policies; the translation programs set up and executed; the patronage networks involved; and the lives and works and ambitions of such leading men in the movement as Cardinal Bessarion and Pope Nicholas V (humanist patrons), Tartaglia, Maurolico, and Commandino (translators), and Regiomontanus, Copernicus, Tartaglia, Benedetti, Guidobaldo dal Monte, and finally Galileo (the men who set out to exploit the Greek legacy thus recovered by all of them together). Third, Rose adds relief to his treatment by embarking on a sustained comparison with that earlier effort at recovery of the Greek legacy in pure and applied mathematics that took place during the Middle Ages. Here his leading question is why the medieval movement came to a dead end whereas its Renaissance counterpart ultimately led to the Scientific Revolution. We take up this line of inquiry first.

The medieval and the Renaissance translation efforts compared. Why, so Rose asked, did the effort to translate Greek mathematical texts into Latin that culminated in Willem van Moerbeke in the third quarter of the 13th century remain incomplete, and why was it so ineffectual compared to what happened when the humanist movement took up the job afresh? In particular, how is it possible that the vitalizing effect exerted in the late Renaissance by the publication of Latin translations of Archimedes' works in mathematical physics (*On Floating Bodies, Equilibrium of Planes*) eluded their medieval counterparts? Even more strikingly, how is it that the very same Moerbeke translation of *On Floating Bodies* which acquired "a new lease on life" when published by Tartaglia and by Commandino in 1543 and 1565 had only "led a pale existence throughout the middle ages"?[79] Here is how Rose (with considerable support from Marshall Clagett's painstaking researches into the fate of the works of *Archimedes in the Middle Ages*)[80] goes about answering these questions, with their profound implications for our understanding of how the Scientific Revolution was prepared:

> Before Moerbeke only two complete treatises of Archimedes were known in the Arabo-Latin tradition. William managed to recover directly from the Greek most of the remaining works including the crucial writings on mechanics, yet his translation remained almost unknown for two centuries, making scarcely any impression in scholastic geometry and exercising no influence at all on scholastic mechanics or physics. In fact, a strange reversal of influence took place. Such of Archimedes' works as were known in the Arabo-Latin tradition were 'scholasticised'. Scholastics like Bradwardine and Albert of Saxony elaborated Archimedes' original geometrical proofs into new logical forms, introducing quite alien philosophical and physical arguments. ... Why were Moerbeke's mathematical translations neglected? [True,] there are indications that Moerbeke was not at home with the mathematics of his subject. Yet the reason for the neglect lies not with the quality of the translation, but rather with the failure of medieval scholars to take up the tradition. Those responsible included scholastic philosophers who found a little Arabo-Latin Archimedes and a lot of Adelardian Euclid sufficient for their purposes. Equally to blame were the mathematicians including those who had perhaps encouraged Moerbeke with his project in the first place.[81]

The conception of Rose's that underlies the explanation given above is that of two distinct traditions in early Western science, one mathematical, one scholastic, which should neither be confused nor taken as natural successors to one another:

> Fourteenth century physics and kinematics do not provide the missing link between the two renascences of mathematics. The proof of this is to be seen in

the elaborate publishing programmes compiled by Renaissance mathematicians, programmes in which the writings of the scholastics find little or no place. In fact, it was precisely because scholastic physics and the mathematical renaissance were not the same thing that the two movements flourished independently of one another throughout the sixteenth century. One was philosophy, the other mathematics, as contemporaries clearly saw.[82]

Thus, the first European effort to restore the Greek corpus in mathematics and, in particular, in mathematical physics was both prematurely halted and prevented from exerting its wholesome effects because at the time the scholastic approach to physical questions proved to be the more appealing alternative. The second effort, which started in Italy in the early 15th century and which forms Rose's proper topic, must therefore be seen as originating in "the necessity of a restoration of mathematics [felt by] our later mathematicians because the medieval renascence had lost its impetus by 1300."[83] Now how did this later effort succeed in moving from full recovery to cautious extension to wholesale transformation?

The limitations of the Archimedean tradition and how Galileo overcame them. The principal motor recognized here by Rose is the sharp contrast, felt by virtually all humanist mathematicians and phrased by many of them from Regiomontanus onward, "between the immutable certainty of mathematics and the perpetual disputes of philosophy."[84] The notion that nothing but the Greek language and hosts of textual corruptions stood between them and the original meaning of their Greek forebears with their infallible certainty guided the humanist mathematicians on a gliding scale from translation (with a growing expertise in the Greek as well as in the mathematical languages) and careful emendation to elucidation and the imaginative supply of missing proofs, and from there on to the independent reconstruction of works that appeared definitively lost. One remarkably pointed specimen of this overall drift of Renaissance mathematics is to be found in the fate of the Archimedean tradition in mathematical physics. It centered in the Urbino school, of which Commandino, Guidobaldo dal Monte, and Bernardino Baldi may be counted as the principal members. What to do with the Archimedean legacy once restored? Both Guidobaldo and Baldi realized that Archimedes' achievement in statics could be made productive in other domains, yet they were unwilling to go beyond "Archimedean purity"[85] and to seek guidance from dynamical principles to be found in Aristotelian statics:

> When Galileo became a mathematician in the 1580's he joined the renaissance of mathematics at its high tide. The works of Tartaglia and Commandino were circulating widely; Benedetti's *Liber Diversarum Speculationum* was issued in 1585; and Guidobaldo dal Monte's *Liber Mechanicorum* had just appeared in both Latin and Italian, with the paraphrase of Archimedes' *De Aequeponderanti-*

bus soon to follow. Yet for all its apparent flourish mechanics in the 1580's was passing through a crisis. While the science of statics had been set firmly on an Archimedean frame, dynamics had resisted such a transformation from the time of Archimedes himself.[86]

The man in whom the entire renaissance of mathematics in Italy culminated, then, was Galileo. Enjoying the patronage of Guidobaldo (who secured for him the mathematics chair in Pisa) and adopting his program, Galileo soon went beyond its rigid purism and ultimately learned to infuse it with dynamical ideas which, for all their unmathematical vagueness, proved suitable to mathematization in the Archimedean manner. Here was the culmination point of a movement in course of which "the idea of advancement was transformed into the idea of progress"—"the spirit of the scientific revolution had arrived."[87] It must be recognized, so Rose ends his book, that two forces, in particular, both of which were external to science, contributed greatly to the success of the movement where its predecessor had largely failed. One was the system of patronage, which not only lent financial support to the men who did the work but also tied them to the life of the courts and thus helped integrate the mathematical sciences into Italian culture at large. The other was the newly invented printing press, of which Regiomontanus already recognized the key significance for the propagation of mathematical knowledge. "Printing was one of the critical factors which distinguished the Renaissance from earlier renascences and translated cultural experience into an altogether new dimension."[88] Yet paramount was the urge for restoration itself, and it is not by chance that this was the time when, with Guidobaldo and Baldi, a first effort was made to write a history of the mathematical sciences—so as to "justify innovation in terms of renewal and restoration."[89]

A provisional conclusion about the Archimedean problem.   In the historiography of the Scientific Revolution there has been a notorious difficulty with what might be termed the idea of 'the Archimedean origins of early modern science' as originally upheld by Koyré, in particular. While surveying Greek science and its 'failure' to bring forth the Scientific Revolution (section 4.2.2) we defined the problem as follows: If the 16th-century impact of Archimedes' work in mathematical physics was so revolutionary, why, then, was its original impact in the 2nd century B.C. so negligible? When, in the sixties through eighties, Marshall Clagett went ahead to demonstrate that the Archimedean tradition in mathematical physics had already been restored in the later Middle Ages and had lived on ever since, this served only to exacerbate the problem. It appeared now that the question should be reformulated as follows: Why is it that the immediate and revolutionary impact which, according to Koyré, to Dijksterhuis, and to many other historians of science, the 16th-century printing of Archimedes' work had on subsequent scientific thought previously eluded, not only his own work in the time when he wrote it and made it known, but also the manuscript tradition that now appears to have lived on throughout the Middle Ages?

One remarkable thing about Rose's book is that, without raising the problem to such a level of explicitness, he nonetheless brought it a couple of steps closer to a solution. His contribution consists essentially of two points. First, Rose moved beyond Koyré's position by demonstrating that the Archimedean mode of thought, if rigidly applied, could and did lead to a dead end no less than Aristotle's kinematic principles could and did. In treating Galileo as the culmination point of Archimedean mathematical physics restored to, and cultivated in, its original purity, Rose put himself in a position to gauge the width of the creative step needed to transform the Archimedean legacy into early modern mechanics. This does not, of course, show that that creative step could not possibly have been made right after Archimedes himself. It does show, however, that pure Archimedeanism was not enough, but that an admixture of Aristotelian modes of though was required to cross the threshold between ancient Greek and early modern science.

Second, Rose's point that the mathematical tradition could be, and effectively was, 'scholasticized' in the thirteenth century goes a long way toward explaining why the Archimedean legacy remained ineffectual during the medieval period.

It is clear that these two answers at once call forth new questions in their turn. Such novel, more pointed questions appear to turn around the relations between Aristotelian thought and the Archimedean approach from the thirteenth century to Galileo's time. Rose's study strongly suggests that a one-sided grounding of Galileo's work in either a purely Archimedean or a purely Aristotelian tradition is insufficient for understanding how he could arrive at his most creative steps, because by themselves, both had run into a dead end. Rather, the question of how the mutual interaction in Galileo's mind between these distinct intellectual traditions could lead to something almost entirely new is what constitutes the true problem. This problem gains further urgency from the consideration that in the Middle Ages the Archimedean tradition could apparently be submerged by scholasticism, whereas three centuries later the former succeeded in holding its own against the latter. Having so far examined the Archimedean side of our problem, we are ready to move on to a survey of that particular portion of the literature on the Scientific Revolution in which the key to the event is sought in the sixteenth-century reform of Aristotelianism. We begin by putting the pertinent literature in its proper frame, which is that of the historiographical tradition of explaining the Scientific Revolution through one or another aspect of Renaissance philosophy.

Three theses on Renaissance thought and the birth of early modern science. To the extent that causes of the Scientific Revolution have been sought in currents of thought during the immediately preceding period, the recovery

and the subsequent extension of ancient texts in Greek mathematical science
have altogether made rather less of an impact upon the literature of the past
two or three decades than certain philosophical currents broadly inspired by
the humanist effort at recovery. Historians of science have traced the influence
of such currents in considerable detail.[90] Thus, the recovery of Epicurus' and
Lucretius' atomistic ideas in the course of the Renaissance has long been recognized as an important ingredient that went into the making of early modern
science,[91] whereas long-neglected Stoic elements in the thought of protagonists of the Scientific Revolution have recently become the focus of much historiographical attention.[92] However, in the cases of three other currents of
thought equally inspired by ancient sources, such historical reconstructions
have taken the format of theses on the causation of the Scientific Revolution.
Thus, the 16th-century reform of Aristotelianism has been claimed to have
helped cause early modern science; so has the dissemination of Hermetic patterns of thought; so has the revival of scepticism. We shall consider these respective claims in the three sections that follow, starting with the astonishing
idea, advocated in 1940 by J. H. Randall, Jr., for the first time, "that the most
daring departures from Aristotelian science were carried on within the Aristotelian framework, and by means of a critical reflection on the Aristotelian
texts."[93]

4.4.3. *The Reform of Aristotelianism*

Randall's article (entitled "The Development of Scientific Method in the
School of Padua") may conveniently be divided into three parts. The three
points the author sought to demonstrate are (1) a particular school of Renaissance Aristotelians, located at the University of Padua, constructed a very sophisticated methodology for experimental science; (2) Galileo knew this
school of thought and built upon its results; (3) all this goes a long way toward
explaining the birth of early modern science.

In the empirical portion of his article, Randall called attention to a tradition among Aristotelian philosophers at Padua from the 15th century onward
that was aimed at the improvement of the *Analytica posteriora* where the role
of experience in demonstrative science was concerned. Rather than seeking the
empirical starting point of scientific inquiry in self-evident truths about nature
acquired through observation pure and simple (as Aristotle had upheld), these
philosophers gradually worked out a methodological cycle (called the *regressus*) which went more deeply into the question of how empirical phenomena are connected to the cause or causes invoked to explain them. Randall
showed how, toward the end of the 16th century, the movement culminated in
the logical works of Jacopo Zabarella, in particular:

> The weakness of the logic of the Schoolmen had lain precisely in their accep-
> tance of first principles by mere common observation. In contrast, Zabarella, *and*

with him the whole new science, insisted that experience must first be analyzed carefully to discover the precise 'principle' or cause of the observed effects, the universal structure involved in them. After this analytic way of discovery has been pursued, we are then in a position to demonstrate deductively how facts follow from this principle or cause: we can pursue the way of truth. Scientific method, that is, proceeds from the rigorous analysis of a few selected instances or illustrations to a general principle, and then goes from that principle back to the systematic and ordered body of facts, to the science itself formally expressed. Zabarella calls this the combination of the resolutive and the compositive methods; and such were precisely the procedure and the terms of Galileo. [My italics][94]

Thus we slide naturally into the second step of the argument, which is aimed at portraying Galileo as the culmination point of the Paduan tradition thus laid bare. Readers of Galileo's works had encountered the expressions *metodo risolutivo* and *metodo compositivo* at many places; Randall now showed both the terms and the refined methodology behind them to stem from a long-standing tradition at the very university where Galileo was appointed in 1592, only three years, that is, after Zabarella's death. Hence, it did not seem far-fetched to Randall to suppose that Galileo's accomplishment had essentially consisted in taking up the true method of science that lay ready for him, and in going ahead and applying it to natural phenomena in the manner prescribed by his immediate predecessors. Thus, to survey the entire movement is to see how "the science of the Italian universities could progress steadily in self-criticism to the achievement of a Galileo."[95] And this is how Randall could round off his article with the daring flourish (in evident allusion to Dante's qualification of Aristotle in the *Inferno*):

> The 'father' of modern science, in fact, turns out to be none other than the Master of them that know.[96]

Renaissance Aristotelianism as a budding scholarly subdiscipline. In retrospect, Randall's article can be seen to have served a dual purpose. It advanced a particular thesis to explain the Scientific Revolution along a line of thought that has been taken up again in later years; it also heralded a new, specialized academic subdiscipline which was to acquire its full vigor in the seventies and eighties. This subdiscipline, which also counts the historian of Renaissance philosophy P. O. Kristeller among its founding fathers, was energetically demarcated and pursued by Charles B. Schmitt (1933–1986), in particular. In a first attempt at synthesis, entitled *Aristotle and the Renaissance* (1983), Schmitt strove to eradicate a number of common misconceptions about Aristotelian doctrine during the period between the scholastics and the onset of early modern science. The doctrine was not static but can rather be seen to display a considerable development, which stemmed from internal motors as well as from such external influences as the restoration of both Aristotle's original texts and those of his commentators through humanist efforts. Nor was Aristotelianism an isolated body of teachings amid a proliferation of flourishing Renaissance philosophies: It both exerted an influence upon, and underwent influences from, thinkers working chiefly in other traditions. Further, Aristotelianism was not one homogeneous doctrine but rather encompassed quite diverse approaches and methods and opinions, which ranged from the unimaginative expounding of the letter of the master's teachings as practiced by Galileo's Paduan colleague Cesare Cremonini, to considerable liberties taken by his colleague at the same university Jacopo Zabarella. In short, Aristotelianism was quite a living and flexible philosophy until well into the 17th century, capable of considerable refinement and of adaptation to novel intellectual currents.[97]

In Schmitt's view, only one novel intellectual current appeared capable of robbing Aristotelian doctrine of the vitality it had enjoyed unabated since the 13th century. The culprit was, of course, early modern science. Although not so rigid and not so incapable of further intellectual growth as portrayed by such opponents as Galileo, Bacon, and Hobbes, Aristotelian doctrine ultimately had little to offer in comparison to the insights the new science introduced on the European scene. Schmitt did not fail to point out enduring Aristotelian loyalties among such pioneers in the life sciences as William Harvey; still, to Schmitt the essential connection between Renaissance Aristotelianism and the Scientific Revolution was that the latter movement radically and altogether rather quickly vanquished the former. He ominously put aside Randall's thesis that, on the contrary, the former caused the latter, thus leaving to another student of features of Renaissance Aristotelianism the task of reviving it. This student is William A. Wallace, and it is to his mode of explaining the emergence of early modern science through the reform of Aristotelianism that we turn now.

Randall's missing link discovered. Wallace's chosen domain of expertise is the history of Aristotelian doctrine—in particular, its scholastic and Jesuit varieties. In his *Causality and Scientific Explanation* (1972) he was concerned to show that Aristotelian methodology as set forth in the *Analytica posteriora* and in countless commentaries shares with the methods advocated and applied by the proponents of early modern science a realist bent, the search in the one as in the other being for the true causes of things. The basic continuity thus posited has never ceased to provide the framework for Wallace's subsequent researches, which have increasingly centered on laying bare Galileo's roots in Aristotelian thought on the nature of science. To Wallace's mind, the one weakness in Randall's thesis was that the connection between Zabarella and Galileo, although plausible, had remained inconclusive.[98] True, among the unpublished early papers left by Galileo was a collection of transcriptions and/or excerpts from obviously Aristotelian authors, but a detailed comparison with Zabarella's works left it unclear whether Galileo had been familiar with that particular corpus. However, Wallace was instrumental[99] in spotting the missing link, which appeared to be located at an unexpected spot: the Collegio Romano, that is to say, the center of Jesuit learning at Rome. By means of extensive collations and other diplomatic techniques Wallace established two particular points beyond reasonable doubt. One is that Galileo knew Zabarella's work at one remove—through a textbook by the Jesuit teacher of philosophy Paulus Vallius. The other is that the date when Galileo sat down to excerpt chapters by Vallius on such topics as the *regressus* is an astonishingly late one. Galileo did so in the years 1588–1590, that is to say, just before he went on to write his iconoclastic, heavily Archimedean, incompleted treatise *De motu* in 1591.

All this was set forth by Wallace in exhaustive detail in his book of 1984, *Galileo and His Sources: The Heritage of the Collegio Romano in Galileo's Science,* and elaborated further in a number of articles.[100] If one focuses on how, in these writings, the argument is set up that is meant by its author to construe a causal link between the reform of Aristotelianism and the onset of the Scientific Revolution, one finds a number of critical points being addressed that, in brief asides to be sure, had occupied Randall as well. For it is by no means self-evident that Randall's thesis, even if fortified by the solid connection between Zabarella and Galileo now established, is strong enough to bear the explanatory weight it has been charged with by both its original propounder and its latter-day advocate. The principal critical issues which the thesis must confront, and which were in fact raised soon after Randall's original publication,[101] are three time-honored ones. They are about the role of mathematics in Galileo's work, the role of method in the emergence of early modern science, and the extent to which Galileo's work can stand for the Scientific Revolution as a whole. Let us survey how Randall and Wallace attempted to confront these issues.

Three more links missing from Randall's argument. First, there is the mathematical element in Galileo's work, of which so much was made during the formative years of the Great Tradition. Randall acknowledged in passing that in Zabarella's methodological treatises no trace is to be found of a notion that the principles to be sought in science should be mathematical. To Randall, rather than compelling a substantial modification of his original thesis of the Aristotelian origins of early modern science, this point elicited an adaptation as minimal as he could make it. As the two missing elements which, somehow, infused the new, Paduan methodology, he invoked an independent strand in Renaissance Aristotelianism which he promised to treat elsewhere but never did[102] and the recovery of Greek mathematical texts by Archimedes and others. After the cursory invocation of these elements, he still felt able to proclaim:

> With this mathematical emphasis added to the logical methodology of Zabarella, there stands completed the 'new method' for which men had been so eagerly seeking.[103]

Which at once raises our familiar question (section 3.1.1) of the extent to which the revolution in science is to be regarded as one of *method*. It is quite clear that neither Zabarella nor any of the other Padua Aristotelians ever applied his advocated method to actual scientific problems. Is it, then, not much more natural to take its application for the key novelty, rather than positing a deep continuity between the men who refined the method and the man who actually went ahead to apply it, and on occasion to invoke it? On this issue, which is crucial to his thesis, Randall was less than fully candid. In one side remark he acknowledged "that in method and philosophy *if not in physics* [Galileo] remained a typical Padua Aristotelian" [my italics].[104] But elsewhere he stated, without adducing one shred of evidence, that

> the Aristotelian physics and a nascent 'Galilean' physics were in definite and conscious opposition at Padua, and this critical conflict contributed greatly to the working out of the latter.[105]

Finally, we note that in Randall's article Galileo's work is equated without more ado with early modern science: A thesis about the origins of Galilean science is being identified here by none but rhetorical means with a thesis on the causes of the Scientific Revolution. This is one particular feature of efforts to explain the birth of early modern science that we are to encounter time and again. Whether a purported explanation is focused on Italy (as was equally the case with Rose's analysis of the Archimedean tradition) or on any other center of the Scientific Revolution, in every such case we shall note in what follows its geographical limitations, find out how the author in question dealt with them, and, in our concluding sections, offset their approach against a minority of explanations (such as the one from Copernicanism) in which the Scientific Revolution is addressed in its full geographical extension.

How Wallace dealt with the very same issues. In returning now to the case of the Aristotelian origins of early modern science, we find that Wallace, too, confined his Aristotelian connection to Italy. In one remarkable aside he tried to fortify his claim to have laid bare a vital connection between 16th-century Aristotelianism and the Scientific Revolution by moving forward the dating of the latter event beyond the appearance of the *Discorsi*, that is to say, almost halfway into the 17th century:

> . . . Galileo's unique contribution to the science of mechanics, which indeed transformed that science into a *nuova scienza* on which the Scientific Revolution of the seventeenth century was soon to be erected.[106]

I do not think that this expedient can save the original claim. Even if otherwise wholly accepted, claims on behalf of the Aristotelian origins of early modern science are by nature of their case confined to digging up Italian connections and, hence, to determining the intellectual heritage of Galileo to the virtual exclusion of any other pioneer of the Scientific Revolution. Still, the very passage just quoted shows that Wallace was much more candid than Randall in coming to grips with the principal difficulties facing the thesis: What about Galileo's effort at mathematization? And where is the key novelty to be located: at the point where traditional Aristotelian methodology was refined by his Paduan followers, or rather at the point where the surely sophisticated yet still abstract methodology of Zabarella gave way to Galileo's radical decision to go ahead and actually create a new, largely non-Aristotelian physics?

On these issues we find Wallace ensnared in ambiguities. He readily acknowledges that historians will continue to diverge over "whether or not [the] 'new science' stands in essential continuity with the ideal of *scientia* whose norms are prescribed in Aristotle's *Posterior Analytics*."[107] He also recognizes that "these innovations, clearly more than the seedbed from which they sprang, are what earn for [Galileo] the title of Father of Modern Science."[108] Such considerations notwithstanding, Wallace prefers to conceive of Galileo's new mathematization of motion as fitting in with the Aristotelian ideal of demonstrative science and as compatible throughout with the refinements proposed at the Collegio Romano. But, if one methodology is apparently capable of accommodating two radically different types of science, is not then the obvious inference that the issue of method is foreign to the question of what caused the Scientific Revolution? True, Wallace is of course well aware of the distinction between method and its application, yet he tends to minimize the significance of the distinction as much as he can.[109] Does not, then, the basic fallacy of the approach come from the preconceived desire to define the new science of the 17th century formally rather than in terms of its actual content and of the changes in world-view that made such enterprises as the mathematization of nature feasible in the first place?

In one sense, Wallace cannot but agree. Having exhausted, at the end of *Galileo and His Sources,* available opportunities to connect Galileo to contemporary efforts at refining and reforming Aristotelian conceptions, Wallace feels that the moment has come to ask what remains that was new in Galileo's science. And it is remarkable to see how, after such a much more thorough investigation than was ever made before of the roots of Galileo's thought, Wallace comes up with precisely those crucially new features that had been singled out by historians of science from the time of Burtt and Dijksterhuis and Koyré onward: Galileo's mathematization of physics under the primary inspiration of Archimedes and his devising of experiments to find out whether the mathematical supposition, through its consequences, actually holds in an idealized reality taken to be devoid of accidental impediments. The "Iuvenilia" appear to come far closer in time of writing to *De motu* of 1591, and they testify to the presence in Galileo's mind of marks of 'progressive' Aristotelianism up to a much later date than had been held possible before, yet the year 1591 still marks Galileo's departure, under the aegis of Archimedes, from orthodox positions regarding the problems of local motion. In Wallace's new picture no less than in the received one, the impetus physics of *De motu* has failed to stand up to the process of incipient mathematization to which Galileo subjected it there, thus giving occasion to the radically new conceptualization of the problem undertaken in his Padua years.[110] At this very point is still located one of those critical junctures in the origins of early modern science that form the true issues requiring explanation. Efforts spent on behalf of Aristotelian methodology have not brought us closer to such an explanation. What they have done is to clear up a significant portion of the intellectual heritage of one of those men who went on to transform that heritage almost beyond recognition. The riddle of what made him do so continues to keep us occupied.

4.4.4. *Hermeticism and Neo-Platonism*

From the twenties through the early sixties of the present century, when the concept of the Scientific Revolution was forged, Renaissance neo-Platonism was often granted a couple of pages in brief sections devoted to explaining the origins of the new scientific movement. Thus, in the strongly 'mathematizing' accounts of Burtt, Dijksterhuis, and Koyré, neo-Platonism figured as a prelude to, and as a source of inspiration for, the higher status of mathematics indispensable to the mathematization of nature carried out by such men as Copernicus, Kepler, Galileo, and Descartes. In such less mathematically committed surveys as those of Butterfield and the Halls, neo-Platonism found a place as one of many signs of an altered climate of thought during the early, 16th-century stage of the Scientific Revolution.

Meanwhile, some difference of opinion remained over the exact contribution made by those neo-Platonic currents that had come up from the late 15th century onward in Italy. For example, Koyré reproached Burtt for failing to distinguish between two kinds of neo-Platonism: a scientifically fruitful respect for mathematics as the model of exactitude and a sterile belief in number symbolism mixed up with magic and mysticism.[111] The distinction between potentially scientific and 'mystical' currents in Renaissance thought was drawn in much sharper lines in various books by the Halls, where it pervaded the entire account far beyond the subject of neo-Platonism and mathematics, so as to present the Scientific Revolution as the liberation of rational, scientific thought from the many obscurantist currents still largely dominant in the 16th century. Again, on a more specific level, the attribution of Copernicus' search for an ordered, harmonical arrangement of the planetary system to neo-Platonic influences he underwent in his student days has become rather a commonplace in the pertinent literature.

Altogether, such invocations of 15th- and 16th-century neo-Platonism tended to remain somewhat loose and general. This changed considerably in 1964, when a specific set of texts associated with neo-Platonism was both singled out as the vital core of Renaissance thought on nature and linked in a much more straightforward manner to the onset of the Scientific Revolution. The set of texts was the Hermetic corpus that had been translated from Greek into Latin by Marsilio Ficino in 1463. The connection with the rise of early modern science was principally made through the Hermetic conception of man as a magician. The argument built up to make this remarkable case has become known in the literature as the Yates thesis, in deserved honor of the English historian Frances Yates, who was its principal originator. We have dealt before with part of her work on European thought in the 16th and 17th centuries. In sections 3.3.1–5 we focused on reinterpretations of the Scientific Revolution that followed in the wake of her original thesis as set forth in her 1964 book, *Giordano Bruno and the Hermetic Tradition*. In the present section we take on this thesis itself, claiming as it does that vital aspects of the 17th-century Scientific Revolution can only be explained through certain features of magical thought and practice in the Renaissance that go back to the late 15th-century acceptance of the Hermetic corpus and to its subsequent elaboration.[112]

The Yates thesis. Like all great books, *Giordano Bruno and the Hermetic Tradition* carries a message on various levels. At one level, the one indicated by the title, the book aims to show that the writings of the important Renaissance philosopher Giordano Bruno have been misunderstood. His stream of publications as well as his travels through Europe up to the moment of his arrest in 1592 should be regarded as the messianic call, directed to a religiously divided Europe, to reunite under the banner of a magical and cabalist message derived from an extreme interpretation of the Hermetic corpus. Such usage of the Hermetic texts, so Yates contends, was made possible by their radical misdating. From Ficino's translation onward the Hermetic corpus was not recognized as the 2nd-century concoction of Greek, Roman, Egyptian, Jewish, and possibly early Christian elements Isaac Casaubon showed it in 1614 to be, but, rather, as texts of venerable antiquity, contemporary with Moses, in which both Platonic and Christian truths had supposedly been foreshadowed.

The main body of Yates' book consists of the story of how the Hermetic tradition, with its strongly magical overtones, first, with Ficino, elevated magic from an obscure, medieval conjuring business to a lofty philosophical plane; next, through Pico, was connected with Cabala as well; and finally, with Bruno, was brought to full fruition in a new world-view of strongly activist magic. In this conception, the accepted view of Bruno as the early anti-Aristotelian who helped usher in early modern science by boldly extending Copernicus' limited reform into the modern notion of infinite space gives way to a new picture which, on this level, seems to remove Bruno as far as possible from any current of thought having to do with the rise of early modern science.[113] As one striking example Yates adduces Bruno's usage of Copernicus' diagram of the heliocentric universe. In Bruno's hands this is turned into a magical hieroglyph, serving no longer as the illustration of a mathematical argument but rather as a symbolic sign or talisman.

However, according to this account, on another level Bruno surely deserves the pivotal position in the history of science accorded to him earlier on the wrong, anachronistic grounds, for there is indeed a vital connection between Hermetic thought in the Renaissance and the emergence of early modern science. In the final chapter of her book Yates carefully delineates where, and how, such a connecting line ought to be drawn. It is not in the history of scientific thought proper: "With the history of genuine science leading up to Galileo's mechanics this book has had nothing whatever to do," she cautions explicitly.[114] To Yates in this book (as distinguished from her later, 'Rosicrucian' excursions) an absolute watershed separates premodern from modern European thought on nature:

> No one will deny that the seventeenth century represents that momentous hour in the history of man in which his feet first began to tread securely in the paths which have since led him unerringly onwards to that mastery over nature in modern science which has been the astonishing achievement of modern European man, and of him alone, to this extent, in all the annals of mankind.[115]

The very context of Yates' book enables her to mark with precision the watershed in question: It becomes almost tangible in the controversies fought by Mersenne (aided by Gassendi) and Kepler with that late representative of Hermetic thought Robert Fludd. The early propounders of a radically altered mode of thought, although thoroughly familiar with (Mersenne) or deeply immersed in (Kepler) Hermetic patterns of thought, nevertheless marked with great precision what distinguished their own, truly mathematical approach from Fludd's Magia and Cabala:

> Mersenne is a modern; he has crossed the watershed and is on the same side of it as we are; belief in the power of magic images of the stars seems to him quite mad. . . .
>
> [Kepler] saw with the utmost distinctness that the root of the difference between himself and Fludd lay in their differing attitude to number, his own being mathematical and quantitative whilst that of Fludd was Pythagorean and Hermetic. . . .
>
> The seventeenth century is the creative period of modern science, and the Fludd controversies come at the crucial moment when the new turn begins to be made, when the mechanical philosophy of nature provided the hypothesis and the development of mathematics provided the tool for the first decisive victory of man over nature.[116]

On this level early modern science and Hermetic modes of thought are radical opposites. However, on another they are not. And the move from the one level to the other takes place when we leave the domain of historical description and analysis and begin to search for causes:

> The history of science can explain and follow the various stages leading to the emergence of modern science in the seventeenth century, but it does not explain *why* this happened at this time, why there was this intense new interest in the world of nature and its workings.[117]

Yates goes on to observe that historians of science have been aware of the explanatory gap; she does not, however, seem to be aware herself of how numerous and varied their attempts have been to fill it. Rather, she goes ahead to present her own contribution, in beautiful, stirring prose worth being quoted at some length. She starts from an overall conception of the deepest layers of human thought and feeling in which such intellectual upheavals as the Scientific Revolution find their origin, and then passes on to a more specific frame of analysis:

> It is a movement of the will which really originates an intellectual movement. A new center of interest arises, surrounded by an emotional excitement; the mind turns whither the will has directed it, and new attitudes, new discoveries follow. Behind the emergence of modern science there was a new direction of the will towards the world, its marvels, and mysterious workings, a new longing and determination to understand those workings and to operate with them.
>
> Whence and how had this new direction arisen? One answer to that question suggested by this book is 'Hermes Trismegistus'. And under that name I include the Hermetic core of Ficinian Neoplatonism; Pico's momentous association of Hermetism with Cabalism; the direction of attention towards the sun as the source of mystico-magical power; the magical animation throughout nature which the Magus seeks to tap and to operate with; the concentration on number as a road into nature's secrets; the philosophy, present . . . in the philosophical Hermetic writings, that the All is One, and that the operator can rely on the universal validity of the procedures which he uses; finally, and this is in some ways the most important point, those curious historical errors by which 'Hermes Trismegistus' was Christianised, so that it was lawful for a religious Hermetist to speculate on the world in his company, to study the mysteries of creation with his assistance, and even (though not all were willing to stretch the point thus far) to operate with the world forces in magic.
>
> The reign of 'Hermes Trismegistus' can be exactly dated. It begins in the late fifteenth century when Ficino translates the newly discovered *Corpus Hermeticum.* It ends in the early seventeenth century when Casaubon exposes him. Within the period of his reign the new world views, the new attitudes, the new motives which were to lead to the emergence of modern science made their appearance.[118]

And then Yates faces the crucial issue with which her thesis stands or falls:

> The procedures with which the Magus attempted to operate have nothing to do
> with genuine science. The question is, did they stimulate the will towards genu-
> ine science and its operations?[119]

Portions of her answer to this question had been prepared by Yates in earlier
chapters. We can glean these chapters and come up with five distinct (though
obviously connected) claims. They may be labeled as follows: from number to
mathematics; universal harmony; the sun at the center; man as operator; and,
finally, a Scientific Revolution in two stages. These claims are now to be discussed in the order in which Yates brought them up.

Cabala and angel-summoning; or, From number to mathematics. The two
levels of analysis with which Yates operates in the main argument of her book
can also be discerned in many of its subthemes. So it is in her contention—
to become ever more central in her later work—that magical operations with
numbers contributed to the emergence of 'genuine mathematics' as applied in
the Scientific Revolution. On one level, she acknowledges that

> neither Pythagorean number, organically wedded to symbolism and mysticism,
> nor Cabalistic conjuring with numbers in relation to the mystical powers of the
> Hebrew alphabet, will of themselves lead to the mathematics which really work
> in applied science.[120]

And yet, she continues, not only do such magical operations leave room for
'genuine mathematics', but also does a figure like John Dee testify to a Renaissance conception of the world in which his summoning of angels through magical signs and numbers and his advocacy of the cultivation of mathematics
appear as two sides of one and the same coin. In the outlook of the Renaissance
Magus, of which Dee, in Yates' view, represents the prototype, activities were
united which to us seem utterly distinct. Their inner connectedness was felt at
the time to be rooted in the existence, in the magical view of things, of three
levels of reality, succinctly distinguished in Cornelius Agrippa's 1533 textbook
of magical theory and practice. Yates sums up the connection as follows:

> The universe is divided into three worlds, says Agrippa, the elemental world, the
> celestial world, the intellectual world. Each world receives influences from the
> one above it, so that the virtue of the Creator descends through the angels in the
> intellectual world, to the stars in the celestial world, and thence to the terrestrial
> elements and all things composed of them. *Magicians think that they can make
> the same progress upwards, and draw the virtues of the upper world down by
> manipulating the lower ones.* They try to discover the virtues of the elemental
> world by medicine and philosophy; the virtues of the celestial world by astrology
> and mathematics; and in regard to the intellectual world they study the holy cere-
> monies of religions. [My italics][121]

In later books of Yates, Dee was to be accorded an ever more pivotal position in her evolving views on the transition from the world-view of the Renaissance Magus to what she called the Rosicrucian stage of early modern science. Having dealt before with the 'Rosicrucian' stage in Yates' own thought, we must pay some critical attention here to the mathematical part of the original Yates thesis, which, in my view, is among its weakest. Not so much because, as Hall was to phrase it, John Dee "simply discovered nothing new,"[122] or because later research has called into serious question Yates' interpretation of Dee's world of thought, but principally because of her persistent confusion between the furtherance of the discipline of mathematics, which had been an ongoing concern for centuries, and the mathematization of nature as a key feature of the Scientific Revolution. Here we touch on one notable shortcoming of the Yates thesis, to be met again in other contexts: her understanding of what the Scientific Revolution was about may surely be called less than perfect.

One little, though significant, symptom is her naive characterization of Robert Hooke as "one of the best mathematicians in the Royal Society," a view which may well in heaven bring Dame Frances to posthumous blows with no less an authority on Hooke's mathematical abilities than Sir Isaac himself.[123] Obviously, the cultivation of mathematics is to be kept distinct from the insight that certain portions of mathematics can fruitfully be applied to certain natural phenomena like free fall and projectile motion or to the realities of the trajectories of the planets. To the extent that Yates' argument is about a heightened respect paid to mathematics in the magical world-view, the contention is more or less equivalent to the traditional attribution to neo-Platonism of such a function. But Yates goes much further than that:

> The Renaissance magic was turning towards number as a possible key to operations, and the subsequent history of man's achievements in applied science has shown that number is indeed a master-key, or one of the master-keys, to operations by which the forces of the cosmos are made to work in man's service.[124]

Even if allowance is made for the serious overstatement of the extent to which the 17th-century mathematization process yielded immediately applicable results, Yates' claim is valid only if the term 'number' is taken as a catchall for whatever is measurable. With number taken in the Cabalist sense of involved symbolisms or in the strictly Pythagorean sense of the abstract entity of which the world is made, her claim misses its mark—the rejection of numerology forming, after all, one crucial element in early modern habits of thought that made the mathematization of nature possible in the first place.[125]

Universal harmony. Another idea that pervades Hermetic thought is the close link between the microcosm and the macrocosm—a link established by the harmony which rules throughout the universe. Yates touches this theme only very briefly, claiming without much substantiation that here another motive force is to be detected "for genuine mathematical thinking about the universe."[126]

No doubt the theme of universal harmony figured in Hermetic patterns of thought besides many others. Still, the theme really constitutes a thought complex all by itself, narrowly connected to musical theory, which provides the proper context for its further examination. The place of the idea of harmony in Renaissance and early modern thought has begun to be turned into a matter of some historiographical concern over the past decades.[127] I believe indeed that our understanding of the Scientific Revolution and of its causes cannot but gain from a closer attention to this huge topic than has so far been given to it. For example, the well-known question of what made both Copernicus and Kepler value the 'harmonic' arrangement of planetary orbits in the heliocentric hypothesis so highly as to accept for the time being its many apparent absurdities may well come closer to an answer through a deeper insight into the place of the concept of harmony in contemporary thinking. All this, however, has little to do with the Yates thesis, which, to the extent that it touches upon matters astronomical, rather entails claims of a very different order.

The Sun at the center. The Sun-worship that shines through many texts of the Hermetic corpus naturally provided a tempting opportunity for Yates to provide one more link between Hermetic thought and the Scientific Revolution. One obvious point of connection was Copernicus' well-known invocation of Hermes Trismegistus at the well-known place in Book I of *De revolutionibus* where the dry mathematician gives way to the inspired poet of the Sun's exaltation. The other was her attempt to demonstrate that Bruno's Copernicanism was closely bound up with his Hermetic world-view. Here is how Yates turned the connection into a cause of the Scientific Revolution:

> Bruno's world view shows what could be evolved out of an extension and intensification of the Hermetic impulse towards the world. . . . Drained of its animism, with the laws of inertia and gravity substituted for the psychic life of nature as the principle of movement, understood objectively instead of subjectively, Bruno's universe would turn into something like the mechanical universe of Isaac Newton, marvelously moving forever under its own laws placed in it by a God who is not a magician but a mechanic and a mathematician. The very fact that Bruno's Hermetic and magical world has been mistaken for so long as the world of an advanced thinker, heralding the new cosmology which was to be the outcome of the scientific revolution, is in itself proof of the contention that 'Hermes Trismegistus' played some part in preparing for that revolution.[128]

In other words, even though Bruno's world-view was the very opposite of the world-view embodied in early modern science, and even though a radical alteration of all crucial parameters of the former was required to arrive at the latter, still the former may be taken to have helped bring forth the latter, through the agency of an enhanced "will towards the world." Even if we accept the conclusion, it still leaves wholly unexplained the question of how these parameters were altered in the first place, the primary problem being rather to understand what made it possible for early 17th-century thinkers to 'drain the universe of its animism'.

Now this is largely an *a priori* point, aimed at the logical structure of Yates' argument. A close scrutiny of the factual evidence underlying her pertinent claims was made in 1977 by the historian of Copernican thought Robert S. Westman. In his essay "Magical Reform and Astronomical Reform: The Yates Thesis Reconsidered" he demonstrated in convincing detail the following three points. (1) Hermetic Sun-worship was no less compatible with the traditional position of the Sun taken as the middle one of seven heavenly spheres orbiting a stable Earth located at their common center than with Copernicus' heliocentric hypothesis. (2) With the sole exception of Bruno, no Hermetic philosopher of the Renaissance added any original idea to what, during any given decade of the 16th century, contemporary professional astronomers had to say on Copernicanism. (3) Bruno's realist adoption and subsequent infinitist elaboration of the heliocentric viewpoint does not so much fit in a Hermetic context as, rather, in a long-standing tradition of thought about the infinite. These points, taken together, dispose rather effectively of Yates' contention that "Bruno's use of Copernicanism shows most strikingly how shifting and uncertain were the borders between genuine science and Hermetism in the Renaissance."[129]

Man as operator. A more promising connection established by Yates between the Renaissance outlook and the Scientific Revolution is to be found in a radically altered conception of man that was inspired by the Hermetic corpus. This involved

> a view of the cosmos as a network of magical forces with which man can operate. The Renaissance magus had his roots in the Hermetic core of Renaissance Neoplatonism, and it is the Renaissance magus, I believe, who exemplifies that changed attitude of man to the cosmos which was the necessary preliminary to the rise of science.[130]

In the Hermetic corpus, man is conceived as 'a great wonder', divine in origin, and endowed with powers to dominate nature:

> What has changed is Man, now no longer the pious spectator of God's wonders in the creation, and the worshiper of God himself above the creation, but Man the operator, Man who seeks to draw power from the divine and natural order. . . . By Magia man has learned how to use the chain linking earth to heaven, and by Cabala he has learned to manipulate the higher chain linking the celestial world, through the angels, to the divine Name.[131]

This new, activist picture of man, so Yates claims, was crucial to the emergence of early modern science in two distinct yet closely related respects. For one, it stands at the origin of the idea, powerfully expressed by Francis Bacon, that the vocation of the scientist is not confined to the passive contemplation of nature. Equally required is man's active involvement in tapping nature's hidden forces—to be revealed through an improved knowledge of its laws—and their subsequent harnessing for man's own good. This activist conception of science, Yates maintains, is exactly analogous to the activism of the Magus, with only the conception of nature that lurks behind the activism altered in the meantime.

Evidence for this historical rooting of Baconian utilitarianism was found by Yates in a book on *Francis Bacon* with the significant subtitle *From Magic to Science*. It had been written in 1957 by Paolo Rossi. His leading conception here was that Bacon's "idea of science as the servant of nature assisting its operations and, by stealth and cunning, forcing it to yield to man's domination; as well as the idea of knowledge as power," both go back to the tradition of magic and alchemy.[132] This is not to say, Rossi adds, that Bacon himself still adhered to the tradition from which his activist conception sprang. Rather, he gave these very ideas a decisively new twist in rejecting the magician's secretive restriction of his discoveries to the initiated few, wishing instead to turn science into a collaborative effort open to all and directed to the benefit of all. Also, Bacon condemned magic for its inflated individualism and for the rashness with which insights were claimed that could only be attained through much patient labor. The overall picture thus emerging from Rossi's case study fits in neatly with Yates' own, with her conception, that is, of Renaissance magical thought during the decades around 1600 being modified in the direction of key features of early modern science.

Such a conception can be extended further. After all, what underlies Bacon's activist idea of the scientific enterprise is his notion that nature may be subjected to manipulation. This notion, in its turn, was closely bound up with his advocacy of the experimental approach to nature (sections 3.4.1–3). Hence, it seems natural, in the Yatesian view of things, to regard the magical mode of operating as a natural precursor to scientific experiment in which nature is manipulated so as to display its hidden properties. Curiously, this is the line of thought about which Yates herself was least explicit, even though it is easily the most plausible of the various causal connections she adduced between the Hermetic movement and the rise of early modern science. It is also the one that has raised the least controversy. It may well be the most enduring contribution made by the Yates thesis to our understanding of those altered attitudes in Western thought (as compared to the ultimate scientific stagnation of the Hellenistic world) that made the Scientific Revolution possible.

Still, we should be aware of the limitations of the connection thus established. Experiment, after all, fulfilled a variety of functions in the 17th century, which Yates did not sufficiently distinguish. She certainly associated 'science' in too one-sided a fashion with 'operations'. It is not perhaps wholly accurate to say, with Hall, that "Dame Frances has a vision of modern science as being 'theoretical technology' ";[133] rather, she viewed science as in the nature of a 'gnosis', as we noted in section 3.3.5. Yet it certainly goes too far to link the new activism, with its magical origins now revealed, to the Scientific Revolution in its entirety.

At this point it becomes useful to recall Kuhn's idea of the Scientific Revolution as composed of two distinct intellectual movements: the transformation of the 'classical, mathematical sciences' and the birth of the 'Baconian sciences' (section 2.4.4). Kuhn recognized as one key virtue of his distinction that, with its help, a number of standing issues regarding currently accepted explanations of the Scientific Revolution may be cleared up. For example, beyond the long-standing idea of neo-Platonism as a contributor to the enhanced status of mathematics, the mathematical, classical sciences of the late 16th and early 17th centuries stood to gain very little from Hermetic patterns of thought. However, so Kuhn argued, the explanatory power of Hermeticism in relation to the emergence of early modern science is greatly enhanced if the thesis is confined principally to the birth of the Baconian sciences:

> Recognizing Francis Bacon as a transition figure between the magus Paracelsus and the experimental philosopher Robert Boyle has done more than anything else in recent years to transform historical understanding of the manner in which the new experimental sciences were born.[134]

If Yates' claims on behalf of Hermeticism as the cause of the Scientific Revolution appear greatly overdone where mathematical science, universal harmony, and a Sun-centered conception of the universe are concerned, and also seriously diminished in import by being confined to the furtherance of experimental science in the Baconian, mostly heuristic sense, the question arises of whether this must be the end of what the Yates thesis has to offer to our understanding of how the Scientific Revolution came about. Or is perhaps a wider truth to be discerned here?

A Scientific Revolution in two stages. For Yates the period of crucial transition, when the new conception of man and the world embodied in the Hermetic world-view gave way to early modern science, was the first decades of the 17th century. Despite the radical opposition between magical and scientific modes of operating which she recognizes, the transition itself took place in quite gradual a fashion:

> The figure of Bacon is a striking example of those subtle transformations through which the Renaissance tradition takes on, almost imperceptibly, a seventeenth-century temper and moves on into a new era. . . . It is perhaps somehow in these transitions from Renaissance to seventeenth century that the secret might be surprised, the secret of how science happened.[135]

One other illustration of how we must conceive of this subtle yet decisive transition is the well-known story of Descartes' mystical revelation in 1619. In the turn given to the event by Yates, the experience took place in an atmosphere saturated with Hermetic notions in their later, Rosicrucian variety, whereas the direction into which the experience ultimately guided Descartes was the very opposite, namely, the path of exact science:

> A transition has been made to an epoch in which what is still a Hermetic, almost a 'Rosicrucian', impulse towards the world results in valid scientific intuitions. But may not the intensive Hermetic training of the imagination towards the world have prepared the way for Descartes to cross that inner frontier?[136]

Having thus at long last arrived at the highest level of Yates' thesis, we find that the radical opposition between Hermetic and genuinely scientific patterns of thought, already made more and more elusive by the subtlety of the transition between them, evaporates entirely and gives way to an altogether novel conception of the Scientific Revolution, never to be set forth by Yates in more than the following sentence:

> It may be illuminating to view the scientific revolution as in two phases, the first phase consisting of an animistic universe operated by magic, the second phase of a mathematical universe operated by mechanics.[137]

Underlying such a view of what happened to science at one of its dramatic turning points in history is Yates' ultimately gnostic conception of science, which we know to have colored in a decisive manner her approach to the history of thought. For those to whom the idea of science as a revelation proceeding in successive stages means nothing, much of her Hermetic explanation of the rise of early modern science cannot but remain chiefly (even if not necessarily wholly) barren. For those who share it to a greater or lesser degree, much that otherwise would seem no more than vaguely suggestive and insufficiently substantiated in her enthusiastic prose may acquire a meaning and a significance less than fully expressed in the sentences she employs.

Frances Yates has more than once been identified, with greater or lesser caution, with the Hermetic views she wrote about. To this she used innocently to reply that, rather than being an "occultist" or a "sorceress," she was just "a humble historian whose favourite pursuit is reading."[138] Yet the questions raised about her personal views were not altogether unjustified. There remains something mysterious in her writing about these subjects. In reading her work one feels that she tries to define some ineffable core by circling around it and approaching it from all kinds of different viewpoints—her ultimate message is left to be guessed by the reader. This approach is quite appropriate to her subject, which is itself about things that lend themselves better to an intuitive grasp than to logical analysis. All this is why academic criticism of the usual type, such as has been indulged in throughout the preceding pages, too, may in the end prove inadequate. It is clear that at the original time of writing, in the sixties and seventies, Yates' work struck a responsive chord with many whose sensibilities have since shifted with time. Precisely because Yates' thesis is bound up to such an unusual extent with a peculiar, and also hard to define, conception of science did it seem important to sift out in somewhat nitpicking detail what in her thesis, if viewed from the perspective of more-conventional conceptions to which I myself by and large subscribe, appears worth preserving. There remains the lingering suspicion that Frances Yates may have glimpsed truths about the origin of early modern science whose full import at present still eludes us.

4.4.5. *The Revival of Scepticism*

One more contribution to the rise of early modern science said to have come from the Renaissance recovery of ancient philosophical thought concerns the 16th-century revival of scepticism. In a book first published in 1960 that has since been entitled *The History of Scepticism from Erasmus to Spinoza,* Richard H. Popkin, a historian of religious and philosophical ideas in early modern Europe, assembled a great deal of earlier and of fresh historical research on the thought of individual sceptical thinkers, held together by an overall thesis on the peculiar features of the period covered in his survey.[139] It is his contention that, whereas ever since Spinoza sceptical thought has been identified chiefly with doubts about revealed religion, during the period demarcated in his book sceptical arguments fulfilled a key role in the theological debate between the Reformation and the Counter-Reformation, with consequences spilling over from the domain of religion proper to create a profound 'sceptical crisis' in philosophy and science, too. Out of the resolution of this crisis as carried out in the first half of the 17th century, "modern philosophy, and the scientific outlook finally emerged."[140]

It all started, so Popkin begins his argument, with two distinct, quite unrelated events which happened to coincide fairly closely in time. One dates from the early twenties of the 16th century, when Luther widened his critique of certain practices of the Church into an all-out theological attack on the accepted sources of Christian faith: the tradition of the Church Fathers and the authority of the Pope and the Councils. Claiming instead that the conscience of the believer in reading Scripture provides the only valid criterion of religious knowledge, Luther opened a veritable Pandora's box. What over the centuries had been taken for granted suddenly became a matter of dispute: What criterion may validly be applied in judging the grounds of religious belief? Adherents to the Reformation and to Roman Catholic orthodoxy set out to undermine each other's proclaimed certainties. Protestants railed at the ingrained fallibility of humans presuming to tell us for certain what we must take God's word to mean, whereas Roman Catholics pointed out with relish the inevitably subjective nature of a man's conscience: to know a thing because one's conscience, or one's sense of personal illumination, tells one that it must be true can hardly be called a solid criterion for the objective truth of what are, after all, subjective convictions.

Into the turmoil of this dispute over the foundations of theological knowledge fell the humanists' discovery of the sceptical tradition in ancient Greek philosophy. The 'Academic' variety of scepticism, which holds that no certain knowledge is possible, just began to spread at the time, chiefly through the intermediary of Cicero's exposition of the doctrine. Arguments taken from it were employed on occasion by theologians to demonstrate that the opponent's stance inevitably led to scepticism (taken as a position to be avoided at all costs). In a few cases the sceptical position was used to underscore that in religious matters no certainty can be attained, so that one had better stick to tradition (as was often concluded) or leave one's fellow man at peace over his religious preferences (as a dissenting few suggested). The sceptical trickle turned into a flood, however, when, in 1562, the works of Sextus Empiricus (2nd century A.D.) began to be printed. Sextus had written out a systematic presentation of arguments developed in the other variety of scepticism: Pyrrhonism. The point of departure here is that Academic scepticism is inconsistent in holding that nothing is certain, since the affirmation inevitably contaminates itself. Thus, Academic scepticism can be seen merely to exchange the positive dogmatism of its opponents for a negative dogmatism, whereas the true opposition is between those—the dogmatists at large—who think that we can know some things with certainty (even if such knowledge is confined to the idea that nothing can be known with certainty) and those who are resolved to suspend judgment on everything, as the Pyrrhonists feel compelled to do. Hence, an exposition of Pyrrhonism as given by Sextus comes down to the systematic demonstration that, and why, all presumedly certain, dogmatic knowledge is inherently uncertain. Sextus was hardly original in adopting this position, but his pertinent writings happen to be the only ones to have been preserved. This is why a thinker who was of the second rank at best in his own time and almost entirely unknown during the Middle Ages could turn into 'the divine Sextus' within decades of the first printing of one of his works in 1562. The man to employ the battery of Pyrrhonist counterdogmatic argument as expounded by Sextus and to turn it into a formidable intellectual challenge to theology, philosophy, and science alike was, of course, the French humanist Michel Montaigne.

In a loose, rambling, yet highly persuasive literary style Montaigne gave reasons for doubting any statement that goes beyond one's personal sense impressions. Many of his reasons were taken from the Pyrrhonist arsenal, such as the unreliability of sense experience, the proneness of reason to error, the fact that the search for a criterion by which to judge the false from the true is bound to lead to an infinite regress, and so on. The Voyages of Discovery have made the last quandary even more pressing, Montaigne argues, because we know now that there is a New World where people have been found to apply quite different standards of reasoning and judging—who shall choose between theirs and ours? To all this and much more Montaigne added the sheer multiplicity of philosophical doctrine which had become so striking a feature of his own time. Now that, besides Aristotle, so many rival doctrines have been recovered by humanists' efforts, who is to decide which of these carries truth? No statement can be found in the writings of the learned for which a counter-statement is not to be found elsewhere (and here Copernicus is advanced as a striking example of how not even the seemingly self-evident truth of the stability of the Earth at the center of the universe has gone unchallenged). What, then, can we securely know?

As far as religion is concerned, Montaigne concludes that, because of the apparent impossibility of finding a reliable standard for the true faith, it is best to stick to tradition, which is embodied in the Catholic Church. For our knowledge of the natural world, so Popkin contends, Montaigne had created a veritable 'Pyrrhonian crisis':

> The occurrence of Montaigne's revitalization of the Pyrrhonism of Sextus Empiricus, coming at a time when the intellectual world of the sixteenth century was collapsing, made the 'nouveau Pyrrhonisme' of Montaigne not the blind alley that [certain] historians ... have portrayed, but one of the crucial forces in the formation of modern thought. By extending the implicit sceptical tendencies of the Reformation crisis, the humanistic crisis, and the scientific crisis, into a total *crise pyrrhonienne,* Montaigne's genial *Apologie* became the *coup de grâce* to an entire intellectual world. It was also to be the womb of modern thought, in that it led to the attempt either to refute the new Pyrrhonism, or to find a way of living with it.[141]

Let us find out how 'modern thought' was born out of efforts to come to grips with the sceptical crisis created by Montaigne.

The 'sceptical crisis' of the early 17th century. In Popkin's presentation, matters came to a head in France during the second decade of the 17th century. Sceptical arguments taken from Montaigne (mostly through Charron, who had systematized his teacher's literary ramblings into solid doctrine) were employed by a group of learned men who called themselves 'libertines'. Their true, as distinct from professed, allegiance to the Catholic Church, or to any religious belief for that matter, was (and still is) a debatable matter to say the least, and the suspicion that scepticism led to atheism gave rise to vehement disputes over the merits of Pyrrhonian thought. In this hot-house atmosphere, saturated with accusations and with polemical blasts and counterblasts, three ways were devised to come to grips with Pyrrhonism. One was to choose Aristotelian doctrine for one's secure bulwark: Since Aristotle had been right, his philosophy was above doubt and could not therefore be tainted by even the cleverest of arguments from the stock of the sceptics. Another, more imaginative approach was to out-sceptic the sceptics: to adopt all their considerations without exception, to carry these to extremes hitherto undreamt of, and then to show that nonetheless a principle of unshakable certainty could be found, upon the basis of which a novel philosophy could then be constructed. That is to say, we are talking here of the path along which Descartes arrived at his 'cogito' and of how he went on to employ it. The third approach, which preceded Descartes' by a decade or so, was rather to grant the irrefutability of the sceptical arsenal while upholding that, if our idea of what constitutes knowledge is made less rigorous, we find a middle way between scepticism and dogmatism, which Popkin calls 'mitigated' or 'constructive' scepticism. This, he argues, was the road taken by Gassendi and Mersenne. They arrived at it from opposite ends.

Gassendi was the one among the 'erudite libertines' who moved from the purely sceptical position (which he had used for a vehement attack on Aristotelianism) toward the view that, if we give up the ideal of proven knowledge and exchange it for hypothetical knowledge which takes care not to go beyond the appearances of things and their mutual connections, we arrive at a form of science which is no longer open to sceptical devastation while moving a step beyond the barren hopelessness of a perennial suspension of judgment on everything. The sceptics are right: It is not given to man to gain knowledge of the essence of things, and nature is not necessarily wholly transparent to our understanding. But the sceptics are wrong, too, for the inescapable limitations of human reason and sense experience do not condemn us to ignorance. Rather, we can construct a science of how phenomena appear to us, with our experience serving as a guideline and the verification of predicted experiences as a criterion.

Gassendi's friend Mersenne arrived at by and large the same position from the other end. Invoking *La verité des sciences contre les septiques ou pyrrhoniens* (The Truth of the Sciences against the Sceptics or Pyrrhonists), Mersenne sought to circumvent Sextus' compelling arsenal by granting that the sceptic has a point against traditional dogmatists. However, the arsenal falls flat before the sciences, which are given to another ideal of knowledge: the same type of hypothetical knowledge of sense phenomena and their mutual connections that Gassendi had come to recommend. In filling the remaining 800 pages of his book with an outline of what the sciences had meanwhile taught humanity Mersenne demonstrated that there is room for a middle way between scepticism and dogmatism.

This was how, according to Popkin, the 'outlook of science' was born out of the sceptical crisis of the early 17th century. Its triumph was delayed, though, because the very same sceptical crisis equally gave occasion to the construction of a reinforced dogmatism—Cartesian philosophy—which first had to be shown to be unable, after all, to slay the sceptical dragon. Either

Descartes had produced a novel dogma no less open to sceptical doubts than its predecessors (as Descartes himself at one point seemed ready to grant in acknowledging that his system ultimately rested upon a sense of personal illumination) or, if not, he appeared to be none but a 'sceptic despite himself'. In either case, scepticism ultimately came out victorious; the Cartesian onslaught had come to naught, and the time became ripe for the definitive triumph of science as a form of 'mitigated scepticism':

> Gassendi, perhaps even more than Mersenne, had accomplished one of the more important revolutions of modern times, the separation of science from metaphysics. Building his new outlook on a complete Pyrrhonism with regard to any knowledge of reality, or the nature of things, he was able to develop a method, and a system of the sciences, which, of all those of the seventeenth century, comes closest to the modern anti-metaphysical outlook of the positivists and pragmatists. . . . [He was] a most important link between Galileo and Newton, in moving from a conception of the 'new science' as the true picture of nature to one wherein it is seen as a hypothetical system based solely on experience, and verified through experience, a conception in which science is never thought of as a way to truth about reality, but only about appearance.[142]

Popkin's account of scepticism during the Renaissance and the early 17th century has been outlined here to the extent that it seems to hold a message about the causes of the Scientific Revolution. The question to be asked is: What exactly is that message? What can Popkin's thesis be taken to explain?

The 'sceptical crisis' and the Scientific Revolution. On the one hand, Popkin's book leaves the reader with the impression that the author has not been out to 'explain the Scientific Revolution'. Rather, he seems to take the Scientific Revolution as a contemporaneous event which contributed its share to the culmination of the *'crise Pyrrhonienne'* and for the rest went its own, independent way. Thus, in one typical aside Popkin talks about "the abyss of doubt that the crisis of the Reformation and the scientific revolution had opened up."[143] Further, he is well aware that a form of 'mitigated scepticism' as advocated by Gassendi, in particular, inhibited, rather than stimulated, the latter's powers as a creative scientist, in that it made him overcautious and conservative. Again, Popkin makes it clear that for Galileo there is no place in his account:

> In the view of Galileo and Campanella, God has given us the faculties to attain knowledge of the nature of things. However, our knowledge is only partial, unlike His complete knowledge. Nonetheless, we have no reason to question or doubt what we know, and we have no reason to restrict our knowledge to appearances, rather than reality. *The sceptical crisis seems to have bypassed these thinkers,* and left them only with doubts about the Aristotelian quest for certainty, but not with the quest itself. [My italics][144]

Considerations like these evoke a picture of the Scientific Revolution as one *agent* of the *crise Pyrrhonienne* among others, rather than as its *product*. But on the other hand, if one seeks to digest Popkin's full message one wonders whether we have not been granted an illuminating glance at the making of early modern science after all. True, Popkin's thesis is narrowly confined geographically. In surveying the *crise Pyrrhonienne* and efforts to resolve it he deals virtually exclusively with ideas of Frenchmen. Still, the French scene in the early decades of the 17th century surely constituted one of the focal points of the Scientific Revolution, with the very men on whom Popkin's account centers—Gassendi, Mersenne, Descartes—as the key pioneers, even though Popkin tends to treat these men as formed by, rather than as contributing their share to, the Scientific Revolution. Still, it is hard indeed to leave Popkin's book without gaining the distinct impression that an urge to resolve the quandary posed by the revival of ancient scepticism was instrumental in directing the thought of those protagonists of the Scientific Revolution whose work centered in France.

This is not, of course, to say that no other formative influences can be detected here. The point is rather that, while equipped with quite diverging intellectual backgrounds and habits, each of these men came to focus his constructive thought upon one and the same project—the renewal of science in one mode or another—by meeting the sceptical challenge head-on. The question is whether this partial explanation of the Scientific Revolution (partial in geographical scope; partial, too, in that the 'sceptical crisis' was hardly the sole agent here) can be widened further.

In one sense, Popkin certainly widened it. After all, his ultimate claim is that no less a thing than 'the scientific outlook' was born out of the sceptical crisis. By this he means that scientists began to learn to give up the quest for certainty; that metaphysics was cut loose from the pursuit of science; that sceptical objections to knowledge claims were met by confining such claims to the appearances rather than extending them to the reality of things.

One difficulty with these claims is that they are by no means equivalent. In equating without more ado the 'scientific outlook' with such a positivist conception of science, Popkin allows too much to fall by the wayside that is integral to the Scientific Revolution however conceived—Galileo's and Kepler's achievement in the first place. One may very well conceive of science as a 'middle way' between scepticism and dogmatism without necessarily renouncing the idea that science deals with reality—the latter position forming a source of controversy that is unlikely ever to dry up for good. Is it not sufficient here to claim that, whether or not science informs us about the reality of things, its inherently hypothetical nature moves us beyond dogmatic quasi certainties toward that middle ground where knowledge claims are no longer open to sceptical assaults?

Recall here McMullin's account of method in the Scientific Revolution as discussed in section 3.1.2. The transition depicted there from a 'conceptualist' idea of scientific method as still partly upheld by Galileo and Kepler toward the notion of 'pragmatic adequacy' to be expressed in the work of such men as Huygens and Newton now gains further relief through Popkin's thesis. It enables us to see how this transition was prepared ideologically, so to speak, by our early French empiricists in search of a way out of the sceptical crisis. In this sense can we fruitfully talk of the revival of scepticism as a cause of one important aspect of the Scientific Revolution.

Recall also Rabb's thesis on early modern science as one principal agent in both the sharpening and the ultimate resolution of an all-European 'crisis of authority' (section 3.6.1). Again, this thesis gains further relief from Popkin's demonstration of the extent to which the 16th-century sceptical onslaught expresses an acute sense of loss of intellectual certainty, with the new science eventually coming to the rescue as the middle ground upon which intellectual disputes may peacefully be settled.[145]

One further consideration seems to impose itself. Science does not deal in certainties but rather in provisional truths, subject to revision over time. Having noted, in a previous chapter, how modern notions of the temper of science came to be worked out over the course of the 17th century—the provisional nature of its hypotheses; the abandonment of authority as a source of reliable knowledge; a reorientation toward the future—is it not tempting to suppose that such a revolutionary conception of human knowledge may only be attained through the intermediary of a sceptical stage in human thought? Such humanists as Erasmus and More and Rabelais reacted against scholastic dogmatism by adopting that humorist sceptic of the 2nd century A.D. Lucian of Samosata as their model of how to poke fun and jeer at all knowledge claims (not counting revealed knowledge, to be sure). To write tongue-in-cheek; to seek an escape in the imagination; to mock and contradict oneself and others—such were the expedients these men employed when facing the same epistemic emptiness to be expounded in a more serious vein by Montaigne.[146] A plausible case can be made for the idea that a civilization has to go through such a phase of utter scepticism to rid itself of the spell of dogmatic thought and to hit in the end upon a more constructive way out of the resulting abyss of uncertainty—the way out embodied in early modern science.

Such a widened case has its limitations, too. As Popkin observed, there is no evidence that Galileo was moved by a 'sceptical crisis', even though it remains true that he was steeped in the thought of the humanists and may very well have taken from their example the urge to go against dogmatic claims while also giving the urge a more constructive twist than had been within their reach. Also, the Hellenistic world provides an antecedent example of how a sceptical mode of thought—in fact, the birth of scepticism itself—came up with the purpose of laying to rest confusion over a multiplicity of philosophical doctrines. Sceptical thought even coexisted for a time with a flourishing move-

ment in science—the Alexandrian variety. The latter petered out, whereas its
17th-century counterpart was to go on from there and conquer the world. In
the 3rd and 2nd centuries B.C. scepticism was not given the constructive twist
it was given in France in the early decades of the 17th century, and we are left
wondering why this was so. Once again, western European thought during the
Renaissance appears as little more than a replay of a performance originally
staged by the ancient Greeks. What, then, caused those characteristic twists to
occur at the end of the later period? It seems about time for us to collect the
efforts at explanation of the riddle of the birth of early modern science that we
have surveyed so far and consider where they have got us.

4.5. THE HARVEST OF THE 'INTERNAL' ROUTE

In making the effort that follows now at determining what our first collection
of purported explanations of the Scientific Revolution has actually taught us, I
wish by all means to refrain from any attempt at imposing my own personal
preferences upon the reader. Inevitably, one person is struck most forcefully
by this explanation, another by that, and it seems presumptuous to interfere
unduly with such personal judgments. Still, the historian of historiography can
perform a possibly useful job in suggesting guidelines which may assist the
reader on a path toward a judgment that, in the end, cannot but be a matter of
personal taste. I hope to provide such sustenance through a brief yet systematic
effort at comparison between our explanations. The comparison takes place
along a variety of dimensions: of time, of place, of explanatory scope, and of
susceptibility to restatement in terms of a sharpened definition of the original
problem of explaining the Scientific Revolution.

Before embarking on the comparison I wish to point at one striking feature
our explanations surely have in common. This is that each and every one of
their authors can be seen to have undertaken a bold effort to come to grips with
one of the major riddles of history by marshaling large quantities of erudition,
imagination, and orderly thinking into the service of such an overarching objective. We have been surveying here history writing in the grand style, and
whatever shortcomings we may go on to point out should never be seen to
detract from that principal merit. No less a merit of these efforts at explanation
is that they have taught us much that is new about the overall climate of thought
in which the protagonists of the Scientific Revolution grew up. Whether or not
we are entirely satisfied with the explanatory power of our various theses, we
have learned a great deal about mathematical, Aristotelian, Hermetic, and
sceptical modes of thought during the period immediately preceding the Scientific Revolution that otherwise might have remained obscure.

Still, there are some difficulties with our theses. Of these, problems with
time are the least important. Claims for causal continuity between Renaissance

thought and early modern science are not, by their very nature, affected by a difficulty facing similar allegations of causal continuity between the Middle Ages and the early modern period. This difficulty is simply one of time lag. 'Medieval' claims have always had to construe connecting links stretching several centuries. Only the Duhem thesis, in taking the Revolution in Science to have occurred in the 14th century, has been exempt from such a requirement; Hooykaas' thesis of the wholesome idea of the contingency of nature and of its furtherance by Tempier's decree, however, must face the temporal gap. The author has met the difficulty by invoking an overall "favourable climate for science" through which the idea was preserved (more on this in section 5.2.8).[147] None of the other explanations treated in the present chapter is affected by the difficulty, since for Renaissance continuities a flawless connecting point presents itself in the smooth temporal transition between Renaissance and early modern science.

Problems of place and problems of explanatory scope are another matter, however. Even if we were to accept every single explanation *qua* explanation, it can still be seen that each and every one of our explanations is incomplete in that it addresses only part of the problem at issue. Both the 'Archimedean' explanation and its 'Aristotelian' counterpart picture Galileo as the culmination point of a particular tradition in Renaissance thought. Both are devoted exclusively to Italy; neither has anything to say about other centers of the Scientific Revolution. By contrast, the 'sceptical' explanation, while capable of being extended to the whole of Europe, is chiefly about France, with Gassendi, Mersenne, and Descartes at the center of attention. Again, the 'Hermetic' explanation, while addressing Kepler, Mersenne, Descartes, and Bacon, in particular, is conspicuous in leaving Galileo out of the picture altogether. Only the 'Copernican' explanation covers the Scientific Revolution in its full geographical scope, in that the effort to meet objections to the Copernican hypothesis led to a reconceptualization of standing scientific ideas by pioneers all over Europe. However, while the 'Copernican' explanation covers the revolution in the domains of astronomy and of mathematical physics, it has nothing at all to say about the emergence of the Baconian sciences. Conversely, the 'Hermetic' explanation, which works so poorly in the domain of mathematical science, has appeared strongest if its explanatory sphere is confined to the Baconian sciences.

In face of such apparently damaging difficulties we cannot avoid the question of whether they are inherent to our original problem of seeking causes of the Scientific Revolution or not. Can they perhaps be overcome? If so, how?

Before seeking a way out of our quandary, let us exacerbate it further. Thus far we have only addressed problems arising from explanations whose validity we have accepted without questioning. But what is it that their authors have set themselves to explain? Here we meet one final problem—possibly the greatest of all. Consider what our survey of explanations of the 'failure' of the

Greeks to make a Scientific Revolution implies for our appraisal of what, at the time of the Renaissance, grew out of the humanist effort at recovery.

The principal result of the effort was that the overall state of science in western Europe attained a level roughly equivalent to the state in which the Greeks had left it (and in which the Muslim world had recovered it many centuries earlier). From the vantage point thus reached, three possible consequences might in principle have ensued. One possibility is that the Greek corpus might have remained stagnant. This is what happened twice in history: at the end of the fertile stage of science in the Hellenistic world and again during the entire Byzantine period. Alternatively, the legacy might have been somewhat enriched and extended. This, too, happened twice. It is what was done to the Greek legacy in the Muslim world; it also happened in Renaissance western Europe. Finally, such extensions as did ensue might, in their turn, usher in a novel type of science: the science of the Scientific Revolution. This, as we all know, happened only once. Hence, it is here that our true problem is situated— it is in this particular context that the search for causes of the birth of early modern science out of Renaissance thought acquires its full significance.

It follows from the considerations just advanced that for an explanation of the Scientific Revolution taken from aspects of Renaissance thought it is not sufficient to pile up data from such-or-such a current in Renaissance thought and seek a connection with elements in the thought of the scientific pioneers of the 17th century. Rather, the search must be for those characteristic twists in the tradition that constitute as it were the critical junctures, the truly creative steps which, together, constitute the onset of early modern science. For example, it was one thing, for Rose, to point at the indispensability to Kepler of the antecedent printing of Apollonios' work on conic sections; it is something else to go on to explain what enabled Kepler to employ Apollonios' ellipses for a purpose to which no one else had ever subjected them.[148] Similarly, the problem is not so much to show that Galileo worked in an Archimedean tradition and was acquainted with refinements in Aristotelian methodology, nor is it to lay bare Kepler's preoccupation with Pythagorean thought or Descartes' and Mersenne's and Gassendi's concern with scepticism. The problem is not to explain how these men came to use the Greek legacy (however interesting a subject this is in its own right) but rather how they managed to transcend it.

If we review our list of explanations once again, while focusing now on their respective capacities to meet this particular challenge, we do indeed come up with a few striking insights. Kuhn has explained to us why Copernicus' reform could hardly have been made in the immediate wake of Ptolemy, but rather required intervening centuries for the weaknesses of the latter's solution to the problem of the planets to become manifest. He also showed how this reform, kept so limited by its originator, could nonetheless be elaborated into vital portions of early modern science through efforts to meet seemingly inescapable objections. Similarly, Yates pointed the way toward realizing how the activist twist given to the Hermetic legacy by Bruno and others could inspire the onset of scientific experiment, heuristically employed.

These are no mean results, and we shall keep them in mind once we make up our final reckoning. Meanwhile, we go on to note that remaining explanations have yielded less valuable results in this respect.[149] With Aristotelian methodology it has appeared hard to find how it might have contributed so much to Galileo's creative steps—the less so in view of crippling habits of thought in the Aristotelian tradition as outlined by Maier and by Grant. By contrast, with the sceptical heritage the twist indeed manifested itself, even though it has remained unexplained. Owing to Popkin's work we can now see that Gassendi and Mersenne did seek in science a middle way between scepticism and dogmatism—yet we are left to explain what enabled them to make this forward step. Equally, Rose made it clear that the Archimedean legacy was still in need of further enrichment before it could yield Galileo's science—we do not yet understand what made Galileo go ahead and do it.

All this means that we have now reached from the vantage point of Renaissance thought the very same outcome arrived at earlier when we surveyed the dead end into which Greek science had run. There we concluded that the Greek legacy stood to gain from a cultural transplantation, and we went on to list three respects in which, as a range of historians of ancient science suggested, the civilization of western Europe differed in a possibly relevant manner from its Hellenistic counterpart. We encountered the suggestion that, in the course of the late medieval and the Renaissance periods, western Europe came to offer a fertile environment for a Scientific Revolution to occur because of (1) formative influences emanating from the biblical world-view; (2) an atmosphere of technological innovation and vitality; (3) the presence of social layers responsive to the justification of science as a pursuit with a dignity and an autonomy of its own. These three suggestions enjoy one great advantage over most of those discussed so far in that they cover the entire subcontinent and, hence, may be expected to address the Scientific Revolution in its full geographical scope. After all, the Scientific Revolution happened all across western Europe—is there not, then, good reason to seek explanations of the Scientific Revolution in events that affected equally all of western Europe?

In order to follow such an approach we must leave the domain of thought on nature proper and go on to inspect a great variety of historical developments which, in one way or another, were specific to western Europe and thus may be invoked for an explanation of how it was that what had stagnated in ancient Greece came to full fruition in western Europe of the early 17th century. Over the past sixty years or so, historians have been quite vocal in suggesting how one or another of such events may have caused the emergence of early modern science. The next chapter is devoted to finding out how they made their respective cases. It will take us from the Reformation to the Voyages of Discovery, from the mechanical clock to mercantile capitalism, from the printing press to the arts and crafts, from the practical needs of the gunner and the sailor to the rise of an ideology for science; and when such a widely cast panorama of western European history has fully passed before our eyes, we shall ask once again what we have learned about the original riddle of the rise of early modern science.

✺ FIVE ✺

THE EMERGENCE OF EARLY MODERN SCIENCE FROM EVENTS IN THE HISTORY OF WESTERN EUROPE

5.1. RELIGION AND THE RISE OF EARLY MODERN SCIENCE

T HE GENERAL SUBJECT OF THE RELATIONS, in past and present, between science and religion has yielded an enormous literature. Not many other historical issues are capable of arousing quite so deep-seated emotions. To find the proper balance between detachment and involvement seems almost beyond reach when this particular subject comes up for discussion—partisanship and irrelevance being the Scylla and the Charybdis on which more than one historiographical ship has been wrecked.

In its vastness, the subject covers a broad spectrum of widely diverging topics. One may, for example, examine present-day struggles over evolutionary theory versus creationism or 17th-century tracts refuting the notion that the cultivation of science leads to atheism, and in both cases find oneself to be a— perhaps unwilling, or even unwitting—participant in a far larger debate, with views from one domain spilling over into others.[1]

Even in confining oneself to questions about religion and the Scientific Revolution a number of really distinct topics comes up. For example, it has been argued by Amos Funkenstein that the 17th century witnessed the rise of a new, rather short-lived genre of writing on religious affairs in relation to scientific matters. He calls the genre 'secular theology'. It may be taken to consist of the numerous treatises devoted by such early modern scientists as Galileo (his "Letter to the Grand-Duchess"), Pascal, Boyle, Newton, and many others to religious and theological issues which come up in connection with the new science and their own stakes in it. Quite a few such treatises in 'secular theology' have been examined by historians—as a genre with distinct features of its own it awaits its students.[2]

Another issue that comes up in this connection is how early modern science led to the alteration of vital aspects of Christian belief. Westfall, in his earliest book, *Science and Religion in Seventeenth-Century England,* summed up the transition involved here in a succinct manner:

In 1600, Western civilization found its focus in the Christian religion; by 1700, modern natural science had displaced religion from its central position.[3]

Copernicus' reform in astronomy, so Westfall goes on, had already called into question the interpretation of significant Bible passages. Furthermore, the advent of early modern science raised big questions about revealed truths as compared to findings attained through the methods of science; about man's place in the cosmos—a subject on which both theologians and scientists possessed or developed views of their own; and, perhaps most disturbing of all, about the position and function of God in a mechanized universe. Here is how Westfall brought out the ultimate paradox that lies hidden here:

> Despite the natural piety of the virtuosi [English 17th-century scientists], the skepticism of the Enlightenment was already present in embryo among them. To be sure, their piety kept it in check, but they were unable fully to banish it. What else can explain the countless dissertations on natural religion, each proving conclusively that the fundamentals of Christianity are rationally sound? They wrote to refute atheism, but where were the atheists? The virtuosi nourished the atheists within their own minds. Atheism was the vague feeling of uncertainty which their studies had raised, not uncertainty of their own convictions so much as uncertainty of the ultimate conclusions that might lie hidden in the principles of natural science. With wonderful certainty and assurance each virtuoso proved the existence of God from the creation; yet repeated too often, the assurance acquired an odor of insecurity. With Newton the insecurity was growing toward open fright. . . . Following the birth of modern science the age of unshaken faith was lost to western man.[4]

Here the Scientific Revolution is seen as the decisive event to turn the relation between science and religion into a deeply problematical one ever since, with the upholders of Christian belief clearly on the defensive. Frequently, this problematical relationship has been projected back onto the days before the Scientific Revolution. Thus the rise of early modern science was described as one particular stage in what Andrew D. White, in 1896, called the *History of the Warfare of Science with Theology in Christendom*. In such a—once widely shared—view the emergence of early modern science appeared as the liberation of independent scientific thought from the fetters of dogmatic theology. Rather one-sided interpretations of Galileo's trial figure prominently in such accounts.

Crude notions like White's have few scholarly adherents anymore. Weaker versions, however, are still around with good reason. After all, it remains an incontrovertible fact of history that, to say the least, the new science was accorded a less than enthusiastic acclaim by many religious authorities of the time. There is therefore more than a tinge of a paradox in the assertion that, rather than having been gained through prolonged battles with religious forces, the rise of early modern science was due at least in part to developments in Christian thought—in particular, to certain aspects of Protestantism. That this is so has been maintained on two distinct levels of argument. The first effort to make such a case (1935/1938) has entered the historiography of science as the 'Merton thesis'. Merton's original thesis holds that the more than proportional participation of Puritans in 17th-century English science was at least partly due to a specifically Puritan ethic which tended to foster the cultivation of science. In subsequent debates many historians have rightly or wrongly taken this thesis for an effort at explanation of the Scientific Revolution, and it is in this capacity that we shall discuss it in section 5.1.2. Treatment is preceded by a rival conception put forward by Hooykaas. He argued in 1972 that the biblical world-view, which was brought to the forefront of western European thought through central tenets of Reformed belief, supplied natural science— hitherto cultivated in a predominantly Greek spirit—with precisely such vital elements as were needed for early modern science to arise. This particular attempt at explanation of the Scientific Revolution shall hold our attention first.

5.1.1. *Hooykaas and the Biblical World-View*

For the Dutch historian of science R. Hooykaas (born 1906), nature is first and foremost made up of *things,* concrete and tangible rather than ideal and mathematical. To be sure, he has duly acknowledged the element of quantification in the changes in 16th- and 17th-century science that together make up the Scientific Revolution, yet neither Koyré's mathematical realism nor Dijksterhuis' mathematical instrumentalism has been able to satisfy his conception of nature. For him, in an important sense rational mechanics is almost foreign to the realm of the sciences of *nature,* and the emergence of this particular discipline in the early 17th century cannot therefore be allowed to stand for the core of what the Scientific Revolution was about.

Rather, in Hooykaas' view the principal key to the Scientific Revolution resides in the historical process through which humanity gradually learned to curb its inborn aptitude to overextend its powers of reasoning. Not until reason was tamed into allowing itself to be guided wherever the facts of nature led could early modern science be born. A humbler attitude toward what nature has to say to us than displayed by the ancient Greeks and most of their medieval successors was therefore required. Not that Hooykaas has ever advocated a pure and simple empiricism: It is the right balance between reason and experience that has to be struck again and again. Yet before the onset of early modern science reason had generally been given far too much weight in the balance (section 4.2.1). The two principal events that contributed most (one directly, one indirectly) to granting to experience its right place in the balance were the Voyages of Discovery and the Reformation.

In a later section (5.2.8) we discuss how, according to Hooykaas, the Voy-

ages of Discovery helped produce a new climate of thought in which nature
was allowed to speak with its own voice rather than being squeezed into the
Procrustean bed of rational *a priori* constructions. In the present section we
examine Hooykaas' account of how, in his view, the Reformation contributed
greatly to the same effect. While pertinent notions run across many of his
books and articles from the late thirties onward, they were most cogently set
forth in his little book of 1972, *Religion and the Rise of Modern Science*.

The biblical world-view.   The Bible, Hooykaas allowed, does not present a
world-*picture* in the sense in which one can speak of the world-picture expressed by Greek science and natural philosophy, for there is no biblical science, nor does the Bible claim to present one. Yet it is possible to speak of a
certain biblical world-*view*—an overall outlook on nature and on how nature
is related to its Creator. In several respects, Hooykaas maintained, this biblical
world-view contains all the necessary antidotes to Greek rationalism. In Greek
science no less than in other, more overtly religious conceptions of nature,
nature belongs to the domain of the numinous. The biblical world-view is
unique in that, owing to its conception of an absolutely sovereign God, nature
is radically 'de-deified'. When applied to the scientific investigation of nature,
this leads to an attitude at once more humble and more forceful to nature—to
acceptance as well as to dominion. God's acts are inscrutable to man: No human *a priori* construction can ultimately fathom His ways, and this means that
we must accept as divine gifts whatever we find in reality to be the case:

> In this voluntaristic way of thinking, the order of nature was not *our* logical order,
> but that *willed* by God.[5]

It equally follows that, opposite to the Greek view, it is not impossible for
Art—that is, for human intervention in natural affairs—to imitate, or even to
surpass, Nature. Also, the Greeks despised manual work, thus barring a truly
experimental science, whereas in the biblical view of things it is an honor to
work with one's hands, since labor in general is seen as a service to God and
as sanctioned by such service.

Common to these various features of the conception of God and nature
contained in the Bible is precisely what was needed to produce a consistently
fruitful interaction between reason and experience. Surely the Greek powers
of abstract reasoning and of thinking up idealized constructions were necessary for mature science. But the biblical humility toward accepting the facts of
nature as they are, combined with a view of man as fitted out by God with the
power to take nature on, was equally indispensable for early modern science
to arise:

> Metaphorically speaking, whereas the bodily ingredients of science may have
> been Greek, its vitamins and hormones were biblical.[6]

Throughout large portions of his book Hooykaas is concerned with illustrating these general tenets. He shows how, in the 16th and 17th centuries, many devoutly Christian scientists, in setting forth their views on the peculiarities of their pioneering science, displayed these very features of the biblical world-view, now for the first time applied to the scientific investigation of nature. The question that naturally arises at this point in the argument is why the biblical world-view, which had been around for a very long time, did not affect the cultivation of science in such a decisive manner earlier than the 16th and 17th centuries A.D.

The priesthood of all believers. Although this question is not specifically handled at one definite place in Hooykaas' book, two distinct, mutually supplementary answers may be gleaned from his densely presented argument.[7] On the one hand, so he argues, we must understand that, when from the 13th century onward natural philosophers became familiar with the immensely impressive achievements of the Greeks, they lacked the boldness at first to confront these squarely with the world-view implicitly present in the Bible:

> In the Middle Ages . . . the biblical view was only superimposed on, and did not overcome, the Aristotelian conception.[8]

The uneasy compromise struck in varying degrees by successive generations of scholastics, however, gave way to a much fuller confrontation of Greek rationalism with biblical voluntarism—of 'Athens' with 'Jerusalem'—as a result of the Reformation. The key concept Hooykaas introduces here into the argument is

> the typically Protestant (perhaps even 'Reformed') emphasis on the 'general priesthood of all believers'. This implied the right, and even the duty, for those who had the talents, to study Scripture without depending on the authority of tradition and hierarchy, together with the right and the duty to study the other book written by God, the book of nature, without regard to the authority of the fathers of natural philosophy.[9]

In other words, the conventional idea that God revealed Himself in two books—the Bible and nature—thus acquired a new import through a fundamental tenet of Protestant belief. Both revelations had to be studied with the same attitude of devout and humble acceptance of the immediately given, unperturbed by the intervention of authority or tradition. In the case of God's other book this inevitably entailed a deep respect for the hard and often unwelcome *facts* of nature over and above the *a priori* constructions our fallible reason might wish to make in order to explain them away.

Now this is precisely the approach that fits in best with Hooykaas' conception of nature as the domain of the hard and stubborn fact. Hence his key examples of the effect of the new, biblical respect for nature as a given are those very occasions in the history of 16th- and 17th-century science when the investigator of nature, once having made a beautiful construction, gives it up in the face of the conflicting fact. The crown witness here is Kepler, with his celebrated rejection of his own *hypothesis vicaria,* seemingly matching the data so well and reached after so much tortuous calculation, because of a misfit of 8 minutes of arc. Here, at this key juncture on the road to early modern science, we find the triumph of the biblical conception of nature over Greek rationalism, expressed in Kepler's own vision of 'the astronomer as a priest of God to the book of nature':[10]

> He submitted to given facts rather than maintaining an age-old prejudice; in his mind a Christian empiricism gained the victory over platonic rationalism; a lonely man submitted to facts and broke away from a tradition of two thousand years.[11]

Throughout his book Hooykaas is concerned to adduce examples of this novel attitude of reverence—especially but by no means exclusively among Protestant scientists—toward the facts of nature.[12]

Qualifications of the thesis. In all this Hooykaas takes care to qualify. In the first place, he does not assert that the new respect for the empirical came *solely* from the biblical world-view as expressed in the Protestant idea of the priesthood of all believers. Hooykaas thinks that 14th-century nominalism, the Voyages of Discovery, and a Christian-inspired, heightened respect for manual labor had already begun to turn European minds in the same direction.

Also, Hooykaas is well aware that there certainly existed a narrow-minded antiscientism among leading Protestant theologians. On this score a large part of his pertinent argument is directed toward showing that the theme of their literalist clinging to the Bible has greatly been overdone in the literature, to the point even of historians' invoking wholly spurious quotations.[13] Furthermore, even where biblical literalism or antiscientism indeed prevailed, this did not prevent the healthy ingredients of the biblical world-view from doing their work in the minds of a great number of Protestant scientists, who were not constrained by Church authorities as their Roman Catholic colleagues were.

Finally, and most importantly, Hooykaas does not of course claim that the Scientific Revolution was exclusively the work of Protestant scholars. The road toward the new respect for nature was trodden by Catholics and Protestants alike.

Nevertheless, there remains a question here. Nowhere does Hooykaas face squarely the issue of how, if indeed the biblical world-view provided a key ingredient to early modern science, this new ingredient began to affect Catholic scientists in the 16th and 17th centuries, after rather consistently having failed to do so during the reign of scholasticism. At this point two paths seem open, both in fact taken by Hooykaas. One is to stress the relative nature of the thesis by pointing out that the biblical world-view was only one among a variety of historical factors contributing to the new respect for the facts of nature. Another is to shift the argument from direct causation to statistical regularity. Already in the late 19th century the Swiss naturalist Alphonse de Candolle had pointed at the disproportionally high contribution of Protestants to science from the 17th century onward.[14] One other scholar to draw inspiration from de Candolle's pertinent calculations was Robert K. Merton, whose thesis on Puritanism as a boon to the scientific vocation we shall now subject to closer scrutiny in the next section.

5.1.2. *The Merton Thesis*

The Merton thesis was not originally set up as an explanation of the Scientific Revolution; rather, its adherents and critics turned it into one. Nor is this 'Merton thesis' more than part of the book that in reality embodied the full Merton thesis, namely, the doctoral thesis defended by him in 1935 and published in 1938 under the title *Science, Technology and Society in Seventeenth-Century England.*[15] The 'Puritanism and science' hypothesis which, among historians of science, has come to stand for 'the Merton thesis' in reality takes up only three to four chapters out of eleven, and in his preface to the 1970 reprint of the book Merton complained with justice about the relative neglect of the other elements he had deemed relevant to his large-scale inquiry into how science and technology were related to the environing society of 17th-century England.

The author's intention with the book can perhaps best be described as 'an essay in the historical sociology of vocations'. His starting point is the observation—underpinned with a statistical analysis of 17th-century careers culled from the *Dictionary of National Biography*—that in the course of the century a strongly increasing number of Englishmen devoted their careers to science. Given that science, at the beginning of the century, had not yet been institutionalized or been widely regarded as a legitimate activity in its own right, there appeared reason to inquire into the social setting of such a notable concentration of talent in the realm of science and, more specifically, into the how and why of the various foci of scientific activity that more detailed prosopographical research additionally revealed.

Taking his cue from the celebrated Weber thesis—which aims to demonstrate a crucial link between the Protestant ethic and the rise of the spirit of capitalism—Merton elaborated Weber's one-line suggestion that a similar link might exist between the Protestant ethic and the rise of empirical science into a full-fledged inquiry into precisely such a causal connection, expressly limited, however, to the English case. To his thesis that English Puritanism and science had certain important values in common which helped explain the contemporary rise of science in that country, Merton devoted three consecutive chapters which were intended to make the following points:

1. Irrespective of doctrinal background and of the original intentions of the Founders, a specifically Puritan *ethic* can be distinguished. It developed during the 16th and 17th centuries and may be characterized, in Weber-like fashion, by its 'intramundane asceticism' in the sense of a practical utilitarianism devoted to the honor of God but directed toward worldly activities rather than deflected into the 'extramundane asceticism' of monastic life. As with Weber, Merton's crown witness for the thrust of this ethic was the *Christian Directory* of the Puritan divine Richard Baxter.

2. An examination of how English scientists themselves justified their cultivation of science reveals the expression of many of these very motives, with a strong emphasis put upon the useful nature of science. Equally, there appears to be a remarkable similarity between the practical orientation of the Puritan ethic and the experimental bent of the new science. Among these invocations of scientists' efforts at justification, Merton's principal (though by no means exclusive) witness was Robert Boyle. Merton's main conclusion at this point was:

> Empiricism and rationalism were canonized, beatified, so to speak. It may very well be that the Puritan ethos did not directly influence the method of science and that this was simply a parallel development in the internal history of science, but it becomes evident that, through the psychological sanction of certain modes of thought and conduct, this complex of attitudes made an empirically founded science commendable rather than, as in the mediaeval period, reprehensible or at best acceptable on sufferance. *In short, Puritanism altered social orientations.* It led to the setting up of a new vocational hierarchy. . . . One of the consequences of Puritanism was the reshaping of the social structure in such fashion as to bring esteem to science. This could not but have influenced the direction of some talents into scientific fields which otherwise would have [been] devoted to callings which were, in another social context, more highly honored. [My italics][16]

3. Having thus established, through an abstract derivation guided by pertinent statements of English scientists, the hypothesis that "there is, to some extent, a community of assumptions in Protestantism and science,"[17] Merton went on to test the hypothesis against statistical material regarding the distribution of scientists over various religious denominations. To begin with, Merton found a remarkably high number of Puritans among the founders of the Royal Society. Extending his research at this point to other times and places, he found that available statistics invariably indicated a far greater proportion of Protestants among scientists than might be expected on the basis of the distribution of Protestants and non-Protestants (primarily, of course, Roman Catholics) among European populations at large. Merton did not take this to mean that, therefore, science could only have flourished among Protestants. Clearly, science had been cultivated in Catholic countries, too. However, so he asserted, this does not refute the thesis—it only serves to underscore its *relative* nature.

This was a feature to which Merton attached much importance throughout his study. The anticipated objection only went to show

> that other circumstances may equally conduce to the espousal of science and that these factors may be sufficiently effective to overcome the antagonism involved in the existing religious system.[18]

Altogether, however, Merton felt sufficiently justified to conclude the pertinent portion of his book as follows:

> The formal organization of values constituted by Puritanism led to the largely unwitting furtherance of modern science. The Puritan complex of a scarcely disguised utilitarianism; of intramundane interests; methodical, unremitting action; thoroughgoing empiricism; of the right and even the duty of *libre examen;* of anti-traditionalism—all this was congenial to the same values in science. The happy marriage of these two movements was based on an intrinsic compatibility and even in the nineteenth century, their divorce was not yet final.[19]

The remaining chapters of the book were devoted to pointing out the mostly utilitarian motives (taken from contemporary usage of science for mining, navigation, gunnery, and suchlike practical pursuits) which, according to Merton, helped explain the specific directions into which the new scientific activity primarily tended to flow. We return to this particular portion of the original Merton thesis in section 5.2.3.

The Merton morass. *Habent sua fata libelli,* and this has been particularly true of Merton's book, which has been taken on by critics more than perhaps any other book on 17th-century science.[20] In 1973—thirty-five years after the book had appeared, that is—the author of *Science and Religion in Seventeenth-Century England,* Richard Westfall, emphatically declined to

> venture into the morass of Puritanism and science, which seems to exist only to swallow all those who attempt to cross it. Although the question was muddy in the early 1950's, at least it was relatively fresh. Twenty years of polemics appear to me to have done nothing to clarify it; they have done even less to keep it fresh.[21]

It is our historiographer's duty to go ahead nonetheless and to find a safe and, if we are lucky, somewhat refreshing path through the morass. The most promising route, for our purposes at least, seems to lie in an attempt to find out what the Merton thesis actually has to do, in the view of author and critics alike, with the causation of the Scientific Revolution—an issue that is unlikely to have appeared from our summary to be particularly obvious. We set out to cross the morass by taking a quick survey of various types of criticism that have been hurled at the thesis. For the most part these concern the tenability of the various links in the chain of Merton's original argument.

Critique of the Merton thesis: What 'Puritans' had to do with science.   There is virtually no link in Merton's argument that has not been subjected to the severest criticism. Thus, Richard Baxter has been declared to be singularly unrepresentative of 17th-century Puritanism. Merton's quite broad conception of 'Puritanism' has been variously criticized as too comprehensive, too exclusive, altogether misconceived, or otherwise unable to accommodate the elusive, continually shifting patterns of religious allegiance throughout the political and religious convulsions of the century. The reliability of Merton's counts among the founders of the Royal Society has been called into doubt, partly on the preceding ground and partly because of the inclusion of so many worthy nonentities among the founders. His theoretical derivation, in Weber's footsteps, of 'intramundane asceticism' from Calvinist predestination beliefs has equally been disputed. Also, it has been argued that Merton handled both science and religion as essentially homogeneous entities, as if throughout the course of a most eventful century no significant changes had occurred in either.

From such criticisms the most varied conclusions have been drawn. These range from alternative causal connections between 17th-century science and Protestantism (Hooykaas' previously discussed account is by far the most closely argued among such attempts), through the idea that Latitudinarianism (or, in the view of another critic, 'hedonistic libertarianism') rather than Puritanism was particularly favorable to the cultivation of science, to the flat denial that there is any link to be detected here at all.

Matters such as these—into the validity of which we are not going to delve here any further—at least had the merit of confronting actual statements and intentions of the author. However, Merton has also, and quite undeservedly, been accused of explaining the rise of English science exclusively through the Puritan ethic, whereas in reality he had throughout his study taken great pains to point out, not only that there was a mutual relationship between the two, but also that other factors might well have been equally or perhaps even more important in accounting for the phenomenon. Generally speaking, throughout this flurry of criticism the original historical problem Merton had set out to solve was frequently lost sight of: the explanation, through changes in social values, of shifts in vocational interest in 17th-century England.[22] But how has Merton's stated aim to solve this particular problem come to be seen as an attempt at explanation of the Scientific Revolution? As far as I can detect, this was chiefly the work of A. R. Hall, when he subjected the Merton thesis to scrutiny in his celebrated article from 1963, "Merton Revisited."

The Merton thesis and the explanation of the rise of early modern science.   Nowhere in his book had Merton claimed that he was going to explain, or had explained in fact, the rise of early modern science. This appears unmistakably from his quite conscious limitation to science in England. On the very first page of his preface, science on the Continent is explicitly excluded from treatment. (Yet it is true that Merton could not refrain from adding that "the connections between science in England and on the Continent . . . leave the picture materially unaltered.")[23] Considering the time when the book was written, this is not to be regarded as quite so serious a lacuna as it has become since the concept of the Scientific Revolution was created as an analytical tool meant to give a closer unity to the range of phenomena which together constitute the rise of early modern science. This, however, we know to have been the work of Koyré and Butterfield rather than of Merton's godfather in the history of science, George Sarton.

In fact, the subsequent extension of the scope of Merton's thesis by his supporters and critics is one of the clearest signs of the analytical gains attained by the coining of the new concept. So persuasive was the notion of the Scientific Revolution that in retrospect Merton's thesis was projected forward onto it. The extension, then, can be traced to Hall's article, in which Merton is taken to uphold that

> sociological history provides (if I follow Merton's view of 1938 correctly) principles sufficing to explain that crucial event, the scientific revolution of the seventeenth century.[24]

Remarkably, in the same article by Hall the onset can be found of a related historiographical pattern, namely, to find fault with the thesis because, by its effective limitation to the case of England, it was not really capable of explaining the Scientific Revolution in its entirety, as it was supposed (by the critic, not by the author) to have done. The next link in this historiographical comedy appears to have been the adoption of the Merton thesis by the English historian Christopher Hill, who, in his *Intellectual Origins of the English Revolution* (1965), took Merton to have established, once and for all, a crucial link between 17th-century science and Protestantism. After which wholesale adoption Hill, too, readily admitted that he confined his own argument to the case of England!

So much for the odd distortions through which Merton's thesis has simultaneously come to be held for an explanation of the Scientific Revolution and to be castigated for its failure really to provide one. The net result has been that the Merton thesis—whether or not taken in conjunction with Hill's defense of it—has customarily come to stand for one purported explanation of the rise of early modern science. It remains for us to find out how well (or how badly) it has done among historians of science in this mostly unintended capacity.

Part of Hall's critique held pace with the limited scope of the thesis, in that Merton had stressed throughout the Baconian aspect of early modern science at the detriment of the conceptual overhaul accomplished by Galileo and Kepler and many others. For the rest Hall's objections derived in large measure from his aversion to any attempt at 'external' explanation of events in the history of ideas (on which more in section 5.2.5). A more specific rebuttal was provided in 1965 by Theodore K. Rabb, who argued that, if the concern at hand is the explanation of the *rise* of early modern science, then the decisive period to look at is the decades *before* 1640. By 1640, so Rabb asserted, the Scientific Revolution was largely over. The crucial discoveries had been made, and what remained to be done was rather in the nature of elaboration and consolidation. Now it might well be true that Protestants had been preponderant in the consolidation process of the new science. (After all, it was only at this time that prominent upholders of the Counter-Reformation turned against the scientific movement.) But, obviously, this did not mean much for the onset of the Scientific Revolution, which had of course been as much an affair of Catholics as of Protestants:

> It is only by ignoring the enormous Catholic scientific activity of the period that major claims can be made for the importance of protestantism, let alone puritanism. . . . Evidently orthodox Catholics somehow managed to develop a great interest in science during this century [from the 1540s to the 1630s] despite their want of religious encouragement. Could it be that religious stimuli were not enormously important to the *rise* of science, or did the Catholics have the same advantages as the Protestants?[25]

At this very same point, we already know, is located the most vulnerable spot of Hooykaas' later thesis, and of any attempt generally to explain the Scientific Revolution through developments in contemporary religion. Although there have undoubtedly been specific networks of scientists with strongly similar religious convictions and motivations in specific areas and during specific periods, the Scientific Revolution, when taken as a whole, was simply too variegated an event to allow for any overall explanation through the dominant contribution of any specific denomination of Christian belief.

Such a tentative conclusion leaves room for taking up J.L. Heilbron's recent suggestion that what we need to work on is an 'ecumenical Merton thesis' stating that "where science prospered in early modern times, it derived important support and reinforcement from organized religion"—whether Protestant or Catholic, that is.[26] Nor does the conclusion exclude the possibility (pointed out most cogently by Hooykaas and by Farrington) that the Scientific Revolution may have been due at least in part to one or more features common to the Christian religion as such, as distinct from a variety of non-Christian creeds (a possibility to be pursued further in the next chapter, when we deal with the 'failure' of early modern science to emerge in Eastern civilizations). Nor is the possibility excluded that "in the story of the rise of science . . . religion is a peripheral concern"—this being the final conclusion drawn by Rabb from his preceding considerations.[27] Nor is it *a priori* ruled out that the Merton thesis, although clearly defective as an explanation of the Scientific Revolution as a whole, may yet acquire fresh significance if applied to just one portion of the event. Two historians have indeed put forward suggestions along such lines.

The Merton thesis returns to England. Even if we take Merton's connection between Puritanism and English science as established, the thesis has appeared to fail to apply beyond England so long as 'England' continues to stand simply for one geographical area among the many countries where the Scientific Revolution in fact took place. However, once English science is granted a distinct role of its own in the ongoing process of the Scientific Revolution the Merton thesis may gain a new sort of validity. That this is so has explicitly been suggested by Kuhn in his "Traditions" article of 1976, where he split up the Scientific Revolution between the transformation of the classical sciences and the simultaneous emergence of their Baconian counterparts. Here is how Kuhn hoped further to enhance the analytical fertility of his own schema through a rescuing operation on behalf of the Merton thesis:

> The main drawback of this viewpoint [the Merton thesis] has always been [N.B.!] that it attempts to explain too much. If Bacon, Boyle, and Hooke seem to fit the Merton thesis, Galileo, Descartes, and Huyghens do not. . . .
>
> Its appeal is, however, vastly larger if it is applied not to the Scientific Revolution as a whole, but rather to the movement which advanced the Baconian sciences.[28]

Although Kuhn does not go on to say so, his reinterpretation equally makes much more plausible the connection drawn by Merton between the Puritan ethic and the strongly empirical bent of the science it was supposed to foster, a feature that applies much more clearly to the 'Baconian' than to the 'classical' tradition in Kuhn's scheme of things. A revision along such lines still leaves open all the knotty questions raised in connection with the original thesis, to be sure. It only shows that, *if* these can be resolved, the result might indeed be relevant for the causation of the Scientific Revolution.

A similar consideration applies to one more historiographical adjustment of the Merton thesis. Just as, in Kuhn's proposed revision, the special position of England as a *locus* of the Baconian sciences is stressed, so England is granted a place of its own in Joseph Ben-David's explanation of the Scientific Revolution. Here Merton's insights into the contemporary justification of science on utilitarian grounds are combined with an analysis of which social layers all across 17th-century Europe were (or failed to be) receptive to such efforts at legitimation on the side of the scientists. Ben-David's resulting thesis, which has been seriously misjudged in the literature, cannot unfold its full potential until we have surveyed the historiographical tradition of which it forms one culmination point. That particular tradition is to occupy us as soon as we have properly completed our treatment of Merton's thesis on Puritanism and science in 17th-century England.

The vagaries of the Merton thesis: A brief recapitulation.   The historiographical adventures of the Merton thesis have been so confusing that, at the risk of
being overly repetitive, I wish to round off the present section with a brief,
stage-by-stage summary of what historians of the Scientific Revolution have
successively done with it.

At its birth in the late thirties, the Merton thesis was intended to explain
something topical about English science (although there need be little doubt
that at the back of the mind of its author an unexplicated notion rested of
the special significance of 17th-century English science for the rise of early
modern science).

Quite independently from what Merton was doing at the time, the notion
of the rise of early modern science was given greater conceptual coherence
than it had enjoyed before through the coinage of the concept of the 'Scientific Revolution'.

When, in the fifties and sixties, the concept gained ever greater currency,
the Merton thesis was taken, by advocates (Hill) and critics (Hall) alike, for an
explanation of the Scientific Revolution.

In that particular capacity its obvious shortcomings were rapidly realized.

Fresh interpretations of the Scientific Revolution made in the seventies,
however, appeared to provide new breathing space for the Merton thesis and
to restore to it the capacity to inspire fruitful thinking on the time-honored
question of how early modern science came into being.

One more source of confusion over the Merton thesis.   One element that has
contributed further to the confusion has still been left out of our account. This
is the fact that the Merton thesis, in its original setup, encompassed a good
deal more than the 'Puritanism and science' hypothesis which has been taken
by most historians to express the full Merton thesis. These neglected elements
of the original thesis concern the practical purposes for which Merton believed
science to have been cultivated during the 17th century. This particular portion
of the Merton thesis was itself part of a larger tradition of purported explanations of the rise of early modern science. It is to this tradition that we turn now.

5.2. THE ACTIVE LIFE OF EARLY MODERN EUROPE

One particular cluster of explanations of the Scientific Revolution is focused
on what might broadly be termed the turn toward a new dynamism in the history of western Europe from about the early 16th century onward. Whether or
not such feats as the Voyages of Discovery, the flowering of the arts and crafts,
the advent of gunnery, the invention of the printing press, the improvements in
mining, etc., were contrasted with the supposed quietude of medieval contemplation, it is in areas such as these that a number of historians have sought for

the decisively novel element in European history that made the emergence of early modern science possible. The earliest serious work I am aware of in which these themes were sounded all together and given an inner coherence all their own is Leonardo Olschki's three-volume *Geschichte der neusprachlichen wissenschaftlichen Literatur* (History of Scientific Literature in the Vernacular; 1919–1927).

5.2.1. *Olschki and Koyré on the Scientist's Response to Europe's Budding Dynamism*

Leonardo Olschki was born in Verona in 1885, a son of the publisher Leo S. Olschki. The main areas of his interest were Italian and French language and literature. He taught Romanic philology at the University of Heidelberg from 1913 to 1933, when he was expelled as a consequence of the enactment of the Nuremberg Laws. After occupying a chair in Rome he left in 1938 for the United States, where he taught at Johns Hopkins in Baltimore and, later, at Berkeley. He died in 1962.[29]

The formal objective of Olschki's 1266-page effort was to present an overview of works on topics pertaining to science and written in the Italian vernacular in between Leon Battista Alberti and Galileo Galilei. His original intention to extend the analysis to Galileo's disciples and to add volumes about French scientific literature from Oresme to Descartes and beyond, as well as about the German counterpart to this genre, was never carried out.[30]

On one level, this work of vast erudition represents an attempt, never since followed up as far as I know in anything like Olschki's comprehensive fashion, to analyze the scientific writings of the time as products of language. That is to say, Olschki examined such topics as the emancipation of the vernacular from Latin; the reasons one or another scientific author may have had for writing now in the one, then in the other, language; the place of all these writings in the literary prose of their time; their stylistic, grammatical, semantic, and literary peculiarities; the gradual adaptation of a language originally quite unsuited to scientific terminology to the formal requirements of standardized scientific reporting. In short, if Olschki had indeed managed to complete his full program, we would have possessed what is still sorely missing in the historiography of the Scientific Revolution: a systematic overview of its pioneering writings considered as products of language and as works of literary art by an author thoroughly versed in the history of language and literature as well as in the history of science.

Olschki did not confine his analyses to the formal aspects of the language used by men like Alberti, Leonardo, Tartaglia, and Galileo. Rather, he was concerned throughout to show how linguistic matters often went far to determine arguments, his treatment of Galileo's great works in the form of dialogues easily providing the culmination point of his entire effort. Now, the principal example of how, in Olschki's view, the issue of language helped determine the direction of scientific thought is to be found in the increasing usage, through-out the Renaissance, of the vernacular as a reflection of a decisive switch from dry and bloodless scholastic erudition toward a mixed scientific/technological literature based upon the experience of the artisan, the practitioner, the trav-eler—in short, of all those elements in Renaissance Europe that gave life a new dynamism. Olschki's ultimate purpose was no less than "to lay bare the cultural preconditions of the development of science."[31] He believed he had found a path leading toward this goal through an analysis of scientific literature in the vernacular. This new genre, after all,

> arose when the secularization of the forms and conceptions of life forced men to draw the sciences, which had removed themselves far from the world, into the sphere of practical and mental activity. . . . This is why scientific literature in the vernacular starts with the applied and the empirical sciences, so as to find, once having arrived beyond the limits of practical necessities, the road toward purely scientific abstractions in an independent way. The end-point of this development, to which this history of the rise and formation of early scientific prose is devoted, is to be found in the work of Galileo and of Descartes, whose creations and discoveries are not the emanation of ancient and medieval methods of inquiry but rather the further development and the triumph of an idea.[32]

Thus Olschki was *not* concerned with demonstrating that Galileo's science was focused more or less exclusively on practical issues. Rather, his point was that what enabled Galileo to transcend the amassed erudition of his predecessors in science was the recently emerged tradition of applying mathematical notions to practical matters of a technological nature, which he adopted from the pre-ceding literature in the vernacular. That is to say, matters of perspective, min-ing, fortification, ballistics, and so on provided the impulse for the turn toward the empirical without which the decisive renewal of science in the 17th century would have been inconceivable. Thus, to mention only one example, Galileo's well-known invocation (right at the start of the First Day of the *Discorsi*) of the Venetian Arsenal and the practical problems encountered there acquires, in Olschki's treatment, a highly programmatic value in announcing what had made the new science possible in the first place. This was a very unusual and new focus at the time. Still, Olschki took care to qualify:

> The problems of the economy of power and of how much machines can accom-plish, of the accuracy of guns, of the resistance of fortifications, are the very same ones that for two centuries had found treatment in the technical literature. Galileo, however, considered the tradition of the workshops, with which he had become acquainted through his teachers, primarily as the area of experience and observation, as suitable for drawing the preliminary lines of the theoretical foun-dations of the mechanical arts. This is why the formulation of those questions is nevertheless fundamentally different, and why their solution is fully independent

from this direct tradition of the workshops and of the theorists, even though his attention is drawn time and again to the possibilities of applying in practice the theories he discovered through speculation and experiment.[33]

Olschki's transcending of Mach's empiricism.   Olschki's views have been sketched here in their barest outlines only. His highly detailed, dense, not very closely argued yet captivatingly written volumes were instrumental in starting a tradition in the historiography of the Scientific Revolution which has remained alive up to the present. Yet his work is rarely cited, let alone discussed in any detail, and I hope one day to be able to devote to it the separate study it so fully deserves. With hindsight one can see that Olschki managed from the outset to avoid a trap into which many of those were destined to fall who went on to elaborate some of the themes he had sounded for the first time. With Olschki there is no question of purely and simply explaining the rise of early modern science through its practical utility. Note, for example, the subtle phrasing in the passage just quoted: While indicating that ultimately Galileo's source of inspiration was located in the domain of technology, Olschki makes it clear that Galileo brought the insights gained from that source to a new, higher plane of theoretical abstraction.

Insofar as Olschki's conception of the importance of the arts and techniques of the Renaissance for the emergence of early modern science relies on the characterization of the latter as empirical and observational, he quite consciously followed the footsteps of Ernst Mach.[34] Even though more than a modicum of Mach's positivist approach to the history of science may equally be found here and there in Olschki's work, his more refined historical sensibility allowed him to a considerable extent to transcend the limitations inherent in Machian historiography. This happened quite independently from Duhem's findings, which were to transform so deeply historians' views on the rise of early modern science during the very period when Olschki was writing his volumes—apparently in considerable isolation. A critical admirer of Galileo, Olschki was put off by what he regarded as Duhem's disparagement of Galileo's independent merits through the attribution of prime originality to the Parisian Terminists. Thus he could not bring himself to attach any significance to what the scholastics had contributed, beyond sheer erudition, to the origins of early modern science. In Olschki's view, Galileo was primarily a mathematical physicist in the Archimedean tradition, yet what enabled Galileo to surmount the combined limitations of classical tradition and of scholastic erudition was precisely the creative elaboration of practical matters for which the newly emerged tradition of the scientific and technical literature in the vernacular had sharpened his eyes.

Such a conception, with its focus on Galileo's achievement, was bound to clash with the more 'idealist' interpretation of the origins of early modern science Alexandre Koyré was to develop more than a decade later. The clash is the more interesting to investigate because it was not in the simple nature of an 'empiricism *versus* mathematicism' opposition.[35] After all, Olschki's personal brand of sophisticated empiricism did leave him sufficient room to call attention, ten years before Koyré, to "the thoroughly Platonic direction of [Galileo's] mind" and to his "Platonic method of considering nature," even though, according to Olschki, Galileo's

> relation to Plato . . . was founded on a sense of mental affinity rather than on any
> awareness of a similarity of views. In a general sense of mental direction Galileo
> is a Platonist, but Plato's literary peculiarities captivated him far more than his
> doctrines proper.[36]

Precisely such a resemblance in their views on Galileo may help us determine the point where, nevertheless, Olschki's and Koyré's views on what had made the new start in science possible went far apart.

Olschki's and Koyré's conceptions contrasted. Alexandre Koyré never set forth his ideas on the causes of the emergence of early modern science in other than scattered, fleeting passages, his main concern in these matters being the correction of other historians' mistaken views in light of his own core conception. Thus, so Koyré argued in *Études Galiléennes,* Leonardo Olschki's error had resided, not so much in his acknowledgment of Galileo's Platonism (even though Olschki had wrongly found this compatible with stressing Galileo's empiricist leanings), as, rather, in his idea that the tradition of the engineering scientists stood at the origin of early modern science.[37] The point was that all those activities in such areas as ballistics, fortification, hydraulics, and the like had not by themselves been sufficient to lead to the required overthrow of received scientific doctrines:

> Bullets and cannon balls brought down feudalism and medieval castles, but me-
> dieval dynamics resisted the impact.[38]

Koyré readily granted that the Middle Ages and early Renaissance had witnessed a splendid and quite unique outburst of technical invention: the foliot balance, the building principles underlying the Gothic cathedral, and so on. Such inventions demonstrated what the arts and crafts, that is, "technical thought of the commonsense variety,"[39] could accomplish, but they also demonstrated the limitations inherently set to such activities without the support of the as yet unborn science of the Scientific Revolution. As soon as this new science came into being through the theoretical efforts of Galileo and Descartes, however, it went on at once to transform these empirical pursuits. In thus being drawn into the newly gained domain of precision, technical arts were, for the first time in history, turned into technology proper. Thus attempts like Olschki's to ascribe to the 'science' of the engineers the decisive turn toward the practical that was needed to set Galileo's revolution going came down to the projection backward of a link between scientific discovery and technical invention that is nowadays taken for granted but that scarcely existed before the advent of the Scientific Revolution. Rather, the crucial impulse had been the introduction of Archimedes' work, and the decisive switch toward an altogether different plane of physical theorizing that immediately followed, although only achieving full fruition in Galileo's work, can already be discerned in the extent to which Benedetti managed to overcome the limitations of the engineering scientists.

I do not think that Olschki and Koyré were as far apart on this score as the latter made it appear. After all, Olschki, too, recognized that the mathematical tradition had been indispensable for the newly emerged tradition of engineering science to overstep its original limits. Yet from this shared insight they drew opposite conclusions, and it is by delving somewhat into Koyré's views on explanation in the history of science generally that we can reach the deepest level on which their divergence is to be situated.

Koyré's approach to the history of science is customarily taken to be the embodiment of 'internalism'. This, although of course largely true, does not yield the complete picture.[40] Lumping together approaches through scientists' social environment and their individual psyches under the interesting label 'psychosociological explanations', Koyré was quite ready to acknowledge that, for science to come into being at all, certain conditions had to be fulfilled, which were met in fact only in the rarest of cases. After all, civilizations can flourish without science, as is witnessed by the examples of China, Carthage, and Persia. Individuals need leisure to devote themselves to scientific pursuits, as well as a willingness to probe the depths of purely theoretical research. For those conditions to come into being an audience is required, or at least some esteem and a recognition in the wider society of the value of research beyond the domain of the immediately practical and useful, which in most cases is the only concern of the bureaucrat. Therefore, bureaucratic and also aristocratic societies tend to value only knowledge that is magical, sacred, or concerned with the exertion of political power. We can understand therefore why science was not born in the huge bureaucracies of China and Persia, and also, conversely, how it *could* arise in Greece, but not *why* the latter event *did* indeed take place.

The same limitations apply to any attempt at explaining the Scientific Revolution of the 17th century. The persistent appeals made at the time by scientific enthusiasts to the applicability of the new science were largely intended for consumption by princes and bureaucrats who had at their disposal the authority to grant leisure for free, theoretical research. This is how scientists could gain the social recognition without which the pursuit of pure science is ruled out *a priori*. But beyond the largely successful search for legitimation of the new science, the appeal to its applicability scarcely played a role. For example, although the problem of finding longitude at sea was indeed a spur to

Huygens' construction of precision clocks, for him the benefits for theoretical astronomy definitely came first.

This very point of the legitimation of science, so Koyré implied rather than asserted, marked the limit of how far 'psychosociological explanations' can carry the historian of science. The structure of the Greek *polis* cannot explain Greek science, just as "Padua or Florence do not explain Galileo."[41] All of which goes to show that "it is impossible . . . to give a sociological explanation of the birth of scientific thought."[42]

Koyré and the limitations of historical explanation. Two sets of comments are in order at this point. The first is that Koyré failed to raise here to the level of explicit discussion a distinction which is basic, not only to his own argument, but to the whole issue of 'external' explanations in the history of science. We have just found Koyré acknowledging (along lines later to be explored less crudely and in much greater detail by Ben-David) that it does make sense to investigate social conditions which made the emergence of early modern science possible. However, he rejected any explanation that threatened to diminish the autonomy of the internal development of scientific thought as a process guided above all by an inherent logic all its own. The curious thing is that, after implicitly distinguishing between these two levels of analysis, he at once conflated them in a wholesale dismissal of the approach as such. I think instead that the distinction is vital, and that explanations through the absence or presence of certain social conditions in given civilizations and time frames may be quite illuminating, whereas attempts at the social *reduction* of scientific ideas are indeed to be withstood. Such social reductionism was surely rampant in the thirties—the flexible period in Koyré's thought on the history of science— so we can understand how, despite his awareness of the distinction, he tended to conflate the two and to throw Olschki's quite sophisticated analysis on the same rubbish heap as the crudest products of historical materialism put forth at the time. We discuss those products in the next section, where we come back to the idea that the failure to keep these two levels of social analysis of the history of science distinct has greatly impaired the discipline for decades. Nor has it in fact been quite overcome yet.

At the present point in the argument I wish to assert only that, when one seeks to learn what enabled 17th-century Europe to accomplish what ancient Greece never managed to pull off, the invocation of the new dynamism of European society as explored in pioneer fashion by Olschki is not as easy to dismiss as Koyré thought. It is true, though, that Koyré was quite consistent here in that, for him, the problem of Greek stagnation did not exist: We know that, as far as he was concerned, Galileo might have followed upon Archimedes, not at the remove of eighteen centuries, but rather as his direct successor (section 4.2.2).

Koyré had another reason for caring so little about probing the causes of the Scientific Revolution. This is his scepticism about what historical explanation can accomplish even in the best of cases. Ultimately, he believed, we do not know why technical invention, or scientific discovery, emerged at all:

> For all explanations, however plausible they may be, ultimately turn around in a circle. Which, after all, is not a scandal for the human mind. It is fairly normal in history—even in the history of the human mind—that there are inexplicable events, irreducible facts, absolute beginnings.
>
> [In short:] It is impossible, in history, to empty the fact and to explain everything.[43]

And thus Koyré felt assured that, with his own revolutionary conception of how early modern science emerged, the prime task was to work out these new insights of his. The Scientific Revolution had first of all to be conceptually shaped, factually described, and interpretively analyzed.[44] To others might be left the business of determining what little could sensibly be said about its causes. We are going to explore now to what extent, despite Koyré's surely wise warnings, some illumination may nonetheless be attained about the possible roots of the new science in the new, European dynamism without relapsing into social reductionism and before falling into the trap of ultimate circularity.

5.2.2. The Hessen Thesis: Early Modern Science and Capitalism

The fate of the themes first touched upon by Olschki has been a curious one. Acknowledgment of his contribution to our understanding of the rise of early modern science has, with few exceptions only, been relegated to the domain of the footnote, so that the full extent of his achievement has remained hidden from sight.[45] Yet the themes themselves did not go away thereby. The broad picture is as follows. During the thirties and forties the overall idea of practical requirements having provided a major boost to the rise of early modern science was given a Marxist twist far removed from Olschki's intentions by the Soviet physicist Hessen and by the Austrian philosopher/scientist Zilsel, whereas the idea was connected by Merton to his own thesis on 'Puritanism and science' with less conspicuously Marxist overtones. In the fifties the entire approach came under heavy attack (notably by Hall) because the newly arisen concept of the Scientific Revolution seemed incompatible with it. When, from the late fifties onward, the theme staged a comeback, it was mostly treated in a more fragmentary way. Such landmarks of European dynamism as the mechanical clock, exploration overseas, or the printing press, which Olschki had just touched in passing, were now being claimed in a more detailed manner to have helped bring about the Scientific Revolution. Authors to be discussed in this connection are Landes (but also Koyré), Hooykaas, and Eisenstein. Since the general theme of the explanation of the Scientific Revolution through aspects

of European dynamism constitutes a somewhat coherent historiographical tradition in its own right, its treatment in by and large chronological order can
help clarify the nature of the debates carried on in its name. Chronology, then,
requires that the stage be now cleared for the appearance of that archvillain of
so many pieces, Boris Hessen.

In his paper of 1931, entitled "The Social and Economic Roots of Newton's 'Principia,'" Hessen was concerned to make three points in succession.[46]
He wished to demonstrate, first, that all subjects handled in the three books of
the *Principia* derive from technological issues that had come up during preceding decades of the century as a result of the needs of incipient capitalism.
Second, he argued that Newton—a true child of his class—had been unable to
draw the materialist, atheist consequences of his own physical world-picture.
Finally, because of his lack of familiarity with the steam engine, the principle
of the conservation of energy in the definitive form given to it by Marx's collab-
orator, Friedrich Engels, had eluded Newton. These three points were preceded
by a short account of historical materialism in the format of a paraphrase sprin-
kled with literal quotations from Marx's classical statement in the preface to
Zur Kritik der politischen Oekonomie. Throughout his paper Hessen alternat-
ingly took for granted and proclaimed loudly the complete, indisputable, and
definitively established truth of these Marxist starting points of historical
analysis.

Hessen's first point has customarily been taken as an argument about the
rise of early modern science rather than about Newton alone because (the title
notwithstanding) that is what it is.[47] Leaving out of consideration from here on
Hessen's second and third points, we do not seriously distort his intentions by
stating that what has both famously and infamously become known since as
the 'Hessen thesis' comes down to the argument that, through the mediation
of technological developments, the emergence of early modern science is to
be explained by the needs of early capitalism:

> The brilliant successes of natural science during the sixteenth and seventeenth
> centuries were conditioned by the disintegration of the feudal economy, the de-
> velopment of merchant capital, of international maritime relationships and of
> heavy (mining) industry.[48]

Hessen's starting point in seeking to make his case is that "the development of
merchant capital" created a set of quite distinct problems in several areas of
technology: transportation over water, the mining industry, and ballistics (in
the sense both of what goes on inside the gun once fired and of what happens
to the bullet outside). He then proceeds to identify and list the full range of
these contemporary problems. One example from each realm is "an increase
in the tonnage capacity of vessels and in their speed," "methods of ventilating
the mines," and "the stability combined with least weight of the firearm."[49]

For each separate item on his list Hessen determines "what physical prerequisites are necessary in order to resolve these technical problems."[50] For example, "in order to increase the tonnage capacity of vessels it is necessary to know the fundamental laws governing bodies floating in liquids, since in order to estimate tonnage capacity it is necessary to know the method of estimating a vessel's water displacement. These are problems of hydrostatics." Similarly, "ventilation equipment demands a study of draughts, i.e., it is a matter of aerostatics, which in turn is part of the task of statics."[51] And so on.

Having thus identified the domains of physics which contain the solution to the technical problems in question, Hessen proceeds to an enumeration of 16th- and 17th-century scientists who worked in these areas of physics and finds that all areas were indeed covered at the time:

> The above specified problems embrace almost the whole sphere of physics.
>
> If we compare this basic series of themes with the physical problems which we found when analysing the technical demands of transport, means of communication, industry and war, it becomes quite clear that these problems of physics were *fundamentally determined* by these demands. [My italics][52]

Never tiring of reiterating a point rather than shoring it up with arguments, Hessen went on to assert that

> we come to the conclusion that the scheme of physics was mainly determined by the economic and technical tasks which the rising bourgeoisie raised to the forefront.
>
> During the period of merchant capital the development of productive forces set science a series of practical tasks and made an imperative demand for their accomplishment.[53]

Then follows an interlude in which Hessen—taking Friedrich Engels for his guide while fully ignoring Duhem—sets off the scientific darkness of the medieval universities against the subsequent revival of science. Finally he takes up the *Principia*. Hessen grants that its three books were written in abstract, mathematical language, "and we should seek in them in vain for an exposition by Newton himself of the connection between the problems which he sets and solves with the technical demands out of which they arose."[54] Rare indeed is the Marxist who allows himself to be put off by such a detail. If the facts of history happen to forbid him to enter through the front door of explicit contemporary statements, entrance through the back door is still kept open by the convenient ploy of 'history happening behind people's own backs'. Hessen first assembled what little Newton, throughout his correspondence, had to say on technical matters (chief witness being his letter to Francis Aston) and briefly condemned all other, idealist historians and their inability to recognize in Newton anything else but the lofty genius, "soaring only in the empyrean of abstract thought."[55] Hessen then proceeds to enumerate the areas of physics addressed in the *Principia*, going on to notice with glee that these are the very ones needed to solve those technical problems with which the rising bourgeoisie found itself grappling at the time. *And the above account is all there is to Hessen's pertinent argument.*

Origin and effect of the Hessen thesis.   If this is so, why devote so much space to an argument whose shortcomings are apparent not only to the present-day reader but scarcely less so from the viewpoint of levels of sophistication reached by the historiography of science way before 1931? The crude anachronism of Hessen's argument is not even matched by Ernst Mach's pronouncements of half a century earlier. By 'crudeness' I do not mean here primarily such factual mistakes as the sudden appearance of a certain "Herique" (allowing for some errors of transliteration from the Cyrillic to the Latin alphabet, done in a five-day translation rush, as a result of which Magdeburg's famous mayor came out in somewhat garbled fashion) or the reference to Galileo's praise of the "Arsenal at Florence."[56] By crudeness I mean above all the crux of the entire argument, which is simply that *our* present-day awareness of the physical background of certain technical problems is projected back by Hessen onto 16th- and 17th-century engineers who in reality either solved their problems empirically or, in almost all other cases, were at a loss or did not bother to solve them at all. It is as if no ship builder in the whole history of the world had ever been able 'to increase the tonnage of a ship' until an expert in hydrostatics had come along and told him how to do it. So, again, why devote so much attention to such a sloppy piece of scholarship?

The answer has almost everything to do with the outer circumstances accompanying the publication of Hessen's paper.

It was delivered at the Second International Congress of the History of Science and Technology, held in the summer of 1931 in London. Hessen was a member of a delegation from the Soviet Union headed by the still-prominent Old Bolshevik Nikolai Bukharin. Already outmaneuvered by Stalin politically, Bukharin was still found ideologically suitable to present to the world the current views of the Party on the respective roles of science in rotten capitalism and in the process of building a socialist society. Boris Mikhailovich Hessen (born 1883) was serving in 1931 as director of the Moscow Institute of Physics. His commitment to the theory of relativity had already caused him increasing ideological difficulties with the authorities, and, as Loren Graham has recently shown, Hessen's paper can be seen as his defense against strident, potentially lethal attacks back home. Still, the machinery of the Great Terror caught him in 1935, and in 1938 he perished together with his patron, Bukharin.[57]

The sudden appearance of such a politically loaded delegation at a meeting of quiet historians of science created quite an uproar. Its lasting effect was threefold. First, through a remarkable effort centered in the Soviet embassy, the delegation managed within five days to have its mostly quite lengthy papers printed in book form under the title *Science at the Cross Roads*. The title refers to the dual state of science pointed out by Bukharin in his opening speech: in deep crisis under decadent capitalism but flourishing in the country where a new society was being built. This idea struck profound chords in the minds of a number of somewhat vaguely leftist British scientists of high promise and prominence, among whom for present purposes J.D. Bernal and Joseph Needham must be singled out. Never mind that, as they well knew, rather than being in a state of crisis science as cultivated in their own hometown, Cambridge, was passing through its golden age. Never mind that the crudeness of Hessen's argument cannot entirely have escaped them—certainly not Needham, who had already won some spurs in the historiography of science. For reasons beyond the scope of the present book to pursue it was Hessen's piece in particular that served as an eye-opener to these men in showing what could be accomplished by an analysis of the history of science in terms of its social and economic context.[58] Bernal was to build upon this particular insight the conception of *Science in History,* which we discussed in section 3.6.2, incorporating Hessen's thesis as one element in its construction. The intellectual course on which Needham was subsequently led is to occupy us in the next chapter.

On other historians of science the effect of Hessen's paper was hardly less momentous, though mostly in the opposite direction. 'If this is what the external approach to the history of science is all about', so many must have thought, 'then down forever with the external approach to the history of science'. The disastrous split in the historiography of science between 'internalists' and 'externalists', which has wasted so many intellectual opportunities and is still not quite behind us, finds its origin in the Hessen thesis. In order to begin to overcome the virtually inborn prejudice, coming so naturally to the practicing scientist, of science being a wholly autonomous product of the inquiring intellect, nothing would have been needed so much as patience, subtle discrimination, and the careful avoidance of crude reductionism. Instead, here was this narrow-minded piece of bigoted dogmatism at its Stalinist crudest. What could anybody who was not already looking toward the Soviet Union as the Promised Land do but reject the message hook, line, and sinker?

Tacitly reject it was the solution almost universally adopted for the time being. Still, two decidedly noncommunist scholars with an interest in the history of science did more than that. While Merton put elements of Hessen's piece to service in arranging the technological chapters of his own thesis (see next section), the economic historian G.N. Clark took up the challenge head-on. He did so in 1936, in a polite yet fairly devastating lecture called "Social and Economic Aspects of Science," which was published one year later in a little book under the strangely inappropriate title *Science and Social Welfare in the Age of Newton.*

Clark's challenge of the Hessen thesis. Clark was quite ready to grant that economic life and warfare had contributed to directing the minds of 17th-century scientists toward specific research topics. But other stimuli to the 'scientific movement' of the 17th century could be detected besides just these two. Here he mentioned in the first place medicine, or "the technology of healing," with its effect on biological research.[59] Next he called attention to the fine arts, including specifically musical theory, which "had much to do with fixing the predominantly mathematical method of seventeenth-century science."[60] However, he insisted, the impulses coming from healing and art should not be reduced to the economic sphere, since both transcend sheer utility.

Clark noted that historians had for a long time recognized the origin of the experimental method in procedures regularly used in industry, in the arts and crafts. Also,

> it has long been recognized that the introduction of rational accounting in business in the later Middle Ages was another result of the habit of quantitative thinking which was married to experimentation in the work of Galileo and Newton.[61]

The point, however, was that these experimental and quantitative methods, originating as they did in the worlds of miners, artisans, craftsmen, and business accounting, in order "to become scientific had to be combined with something else."[62] This 'something else', then, was religion. Clark illustrated the assertion with some examples of contemporary scientists (in particular, Newton), pointing out their religious motives for doing science. These, Clark cautioned, "are to be believed unless there is evidence to the contrary."[63] And then Clark crowned his whole argument by pointing to "the disinterested desire to know, the impulse of the mind to exercise itself methodically and without any practical purpose [as] an independent and unique motive."[64] It was this greatest motive of all which, together with the other five, had produced the scientific movement of the 17th century. This new scientific movement "touched the needs of human life only here and there."[65] After all, many practical problems whose solution was urgently sought at the time resisted such resolution by science for a long time after the century had ended; equally, many useful applications of science resulted from research originally undertaken purely for its own sake. Only with these two qualifications in mind, so Clark concluded his essay in deadpan manner, could Hessen's thesis be accepted.

5.2.3. *Merton on Science and Technology in the 17th Century*

In chapters 1 through 6 of *Science, Technology and Society in Seventeenth-Century England* Merton had sought to establish that in England, in the course of the 17th century, there had been a notable increase in vocational interest in science, and that this could be causally connected to the Puritan ethic of the time. But this was only part of the argument he aimed to make. The remainder of Merton's doctoral thesis was concerned with finding out:

which forces guided the interests of scientists and inventors into particular chan-
nels? Was the choice of problems a wholly personal concern, completely unrelated to the socio-cultural background? Or was this selection significantly limited
and guided by social forces? If so, what was the extent of this influence?[66]

More than in the 'Puritanism and science' chapters of his dissertation Merton
was concerned in the later chapters to make a point, not only about English
science, but about the origins of early modern science on the Continent as
well. Somewhat surprisingly for the man who was to become one principal
representative of mainstream American sociology, both the overall organization of his pertinent account and a substantial number of factual data were
taken over from Hessen, with due reference but rather uncritically. Thus, besides a number of pages where Otto von Guericke is mentioned, we also see
the otherwise unknown Herique turn up, together with the equally spurious
Florentine Arsenal invented by Hessen.[67] Insofar as Merton's argument extended to the Continent, his account relied quite heavily on Hessen. What Merton independently added to the substantive content of Hessen's thesis, besides
the invocation of much additional literature, is chiefly the following.

In the first place, he presented a great deal of empirical material taken
from English science and technology. For example, in order to supply supporting evidence for Hessen's thesis about the *Principia* Merton pointed to a remark
by Newton (apparently overlooked by Hessen) at the end of a theorem about
the resistance of fluid media. Here Newton wrote, "This Proposition I conceive
may be of use in the building of ships."[68] It is characteristic of the extent to
which Hessen's influence on the making of Merton's argument has frequently
been overlooked that this particular find has later been attributed to Hessen, to
be held up to deserved derision in view of the fact that no contemporary ship
builder could possibly have made sense of the highly complex mathematical
reasoning involved here.[69]

More importantly, Merton once again supplemented the products of his
rummaging through the contemporary literature with statistical evidence—a
procedure to which he attached considerable programmatic value. Too often,
so he complained with remarkable foresight into the shape of future debates
over his own thesis, arguments in the history of ideas are "justified by an appeal
to carefully selected cases which nominally bear out one or the other of . . .
conflicting opinions."[70] Hence, Merton went ahead and broke down according
to their subject matter the topics discussed at meetings of the Royal Society
during four selected years. He reckoned these to belong either to "Research
related to socio-economic needs" (a category he subdivided largely in accordance with Hessen's classification) or to "Pure science."[71] Consequently, the
statistical results that ensued from this mode of classification were disfigured
from the start by the basic flaw in Hessen's anachronistic argument. It is true
nonetheless that, if one seeks to reclassify the topics listed by Merton ac-

cording to less anachronistic standards, strikingly many topics (close to half of the number considered) appear to deal indeed with unambiguously practical matters. The remaining problem, which no amount of statistics can of course resolve, was still the old one, namely, to interpret what this result signified.

At this crucial point in the argument Merton was less dogmatic, more careful, and altogether more sophisticated than Hessen. Throughout his pertinent chapters Merton was concerned to make it clear that he did *not* wish to establish a monocausal, one-way relationship, 'explaining' either the 17th-century upsurge in science or the particular direction science had taken at the time by nothing but contemporary social and technological needs. It is not, however, easy to detect, through his maze of qualifications, the precise shape of the thesis Merton was still prepared to defend. I do not even believe that any reading of the chapters in question can yield a wholly consistent picture here. In his most relative mood, Merton wrote that "in the last analysis it is impossible to determine even approximately the degree to which practical concerns focussed scientific attention upon certain problems."[72] Other pages, where he deals with specific cases, often contain much more definite statements. In passing on now to citing Merton's principal qualifications I shall refrain from attempting to determine their final mix in his own mind at the time.

First, there is the issue of conscious motivation versus institutional constraint. Unlike Hessen, whose thesis was too crude even to make it possible to distinguish between the motivational and the institutional level of analysis, and also unlike Clark, who had taken Hessen's thesis to be refutable by citing scientists' outspoken motives for doing science, Merton quite consciously made and applied a distinction between, on the one hand, what a scientist's actual motives may have been and, on the other, what larger social values, on a much less conscious level, guided his research in one direction rather than another. If men like Boyle showed themselves scientifically interested in potentially useful pursuits, this was most often a matter, not so much of satisfying personal needs, but rather of an orientation guided by predominant social values. Rare was a case like that of Robert Hooke, who, in his relative poverty, was one of the few scientists of the time for whom economic gain was one important motive for doing science the way he did it. His case, so Merton argued, does not exclude at all the possibility that the turn toward the application of science widely observable among contemporary scientists might derive from social constraints of a higher order of generality.

Next, Merton had no difficulty in adopting Clark's 'disinterested search for truth' as an independent motive for doing science, his only point here being that, given certain overall social values, this search could still flow in one, rather than in another, direction.[73] The largely utilitarian values of the day, so Merton asserted, helped significantly to further the application of science to the solution of actual technological problems.

The question remains in what precise manner institutionally grown values were supposed by Merton to foster the direction of scientific research. For determining the ultimate value of Merton's argument this was, and is, a decisively important issue, yet on this point Merton was concerned to attack other authors' views rather than to elaborate an account of his own.[74] Still, a few hints can be detected at various places in his book. The most pertinent statement I found is the following:

> With the rise of the modern era, *when science had not yet attained social autonomy,* the emphasis on utility served as a support. Science was socially countenanced, even esteemed, largely because of its potential use. [My italics][75]

The underlying idea appears to be that the applicability of the new scientific enterprise must be seen as an *ideology* required in order to justify and to legitimate it and thus to procure social support. This idea has been widely ascribed to Merton in the literature since, even though it remained almost wholly implicit throughout his doctoral dissertation.[76] It was to be elaborated by Ben-David into a full-fledged account of the emergence of early modern science, an account we are to examine later in section 5.3.

5.2.4. *Zilsel and the Social Origins of Early Modern Science*

No historian of science concealed a heavier indebtedness to Olschki than the Austrian philosopher/scientist Edgar Zilsel (1891–1944). Together with Neurath he formed the 'left wing' of the Wiener Kreis (Vienna Circle). He also served on the board of the Verein Ernst Mach (Ernst Mach Society). After the *Anschluss* he managed to leave Austria, living in the United States from 1939 to 1944, when he committed suicide. His years of exile were marked by an impressive output of articles on various aspects of the emergence of early modern science, all as it were abstracts deriving from a more detailed monograph he did not live to write.[77]

Like Hessen, Zilsel was a Marxist. With this bald and factual statement, however, the similarity ends. Between them, these two men strikingly exemplify the split in Marxist thought that started with the bolshevik *coup d'état* of 1917 and separated by an ever widening gap the many who succumbed to the Stalinist myth from that vanishing number of Marxists in the West who managed to keep free of this intellectual graveyard and its aura of State power. Unlike Hessen's brand of Marxism, Zilsel's variety served, not as a severe, in many respects enforced, constraint upon an intelligent and civilized man, but rather as a stimulus toward fresh thought in areas where sophisticated social analysis had not exactly been abounding. The method of class analysis as applied by Zilsel surely suffered from limitations of its own, as we shall see, yet one can recognize how it is being put to work here in the manner originally intended by Marx—as a novel framework to enlarge one's intellectual grasp of reality.

Zilsel's paper "The Sociological Roots of Science," which appeared in 1941/1942, contains the most succinct exposition of his fundamental thesis. As such it provides an outline of the full study he had embarked upon but never finished. The article opens with a very powerful and attractive argument in favor of the sociological analysis of the origins of early modern science:

> Were there many separate cultures in which science had developed and others in which it is lacking, the question about the origin of science would generally be recognized as a sociological one and could be answered by singling out the common traits of the scientific in contrast to nonscientific cultures. Historical reality, unfortunately, is different, for fully developed science appears once only, namely, in modern Western civilization. It is this fact that obscures our problem. We are only too inclined to consider ourselves and our own civilization as the natural peak of human evolution. From this presumption the belief originates that man simply became more and more intelligent until one day a few great investigators and pioneers appeared and produced science as the last stage of a one-line intellectual ascent. Thus it is not realized that human thinking has developed in many and divergent ways—among which one is the scientific. One forgets how amazing it is that science arose at all and especially in a certain period and under special sociological conditions.[78]

How, then, to set up such a sociological analysis? Here Zilsel's Marxist presuppositions manifested themselves in his taking it for granted that, since "the emergence of modern science . . . took place in the period of early European capitalism," the transition from feudalism to early capitalism and the social changes of the time that went with it "form necessary conditions for the rise of science."[79] A coincidence in time is, without the invocation of any further argument, being turned here into a clear-cut causal connection.

Analysis along such lines yielded results on two distinct levels. The first of these was the idea that "the emergence of the quantitative method, which is virtually non-existent in medieval theories, cannot be separated from the counting and calculating spirit of capitalistic economy."[80] This was an idea that had been around for quite some time already. Although its origins have never yet been tracked down, they are most likely to be found in the circle of German economists and sociologists composed of such scholars as Weber, Sombart, and Simmel.[81] The connection between the quantitative aspect of early modern science and early capitalist accounting habits has been quite variously construed, from the type of subtle intertwinement we encountered when discussing Weber (section 3.6.3) to the straightforward causality of which Zilsel provides one example. When presented in the latter format, all such theories seem to suffer mortally from an elementary objection voiced by Dijksterhuis: that quantitative operations preceded by many centuries the onset of merchant capitalism, since the former go back to early antiquity.[82] What such theories equally seem to have in common is that they are proposed on a largely *a priori* basis rather than being anchored in solid evidence from the history of both early capitalism and early modern science. Zilsel, for one, confined himself here to invoking contributions made by mathematicians like Pacioli and Stevin to bookkeeping techniques—facts of history, no doubt, yet hardly sufficient evidence for connecting causally the rise of merchant capitalism to the rise of early modern science. The lasting value of Zilsel's inquiry into the 'why' of the latter event, then, resides rather in a more original line of thought, which occupies the principal part of his study and which starts with a distinction between "three strata of intellectual activity in the period from 1300 to 1600: the universities, humanism, and labor."[83] The three social layers, so he argued, were characterized by importantly different intellectual habits and capabilities.

The type of rationality to be encountered in the universities during these three centuries was quite as different from the "rational methods of a developed economy"[84] as it was typical of the scholastic pleasure in endless distinctions and classifications to be found wherever priests pass on their traditional, religious doctrines. Secular learning, on the contrary, originated outside the universities in Italy, where it took the form of humanism. This type of scholarship was interested in words rather than in things. Sociologically it may be characterized by its function of gaining prestige both for the patron and for the humanist scholar himself. With the scholastics the humanists shared the classical disdain for the mechanical, as opposed to the liberal, arts:

> The social antithesis of mechanical and liberal arts, of hands and tongue, influenced all intellectual and professional activity in the Renaissance. . . .
>
> [Meanwhile,] beneath both the university scholars and the humanistic literati the artisans, the mariners, shipbuilders, carpenters, foundrymen, and miners worked in silence on the advance of technology and modern society. . . . Having outgrown the constraints of guild tradition and being stimulated to inventions by economic competition, they were, no doubt, the real pioneers of empirical observation, experimentation, and causal research.[85]

These artisans were generally ill-educated men, but a few groups from their midst managed to emancipate themselves to some extent. These were the famous artists/engineers (examples run from Brunelleschi to Cellini; Stevin is counted among them, too),[86] but also the surgeons and the makers of musical as well as of nautical and astronomical instruments. Far removed from the empty verbosity of scholastics and humanists alike, these superior artisans in fact made numerous scientific discoveries. However, and this is crucial, they lacked the analytical skill to systematize these discoveries and raise them from rules of thumb to the plane of exact scientific laws. After all:

> Natural science needs theory and mathematics as well as experiments and observations. Only theoretically educated men with rationally trained intellects were able to supply that other half of its method to science.[87]

But this meant that

> the two components of scientific method were still separated before 1600—methodical training of intellect was preserved for upper-class learned people, for university scholars, and for humanists; experimentation and observation were left to more or less plebeian workers.[88]

This separation manifested itself clearly in two entirely different types of literature, the one in Latin, the other in the vernacular. Authors of the latter type, which began to abound in the 16th century, usually could not read the former; the former consistently ignored the latter.

> As long as this separation persisted, as long as scholars did not think of using the disdained methods of manual workers, science in the modern meaning of the word was impossible.[89]

Having thus with great patience moved each piece of the game carefully to its proper place, Zilsel is now ready to checkmate in one final move:

> About 1550, however, with the advance of technology, a few learned authors began to be interested in the mechanical arts, which had become economically so important. . . . Eventually the social barrier between the two components of the scientific method broke down, and the methods of the superior craftsmen were adopted by academically trained scholars: real science was born. This was achieved about 1600.[90]

Not wasting one more word on the somewhat vexed question of what caused this break in a persistent pattern to occur, Zilsel sought evidence for his thesis in the works of selected scientific thinkers who flourished around 1600. His chief contemporary witnesses were Galileo, Bacon, and Gilbert. With all three he noted their scorn for scholastic as well as humanist modes of argument. In the case of Galileo, in particular, Zilsel pointed to all his background experiences in contemporary technology which had previously been treated by Olschki.[91] In Bacon's works Zilsel found especially significant the call to scientific teamwork, which, so he claimed, obviously derived from the division of labor so common to craftsmen. The background of Gilbert's magnetic theories in navigational problems as discussed at the time by the compass-maker Robert Norman was examined extensively by Zilsel in a separate paper. He also sought to underpin his thesis by demonstrating how the idea of progress in science equally arose in the early 17th century through the tradition of the artists/engineers.[92] Altogether, much of the persuasiveness that his set of articles undoubtedly exudes derives from the wide scope of the historical data he collected and from the skill with which he arranged them and made them serve his theoretical ends.

Strengths and weaknesses of the thesis.   The empirical material assembled by Zilsel surely went far enough to reinforce the principal point made before by

Olschki, to wit, that there is a technological and manual labor background to early modern science which may well have helped bring the latter about. What Zilsel added to this are chiefly four claims: (1) Technology and the arts were bound up with the rise of early capitalism during the centuries preceding the decades around 1600, when early modern science was born; (2) natural philosophy, on the one hand, and the arts and crafts, on the other, had been frozen in different social strata whose mutual distance did not begin to melt away until that very date; (3) early modern science was effectively prepared by empirically skilled artisans, in the guise of rules of thumb discovered through experience; (4) the breakdown of the social barrier enabled logically skilled thinkers to turn these rules into scientific laws for the first time, thus producing early modern science.

Zilsel's greatest merit, I think, is that he gave to the 'technological background to early modern science' theme, which had remained just a bit vague and indistinct with Olschki and Merton, and had been intolerably narrowed down by Hessen, a new conceptual clarity. However, here was also the main liability of his thesis; perhaps Olschki and Merton had been wise after all in leaving their respective theses just a trifle undetermined. Somehow it seems that, as soon as explanations along such lines are specified beyond a certain, not so easily located, borderline they display a disheartening habit of losing in plausibility what they have just gained in sharp, conceptual definition.[93]

Or so at least it is with Zilsel's thesis. One obvious drawback is that he fails to tell us how the decisive change came about. How is it that between 1550 and 1600 the social barrier between the skilled artisan and the Latin-writing intellectual suddenly began to break down? Was this part of a more general shift in the social history of western Europe? Or was it just a singular event? If so, how did it come about? From Zilsel we hear no more about it.

A closely related defect, already noted, lies in Zilsel's largely *a priori* approach to how 'early capitalism' was related to the rise of early modern science. Besides the mathematicians' contribution to early capitalist accounting methods little more is to be found here than a reference to the dissolution of the craft guilds—another highly disputable statement to make about the social history of 16th-century Europe.

Finally, and most importantly, Zilsel's thesis stands or falls with a quite definite view of what constitutes early modern science. True, quantification has a part in science-as-conceived-by-Zilsel, yet his attachment to Mach's conception shines through the prime importance given to observation and experiment as the decisively novel characteristics of early modern science. And indeed, it is easier to see what manual labor had to do with early modern science when the experimental features of the latter are stressed than when the emergence of a mathematical world-picture is at stake. But is it true to say that the artists/engineers did in fact perform scientific work, with the only thing still required being the application of a certain amount of logical rigor and systematic thought? In other words, are the skilled artisans' "quantitative thumb rules" truly "the forerunners of the physical laws of modern science," as Zilsel upheld in so many words?[94] What possible rule of thumb, even if married to a capacity for logical, systematic, theoretical thinking, might even come close to the principle of inertia or to Kepler's laws or to the establishment and proof of the tautochrony of the cycloid?[95] It is in questions like these that the most problematic aspect continues to reside of a line of approach to the riddle of the Scientific Revolution that at least has the merit of seeking to anchor the event in the historical peculiarities of the subcontinent where it took place.

The road to the comparative approach.   Zilsel undoubtedly took his own considerations to have established, albeit in outline, the case he had set out to make, but for one important reservation which he took pains to spell out. He readily granted that his conclusions left open the question of how it is that certain other civilizations with money-based economies and with social strata of skilled artisans and secular scholars did *not* produce something like early modern science. Zilsel's initial reply to his own question focused on classical antiquity. Here we see our familiar mirror imaging at work once again: Zilsel argued that manual labor, being done as it was at the time by slaves, was held for a most humble occupation, so that the social barrier between 'hand and tongue' could not in this case be overcome. It follows

> that science could fully develop in modern Western civilization because European early capitalism was based on free labor. . . . Lack of slave labor is a necessary but not a sufficient condition for the emergence of science.[96]

At this point as at several others Zilsel drew a careful distinction between necessary and sufficient conditions for the rise of early modern science. It is not clear whether he thought that sufficient conditions could ever be discovered by means of historical investigation. However, he did think that historians have at their disposal one method which enables them to discover necessary conditions. This is the comparative method: the sole counterpart to experimentation available to historians. For example,

> further necessary conditions would be found if early capitalistic society were compared with Chinese civilization. In China, slave labor was not predominant, and money economy had existed since about 500 B.C. Also there were in China, on the one hand, highly skilled artisans and, on the other, scholar-officials, approximately corresponding to the European humanists. Yet causal, experimental, and quantitative science not bound to authorities did not arise. Why this did not happen is as little explained as why capitalism did not develop in China.[97]

Zilsel's article ends with a plea for research along such lines. He takes it for granted that comparative research must be done the sociological way. That is to say, he assumes without more ado that differences between one civilization and another, to the extent that these differences can account for the emergence or nonemergence of science, cannot but be of a social, rather than of a mental/intellectual, kind. Another unargued assumption of Zilsel's is that the admittedly 'approximate' nature of the 'correspondences' historians inevitably work with (unlike the exact, one-to-one correspondences of the scientist) do not impair the validity of their results. Even though he ignored these and other methodological difficulties involved in the procedure of cross-cultural, historical comparison, Zilsel deserves great honor for being the first to grasp the exciting intellectual opportunities that lie in store here for the student of the birth of early modern science. Or was he?

At the end of his plea for the method Zilsel noted that "it is strange how rarely investigations of this kind are made."[98] He had every right to make the point. How could he possibly know that, as he was writing these very words in New York, a British biochemist in Cambridge, on the verge of a trip to China on behalf of the Royal Society, was contemplating 'an investigation of this kind' and, what is even more, was busily learning Chinese to give substance to the investigation?[99] In due time we shall learn more about the—still unfolding—outcome of this Chinese adventure (ch. 6).

5.2.5. Hall against 'External' Explanations

The complex of ideas we have been discussing—about technology, the arts and crafts, capitalism, and their possible relations to the birth of early modern science—was very much a matter of the thirties and forties. The whole issue rapidly receded to the background during the fifties and early sixties.[100] At that time Rupert Hall was one of the few to take up these questions, not so much to add further ammunition as to weigh the pros and cons in a very critical manner. He did so in two programmatic, highly influential articles: "The Scholar and the Craftsman in the Scientific Revolution" (1959; a fine piece of subtle analysis along typical Hall lines) and "Merton Revisited, *or* Science and Society in the Seventeenth Century" (1963; here party colors are much more in evidence). The main objective of the former piece was to delineate the limited significance for science of 16th- and 17th-century technology; of the latter, to show up the limitations of the Merton/Hessen/Zilsel approach in light of the promising historiographical opportunities offered by a research program that centered on the novel concept of the Scientific Revolution with its built-in notion of the autonomy of scientific ideas.

The scholar and the craftsman.   Hall's interest in the history of technology has always been great, one of his persistent concerns being to keep the area distinct as well as apart from the history of science. His doctoral dissertation of 1952 was *Ballistics in the Seventeenth Century,* and one of its main conclusions was that the significance of the rise of gunnery for the science of ballistics—or for any other physical science of the time, for that matter—was negligible. This research put him into contact with the work of Merton, who, in Hessen's footsteps, had made much of the importance for science of gunnery. The friendship Hall subsequently struck up with Merton[101] neither prevented him from distinguishing sharply and fairly between different shades of dogmatism in the respective theses of Hessen, Merton, and Zilsel nor from expounding his objections to the general line of research undertaken by these three men.

The common theme of his objections is the mostly independent role of the scientist throughout the Scientific Revolution. Hall was quite ready to grant that during the Middle Ages and the Renaissance technology had made remarkable progress. However, so had technology almost anywhere almost anytime, and if, therefore, in the 16th century scholars began to pay attention to it for the first time, "the change was in the eye of the beholder."[102] More generally speaking, to regard the scientist of the 17th century as "a sort of hybrid between the older natural philosopher and the craftsman"[103] was grievously to misjudge the kind of intellectual problems the protagonists of the Scientific Revolution had to grapple with. This is not to say that there was no connection at all; the point, however, is that nothing but proper qualification entitles the historian to speak of the significance of technology and the arts and crafts for the Scientific Revolution. Such proper qualification, Hall averred, showed that their contribution was in fact limited to no more than four possibilities. Technology and the arts and crafts might (1) present problems and information suitable for analysis and inquiry by men with a scientific training; (2) produce "techniques and apparatus adaptable from the purpose of manufacture to those of scientific research"; (3) provide scientific instruments; (4) develop "topics not embraced in the organization of science proper."[104] Examples of these various functions of contemporary technology, so Hall maintained, may be found throughout the Scientific Revolution. Still, here again one must qualify carefully in that some of the four possibilities were realized to a far larger extent in one particular domain of science, such as chemistry, than in another, like astronomy, which but for its instruments owed nothing whatever to the crafts.

Which leads us straight to Hall's next principal point: the extent to which the entire approach failed to fit the concept of the Scientific Revolution in both its intellectual and its geographical scope.

The Scientific Revolution: A more fruitful research program.   The core idea in "Merton Revisited" is that the Hessen/Merton/Zilsel line of research had become outdated as a result of the rise, in the meantime, of a much more promising research program for the rapidly professionalizing discipline of the history of science: the one embodied in the notion of the Scientific Revolution. The framing, by Koyré and his disciples, of this concept of the Scientific Revolution in the sense of a grand intellectual upheaval occurring all across the sciences of the 16th and 17th centuries and involving a new picture of the world made readily apparent the shortcomings of the older line. The conceptual overhaul that had been carried out by Copernicus and Kepler and Galileo *cum suis* was explicable only in terms of the autonomous development of scientific ideas. The defense of this particular viewpoint is what moved Hall almost imperceptibly toward taking Merton's two theses—the one on Puritanism and the one on the uses of science—as meant 'to explain the Scientific Revolution'. This misunderstanding (to which Merton had surely given some occasion, chiefly by his uncritical reliance on Hessen's data and mode of categorizing) was compounded further by Hall's peculiar conception of 'explanation' in the history of science. This conception comes clearly to the fore in a complaint leveled by Hall against Merton's 'Puritanism and science' hypothesis. Hall regretted

> the reluctance of Merton to declare precisely what the relation between religion and science was. If Puritanism was not the 'ultimate cause' of (say) the *Principia*, was it *a* cause? Was the religious encouragement of science the decisive factor or not?[105]

This is a striking misconception of Merton's original problem, as if the latter had ever set out to 'explain the *Principia*'. To point at, and to account for, social values which helped make it possible for a young man of Newton's upbringing and talents to take the route of science rather than of another, socially more recognized intellectual calling is what Merton claimed at best to have accomplished by means of his Puritanism and science hypothesis. But Hall demanded more clear-cut theses, which for him meant that they must address outspoken *motives* for doing science. Hence, he went on to complain in a similar vein that it was not clear what precisely Merton was claiming in his chapters on contemporary technology:

> As with the religious issue, the matter must be clearly and definitely put if it is to have significance. When and in what circumstances is one entitled to infer that a particular piece of scientific work was done for some extra-scientific reason? Those who maintain that often or in some telling way this is the case should lay down their principles of inference.[106]

Interestingly, when a particular scholar, such as Zilsel, indeed took a more definite position, Hall was not satisfied either. It is evident that a deeper objection underlay such appeals to his adversaries to stand up with an unambiguous thesis and let themselves be counted and then cut to pieces. In what follows in the article Hall made no secret of what his chief objection was directed at. The remainder of "Merton Revisited" is one sustained attack on the external approach to the history of science *tout court*. From what Hall has to say in these pages it transpires that for him the external approach to the history of science is *ipso facto* reductionist, or, in his own words, comes down to "the treatment of scientists as puppets." [107] In thus identifying, for all practical purposes, the technology and arts and crafts issue with the external approach to the history of science, and by showing his lack of comprehension for what an external (more particularly, a sociological) point of view can do to help explain the rise of early modern science, Hall contributed his share to keeping dormant the issues Leonardo Olschki had put forward forty years earlier. However, the next quarter-century was to prove that they have a way of coming back, even if in somewhat altered guises.

5.2.6. *The Craft Origins of Early Modern Science: An Interim Assessment*

What Olschki started in the twenties had by the early sixties become a multi-layered and, on occasion, quite confusing complex of ideas and counter-ideas. Four successive sources of complication of the original theme of the crafts background to early modern science may be distinguished. First, the introduction of capitalism as an independent causal agent (compounded by the manner in which it had been introduced) burdened the issue with an ideological load it did not previously possess, and which it was never quite to lose. Then the shaping of the concept of the Scientific Revolution served to highlight the key problem of the extent to which contemporary craft practice could account for a revolution in the domain of creative thought; it also focused attention upon the need to define the proper geographic area to which any thesis along such lines should be taken to apply. Finally, contributors to the debate entertained quite different ideas on what a historical explanation is about and what it can accomplish—differences all too easily dichotomized alongside the pernicious 'internal' versus 'external' boundary line.

It seems to me that as long as the topic of the craft origins of early modern science remains entangled in such multilayered complexity we must despair forever of moving closer to its resolution. Let us therefore seek to disentangle things somewhat. In the first place, a great deal can be gained by excluding, at least for the time being, possible relations between the onset of early capitalism and the emergence of early modern science. Undoubtedly there are tangible connections, which have been noted in scattered passages all across the literature. Equally undoubtedly, no convincing case has ever been made for turning such connections into a solid, causal relationship.[108] It is one thing to recall, for example, mathematical contributions made to capitalist accounting practices; quite another to argue that early modern science arose owing to the increasingly capitalist organization of the economy at the time. Pending a persuasive case possibly to be made for such a thesis once upon a time in the future, it seems wiser to confine the matter for now to the notion that the two may well go back to a common origin—to certain peculiarities of preceding Western development which gave rise both to merchant capitalism and, somewhat later, to early modern science among a range of other, uniquely Western phenomena.[109]

Such a move might be of help, too, in releasing the issue from the bonds of the internal versus external dichotomy. Hall surely had a point; no effort to explain the Scientific Revolution ought to lead to, in effect, 'turning scientists into puppets'. But he was quite mistaken in thinking that *only* an internal point of view can avoid such reductionism, as even Koyré was aware before he went on to forget the distinction between the search for social *conditions* of science and the *reduction* of scientific ideas to whatever agency one might search for in their wider environment. The inner autonomy of ongoing scientific thought (which historians ignore at their peril) is not without its limitations (which it is the historians' business to trace down as best they can). Some very creative thought along such nonreductionist, external lines has occurred in recent decades; it is to occupy us in following sections.

The matter of geography, finally, is comparatively the easiest to deal with. If a given study of features of early modern science is intentionally and/or factually confined to one particular country or region, the important thing is to recognize so much—not to extend without more ado its conclusions either by hindsight or by foresight to the entire geographical domain covered by the Scientific Revolution. Stated this baldly the rule sounds utterly trivial, yet we know both from the vicissitudes of Merton's theses and from what we noted in section 3.7 that it has been less than self-evident in historiographical practice.

Rules of thumb and laws of nature, and the gap between them. Having now either bracketed or wholly removed three important accretions to the original thesis, we can at long last return to what I hold to be the principal bone of contention in the historiographical debate that keeps us occupied here. This is the question of whether and, if so, to what extent empirical discoveries in preceding and contemporary arts and crafts practice helped prepare the advent of the new scientific world-picture that constituted the principal product of the Scientific Revolution. Or, in the apt vocabulary chosen by Zilsel to nail down the issue: What was the precise relationship between empirically found rules of thumb and intellectually expressed laws of nature? Was the difference between the two just a matter of more or less rigorous *formulation,* or did a wider gap yawn between them?

Phrased thus, our dilemma provides the stuff case studies are made of. One such study was undertaken by the historian of medieval technology Lynn White, Jr. In 1966, in an article "Pumps and Pendula: Galileo and Technology," in which, exceptionally, Olschki was given full credit for his pioneership, White focused on two products of late Renaissance craftsmanship: the two mentioned in his title.[110] The issue at stake is what Galileo did with pumps (the onset of theorizing on the void as a practicable matter) and with pendula (the discovery of isochronism and its possible application to accurate timekeeping).

Here White introduces his finding that both devices, which provided the raw material for Galileo's pertinent theorizing in the *Discorsi,* were rather new in his own time. The suction pump, which provided the only available means at the time for wittingly or unwittingly displaying a vacuum in actual practice, was invented in Italy in the second quarter of the 15th century. Similarly, no earlier description of how a pendulum can be used for governing reciprocating motion is known than one found in a book that dates from 1569. Thus, so White argued, two substantial pieces of what we have called 'artificially produced nature' (see section 3.4) lay in waiting for Galileo only through fairly recent developments, which leads him to conclude

> that the rapidly expanding mechanic arts of Galileo's age—in his own metaphor, the Venetian arsenal—provided novel controlled situations, almost laboratory situations, in which he could be among the first to observe natural phenomena, like isochronism or the breaking of a column of water, which are not easily perceived in a pure state of nature. It is exactly Galileo's environment of technical innovations like suction pumps and pendula which makes the tonality of his new sciences historically intelligible.[111]

Modestly phrased as the conclusion is, its final sentence still seems to overstep a limit carefully respected by Olschki, and the question is whether the step is justified. Olschki's position on Galileo (section 5.2.1) may be summed up in the following three points: (1) He used the tradition of the workshops with which he had grown up as a gold mine for mechanical problems; (2) he handled these in a fully independent manner, reaching far beyond the capacity for theoretical reflection inherent in that tradition; (3) he persistently kept an eye open to the application in practice of his theoretical findings. The key point for us here is the second. At the Venetian Arsenal Galileo observed a phenomenon which other natural philosophers of the day were quite as unaware of as it was common knowledge to the theoretically unskilled engineers and workers who handled the pumps. (This is Zilsel's point, too: Galileo was personally bridging a social gap here.) White's amendment that pumps had not been around for a long time does not alter the principal situation at all, which is that Galileo carried to the plane of theory a phenomenon never before taken there. Why did the column of water 'break'? Casually ascribing the phenomenon to deficient material or to similar evasions, engineers hardly cared for any further explanation, meanwhile accepting the phenomenon as an inescapable reality one has to live with. Galileo, on the contrary, had been prepared for the situation because of his preceding immersion in Archimedes' thought and because of the critical stance toward Aristotle that followed in its wake. Hence, he could see what no one for over half a century had seen: that the void was somehow involved. From there he set off a chain of reasoning and experimenting which, within twenty years, was to culminate in Pascal's "Treatises on the Equilibrium of Fluids and on the Weight of the Mass of Air." Was *really* nothing going on here but rules of thumb being formalized into natural law?

I have lately examined a similar case: that of Isaac Beeckman's treatment of beats.[112] On 8 May 1628 Beeckman was told by an organist of his acquaintance that organists are wont to tune the instrument by using beats. (A 1511 description of the procedure specifies that the critical interval must not be tuned as perfectly pure, but "hovering somewhat lower, as much as the ear can stand.")[113] Beeckman at once devised a physical theory of beats (in which he managed to capture the essence of the phenomenon). What is going on here? The organists' procedure is clearly a rule of thumb (a 'rule of ear') if ever there was one. Now the notion of the craftsman's rules of thumb being refined into the physicist's laws of nature would lead one to expect that Beeckman, once made sensitive by the organist to the interest phenomena of musical sound hold to the theorist, would elaborate the former's unsophisticated 'as much as the ear can stand' into a full-fledged theory. Partly, this is what happened. Beeckman did introduce a degree of quantitative precision hitherto never sought into a novel, physical account of the phenomenon. However, the signal given by the organist had hardly caught him unprepared. In 1614/1615 Beeckman had devised a physical theory of consonant musical sound. Around 1617 he had gone on to speculate about what happens when a consonant interval begins to deviate from purity—not yet seeing that this expresses itself in beating. In 1619 he had written down a brief note on beating—not yet seeing the link with the nature of consonance. Nine years later, one hint from his friend, the organist, sufficed to make him see at once how these things really hang together. The upshot, then, is that Beeckman owed the very fact that his thinking on the topic of beating could bear fruit to his preceding theory. And this theory, in its turn, owes nothing at all to any technological insight but is rather 'the fruit, and the expression', of a wholly novel, theoretical approach: the application of a new, non-Aristotelian principle of motion to the problem of consonance, which over the centuries had consistently been treated as a problem in abstract arithmetic.[114] What in the present case was contributed by craft experience is important yet secondary; it helped concentrate wonderfully a mind already well prepared through its own efforts.

Beeckman's work in its entirety offers a perfect test case for the historiographical theory we are dealing with here. Equipped with an academic training, yet thoroughly immersed in craft practice and equally well conversant with practitioners from the most diverse walks of life, Beeckman is ideally suited for the historian to ask what, if anything, his technological background contributed to his considerable (although at the time unpublished) scientific achievement. This question forms the subject of a persuasive chapter in K. van Berkel's as yet untranslated, pioneering book of 1983, "Isaac Beeckman (1588–1637) and the Mechanization of the World Picture."

In this extensive study of large portions of Beeckman's collected notes van

Berkel observes that, for the solution of specific problems of a scientific nature, Beeckman frequently resorted to analogies and metaphors taken from craft practice, more often than not from his own professional experience with candles and water conduits. But this, van Berkel asserts, is also where the uses of rules of thumb came to an end. Beeckman's scientific diary shows that he was much too aware of the conservative, theoretically unskilled bent of mind of craft practitioners as well as of the limitations inherent in purely empirical procedures to expect from this domain a real contribution to our understanding of the world around us. Just as the improvement of contemporary devices ought to be left to the practitioners themselves, so, Beeckman felt sure, the theoretician is indispensable when it comes to formulating general insights about nature:

> Beeckman certainly did not belong to those who were of the opinion that practice without theory was the only true thing, that theory was nothing but the systematic formulation of what was already being applied in practice, and that for this reason the theoretician should always consult the practitioner.[115]

Two conclusions from all this seem to impose themselves: one already drawn by Olschki, the other by Hall. The first concerns Galileo, in particular. We gathered from Rose's account of mathematical humanism (section 4.4.2) that Galileo may fruitfully be seen as the culmination point of a tradition in Archimedean thought which, by itself, had run into a dead end. What enabled Galileo to overcome its limitations appeared insufficiently explicable by means of contemporary refinements in Aristotelian methodology (section 4.4.3). Yet it seems easily explicable upon considering Galileo's background in the arts and crafts, to which Olschki has been the first to call attention. Yet—and this is the second conclusion—it is of paramount importance to realize that enormous creative steps were still involved here and that these are by no means reducible (as Olschki was well aware) to that earlier tradition in empirical technology. This conclusion, which finds support in an examination of two pioneers of the Scientific Revolution with an exceptionally deep involvement in the world of empirical technology, bears generalization pending possible counterexamples to be proffered in future. It is strengthened further by the consideration that it forms nothing but the counterpart to the conclusion about the overall absence of a science-based technology in the 17th century which we reached in section 3.4.3. It is strengthened, too, by a striking summing-up of the overall relation between scientist and technician due to J.D. Bernal:

> The strength of technical tradition is that it can never go far wrong—if it worked before, it is likely to work again; its weakness is that it cannot, so to speak, get off its own tracks. Steady and cumulative improvement of technique can be expected from engineering; but notable transformations, only when science takes a hand. As J.J. Thomson once said, "Research in applied science leads to reforms, research in pure science leads to revolutions."[116]

For the time being, then, we seem committed to the conclusion already drawn by Koyré in a general manner and by Hall more specifically: Contemporary performance in the arts and crafts gave a useful impulse to the revolution in providing food for thought and thus in helping determine the direction it took—it did not, however, implicitly contain that revolution.

The reform of science sought in the streets. Still, we may ask whether this is where, perforce, the matter ends. As always when in the end we stumbled upon an unbridgeable gap between the early 17th-century construction of the beginnings of a new, mathematically and experimentally inspired picture of the world, on the one hand, and a putative cause on the other, it may be worth inquiring, with Kuhn, whether his distinction between the classical and the Baconian sciences may once again be profitably invoked.[117] Additional reasons for suspecting as much reside in one persistent theme in the writings of certain Renaissance thinkers with an interest in the understanding of nature. This is their urgent call to move away from armchair philosophy and to find on the streets and in the workshops a kind of practical wisdom which, so men like Ramus and Vives enthusiastically suspected, might be the very ingredient needed for a hoped-for reform of the sciences. This theme was taken up in the late fifties and sixties by Hooykaas and by Rossi, in particular. When taken together with the tradition previously analyzed by Olschki and now extended by both Hooykaas and Rossi beyond Italy to such literate craftsmen as Palissy, Norman, and Agricola, all this material goes a long way toward establishing the general point Zilsel made about 'hands and tongue' finally coming together in the course of the 16th century. For Hooykaas the primary point of interest was the heightened respect for manual labor that appeared from these writings—a deep historical change he attributed above all to the biblical worldview, which was beginning to make itself felt.[118] In Rossi's book of 1962, *I filosofi e le macchine* (translated in 1970 as *Philosophy, Technology, and the Arts in the Early Modern Era*), the main point was to show how the tradition culminated in Bacon, who drew considerable inspiration for his ideal of a collaborative science from workshop practice.

All of which forcefully suggests that the causal potentialities inherent in the broad idea of the crafts origins of early modern science may be heightened considerably once the idea is focused upon the rise of the Baconian sciences— in particular, on that preparatory period which Butterfield and the Halls had loosely described as some kind of 'outer Scientific Revolution' (section 2.4.3), and which we shall seek to conceptualize further in our concluding chapter 8.

Other manifestations of Europe's budding modernity. Olschki had embedded his topic of the craft background to the birth of early modern science in the general subject of the active life that began to mark Europe in the early modern period. Other markers of that heightened activity, which he mentioned more in passing, are the mechanical clock, the Voyages of Discovery, and the printing press. Each of these manifestations of Europe's budding modernity has since been connected causally to the Scientific Revolution. These respective claims form the subject of the three sections that follow now.

5.2.7. Revolution in Time: Landes and Koyré

The mechanical clock—a uniquely European invention dating from the late 13th century[119]—became in the 17th century the metaphor *par excellence* for the new picture of the universe. But the clock played its part in the Scientific Revolution in a more tangible manner, too. Take the problem of geographical longitude—known from 1522 onward to be potentially soluble by means of accurate timekeeping. Although, as noted in section 3.4.3, the gap between science and its fruitful application was too wide yet to be bridged in the course of the Scientific Revolution, it is nonetheless clear that in this domain more than in any other at the time straightforward results of the new science were intimately linked to urgent practical needs. Moreover, money was involved—heaps of it. Thus, the clock-plus-longitude complex provided irresistible bait to the authors discussed in the present chapter, with their mixed 'technological and socioeconomic origins of the Scientific Revolution' theses. Olschki took up the issue mainly in connection with Galileo's concern to have his discovery of Jupiter's moons exploited for the determination of longitude.[120] Hessen found in this domain one more instance of a practical problem arising from capitalists' needs and leading to new scientific discoveries, particularly by Huygens and Newton (curiously so for a Marxist in search of the 'earthy core' of science, he entirely ignored the high money rewards).[121] Merton characteristically followed in Hessen's footsteps, with the equally characteristic qualification that the connection was "complex, indirect, but none the less clear-cut." [122] Zilsel referred to the issue only once, in showing how much Gilbert had owed to the sailor and compass-maker Robert Norman. Here the connection resides in contemporary efforts to find longitude through the declination of the compass.[123]

The problem of longitude was by no means the only way in which horological matters impinged upon early modern science. By far the best unified treatment known to me of the heady stew involved here of clocks, compasses, backstaffs, telescopes, pendula, ships, harbors, astronomical observations and equally astronomical rewards, governments of seafaring nations, scientists' motives, and craftsmen's achievements is by the economic historian and lover of clocks David S. Landes. In his book of 1983, *Revolution in Time,* Landes undertook to analyze in a precisely argued manner, and with irresistible gusto, what in the subtitle he promised to examine: *Clocks and the Making of the Modern World.* Relying on a large amount of specialized literature—most of it narrowly horological—Landes examined in a broad framework of intellectual and socioeconomic history how the mechanical clock has impinged on the life of western Europe during the almost 700 years of its existence. The book is very rich in showing both in large brushwork and in much detail what contemporary labor arrangements, working habits, commercial opportunities, arithmetical capabilities, transportation, warfare, and so on owed to the mechanical clock. It also presents the story of what the mechanical measurement of time did for the problem of longitude during the period of the Scientific Revolution (and, of course, after). In doing all this, and much more of lesser concern to us here, the book points out, in an unforgettable way, in what senses modernity has relied on accurate timekeeping. But there is one thing it does scarcely, if at all—to examine what, if anything, the clock did, in a more specific sense, for the emergence of early modern science.

The precision clock and the Scientific Revolution. On this score Landes leaves us with one hint only: the truism, for the historian of science, that increasingly accurate timekeeping opened up new opportunities in observational astronomy. This was first realized in the second half of the 15th century, and later exploited in the grand manner by William IV of Hesse and by Tycho Brahe. For the rest, the horological literature seems to have little more to offer than the surely important point—made by Lynn White and by Silvio Bedini, in particular—that the rapid establishment, in the 14th century, of a professional group of clock- and watchmakers provided a vital training ground on which to exercise the mechanical skills indispensable for the construction of scientific instruments in the course of the Scientific Revolution.[124]

Still, there may be more to the issue than experts in historical horology, whose attention is directed mostly to the ins and outs of the clockmakers' craft, are likely to focus upon.[125] We may ask whether the quest for increasing precision in timekeeping had anything to do with the kind of precise scientific knowledge that was such an important ideal of the Scientific Revolution.

For Alexandre Koyré there was indeed a relationship to be detected here, and it is a clear-cut and unambiguous one.[126] The mechanical clock—already a unique and splendid invention by itself—had indeed, in its developed, late 16th-century state come to represent the acme of what the very best of craftsmen, unaided by scientific principles since there were none yet, could achieve under their own steam. Yet, despite their collective efforts over more than two and a half centuries, their clocks were *still* not accurate enough to do without regular checking against readings of the sundial (a fact mentioned by Landes, too, but in a significantly different context).[127] For the purposes of daily life, this was surely good enough (this statement, too, would not be disputed by Landes). A decisive change in the degree of accuracy desired, however, set in as soon as scientists made their entrance on the horological scene. Whatever their motives were in their quest for horological precision (Koyré decidedly opts in favor of astronomical observation and against the problem of longitude, but there is nothing in his pertinent argument that compels him to make a choice here), it is *they* who made the crucial improvements possible. The craftsmen, obviously, continued to play a part, but, in the new universe of precision, their efforts became subordinate at once to the ideas of the members of that new species for whom the quest for precision was part and parcel of their enterprise to understand nature in a novel fashion. That is to say, what Landes treated as a sustained, seven-century-long effort to increase the accuracy of the measurement of time signified for Koyré a two-stage affair, with the intervention, around 1600, of the Scientific Revolution as the crucial spur toward previously inconceivable heights of precision.

Two questions must be asked here. We are meanwhile well acquainted with the first of these, which is simply, What of the craftsmen? Would the clockmakers' guild have been powerless to continue its advance on the road toward further accuracy if no Scientific Revolution had intervened? If one considers—to take the most clear-cut instance, naturally invoked by Koyré as his chief witness—that such a crucial idea as the application of a freely swinging pendulum to the clockwork not only was made by a full-blown scientist but also derived directly from a context of highly abstract, mathematical problems,[128] it is hard to arrive at other than an affirmative answer to this question (an answer which, in its turn, tends to confirm the general conclusion arrived at in the previous section). The answer may surely be qualified to some extent by pointing out (as Landes was fully entitled to do) that Huygens and his colleagues had themselves become craftsmen of a kind,[129] or that "the best of the craftsmen [involved in the quest] possessed surprising theoretical knowledge and conceptual power."[130] Still it may be gathered from Landes' overview, too, that around 1600 new protagonists came on the scene, and that they have stolen much of the show ever since.

Which leaves open the question of whether the impulse toward their entrance was perhaps enhanced by what they had so far seen happening onstage. In other words, could it be the case that almost three centuries of living with time measured, for the benefit of the population at large, to a degree of precision matched by no standard of comparison in any other traditional civilization had prepared the European soil for the sudden flowering of a novel science dedicated to making its statements about any subject at all as precise as it possibly could? Or, to state the point more succinctly, may not the very existence of the mechanical clock, as a little pocket of relative yet unique precision in the midst of a still by and large traditional society, have contributed its share to the emergence of early modern science as the embodiment of the quest for precise and reliable knowledge?

This is, indeed, what Koyré appears to have had in mind.[131] His suggestion may be supplemented by an equally passing notion expressed by Lewis Mumford to the effect that "by its essential nature [the clock] dissociated time from human events and helped create the belief in an independent world of mathematically measurable sequences."[132]

To my knowledge, such promising possibilities have hardly been explored yet.

5.2.8. Hooykaas and the Voyages of Discovery

O mar com fim será grego ou romano
O mar sem fim é portuguez.
The sea with end may be Greek or Roman
The sea without end is Portuguese.

—*Fernando Pessoa, 1918*[133]

Os portugueses . . . descobriram novas ylhas / novas terras / novos
mares / novos povos: e o que mays he: novo ceo: e novas estrellas.
The Portuguese . . . discovered new isles, new lands, new seas, new peoples, and, what is more: a new heaven and new stars.

—*Pedro Nunes, 1537*[134]

Alone among historians of science, R. Hooykaas has taken an abiding interest in the history of Portuguese science and letters. Having learned to read and speak Portuguese fluently by the early sixties, and in the absence of 'native' historians of science, it fell upon Hooykaas to study the significance of the Voyages of Discovery for early modern science.[135] To be sure, for the details of those early voyages he could rely on the work, as well as the support, of two eminent historians of geography and cartography, Cortesão and de Albuquerque, yet for the interpretation of the effects the new discoveries had upon contemporary men of science and letters in Portugal, Hooykaas was very much thrown back on his own resources. Unfortunately, his resulting publications, although mostly written in English, have as a rule been confined to circulation in Portugal and the Netherlands and thus have found little, if any, response in the world of historians of science at large. This is the more to be regretted since, among the results of his pertinent investigations, we find a fresh attempt at explaining the rise of early modern science. This was set forth with particular cogency in an article first published in 1966, "The Portuguese Discoveries and the Rise of Modern Science."

The clash between reason and Portuguese experience. Hooykaas' starting point is the intellectual challenge which, from the early 15th century onward, was posed by the discoveries made by Portuguese mariners in their secret probings ever further down the coast of Africa. The digestion, by Portuguese culture, of so much that was new could not have been avoided, yet the process was exacerbated by the near-simultaneous introduction of the values of humanism in Portugal:

> Perhaps there is no literature in Europe that mirrors so clearly as the Portuguese, the painful conflict in the minds of people who, on one hand, by their humanistic education, not only knew better but also more uncritically admired, ancient learning than their medieval predecessors, and, who, on the other hand, in the same epoch, were confronted with abundant proofs of the insufficiency and fallibility of that same Antiquity.
>
> It was the Portuguese who had to cope as the first in Europe with these opposite influences. Their solution had two aspects: *to follow nature wherever it may lead, and, secondly, to respect tradition whenever this is possible* without violating the first directive.[136]

There follows an array of fascinating accounts of, and quotations from, works by contemporary authors who were compelled to face as facts numerous phenomena the ancients had been quite sure could not possibly be observed because they were bound not to exist. Examples are Aristotle's denial that the tropics could be inhabited; Ptolemy's mathematically derived conviction that all dry land is confined to part of the Northern Hemisphere, and so on.[137] In canto V of his *Lusiadas,* Luis de Camões (traveler, scholar, poet) expressed, and took sides in, the dilemma as follows:

> I have seen things which the rude sailors, having long experience as their sole teacher and judging things by appearances, hold for certain and undoubtedly true, but which those who possess fuller powers of discrimination and who, solely through their intellect and their science, know the world's most recondite secrets, judge to be false or misunderstood.[138]

What did this mean for science? In Hooykaas' view we are witnessing here the birth of 'natural history' in the sense of the domain of the hard and given fact, "where the contingency of knowledge (and of nature herself) comes to the fore."[139] The narrow world of sense-data to which the ancient natural philosophers had confined their all-too-rational speculations was now being blown to pieces. And this was not spontaneously being done by fellow natural philosophers, but rather at the urging of scarcely literate sailors! It is their reports that forced such scholars as João de Castro (author of *Tratado da sphaera,* c. 1538; later viceroy of the Portuguese colonies in India) to acknowledge the superiority of experience over and above rational speculation in the realm of the factual.[140] And this is how Hooykaas could state as his fundamental thesis

> that the Portuguese seafarers and scientists of the 15th and 16th centuries made an important contribution to the rise of modern science by unintentionally undermining the belief in scientific authorities and by strengthening the confidence in an empirical, natural historical, method.[141]

Hooykaas did not claim to be the first to point out this connection. On the contrary, Francis Bacon had already perceived it clearly; witness the passage in the *Novum organum* where the opening of the "regions of the material globe" is set off against the disgrace it would be if "the intellectual globe should remain shut up within the narrow limits of old discoveries."[142] Nor, so Hooykaas insists, was Bacon in the dark about the essential simultaneity as well as the causal connection of these two events: the Voyages of Discovery and the imminent reform of the sciences through unprejudiced observation and experiment.

Hooykaas equally agrees with Bacon that the birth of early modern science was "a general and gradual change of the intellectual climate" rather than "marked by a spectacular singular event."[143] There is nonetheless a problem with the temporal linkage between the voyages and the birth of early modern science—a problem Hooykaas faced head-on.[144] In the first place, it was not given to the Portuguese themselves to usher in the new science for which they had laid one indispensable cornerstone. This is accounted for by Hooykaas partly through the well-known, overall decay of Portuguese culture after c. 1540, partly because of insufficient collaboration between scholars and artisans, and also because the Jesuit monopoly over education tended to strangle freedom of scientific thought.[145]

Not only was the Scientific Revolution not wrought by Portuguese, it also did not take place at the time of their great discoveries, but rather one to two centuries later. This "considerable time lag," Hooykaas contends, should be seen as "an incubation period, in which the 'new philosophy' had already arisen, albeit almost noiselessly."[146] The key novelty that links the Portuguese discoveries to the rise of early modern science resides in the "upheaval in sixteenth-century geography," in particular.[147]

Having come thus far, let us recall a couple of explanations for the rise of early modern science we have previously reported to carry the label 'made by Hooykaas'. What is to be made of the present thesis in light of the one he put forward regarding 'religion and the rise of modern science' (section 5.1.1) as well as of his modified endorsement (section 4.3) of Duhem's explanation through Tempier's decree of 1277? How do Hooykaas' three theses on the origins of early modern science hang together?

The contingency of nature and the rise of early modern science. Fortunately, rather than leaving it the historiographer's duty to find out, Hooykaas himself, in an article of 1987 ("The Rise of Modern Science: When and Why?"), took the trouble to weigh his several theses and derive the entire set from a common denominator. Not that it is particularly hard to detect this common denominator, since it forms the principal *topos* in virtually all his writings (already outlined in section 4.2.1).

Crucial for Hooykaas throughout the history of science is the discovery of the *fact,* even if, nay, in particular if, the fact seems to be ruled out by our preordained reasonings. Rare indeed is the article among his numerous publications where he does not quote T. H. Huxley's dictum: "Sit down before fact as a little child, be prepared to give up every preconceived notion, follow humbly wherever and to whatever abysses nature leads, or you shall learn nothing."[148]

For Hooykaas, therefore, the Scientific Revolution (to the extent that it was a revolution at all) was first and foremost the discovery by scientists of an approach centering on the supremacy of facts. His explanation of the Scientific Revolution comes down, therefore, to pointing at those preceding events in history which contributed to the rise of an intellectual climate of humble submission to the fact in full recognition of nature's ultimate contingency. Hence, it is not required to demonstrate that Tempier's decree at once ushered in the new science, for quite obviously it did not, or that the Voyages of Discovery or even the Reformation had this effect, for they did not either, or at least not directly. In each case it is sufficient to point out their contribution to the shift in intellectual climate that constitutes what the Scientific Revolution ultimately was about.[149] The same is true of Zilsel's fundamental point about the meeting of the artisan and the scholar, a thesis substantially endorsed by Hooykaas as well, since the lifting of the social barrier that kept these groups separate equally compelled scientists to abandon their arrogance and submit to the facts of nature as revealed to them by the artisan.

In all these cases the creation of an intellectual climate is the key thing, irrespective of the intentions of those who—in most cases, quite unwittingly—set the train in motion. Among these various stimuli, the one ultimately allotted the greatest significance by Hooykaas is the one that started in tiny Portugal:

> The great change (not only in astronomy or physics, but in *all* scientific disciplines) occurred when, not incidentally but in principle and in practice, the scientists definitively recognized the priority of Experience. The change of attitude caused by the voyages of discovery is a landmark affecting not only geography and cartography, but the whole of 'natural history'.
>
> . . . Henry the Navigator, who organized the first great voyages of discovery, was no scientist, and he had no scientific aims. But it was his initiative that triggered off a movement which, growing into the avalanche of upheaval in sixteenth-century geography, opened the way for the reform, sooner or later, of all other scientific disciplines.[150]

5.2.9. *Elizabeth Eisenstein: Science Goes from Script to Print*

Elizabeth L. Eisenstein is a historian who did research in the past of France until she became fascinated with the history of printing. Her book *The Printing Press as an Agent of Change: Communications and Cultural Transformations in Early-Modern Europe,* which appeared in two volumes in 1979, has received wide and deserved acclaim. With unfailing eye she spotted one of those rare topics of fundamental significance for our understanding of the past which had been left somewhat neglected by all but dedicated specialists (as was true of the revolution in time before Landes came along in 1983). These specialists, however, had consistently failed to pull the fairly narrow field of book-history into the wider problem area of the modernization of Europe. In contrast, Eisenstein went ahead and single-handedly transformed everybody's vague notion that somehow the advent of the printing press changed quite a great deal in early modern Europe into a huge catalogue of contemporary ideas and activities that, so she claimed, had indeed been affected, in a precise and clear-cut manner, by the shift from script to print.

What is even more, in marked contrast to the habits of most other historians, Eisenstein, after having discussed in much detail how our views of the Renaissance and of the Reformation must be altered when considering the effects of printing, did not stop there. She acquainted herself with a good portion of literature on 16th- and 17th-century science and went on to devote a similar, quite extended discussion to what the printing press had wrought in this domain of human endeavor. Vague notions of what printing might have done or failed to do for the Scientific Revolution were now turned by her into precise ideas as to what it actually did. The net effect of her full effort has been for many readers, including me, to make it all but impossible to read anything from, or about, Europe in the 16th and 17th centuries without finding oneself mumbling occasionally, 'Ah, but such or such could only have happened in a print, rather than in a scribal, culture'.

In my opinion, only one substantial liability has come with these formidable assets. In Part III of her book, entitled "The Book of Nature Transformed" (to which the following discussion is confined), one might have expected an independent, running argument in which a number of hypotheses are successively set forth on what precisely print did to science. Such an argument might have been followed by an overview of how her new interpretations affect previous conceptions of the Scientific Revolution, relegating leftover disagreements with scholars dead and alive to the footnotes or, if need be, to an appendix or a bibliographical essay, as entertainment for the cognoscenti.

But unfortunately, ideas seem to come to Eisenstein while disputing those of others, and it is only in setting forth why those are mistaken that she establishes her own point of view. As a result, throughout the 708 pages of her book those points of view of hers are almost buried in one sustained, often strident polemic. Her numerous observations on pieces of pertinent literature remain scattered in the process and are never given the coherence needed for a comprehensive historiographical overview, nor is her own stance ever allowed to make a cogent case in its own right. In this way both are made to suffer from a mix-up that serves no positive purpose either. Or does Eisenstein really be-

lieve that the validity of her own viewpoint stands or falls with establishing the
failure of her predecessors to share hers?

Remarkably, the abridged version she published in 1983 under the title
The Printing Revolution in Early Modern Europe, in failing to remove the scaffolding and to reorganize the entire argument drastically, suffers in proportion
from the same afflictions. In the following discussion of Eisenstein's principal
ideas I shall ignore her somewhat exasperating mode of presentation and do
my best to peel her views loose from the scholarly wrangles in which their
author found fit to keep them immersed.

Scribal versus print culture.   Eisenstein's point of departure as well as the
pivot around which her basic argument is made to turn is the fundamental
difference in the way texts are transmitted between a culture that has printing
and one that does not. With the courage needed to state the seemingly obvious,
Eisenstein discusses here, with much caution because there are so few reliable,
quantitative data about scribal culture, such drastic changes wrought by the
printing press as the incomparably increased number of books distributed and
sold, the amount of standardization within reach now that errors made by
scribes in copying were ruled out or could be corrected authoritatively
throughout an edition, and the sheer preservation of texts available in such
multitude that no fire, theft, or even censorship could fully eradicate them all.
Nor were such novel features of print culture confined to the written word.
The printing press also provided unprecedented opportunities for elucidating
texts—anatomical atlases, mathematical treatises, and the like—by means of
standardized illustrations, figures, diagrams, and other schematic representations. All this taken together meant that no longer was the sheer transmission
of intellectual products through successive generations a major cultural feat in
its own right: The more or less intact preservation of ideas over time in essentially uncorrupted state could henceforward be taken for granted in a way inconceivable before the advent of printing. The implicit appeal Eisenstein directs to her readers throughout her careful discussions of these matters is to
make an effort of the imagination and become aware of what the handling of
texts truly meant in the scribal era—that is to say, during that portion of world
history that is delimited on one side by the invention of the written word, from
about 4000 B.C. on in one center of civilization after another, and, on the other,
by the invention, around 1450 in western Europe, of a reliable method to reproduce, preserve, standardize, and disseminate it through movable type.

Science before and after the advent of printing.   What did this major change
signify for science? Eisenstein's farthest-reaching claim is that "the printing
press laid the basis for . . . modern science."[151] True, she does admit the merit,
as far as it goes, of an objection made by such historians of science as Sarton

and Thorndike against the supposed importance of printing. This is that printers in the 16th century hampered, rather than furthered, the advance of science
by popularizing precisely the wrong books: the *Almagest* and Sacrobosco's
Sphaera rather than Copernicus' *De revolutionibus,* for example. Eisenstein
counters along two lines of argument. First, "however undiscriminating early
printers may have been, they did not subtract from existing book supplies. They
never narrowed, but continuously enlarged, the range of titles available in a
given library."[152] Second, and much more important, precisely by making
available discordant texts on the same topic, printers enabled scientists for the
first time to carry out extensive surveys of the literature, to note inconsistencies
that inevitably had eluded their counterparts in the scribal era, and to build
novel theories upon the shortcomings of their predecessors. This, in
Eisenstein's view, is precisely what made the Copernican revolution possible.

It is a mistake, she insists throughout her book, to accept at face value the
idea that the rise of early modern science is characterized by a determination
to read, no longer books, but 'the Book of Nature'. Rather, the Book of Nature,
too, was apprehended primarily through books, nor could a scientist's findings
about the contents of the Book of Nature be submitted to collegial scrutiny
other than, once more, being set down in printed form. Thus the truly novel
thing about Copernicus is that he was among the first astronomers to be in a
position to scrutinize the entire *Almagest* at his leisure, just as Tycho and
Kepler could choose between competing planetary theories as no predecessor
could:

> By the time Kepler was a student at Tübingen, astronomers had to choose be-
> tween three different theories. A century earlier, when Copernicus was at Cra-
> cow, students were fortunate to gain access to one.[153]

A closely related, novel thing about the opportunities made available by the
printing press in this respect was the leisure gained by not having to copy texts
and establish their reliability:

> As sixteenth-century astronomers they may be distinguished from their prede-
> cessors not so much because they were influenced by one or another Renaissance
> current of thought but rather because they were freed from copying and memo-
> rizing and could make use of new paper tools and printed texts.[154]

Of Copernicus, in particular, it can be said that

> the northern student who went to Italy for graduate studies or to learn Greek was
> a less remarkable phenomenon than the astronomer who could spend thirty-odd
> years studying a vast range of records without leaving home.[155]

Thus, Eisenstein asserts, we begin to understand how a number of celebrated
key events in the history of early modern science had a cause much more concrete than is suggested by the less tangible explanations customarily invoked.

For example, Kuhn attributes Copernicus' ability to recognize in Ptolemy's planetary theory the 'monster' that had escaped his predecessors (section 4.4.1) to a new mode of thought in the nature of a gestalt-switch made possible by the sheer passage of time. Tycho, in observing the nova of 1572, is credited with a new mode of seeing things in the sky. Butterfield finds in these and many more manifestations of subtle shifts of meaning a new thinking-cap suddenly put on in the course of the 16th century (section 2.4.3). All this, so Eisenstein contends with her aptness to take for woolly what is only subtle, becomes much more readily understandable when we compare conditions in the scribal age with those in the new era of printing. Scientists did not so much begin to see familiar things in a new light; rather, old and new data alike were being put before their eyes by the new printing industry.

A number of standing issues in the historiography of the Scientific Revolution are resolved by Eisenstein along such lines. Take the Middle Ages. Many promising new venues were explored, such as magnetism by Pierre de Maricourt or the rainbow by Theodoric von Freiberg, but they were invariably allowed to peter out. There is a lack of continuity here with what happened in the 16th and 17th centuries, and this has everything to do with the shift from script to print. Take the invention of decimal fractions. Stevin's invention had been preceded by that of Bonfils in the 14th century. The difference is that Bonfils' innovation, copied by a Hebrew scribe who failed to appreciate its meaning, remained buried in manuscript until a historian, Solomon Gandz, published it in 1936, whereas Stevin could at once make his invention known to the world in book form. Thus,

> if promising beginnings traced in manuscripts seemed to peter out with negligible results, if significant ancient work 'lay dormant' as did Archimedes' more difficult treatises, the conditions of scribal culture ought to receive at least some share of the blame. The many defects that plagued medieval investigations: a low level of mathematical skills, an absence of observational instruments, a tendency towards 'unbridled speculation' and so forth—may also be partly attributed to the state of communications.[156]

In this respect, our modern, sophisticated editions of manuscripts tend to hide from the sight of the present-day historian precisely those features of scribal culture that most hampered the sustained advance of science—a phenomenon inconceivable before the advent of printing anyway.[157]

Or take the relations between the scholar and the artisan. Here we come across a curious characteristic of Eisenstein's book. She grabs opportunities to disagree with other historians so readily as to overlook completely those many cases where there is in fact much common ground between her position and theirs, and much for her actually to learn from them. So it is with Zilsel. Rather than seeing that she has an important ingredient to add to Zilsel's thesis, she castigates him for believing that the scholar and the craftsman were far apart.

She does not see that this is only half of Zilsel's thesis, the other half being his idea that early modern science could not arise until, around 1600, the twain had actually met. True, the conditions of this meeting remained somewhat nebulous in Zilsel's treatment, and here Eisenstein has an interesting point—where they actually met was in the printers' shops, which "served as gathering places for scholars, artists, and literati; as sanctuaries for foreign translators, emigrés, and refugees; as institutions of advanced learning, and as focal points for every kind of cultural and intellectual exchange." [158] She goes on to qualify her assertion of communication between artisan and scholar with the reminder that technology, in going to press, *ipso facto* overstepped limits carefully guarded by the arts' guilds during the scribal era: The jealously kept ban of secrecy surrounding their typical craft procedures was now being lifted. Thus the scholar and the artisan were both transformed in the very act of serving printers' needs.

Let me interrupt my report of Eisenstein's ideas for a moment with the remark that a point like the one just made sounds very well in the abstract, but that it still stands in need of confirmation by actual examples of actual protagonists of the Scientific Revolution actually meeting actual artisans in actual printshops. Is that where Galileo met them? No, he rather mentions the Venetian Arsenal. Is that where Beeckman met them? No, he was one himself. Is that where Mersenne met them? No, he looked them up in their own workshops. The issue at stake in Zilsel's and others' theses is not so much whether these distinct groups indeed began to meet in the 16th and 17th centuries, for there is no room for doubt left that they did, and Eisenstein surely supplements this well-received insight with a fresh meeting place of her own. The question is rather how far this meeting went in achieving the intellectual reorientation whose explanation is at issue here. To that deeper question, however, Eisenstein fails to respond.

Back now to her contributions to other theses on the origins of early modern science. We find in the first place what she makes of the disputes regarding the Merton thesis. In her treatment these center chiefly around the issue of whether Catholic or Protestant influences were preponderant in the making of early modern science. Here she makes a number of important points on the effects on scientific thought of censorship in Catholic countries, particularly but not exclusively in Italy. Protestant printing houses are shown to have possessed far better business opportunities and behaved with far more daring in publishing risky texts such as Galileo's *Discorsi* than their Catholic counterparts. The Netherlands, in particular, became a free haven for the publication of texts that Catholic authors could not have seen through the press at home. Thus the Inquisition, in discouraging local printers, in forbidding publications, and consequently in stifling much daring thought, contributed its share to the preponderance of Protestantism throughout large portions of the Scientific Revolution.

Finally, there is the issue of Greek stagnation. In discussing Greek science and why it stagnated, Eisenstein focuses on Alexandrian science, the flowering of which she attributes largely to the availability, in the Museum, of a library of unique contemporary scope. It is essential, however, to be aware of how exceptional such a "major scribal message center" was.[159] On the one hand, she cautions, it made possible the unique performance of Greek science; on the other, the reason this came to an end resides in the fact that even such an exceptional research tool suffered from limitations to be overcome only by the advent of printing. Here Eisenstein makes a very illuminating comparison with the uses the astronomer Regiomontanus made of a similarly huge collection of manuscript books collected in Buda by King Matthias Corvinus of Hungary at the end of the 15th century:

> Until 1470, Regiomontanus pursued a course similar to that followed by his im-
> mediate predecessors . . . who had also been wandering scholars and versatile
> servants of cardinals and kings. Insofar as he combined librarianship with astron-
> omy, he followed a pattern that had been characteristic of Alexandrians and Ar-
> abs alike. When he left the library in Buda in 1470 to set up his Nuremberg
> press, however, Regiomontanus crossed an historical great divide. He had, up to
> then, served colleagues and patrons in a traditional fashion. By gaining the back-
> ing of a wealthy Nuremberg citizen to set up a press and observatory, by turning
> out duplicate tables of sines and tangents, series of Ephemerides and advertise-
> ments for instruments, by training apprentices to carry on with the printing of
> technical treatises and by issuing his own advance publication list, he served
> later generations in new ways. He died prematurely in 1476. Some of his manu-
> scripts did not get printed until more than fifty years after his death. Others were
> lost so completely that his ultimate plans for reforming astronomy will never be
> known. Moreover the great library at Buda that he had culled had its contents
> dispersed and was sacked by the Turks in 1527. Similar catastrophes had set
> astronomy back in the past. The course pursued by Regiomontanus gained mo-
> mentum instead. The serial publications he launched never stopped.[160]

Here, then, we have arrived at the core of things. During the scribal age, a never-ending effort to prevent and halt as best one could the ongoing corruption, decay, and destruction of texts was the primary concern of scholars. If and when science managed to flower at all, it could not but do so for quite limited periods of time only. When substantial portions of lost information were recovered, as finally happened to the Alexandrian heritage in 15th- and 16th-century western Europe, this would still not have been sufficient by itself to set science on a course of uninterrupted advance. Only the printing press could achieve this:

> Limits set by the very largest manuscript libraries were . . . broken. Even the
> exceptional resources which were available to ancient Alexandrians stopped
> short of those that were opened up after the shift from script to print. The new

open-ended information-flow that commenced in the fifteenth century made it possible for fresh findings to accumulate at ever-accelerated pace.... Just as geographic space stopped short at the Pillars of Hercules, so too did human knowledge itself appear to stop short at fixed limits set by scribal data pools.... With the shift from script to print, the closed circle was broken.... When engaged in comparing the limited Greek oecumene with the ever-widening horizons of the age of reconnaissance or when tracing the cosmological movement from 'a closed world to an infinite universe', it is worth devoting more thought to the appearance of a 'library without walls'.... The flow of information had been re-oriented, and this had an effect on natural philosophy that should not go ignored.[161]

The printing press as an explanation of the rise of early modern science. This fundamental conception of the continuous advance of science rather than intermittent flowering and decay being the key thing to stand in need of explanation shares a striking amount of ground with Ben-David's idea of the discontinuity of science in traditional society (section 4.2.3). It is no less striking how fully this meeting of souls has escaped Eisenstein, busily engaged as she was in taking Ben-David to task for some scholarly disagreement over a side remark he made, in an early article, on the effects of Galileo's trial. There is, to be sure, an obvious difference in the assignment of the causal agent that finally broke the circle (as we shall see in the next section). For now, let us inspect what it is that Eisenstein actually held to have explained, and what the status of her explanation is.

Her own claim is surely that she contributed in a major way to the explanation of the rise of early modern science. But this claim requires qualification. Throughout her pertinent chapters considerable confusion reigns over what actually stands to be explained: the unique Scientific Revolution or scientific revolutions in general. This has to do, partly, with a lack of awareness of the ambiguities involved in Kuhn's terminology (see section 2.4.4), but also with a certain lack of clarity over what the Scientific Revolution ultimately was about. Here we must distinguish between two issues.

First, is it really a common view among historians of science that the Scientific Revolution was in essence the replacement of reading dusty old books by 'reading the Book of Nature'? Although with varying degrees of emphasis historians have held this to be an important ingredient of the Scientific Revolution (see section 3.2.1), few have failed to focus on the conceptual revolution that took place at the time. Too much has been said about this in chapter 2 of the present book to require a repetition here. Further, in Eisenstein's view of how science and scholarship operate a scholar contributes to the advancement of learning chiefly by criticizing predecessors with whom one has acquainted oneself through a survey of the literature. Here and there we have already had occasion to observe how much her own work might have benefited from taking such surveys just a trifle less seriously and from giving one's own thought and imagination somewhat freer rein. In the case of, for example, Galileo, it seems downright absurd to imagine him engaged in a thorough 'survey of the literature'. How could he possibly have advanced more than tiny little steps if he had taken issue with all those many Aristotelian tomes the printing press could provide him with even more liberally than scribal services had already put at the disposal of Jean Buridan two and a half centuries before?

Second, in talking about the Scientific Revolution Eisenstein mentions chiefly the contemporary tools that became important for scientists, and in one sense the principal point of her whole argument is that books were even more important tools for them than telescopes.[162] This, once more, seriously underrates the mathematical, as against the observational, aspect of the Scientific Revolution. Again, two sources for this neglect present themselves. Of one Eisenstein is herself fully aware. One general point in her book is that not only the written word but figures, diagrams, and the like underwent a major transformation through the printing press as well. She also suggests that this had a major effect on subsequent changes in the *quadrivium*. However, having come thus far she declines to elaborate since this would constitute "too formidable a task to be undertaken here."[163] In view of what she actually managed to accomplish in her book, such a decision has of course to be respected, yet one wishes that the author had realized somewhat better how one-sided a view of the Scientific Revolution was to result from this omission. And once more, she was so happily taken up with castigating Burtt and Dijksterhuis for some minor points of disagreement that she missed entirely what there was to learn from these authors plus Koyré, whose message she equally ignores—what other sides there were to the Scientific Revolution than just improved observation and experiment.

Now what is Eisenstein's own view of the status of her explanation? In her opinion, explanations of the Scientific Revolution have tended to go awry in two distinct directions. On the one hand, there is the narrow-minded refusal to look for explanations outside the world of the intellect. On the other, there is a tendency (or at least no inherent limit) to take just about everything that happened prior to the Scientific Revolution as contributing its own bit to its emergence—"to wander too far afield compiling lists of everything that happened and marvelling at the general turbulence of the times."[164] But if one keeps clear of these two extremes, one runs the risk of remaining enmeshed forever in all those old debates over Plato versus Aristotle, the Merton thesis, the scholar and the craftsman, and so on. Happily, Eisenstein has a way out to offer to all of us, and I give the reader one chance only to guess what it is:

> By placing more emphasis on the shift from script to print, many diverse trends may be accommodated without resort to an indiscriminate mélange and in a way that avoids prolongation of intellectual feuds.[165]

Surely this implicit claim to rightness simply because one has arrived as the latest player in the game carries little persuasion, the more so since Eisenstein's awareness of the full range of scholarly explanations of the Scientific Revolution is far from exhaustive. Explanations of the emergence of early modern science ought to be judged on their own merits, entirely irrespective of the amount of scholarly blood that has already flowed in their defense.

Eisenstein also has a somewhat better argument in store for granting privileged status to the print case. Situated just at the boundary line between the 'internal' (the scientist thinking) and the 'external' (the scientist's thoughts being published), the printing press as a causal factor is ideally suited to satisfy everyone with an explanation that connects thought with the outer world in the most natural way, rather than through the somewhat less straightforward connections implicit in Merton's thesis as well as in many others'. And this, presumably, is how Eisenstein can claim, a few pages before reaching the end of her book, that

> one cannot treat printing as just one among many elements in a complex causal nexus[,] for the communications shift transformed the nature of the causal nexus itself. It is of special historical significance because it produced fundamental alterations in prevailing patterns of continuity and change.[166]

This, unfortunately for Eisenstein, is of course what *every* defender of a specific attempt at explanation of the Scientific Revolution would like to claim for his own conception, and I fail to see how such a claim could relieve the dutiful historian of the obligation to weigh its merits in the balance as carefully as those of any other explanation. Eisenstein, in short, has a point, but we still have to find out how far it goes, and hers is not the way to persuade us that we should adopt it wholly.

This is all the more true because she quite consciously declined to tread one route on which considerable clarification of this very issue might be gained: the route of comparison with other cultures. Here is the one, quite tempting, peep she allows us:

> The early presses which were first established between 1460 and 1480 were powered by many different forces which had been incubating in the age of scribes. In a different cultural context, the same technology might have been used for different ends (as was the case in China and Korea) or it might have been unwelcome and not been used at all. . . . The duplicating process which was developed in fifteenth-century Mainz was *in itself* of no more consequence than any other inanimate tool. Unless it had been deemed useful to human agents it would never have been put into operation in fifteenth-century European towns. Under different circumstances, moreover, it might have been welcomed and put to entirely different uses—monopolized by priests and rulers, for example, and withheld from free-wheeling urban entrepreneurs.

. . . Yet the fact remains that once presses were established in numerous
European towns, the transforming powers of print did begin to take effect.[167]

In summing up provisionally, then, we may conclude that, despite quite a few
subclaims which are doubtful to say the least, Eisenstein has certainly made
the case that the advent of printing contributed something of great significance
to the rise of early modern science. Whether more can be said is a question
that has to be kept in abeyance until we tread the comparative route followed
only a short distance by Eisenstein.

5.3. Ben-David and the Social Legitimation of the New Science

The number of sciences is great, and it may be still greater if the public mind is
directed towards them at such times as they are in the ascendancy and in general
favour with all, when people not only honour science itself, but also its representatives. To do this is, in the first instance, the duty of those who rule over them,
of kings and princes. For they alone could free the minds of scholars from the
daily anxieties for the necessities of life, and stimulate their energies to earn
more fame and favour, the yearning for which is the pith and marrow of human
nature.

   The present times, however, are not of this kind. They are the very opposite,
and therefore it is quite impossible that a new science or any new kind of research
should arise in our days. What we have of sciences is nothing but the scanty
remains of bygone better times.

—al-Bîrûnî, c. 1030[168]

In this striking passage, the great Persian astronomer/physicist prefigured in
everyday language what we have meanwhile come to know as the starting point
of Joseph Ben-David's historical-sociological analysis of the scientist's role in
society. This starting point is the inevitable instability of the pursuit of science
in traditional society owing to a lack of social recognition of the scientist *qua*
scientist (see section 4.2.3). Al-Bîrûnî ascribes the resulting pattern of intermittent flowering and decay to the inherent fickleness of social relations based
on patronage, in particular, and he makes it clear that such ups and downs are
the self-evidently normal course of science over the ages. Nothing in previous
history of science had proved him wrong, or was to do so for more than five
centuries to come. His words nicely encapsulate from before the great divide
of the 17th century what for us, who were born after it, still requires a big leap
of the historical imagination: to realize how extraordinary it is that science ever
became a continuous, ever-growing, ever-advancing enterprise. This sense of
wonder over the very occurrence of the Scientific Revolution, then, lies at the
root of Ben-David's effort to explain so contingent an event: "What needs explanation is the fact that science [in the sense of the continuous enterprise to

367

which we have meanwhile become so thoroughly accustomed] ever emerged at all." [169]

For Ben-David this question was not equivalent to seeking causes of Galileo's or Kepler's or anybody else's scientific *ideas*. On the contrary, he was deeply convinced of the principal (though not, to be sure, unlimited) autonomy of the scientific-intellectual realm. The sociologist of science can hope to arrive at an understanding of the specific social conditions which made possible (or enhanced) the pursuit of science; what he cannot do is explain scientists' thoughts in terms of their social environment. The succession of ideas follows a logic all its own, to be grasped only by a careful reconstruction of those very ideas. But whether any ideas are to arise at all and whether they will fall on fertile soil are questions that social analysis may help elucidate.[170] Hence, for Ben-David the problem of the emergence of science as a continuous enterprise came down to asking

> what made certain men in seventeenth century Europe view themselves for the first time in history as scientists and see the scientific role as one with unique and special obligations and possibilities. What made this self-definition socially acceptable and respectable?[171]

Note that two related yet distinct questions are being asked here. One is, What made scientists think of themselves in a novel way; the other, What caused this novel idea of self to gain respect in the wider world? Ben-David's explanation proceeds through the following stages. He shows first how a number of peculiarly western European features concurred in bringing science, as cultivated chiefly in Italy in the 16th and early 17th centuries, one step beyond where the Alexandrians had left off. He goes on to argue that the enlarged social basis available there for science was still not sufficient to keep the movement going, so that it would probably have faded away but for that particular confluence of social conditions in northern Europe that in the end made the creation of a social role for the scientist possible. We go on now to break down Ben-David's rather complicated argument into its successive components.[172]

Science, the arts, and the vicissitudes of patronage.   The first uniquely Western feature from which a somewhat more solid social basis for science sprouted than had been available elsewhere was the rise of the medieval university. Personal and, therefore, inherently transient master–pupil relations on which the transmission of scholarship had largely depended previously and elsewhere were now solidified in an institution which helped enhance the continuity of scholarship over time. Being defined as natural philosophy, scientific topics stood a good chance of being studied at many universities in the framework of the ubiquitous arts faculties; philosophers with a naturalist bent of mind were free to cultivate such personal interests. The very arrangement which made possible such pursuits, however, also precluded their emancipation from overarching, philosophical concerns. In this way, no dignity could be gained for science as a valuable pursuit in its own right.

A more promising road toward such an objective seemed to be offered by the interest artists began to take in science in the 15th century. Taking his cues mostly from Olschki, Ben-David shows how possessors of mathematical skills, in particular, came to the rescue of architects and artists in search of enlightenment on issues of solid geometry, perspective, and the like. The alliance, which helped to liberate mathematics from the bonds of extreme abstraction and isolation, was a temporary one; fairly soon the artists had learned all there was to learn for them in the domain of science. The lasting benefit of the alliance for science was its introduction, in the company of the artists, at court, especially in Italy. Not that science was taken up in the same patronage system of which the artists had become such a key part from the 15th century onward. The point is rather that, when with the support of princely patronage learned academies were established all over Italy to study matters of humanist concern under the aegis of Plato's Academy, men with an interest in the study of nature could find a place there, too. Certain members of the Italian aristocracy found in such interests one possible source of intellectual refreshment:

> In Italy science was espoused only by an upper class intellectual clique that was trying to displace the official university philosophers and modernize the intellectual outlook of the Catholic Church.[173]

The principal difficulty in this arrangement, with science finding its basis in academies supported by a minority among the clerical and lay aristocracy in search of intellectual renewal, was its lack of permanence:

> Science and scientists remained dependent on the narrow circles within the upper classes that ruled both the country and the church and were interested in learning. These were the circles which had to be convinced that natural science was important and worthwhile enough for them to give to it the full blessing of public recognition and freedom of communication notwithstanding the weighty doctrinal difficulties that might arise from such recognition. The circles were not convinced however. Indeed, it could not have been otherwise at a time when the argument in favor of the Copernican theory was still inconclusive, and when science could offer no more than a few bits and pieces of intellectually interesting astronomical and mechanical theories and the unbounded confidence of Galileo's prophetic genius. In contrast to natural science was the vast body of learning, wisdom, and beauty represented by contemporary humanism and theology. As long as those who had to be convinced by the proponents of science were men who believed in the main depositories of traditional learning, the scientific movement was doomed to failure. For those in the upper class circles who mattered, and probably for the majority of those who did not, science was an intellectually and aesthetically second rate activity as well as a morally and religiously potentially dangerous one.[174]

This, then, is in brief outline Ben-David's sketch of one of those rather rare, temporary institutional arrangements in history under which science had known a period of flowering. Among its aristocratic supporters science could impress when turned into brilliant literature (as Galileo did); it could command respect when called in to assist large engineering projects; it could be honored when cultivated by men of great creative imagination—it could not gain more lasting allegiance as long as it remained tied to social layers like those it was attached to in Italy, with its rigid social composition.

We have now arrived at what, for Ben-David, was the key turn in the history of science:

> Had the fate of science depended everywhere in Europe on the same educated and 'responsible' ruling class as it did in Italy, the emergence of a proud and self-confident body of scientists might have been postponed for a very long period of time, perhaps indefinitely. But fortunately for the development of science, the social structure in Northern Europe was different. . . . there existed in Northern Europe a mobile class whose aspirations, beliefs, and interests—intellectually as well as economically and socially—were well served by their affirmation of the utopian claims made on behalf of science. Furthermore, part of this class found science a religiously more acceptable pursuit than traditional philosophy. *Thus when the ebbing tide of science, which was receding from the scientific circles and academies of Italy, finally touched France and England, its direction was reversed. The changes that took place at that time set in motion a flood which has still to cease.* [My italics][175]

New sources of support for science in northern Europe. What was it that caused northern Europe's uniquely lasting receptivity to the pursuit of science as a by and large autonomous activity?

First and foremost, in Ben-David's view, was its social makeup. In Italy key members of those professions which contributed most to dynamic modes of life—merchants, bankers, entrepreneurs—had been taken up into the city aristocracy, at the price of losing their own outlook on things social and adopting the values of the aristocracy. Despite this infusion of new blood, the Italian class system remained a rigid hierarchy; for scientists in search of support, then, there was nowhere else to go. In England and France, on the contrary, the most dynamic elements in society were seeking a place of their own. Social stratification was much more fluid overall, particularly in England, and this is how what Ben-David calls a 'scientistic movement' could arise in northern Europe which had failed to appear in Italy. That is to say, among these dynamic, un-adapted, and unofficial social layers appeared certain groups whose aspirations coincided with what they perceived as the promise held out by the new science:

> Where there were individuals and groups whose rapidly rising fortunes came from the discovery of new places, new routes, new markets and new kinds of

goods, there was greater readiness to consider the claims of science as a more valid way to truth than traditional philosophy. The recognition came about partly, perhaps, because these claims fitted the new perspective of a socially and materially changing world well; but it also came about more decisively because the interests of these groups were opposed to oppressive assertions of traditional privilege in general.[176]

What, in short, happened—and this is Ben-David's essential point—is that "in Northern Europe . . . science eventually became a central element in an emerging conception of progress."[177] An ideology took shape in the course of the 17th century in which science was being associated with a number of extrascientific concerns. To men who did not necessarily grasp quite what science was all about, the scientific mode of thought came to serve as a symbol for ideals of social and educational reform, in particular.[178] These were the men who helped create and went on to stimulate the informal circles out of which the Royal Society and the Académie Royale were to grow. Important for them was the capacity of the new science—unlike any variety of some all-embracing philosophy—to serve as a neutral meeting ground where not authority but rather the outcome of doctrinally neutral experiment could be called in to settle disputes (a feature much advertised at the time by the Royal Society's first propagandist, Thomas Sprat, and lately employed by Shapin and Schaffer as one cornerstone in a differently shaded historical argument; see section 3.6.1):

> For those interested in changing the world, empirical science was the true prophecy. It produced innovations that contained their own incontrovertible proofs and made all philosophical controversy unnecessary. And not only was it a way to innovation, but also to social peace, as it made possible agreement concerning research procedures to specific problems without requiring agreement on anything else.
>   . . . empirical science symbolized a goal that was still to be attained: the creation of a new social order where things could be changed and improved by rational and objective procedures and without constant violent conflict.[179]

Here was the significance of the emphasis on the empirical procedures and on the practical utility of science, for which key tenets of Bacon's writings could be invoked and turned into a kind of Baconian ideology. Although in reality science could hardly advance by means of the Baconian recipe taken literally, the scientistic movement picked out of the whole of scientific method those elements that conformed most with its own aspirations. If the scientific outlook had been turned into one more rival philosophy, coherent and logically closed like Cartesianism, science could not have broken out of the boundaries traditionally set by the transient support of royal patronage. "Baconism, however, opposed such a closure of the scientific outlook by creating the blueprint of an ever-expanding and changing, yet regularly functioning scientific community."[180]

All this, finally, could take shape only in the absence of a monolithic body of religious authority ready to stifle the fruitful proliferation of competing ideas and thus to destroy the intellectual climate in which the scientific enterprise could prosper. Here, in the domain of religious variety and of consequent religious policies rather than in any shared ethic is where Protestantism became a boon to science in Ben-David's view. Also, the scientistic movement, arising as it did in a mostly Protestant environment, gave to the aspirations it projected upon science a distinctly religious coloring; what took place was "an actual fusion of religious and scientific elements into a new ideology."[181]

Two decisive respects, then, in which northern Europe differed from the much older seats of civilization bordering on the Mediterranean have now been identified:

> The fact that science was turned into such a broad practical outlook in Northern Europe reflected the beginnings of an open system of classes; that this outlook was taken up by intellectuals and developed into an ideology, potentially threatening to traditional authority was possible only as a result of the doctrinal fluidity of Protestantism.[182]

Different destinies for the scientistic movement in England and in France. It may seem as if so far the geographical entities 'northern Europe' and 'England' have been employed almost interchangeably. But this is not really the case. It is Ben-David's contention that a scientistic movement quite comparable to its English counterpart in its social origins and in its social aspirations had arisen in France, too. The get-together circles out of which the leading members of the Académie were to be recruited had come into being largely through the efforts of equally heterodox, nonofficial people with an equally large stake in "the cause of religious pluralism, on which the survival of these groups depended."[183] The shared mentality found a natural expression in lively contacts between such men as Renaudot and Mersenne on the one side of the Channel, Haak and Hartlib on the other. The big difference came when, in the 1660s, the informal groups solidified into regular institutions. In the English case the 'scientistic movement' became part and parcel of the Royal Society, and through the Glorious Revolution it became part of official society at large, too. In centralized, more tradition-bound France, however, Louis XIV succeeded in keeping out "the group for which science had broad social and technological implications" when the Académie was founded.[184] Confining its activity to strictly scientific research, its royal founder and his bureaucracy drove the pursuit of those wider objectives that were symbolized yet not defined by science into the opposition, thus in the end calling into being the 'philosophes' movement which in the next century was to encapsulate every antiauthority, antitraditional intellectual current in France. But by then science proper had already proved its mettle through its own, autonomous merits, no longer re-

quiring support from strata that had been indispensable during the early, formative stages of the Scientific Revolution.

Some difficulties with the thesis.   In weighing the merits of Ben-David's ambi-
tious thesis we must consider first whether he answered the two questions he
set out to examine when seeking to find out how 17th-century Europe accomplished what no traditional society had previously managed to pull off: What
was it that made scientists aware for the first time of the dignity of their own
calling, and what was it that made the environing society support that new-
found awareness, thus giving rise to the social role of the scientist? While the
argument summarized above is clearly directed to answering the second question, it is not so clear how it relates to the first. Unless I have missed something
essential in Ben-David's rather hard-to-digest chapters,[185] the only key resides
in the occasion offered by the intellectual requirements of 15th-century artists to the scientist to move out of the framework of university philosophy
into the open. Or did Ben-David believe that whenever scientists get a
chance to pursue their intellectual interests they take pride in their activity
and begin to seek support wherever they can find it? Although much is
perhaps to be said for such a view,[186] I can find no textual support for it in
Ben-David's treatment.

   Nor is this the only gap in his argument. In effectively restricting southern
Europe to Italy and northern Europe to England and France, not only Spain
and (much more importantly, where science is concerned) Austria and southern
Germany are left out of the former area, but also northern Germany and (again,
much more importantly) the Netherlands are left out of the latter. The logic of
Ben-David's argument requires that the Netherlands, too, should have produced a 'scientistic movement' for which science became a symbol of social
progress. Ben-David attributes the absence of such a movement to an allegedly
monolithic Protestant regime in the Netherlands,[187] but this is just nonsense: If
religious pluralism reigned anywhere in Europe, it was there. Why no ideology
for science developed in this country so rich in scientific talent remains therefore a mystery.[188]

   Despite such weaknesses, the wide geographical scope of the argument
displays a number of impressive strengths, too. The shift of the center of gravity of science from the Mediterranean to the Atlantic—otherwise just hinted
at by a few authors (see section 3.6.3)—is being linked here to the social composition of these distinct parts of Europe, and so becomes a more readily understandable phenomenon. The same is true of the case of science in England. We
have repeatedly noted a tendency in the literature to focus constitutive elements
of the Scientific Revolution upon England. Ben-David's argument provides the
wider picture which to some extent justifies what otherwise appears as a rather
parochial procedure: We now learn what gave England such a special place in
the rise of early modern science. And this conclusion may be generalized fur-

ther. Ben-David's sketch of the social conditions which made possible the Scientific Revolution as an ongoing, rather than a stillborn, event, however much in need of improvement and elaboration, may fruitfully be seen as a kind of synthesis in which almost every putative cause discussed in the present chapter finds a fitting place of its own. In seeking, as we are about to do in the conclusion to the present chapter, to reap the harvest of our external explanations, we shall therefore employ Ben-David's thesis as our framework of discussion.

5.4. *The Harvest of the 'External' Route*

The two basic themes, of which in the present chapter we have pursued a number of variations, are the rise of early modern science from (1) mostly unintended consequences of the Reformation and (2) manifestations of Europe's newly won dynamism. These variations appeared time and again to yield much of great interest, yet we met much less that may serve straightforwardly as an explanation, however partial, of the Scientific Revolution. Now the principal merit of Ben-David's thesis is that, in digesting large portions of the literature on these two themes, he made them serve a fruitful end by the simple expedient of shifting quite consciously and quite explicitly what stands to be explained from science itself to the *legitimation* of science.

Three particular resources enabled him to make this move. The first is its foreshadowing in Merton's seminal *Science, Technology and Society*. The issue of the effective legitimation of science, which lies hidden in Merton's thesis without coming much to the surface, has since been brought out by many, in the most outspoken manner by Ben-David.[189] His second resource is a clear and realist idea of what 'social' explanations in the history of science can and cannot accomplish. Obvious objections from thoroughgoing 'internalists', which are obvious hits when such reductionist external explanations as the Hessen thesis are the target, would become misses if hurled at Ben-David's more modest undertaking. However, the epistemic modesty of his explanatory strategy was not bought at the usual cost of consequent fuzziness. This was the result of the application of his third resource: the introduction of three intermediate concepts to serve as bridges between the domain of science proper and its social environment. Critics have taken Ben-David's explanation for a mere rehash of familiar Mertonian themes. They failed to see how, in Ben-David's treatment, these themes were organized around the powerful unifying concepts of 'the social role of the scientist', 'an ideology for science', and 'the scientistic movement'.[190] Indeed, not only Merton's themes of Puritanism and the alleged utility of science but also Olschki's account of science and the arts in 15th- and 16th-century Italy and Zilsel's notion of the meeting of the scholar and the artisan easily gain a place of their own in Ben-David's explanatory schema, where they are turned into so many pieces of a well-laid jigsaw puzzle. So, too, does Kuhn's move to focus the causal significance of the arts-and-crafts theme upon the rise of the Baconian sciences.

All this is certainly not to say that we can now rest content, in secure possession of a master key to the 'external' explanation of the Scientific Revolution. To begin with, Ben-David's thesis, while both productive and comprehensive, is still too sketchy, and also suffers from too many defects, to serve as such a master key. The thesis surely offers a useful framework for all kinds of further research, of which a deeper inquiry into motives for supporting science in the various geographical domains of the Scientific Revolution (especially those beyond England) seems the most urgent.[191]

Further, not all those purported causes of the Scientific Revolution that we have treated in the present chapter were included in Ben-David's thesis. There is nothing on the influence exerted by the printing press. With the essence of Eisenstein's later thesis residing in the greater opportunity given to science for continuity over time as a result of the shift from script to print, there seems to be no difficulty in including her insight in a widened version of Ben-David's thesis. Also, it would surely go too far to categorize *all* 16th- and 17th-century emphasis on the primacy of experiential facts under the heading of the legitimation of science—one way or another there was an important empirical ingredient to the Scientific Revolution which requires further historical conceptualization. Nor is there a need to take the role attributed by Ben-David to Protestantism of an unintended, occasional source of religious pluralism for its sole contribution to the birth of early modern science. Neither Merton's emphasis on the Puritan ethic nor Hooykaas' pointing at the 'general priesthood of believers' are explained away thereby as possible stimulants to key elements in the Scientific Revolution. In particular, Hooykaas' insistence upon the 'Christian-empirical' element in Kepler's thought serves as a causally powerful answer to the question of what it was that turned this speculative thinker into such a patient checker of the fruits of his own imagination. Something similar is true of Olschki's point about what the technological literature in the vernacular did for Galileo.[192] All such theses seem to contain insights worth preserving when the explanation of the Scientific Revolution is at stake.

Next, there are two important limitations concerning the explanatory scope of Ben-David's thesis. Not only did the author abstain from touching the level of scientific ideas, it is also clear that not so much the *onset* of the Scientific Revolution as, rather, its *persistence* is what his thesis seeks to explain. Not how a new scientific movement arose is the critical question addressed here, but how it could gain ever-increasing momentum.

Finally, the very move by which Ben-David wisely safeguarded his thesis against customary 'internalist' objections precluded his taking another step that is nonetheless indispensable in the end: to strike a balance between external and internal explanations. The previous chapter gave us some idea where fruitful internal causes of the Scientific Revolution have been and may be

6

sought, and the present chapter did the same for their external counterparts. But how do these causes hang together? Is it possible for one comprehensive historical picture to be drawn in which they all find their proper place?

Whether such a picture is feasible at all and, if so, what its contours may look like can only be found out by expanding further the boundaries of our inquiry. Another, no less weighty reason for doing so resides in the suspicion that the two principal explanatory venues pursued in the present chapter— Western science considered in the context of Europe's religion and of Europe's emerging dynamism—have hardly been exhausted yet by historians of the Scientific Revolution. The two themes, which have appeared so often to yield somewhat indefinite and inconclusive results, may well be still much richer in explanatory resources than has come to the fore so far. And we must consider the possibility that both stand to gain further relief when compared to, and contrasted with, the environment and the fate of science in other civilizations.

That is to say, we now take up Zilsel's cue on the possible merit of a cross-cultural, comparative approach to our topic. Having so far examined the literature in which the birth of early modern science in Europe is taken for granted, we begin to follow those rare, daring authors who have sought an answer to that very same question through the detour of asking why the Scientific Revolution did not spontaneously take place elsewhere, in any of the other great civilizations of the past. Our search, after all, has been not only to bring together, sum up, investigate critically, and compare historians' theses on the nature and causes of the Scientific Revolution but also to let the drift of the entire inquiry be guided by a regulative ideal which I have outlined in the Introduction. This ultimate and, as yet, far-off ideal is the integration of the unique event of the Scientific Revolution into history at large. It is to understand what early modern science was about and how it could come into the world at all; why it arose in the West; what special features of thought and states of society in the West were instrumental in leading to this quite unexpected yet hardly fortuitous outcome. In our campaign for viable ways toward such integration and enhanced understanding, the chapter that follows occupies no less a strategic place in the book than its predecessors. As I hope to show by its results—however provisional and subject to improvement as they cannot fail to be—nothing but a comparative survey of science in societies other than the one where early modern science arose can serve to widen our horizon and to sharpen our eye for elements crucial to this ultimate ideal pursued throughout our argument.

Hence, the purpose of the chapter that follows is *not* to report on the present state of knowledge on Oriental science (for which I would lack competence in any case). Rather, I survey a range of literature (with Joseph Needham's writings standing out both in bulk and in perseverance) in which Oriental and Western science have been compared with a view to understanding why the latter, but not the former, ushered in early modern science, and I seek to cull from that survey its apparently most rewarding fruits. Even so, the scholarly risks involved are no doubt considerable, for we are going to tread much less securely laid ground than before, and it will always be possible to argue that the ground is too unstable to tread on at all.

Still, if in response to the inevitable and often quite justified strictures of the specialist we were to forgo asking big questions, it would severely hamper the advance of scholarship, which (as we have seen throughout the present book) thrives on big questions no less than on small answers. All of which is to say that I shall have achieved my nearest aim with the next chapter if the effort in historiography of cross-culturally comparative history of science that follows persuades the reader that the job of seeking to enhance our understanding of the origins of early modern science along this route is worth doing and maybe challenges him or her to do it better than either the few historians who took the route before or the one historiographer who is now to report on their findings.

❧ SIX ❧

THE NONEMERGENCE OF
EARLY MODERN SCIENCE
OUTSIDE WESTERN EUROPE

6.1. FIRST APPROACHES TO A PROPER DEFINITION OF THE PROBLEM

WHEN TAKING UP THE QUESTION of why no Scientific Revolution occurred in any of the other great civilizations of the past, two preliminary sets of issues must be faced. One is to determine what civilizations we are going to talk about, and how we may justify the choice we make. The other is, Are there any meaningful questions to be asked when dealing with cross-cultural, comparative research in the history of the Scientific Revolution, and if so, what are they?

Tough problems are involved in these two principal issues, and they are interconnected, too. For example: Has a thing to be called 'non-Western science' existed at all? Does it make sense to ask questions about events that conceivably could have happened but never did? How is one to show that something could have happened in the absence of its actually having done so? How could one possibly claim that something like the Scientific Revolution lay just round the corner but somehow never materialized? How much 'like the Scientific Revolution' must 'something like the Scientific Revolution' be in order to qualify as sufficiently like the Scientific Revolution to make any comparison viable in the first place? And so on and so forth. Attempts at the careful disentanglement of the really quite fundamental issues at stake here might fill a volume all by themselves. I propose to cut our way through the web in a practicable manner by making two key distinctions which, I believe, in effect underlie a great deal of the existent literature on non-Western science.

To begin with, it ought to be established that we deal only with the state of science in civilizations *before* products of the Scientific Revolution began to be disseminated beyond their European birthplace. This is not to say that before 1600 no transmission of scientific ideas and procedures occurred at all—far from it, and the topic is to receive due treatment farther along in the present chapter (section 6.4). But the point here is that we are concerned with the comparison of relatively *indigenous* sciences rather than with the history of the reception of Western science in other cultures (even though we shall find that certain important insights into the peculiarities of science and civilization in China may be gained from studying the reception accorded to some early results of the Scientific Revolution as passed on by the Jesuit missionaries in Peking).

With this restriction in mind we should, for the sake of a clear conceptualization, distinguish, with respect to science, between four types of civilization. These are, in ascending order, (1) those without science, (2) those with some science, (3) those with advanced science, and (4) the one and only civilization that produced early modern science.

To claim without further ado that each of these four abstract categories is indeed filled with concrete cases manifest in the real world would obviously prejudge the very questions that ought to be asked before one can safely embark on comparative research in this area. The first and last of our four categories seem uncontroversial, however. Civilization no. 4, on which all attention in the present book has so far been focused, evidently exists. Some examples of category no. 1 can also be adduced without fear of raising controversy: Rome and the archipelago of the East Indies, for instance, produced little or no science worth mentioning. Civilizations like these may thus be eliminated from the present inquiry. If there was no science to begin with, we would gain little by asking why no Scientific Revolution occurred.

The West—the sole creator of science?   A more controversial issue is reached by pitting the aggregate nos. 2 and 3 against no. 4, in other words, by asking whether, besides the Western tradition, any other civilization has ever managed to produce authentic science. To be sure, throughout the present chapter our concern is *not* to examine whether and, if yes, to what extent science has existed at all in any other than the Western tradition. Rather, we are dealing with authors addressing the narrower question of how it is that, in that quite limited number of civilizations which seem, in retrospect, to have come reasonably close to the point where something like the Scientific Revolution might have begun, nevertheless such an event happened to elude them. However, it is clear that even the possibility of taking up the latter question depends on an affirmative answer to the former one. Joseph Needham, in particular, has firmly insisted on the distinction between these two issues:

> As the contributions of the Asian civilizations are progressively uncovered by research [so Needham wrote in 1963], an opposing tendency seeks to preserve European uniqueness by exalting unduly the role of the Greeks and claiming that not only modern science, but science as such, was characteristic of Europe, and of Europe only, from the very beginning. For these thinkers the application of Euclidean deductive geometry to the explanation of planetary motion in the Ptolemaic system constituted already the marrow of science, which the Renaissance did no more than propagate. The counterpart of this is a determined effort to

show that all scientific developments in non-European civilizations were really nothing but technology.

[In sum:] Let us take pride enough in the undeniable historical fact that *modern* science was born in Europe and only in Europe, but let us not claim thereby a perpetual patent thereon.[1]

Needham goes on to hammer home his point by showing that the definition of science necessary to uphold such a one-sided view of things is intolerably narrow. Indeed, a restriction of science to the Western tradition can be maintained *only* by definition, not by a somewhat more open-minded willingness to consider the systematic investigation of natural phenomena and the establishment of some measure of correspondence between those phenomena and their interpretation as constituting science in its own right (bearing in mind that much of the Western tradition in science before 1600 would go by the wayside as well if the narrower definition were applied).[2] To round off this point briefly: If fifteen solid tomes to date of *Science and Civilisation in China* cannot convince a reader of the presence, in the past, of at least one living tradition of non-Western science, then, I suppose, nothing can.

On the threshold of a Scientific Revolution. The next question to be asked is whether the distinction between the categories numbered 2 and 3 actually holds water. The empirical basis of the criterion that underlies the distinction is the suspicion that, whereas it may perhaps make sense to ask why China did not have a Scientific Revolution, it is certainly pointless to ask why, for instance, Japan or Korea did not. To be sure, both these and a number of other civilizations displayed in the past some authentic lines of scientific research and achieved a number of results that are far from negligible in their own right (e.g., a Japanese kind of algebra called *wasan* or the Maya calendar with its intricate computations).[3] Yet it does not as a rule occur to experts in any of these sciences to ask the question that is with some frequency taken up among students of Chinese, Indian, and Arabic science—the 'why no indigenous counterpart to the Scientific Revolution' question. The reason is of course that these limited lines of inquiry and these scattered results represent far too little of a comprehensive, coherent whole to make even feasible the idea that early modern science or something like it may have been lying around the corner. Hence, Japan, Korea, and pre-Columbian America do not make their appearance here. The same goes for Byzantium, which is an interesting borderline case between our categories no. 1 and no. 2, in that Byzantium for many centuries sat upon a splendid legacy of scientific achievement, yet beyond sheer preservation did next to nothing creative with it.[4]

Thus we are left with no more than three non-Western civilizations—the Muslim world, India, and China—of which it may perhaps be maintained that their science reached a stage not too far removed from a breakthrough more or less on a par with the one that actually took place in western Europe. But does this category no. 3 exist indeed, or is it no more than an artifact of the conceptual distinction which we made, but which, in historical reality, ought to be subsumed entirely under category no. 2? In other words, is there any sense in which one can say that science in these three civilizations had progressed sufficiently for making something like the Scientific Revolution conceivable as the very next stage of advance but failed, for one reason or another, to cross the threshold? Or, in other words again, does it make sense at all to ask why the Scientific Revolution occurred in western Europe but not elsewhere? Does what Lynn White has dubbed 'Needham's Grand Question' carry any real meaning?[5]

The meaning of the 'Grand Question': A provisional affirmation. We shall find out soon enough that quite a few scholars, both inside and outside the research area of the history of non-Western science, have indeed denied flatly that the Grand Question, negatively phrased as a 'why not' question, makes sense. A first, purely pragmatic response to this denial would be that, if no one had taken the trouble of asking the question, whether ultimately 'meaningful' or not, our actual knowledge of the content of non-Western science would have been infinitesimally smaller than it has become over the past forty years. After all, Joseph Needham's *Science and Civilisation in China* owes both its origin and the guiding thread keeping its many tomes together to the confident expectation that sensible answers can be given to this very question.

For the rest, however, I shall not continue to discuss in a preliminary fashion the issue of whether the Grand Question does or does not make sense. Rather, I beg the reader to join me in assuming provisionally that it does, or at any rate that it can be reformulated in such a way as to make sense. Granted so much for the moment, we shall handle the intricacies of the case as we are, in due course, to find it being taken up by some of the successive main persons in our historiographical overview—Needham, of course, but also the students of Islamic science Sabra and Sayili, and the sinologists Sivin and Graham. I shall present and examine their views on the matter in the context of the issues that have actually come up in scholarly debates and add to the discussion some fresh contributions to be gained from the historiographical perspective taken throughout the present book. Suffice it for now to say that the very debates to which the issues have given rise have contributed greatly, in my opinion, to making the subject matter of the present chapter so exciting: Remarkably, some of the most profound discussions of the causes of the Scientific Revolution have taken place in these fairly remote recesses of historical scholarship.

Excluding India for the time being. Thus, it seems that we are to deal with three civilizations which might have qualified as viable candidates for a Scientific Revolution. However, there is good reason for omitting the case of India from treatment altogether. Just as Byzantium appeared to be a borderline case between categories no. 1 and 2, so it looks as if Indian science straddles our nos. 2 and 3. To be sure, there are historians of Indian science (notably A. Rahman and D. Chattopadhyaya) who have indeed suggested that the question of why the Scientific Revolution eluded the East is worth asking for India, too.[6] But a systematic effort to answer the question—as Needham did for China— is lacking, and one may doubt whether it is ever to be undertaken.

It is certainly true that, ever since the English colonized India, much highly technical work has been done by both Indian and Western scholars in reconstructing the two areas where India is known to have excelled, namely, mathematics and astronomy. A perusal of Chattopadhyaya's collection of *Studies in the History of Science in India* (a sample, published in 1982, of articles in the field, including some very early ones) makes it clear that for the rest (in the physical sciences as distinguished from botany, medicine, and technology), except for a predilection for speculative atomic conceptions, little has as yet been brought to light that would make India a viable candidate for a Scientific Revolution. It is hard for the outsider to decide why this is so.[7] Is it because the history of Indian science has not yet found its Needham—that is to say, a scholar ready to tap treasures of unexamined or misunderstood treatises and to reconstruct from those a full-blown corpus of indigenous science? Is it because what source material has been preserved will never permit such a reconstruction? Is it because of the well-known problem that no reliable chronology of Indian history can be put together? Is it because Indian civilization just was far more oriented toward metaphysical speculation than toward science as a pursuit in its own right?[8]

Clearly, I am not qualified to choose among such alternatives. For present purposes it suffices to note that, whether or not future research is to show that Indian science can in a general way be put to fruitful comparison with the state of its Western counterpart on the eve of the Scientific Revolution, that time has not yet come. There is as yet no historiographical tradition on the topic worth mentioning, and therefore India is to be excluded from our inquiry.

Which leaves us with science in China and in the Muslim world as the sole subjects of historiographical concern to be pursued in the present chapter.

One key difference between comparison along 'Chinese' and along 'Arab' lines. Although before the 17th century some transmission occurred between Chinese and Western science, at bottom Chinese civilization presents us with something unique: A highly developed alternative conception to the interpretation of nature as it has become customary in the West. This is where the great attraction of a comparison between Chinese and Western science comes from; in Needham's eloquent words:

> It is so exciting because Chinese culture is really the only other great body of
> thought of equal complexity and depth to our own—at least equal, perhaps more,
> but certainly of equal complexity; because, after all, the Indian civilization, inter-

esting though it is, is much more a part of ourselves. . . . Chinese civilization has
the overpowering beauty of the wholly other, and only the wholly other can inspire the deepest love and the profoundest desire to learn.[9]

In the case of China, therefore, the comparison takes the form of finding out
similarities and differences between two almost wholly independent traditions.
With Islamic science, comparative research takes a profoundly different point
of departure. Since both Arabic science and Western science were offshoots of
the Greek legacy, the issue of why Muslim civilization did not pull off a Scientific Revolution appears as one variant of the problem of Greek stagnation we
discussed in section 4.2. If indeed the Greeks had come close to the threshold
of early modern science, so necessarily did the Arabs, and the question of why
they did not cross it may be posed with equal cogency. Again, this is quite
different from asking how it is that, of two fundamentally distinct yet equally
coherent scientific traditions, one failed to arrive at the breakthrough accomplished by the other.

It seems to me that this difference in point of departure for comparative
research has not always been made sufficiently explicit by those who have devoted themselves to such research in the grand manner. We shall find throughout our discussions how deeply it informs the kind of answers arabists and
sinologists, respectively, have come up with.

A disclaimer to complete the preliminaries.   We have now sufficiently delimited the subject matter of the present chapter. A word of caution should still
be added about the source material it is built upon. Together with the majority
of my readers, I lack the capacity to trace to their original sources the historiographical issues to be reported here. Since I do not know either the Arab or the
Chinese language, no way is open to me for learning anything about Islamic
or Chinese science other than what their respective comparative historians have
written on the subject in a Western language, and where they disagree, I ultimately lack criteria for adopting one interpretation rather than another. For
example, Needham once claimed that in all inorganic sciences he has always,
together with his collaborators (among whom have been many native Chinese),
been able to find the perfect Western equivalent for a concept found in a Chinese text.[10] Now I may well suspect that this claim betrays a certain overconfi-
dence in the extent to which a member of one civilization, even when guided
by genuine and profound respect for another one, can fully penetrate into the
deepest recesses of thought and feeling (in particular, those of a far-away past)
of that other civilization. (The literary critic I.A. Richards even called the transfer into English of Chinese philosophical concepts "very probably . . . the most
complex type of event yet produced in the evolution of the cosmos.")[11] I may
further suspect that Needham, because he had to do battle with so much mindless condescension toward China, has been overstating his case somewhat. In
support of such sceptical intimations I may point to such widely varying read-

ings as are available of a Taoist classic I have come to love a great deal as a literary gem: the *Chuang Tzu* (traditionally dated 4th century B.C.). Here even such recent and, I take it, highly knowledgeable translators as the sinologists Watson and Graham put quite different meanings to passing phrases as well as to key concepts, both greatly differing in their turn from the meaning given by Needham to certain passages in the book which, in his translation at least, bear on science. (The outstanding example here is Needham's insistence on translating *wu wei,* not by the customary term 'inaction,' but rather by 're-fraining from action contrary to Nature'; in section 6.5.2 we shall see why.) One gathers from the translators' respective comments and prefaces that not only fundamental issues in sinology are at issue here but also divergent conceptions of what mysticism, and science for that matter, are about.[12] Obviously, divergences like these may affect modes of translating a far wider corpus of texts than just the *Chuang Tzu.* Yet, and this is my key point here, in the end I am barred from going back to the original texts and from building up an independent argument to show what I suspect to be the case—that unambiguous translations of Chinese concepts bearing on science are a thing simply not always or in a straightforward manner to be had.

This tentative conclusion, then, completes our preliminaries; let us now go ahead and take a plunge into the deep waters of Oriental scholarship.

6.2. THE DECAY OF ISLAMIC SCIENCE

A brief, authoritative outline of the history of Islamic science was recently published by the historian of 17th-century optics and of Arabic science A. I. Sabra in the *Dictionary of the Middle Ages.* Sabra's survey will now be used to familiarize us with a few basic facts of Arabic science necessary for digesting subsequently what three selected authors in the present century have had to say on the causes of its decline.

6.2.1. *Some Basic Facts about Islamic Science*

"The term 'Islamic science'," so Sabra begins his outline,

> like the alternative term 'Arabic science', refers to the scientific endeavors of individuals and institutions in the medieval world of Islam. These endeavors began in the middle of the eighth century, in the form of a vigorous translation effort centered in Baghdad, which continued throughout the ninth and into the tenth century. They reached their highest levels of achievement at different times and places over a period extending to the beginning of the fifteenth century. Afterward came a period of stagnation and even impoverishment, though faint traces of this tradition were still to be seen in the Middle East when Napoleon

led his expedition to Egypt in 1798, bringing with it some of the early seeds of a new scientific awakening.[13]

These seemingly bland statements of elementary fact really point to certain key notions about Islamic science, subsequently spelled out by Sabra.

One is that Arabic science, which started as a movement to translate Greek science, had not yet come to an end when these translations, in their turn, were passed on to western Europe. For one thing, far from the whole of Islamic science was transmitted in the process; for another, Islamic science itself continued to be cultivated under its own steam for some two more centuries.

Next, two features provided a considerable measure of unity to the otherwise so divergent phenomenon of science in the Muslim world, cultivated as it was in lands as far away as Spain from Persia, and by Arabs, Persians, and Turks, and by Muslims, Christians, and Jews alike. One of these two features that together put a common stamp on the entire enterprise across some seven centuries was "the predominance of Arabic as the language of scientific expression and communication." The other was

the centrality of Islamic religion to the civilization that gave birth and nourishment to the new scientific tradition—a centrality that makes Islam the point of reference (though not always the premise) for all cultural activities, however heterogeneous they may be in origin or character.[14]

It would be profoundly wrong, Sabra insists, to regard the entire effort of translating the Greek scientific and philosophical heritage as no more than a matter of 'reception'—as passive absorption rather than as an act of 'appropriation'. The effort itself, and the grandiose scale on which it was carried out, were already by themselves major feats of the imagination, of organization, and of scholarly excellence. Also, as large portions of the total aggregate of Greek scientific and philosophical thought were put into Arabic, what came out was in one sense something new. What had been worked out by the Greeks over a time span of more than seven centuries and had never quite turned into one coherent, cumulative body of knowledge, now suddenly appeared as one single whole through the sheer act of virtually simultaneous translation. Add to this that at the same time important ingredients from Indian astronomy and arithmetic became available in Arabic, and the conclusion follows that

the result of these efforts over a period of roughly 150 years was that the Arabic language came to acquire a scientific legacy never before equaled in richness and variety.[15]

This scientific legacy was elaborated in a variety of ways. The appropriation was a highly creative one, and important new additions were made, partly on the level of scientific thought proper, partly on the level of the institutionalization of research and education. To take on the latter first, Sabra argues that the world of Islam gave birth to four institutions not previously found in Greek or

Roman civilization, namely, the hospital, the public library, the *madrasa* or school for higher learning, and the astronomical observatory.[16] True, the hospital naturally supported science only obliquely; the libraries were obviously filled with manuscripts, not with printed books; the *madrasa* was meant solely for religious learning; and the observatories were usually quite short-lived. Nevertheless they all fostered science to some extent, even if only in facilitating its speedy transmission all across the Muslim world.

Another uniquely Islamic contribution to the cultivation of science derived from scientific problems posed by faith. One of these was to determine the *qibla:* the problem, that is, of determining the precise direction of Mecca, which one must face when praying. Another was the need, felt in the thirteenth century for the first time, for the accuracy of timekeeping to be guarded by a special official, the *muwaqqit.* Although the *muwaqqit* was connected to the mosque and had no strictly scientific duties, his office nonetheless left him plenty of opportunities to pursue them if he felt so inclined.

Now what was the overall achievement of Islamic science beyond what had already been discovered by the Greeks?

Overall achievement and means of support. In mathematics the combination, achieved in Islam, of Indian computing techniques with the Greek heritage was most productive, leading to improvements in geometry as well as to "setting the science of algebra on a new course."[17] Astronomy, which was cultivated more than any other science, was devoted chiefly to observations meant to enhance the accuracy of parameters within Ptolemaean planetary theory. In optics the Greek tradition was carried in new directions without thereby being placed on a fully new footing. To the extent that the Archimedean tradition of a mathematical, idealized physics was adopted, this was done by engineers and led to

> their extension of the Archimedean investigation of floating bodies, and in their study of weights, of various types of balances, of the determination of specific gravities, and of the proportion of metals in alloys.[18]

Besides all this, alchemical determinations of properties of materials should be mentioned, as well as speculations in natural philosophy, with ideas along the lines of the later, Western-medieval impetus theory as one particular result.

There is no doubt at all that, as a social activity, science received virtually all its funding, encouragement, and general stimulus from the court. The original translation effort had centered around the Baghdad Caliphs. As the Abbasid Caliphate declined and eventually broke up, and the number of independent realms within the Muslim community began to increase, so, *ipso facto,* did the number of opportunities for scholars eager to study or to advance science:

Centers of learning multiplied across the Islamic world following the proliferation of dynastic rules that vied with one another for cultural and intellectual
eminence as well as for political power. Court patronage, now multiple and dispersed, continued to be the predominant form of support under which scientific
activities flourished.[19]

On the one hand, this form of social support was sufficiently stable for sustaining a scientific tradition that was not to die out until the late fifteenth century. On the other, it was obviously dependent on the scientific-mindedness of
individual rulers. Witness the brief life spans of the observatories; the record
in longevity was easily won by Maragheh, which survived for fifty-seven years.
Very little is as yet known of how individual scholars managed to master, and
later to pass on, their science. One thing, however, seems clear: Since paper
was widely used instead of papyrus or parchment, manuscript books were
amply available and reasonably cheap. Here are two examples:

> Early in the ninth century, to buy a copy of the *Almagest* in Baghdad, the young
> Sanad ibn Ali, the son of a Jewish astrologer, who later became a mathematician
> and astronomer of note, sold his father's donkey to raise enough money. In
> eleventh-century Bukhara the young Ibn Sina was able to obtain a commentary
> by al-Farabi on Aristotle's *Metaphysics* for only three dirhams (admittedly an
> exceptionally low price).[20]

The character of Islamic science and the problem of its decline. The actual
process of dissemination of scientific ideas is not the only topic in Islamic
science about which relatively little is known. In fact, Sabra cautions, no less
than "the great majority" of those manuscripts that together constitute the surviving corpus of Islamic science, spreading today from libraries in Tehran and
Cairo to Los Angeles, have never been examined by anybody.[21] This is one
important reason why so much of the efforts of the small cluster of present-
day historians of Arabic science still goes into cataloguing, editing, and commenting on texts. As E. Savage-Smith put it in a recent bibliographical overview, "comprehensive and interpretative histories of Islamic science have yet
to be published."[22] She goes on to define Sabra's particular contribution to the
field as concerned with "reassessing the traditional view of Western scholars
toward Islamic science, namely, that it was merely a holding ground for Hellenistic and Byzantine ideas until they were again encountered firsthand by
Western readers. Sabra is urging a redefinition of terms and a new approach to
early Islamic science that would view it as a phenomenon of Islamic civilization itself."[23]

How, then, is the character of Islamic science to be defined in this new
approach? It is true, Sabra asserts, that form, style, and terms employed by an
Arabic mathematician are often indistinguishable from the earlier Greek
whose tradition he is pursuing. It is also true that "it would be a vain undertak-

ing to look for an Archimedes or a Newton in the Middle Ages."²⁴ Yet such observations, however true, he maintains, are really beside the point. The point is to see Islamic science as an interesting and innovative enterprise in its own right, expressive of a rich and by and large autonomous civilization. And it is only in this particular framework that the question can be asked of how and why Islamic science came to an end:

> It is precisely the high quality and sophisticated content of Islamic science that give poignancy to the problem of decline. The question is not why the efforts of Islamic scientists did not produce 'the scientific revolution' (probably a meaningless question), but why their work declined and eventually ceased to develop after the impressive flowering of the earlier centuries. Why, for instance, did algebra fail to make significant progress after the twelfth century? Why was the work of Ibn al-Haytham and Kamal al-Din in experimental optics not continued along lines already drawn by these two mathematicians? Why did the observatory, once conceived and established as a specialized scientific institution, fail to gain a permanent footing? And why did the long-standing interest in astronomical observations not develop into more sophisticated programs? These and similar questions are forced upon us by the fact that what we have in the extant works of Arabic scientists is not protoscience but science in the real sense of the word.²⁵

Thus, we may gather, asking why no Scientific Revolution occurred is meaningless in Sabra's view because in so doing attention would once more be shifted away from what Arabic science achieved in its own right—once again it would be pressed into service for solving a problem that is inherently foreign to it.²⁶ It is only in seeking answers to concrete, historical questions, Sabra insists, that the true problem, which is that of the decline of Arabic science, can be brought nearer a solution. So far, answers have usually been given in terms of one or another essential character ascribed to Islamic civilization, such as an alleged "propensity of the 'Semitic mind' to gravitate toward the sensible and away from the abstract."²⁷ But "such essentialist explanations are worthless substitutes for genuine historical inquiries."²⁸ According to Sabra²⁹ only two historians have so far approached the question in such a truly historical spirit, namely, the American Arabist J.J. Saunders and the Turkish historian of *The Observatory in Islam* Aydin Sayili. The pertinent writings of these two authors, then, are to form our proper subject of inquiry. I shall add to accounts of their respective essays one other attempt, by the German/American Islamist G.E. von Grunebaum, done in the main along such essentialist lines as condemned by Sabra, yet on a level more than high enough to merit, I believe, our attention.

Of these three historians, it was Saunders who put the problem itself of the decline of Islamic civilization, and of its science in particular, in historical perspective.³⁰

A brief history of the problem of decline.   After western Europe had originally lagged far behind the Arabs, had caught up in the Renaissance, and finally captured a definitive edge through the Scientific Revolution, it was not until the Enlightenment that Europeans began to realize how far their former rival had been outstripped in every cultural area, including science. Montesquieu and Voltaire, in particular, began to search for reasons how this had come to pass, which they found in the tyrannic and corrupt nature of Turkish rule. In the 19th century, explanations were sought along three distinct lines. One was Gobineau's interpretation of world history in terms of race. Another was the area of religion and theology. Here Ernest Renan accused the orthodox clergy in Islam of suppressing and stifling the spirit of free inquiry. Finally, Marx and his followers attributed the decline of Muslim civilization to the economy, in that the growth of feudalism had caused a decline of intellectual life and had encouraged orthodoxy. It is not until the present century, Saunders concluded before venturing on his own explanation, that the full complexity of the problem has become apparent.

One fundamental point of agreement.   Unlike in the debate over Greek science, the three scholars who, in the present century, appear to have contributed most to the discussion of the causes of the decline of Islamic science, while diverging widely over both the exact nature of the problem and key ingredients of its solution, are nevertheless in agreement over the central reason for the decline. Whereas in the case of ancient Greek science explanations of its decline range from its roots in a slaveholding society to its proneness to rational speculation and consequent impairment of the formation of a sound empirical base, concerning Islamic science von Grunebaum, Sayili, and Saunders are all agreed that the root cause of its decline is to be found in the Faith, and in the ability of its orthodox upholders to stifle once-flowering science. This is where agreement ends, though. When it comes to asking what it was that put the orthodox in such a position, we find rather different answers. Very briefly and grossly put, von Grunebaum attributes this primarily to certain key characteristics of the Islamic faith as established during its early, formative period; Saunders to the disastrous effects wrought by the barbarian invasions that plagued the Arab world from the 11th century through the 15th; and Sayili to the failure of Muslim civilization to achieve a reconciliation between science and philosophy on the one hand and religion on the other. Here is how they made their respective cases.

6.2.2. *Von Grunebaum and the Preservation of the Muslim Community under the Law*

In von Grunebaum's book of 1955, *Islam: Essays in the Nature and Growth of a Cultural Tradition,* two chapters bear on the decline of Islamic science. One, entitled "Muslim World and Muslim Science," does so in a straightforward manner. The other, "The Profile of Muslim Civilization," comes in for consideration, too, because von Grunebaum is concerned throughout to show the marginality of science and philosophy both to the dominant values of the faith and to the life of the faithful.

Von Grunebaum's general point of departure is that, in different civilizations, different types of concerns and, concomitantly, of people are valued and honored. He sets forth this point in impressive language:

> Every civilization and every age favor a limited number of human types for whom they will provide the fullest means of self-realization, while denying it to an even larger number for whose peculiar gifts the prevailing pattern affords no socially meaningful use.[31]

The dominant values of Islamic civilization were such that science and philosophy could never appear, to more than a handful of intellectuals, as things of importance to life as it ought to be lived in accordance with the prescriptions of the Prophet:

> The Muslim's apprehension of the purpose of his earthly life as the outreach for felicity, sa'âda, through service, 'ibâda, has shaped the fundamental aspirations of his civilization both on the political and on the epistemological levels.[32]

It was this particular conception of life, von Grunebaum asserts, that led to a radical dichotomy between two types of knowledge. On the one hand, one could gain knowledge of the revelation and of the prophetic tradition, which tells the believer what the purpose and direction of his life are. On the other, there was knowledge of the world in which his life is to receive its proper orientation. This dichotomy corresponds in fact, albeit not in original intention, to the distinction adduced all the time by Muslim scholars themselves between the 'Arabic' sciences and the ancient, or *awâil,* sciences. And the dichotomy was a fundamental one precisely because the Islamic faith leaves no room for "the prevailing attitude of the West, [which is that] research per se, as an effort to widen man's insight into the mysterious ways of the Creator, is . . . experienced as a means of glorifying God."[33] Rather, in Islam knowledge of the world is of value only to the extent that it can serve the community of the faithful. This community is a central category in Islamic thought, or rather, it is the true *locus* of the Muslim's loyalty, over and above the state or the ruler or whatever authority may stand above him other than the Law itself, which has to be observed in order to fulfill the purpose of our life on earth.

Now, beyond the control of the calendar and the computation of the *qibla,* which belong to mathematics and astronomy, and beyond keeping the members of the community in good health, which pertains to medicine, the *awâil* sciences simply had nothing significant to offer the community. As a result, anything that goes beyond these manifest (and religiously justifiable) needs can, and in fact ought to, be dispensed with. No matter how important the contribution Muslim scholars were able to make to the natural sciences, and no matter how great the interest with which, at certain periods, the leading classes and the government itself followed and supported their researches, those sciences (and their technological application) had no root in the fundamental needs and aspirations of their civilization.[34]

This crucial assertion may raise in our minds two equally crucial questions. The first is: If von Grunebaum's analysis of why science was so marginal to the life of the faithful is right, how could any scientific research beyond the *qibla* and the calendar be carried out at all in the context of Islam? And the other is: What could be von Grunebaum's point in raising this issue in specific connection with Muslim civilization? Did science ever really occupy a more than marginal position in any other civilization prior to the Scientific Revolution? Although he does not himself put these questions explicitly, the passage that follows can best be read as an answer to the first one—an answer which, in my view, contains the most valuable insight von Grunebaum has to offer on the question of the causes of the decline of the *awâil* sciences:

> Those accomplishments of Islamic mathematical and medical science which continue to compel our admiration were developed in areas and in periods where the elites were willing to go beyond and possibly against the basic strains of orthodox thought and feeling. For the sciences never did shed the suspicion of bordering on the impious which, to the strict, would be near-identical with the religiously uncalled-for. This is why the pursuit of the natural sciences as that of philosophy tended to become located in relatively small and esoteric circles and why but few of their representatives would escape an occasional uneasiness which not infrequently did result in some kind of an apology for their work. It is not so much the constant struggle which their representatives found themselves involved in against the apprehensive skepticism of the orthodox which in the end smothered the progress of their work; rather it was the fact, which became more and more obvious, that their researches had nothing to give to their community which this community could accept as an essential enrichment of their lives. When in the later Middle Ages scientific endeavor in certain fields very nearly died down, the loss did indeed impoverish Muslim civilization as we view its total unfolding and measure its contribution against that of its companion civilizations, but it did not affect the livability of the correct life and thus did not impoverish or frustrate the objectives of the community's existence as traditionally experienced.[35]

In other words, so we may infer from the above assertions, in Islamic civilization we do indeed find one key divergence from the attitude characteristic of Western Christendom. Not only was, in Islam, science not regarded as a legitimate enterprise by the community as a whole, which, by and large, was no less true of other civilizations prior to the Scientific Revolution, but much more significantly, not even its own practitioners found science a fully legitimate activity, to be pursued without qualms or vague feelings of being engaged in something inappropriate. In quiet, 'liberal' times, this did not matter much. But when bullied by the orthodox, such self-doubts made scientists and philosophers give in easily *because at bottom they shared the hierarchy of values upheld by their adversaries* and the misgivings about the cultivation of the *awâil* sciences that these implied. The exigencies of their faith were such that, in their heart of hearts, they had internalized the orthodox feelings about the ultimate inappropriateness of their own endeavor.

Von Grunebaum does not, however, elaborate this matter beyond the few lines just quoted. Rather, he goes on to reiterate his chief point of the principally suspect, marginal, and easily-to-be-dispensed-with character of science in Islam:

> From the orthodox viewpoint nothing was lost and perhaps a great deal gained when in the later Middle Ages Islamic civilization prepared to renounce the foreign sciences that could not but appear as dangerous distractions. The retrenchment of intellectual scope must have seemed a small price to pay for the preservation of the original religious experience.[36]

He even goes so far as to deduce from the "indulgent curiosity which is the expected response when one confronts derived, creaturely reality" that, in Islam, the scientist's "first impulse will be to trace and marvel at the ways of the Creator rather than to comprehend the particular structure and the immanent value of natural phenomena; pious and somewhat sentimental wonderment replaces the activating astonishment which moved the Greek."[37] From this statement it would seem to follow that there was scarcely any science worth mentioning in Islam at all, simply because for inherent reasons there could not be. And he ends his chapter by comparing the status of philosophy in ancient Greece and in Islamic civilization thus:

> Philosophy as such, *falsafa,* never could attain to a position within Islamic civilization comparable to that held by it in classical antiquity.
>
> [Hellenistic philosophy had] possessed complete freedom of inquiry, not simply in that no organized body wished to restrain it, but in the more profound sense that there was no intellectual barrier to its expanding its systemization to a complete and autonomous interpretation of the universe to account for cosmology and metaphysics as well as for ethics and politics. Its social function was to guide man to felicity, to help him sustain the vicissitudes of fate, and to overcome the fears and uncertainties of life through insight into the structure of the universe and through that virtue which could be cultivated only as a result of a reasoned understanding of the human situation.
>
> [But in Islam] the principal truths had been established and fixed. Speculation could only mean explication. Revelation was autonomous, theology its pri-

mary guardian; philosophy must needs become auxiliary, irrelevant, or heretical.[38]

As a result, philosophy remained a potential competitor to religion. We do not
find an Islamic counterpart to Saint Thomas, who saw philosophy as a welcome addition. It is typical of Islam that, in order to believe, the believer never
felt a need to find support in reason.

> Philosophy in Islamic civilization is a field for the specialist as it was in antiquity,
> but other than in antiquity the educated are no longer expected to be familiar with
> the specialist's results. . . . As *falsafa* could not be fully justified as necessary in
> terms of the fundamental aspiration of Islam, its alien character was never for-
> gotten.[39]

What has been compared to what?   Here von Grunebaum's argument ends.
Since it is really set up as an array of answers to questions he almost never
raises explicitly, I can only hope that the above is a faithful rendition of what
the author wanted to say. I also hope that the reader joins me in finding many
of his points impressive. Whether or not the argument is empirically true—
which is obviously beyond my competence to find out—it exudes an air of
plausibility in light of one's knowledge of the basics of the history of Islamic
civilization. Much of its suggestiveness derives from important things the author has to say about the place of science in society at large (and by now few
readers will have missed how close von Grunebaum's point of view here is to
Ben-David's). But, on the other hand, the author hardly cares to descend from
his lofty level of abstract generalization. Or, to state the point more precisely:
Although the argument makes little sense if not taken as an effort in the comparative study of science in civilization, the act of comparing remains implicit
throughout, and one cannot fail to wonder what it is that is really being compared here. For one thing, one never hears about any concrete achievement
of Islamic science; at one point, as we have seen, the bare existence of such
achievements beyond the narrow domain of science-for-religious-purposes is
even ruled out on *a priori* grounds. For another, what exactly is the *tertium
comparationis,* or that with which Islamic science is being compared, taken to
be? Now we find it placed in Greek antiquity; then science in the Western
Middle Ages seems to bear the brunt of comparison; then it is some implicit,
unspecified, wholly unhistorical entity best labeled 'Western science as such'.
    With all its brilliance, von Grunebaum's argument reveals itself as belong-
ing to a once-flourishing, by now quite old-fashioned, school of German
thought about the *Abendland*—about Western civilization taken as one homogeneous entity and set off as such against the features of other cultures. This
Abendland is not properly differentiated, certainly not in respect to the history
of its science, of which von Grunebaum appears to know next to nothing:
Much of what he has to say on the Muslim tradition in science and philosophy

is equally true of its Christian counterpart before the onset of the Scientific Revolution.

The key question that is lurking at the background of von Grunebaum's entire analysis appears to be whether sacred-book religions with an exclusive claim to the possession of revealed truth can bring forth science at all, and *a fortiori* (so we may add) early modern science. Thus nakedly put, there is of course no way to answer no; witness the science produced by Christian civilization. But if that is so, then why could Islam not have done the same? Obviously, only if, for one reason or another, in Islam the forces behind the exclusive veneration of the sacred book were stronger than in Western Christendom. What, then, does von Grunebaum have to offer us on this score?

There are in von Grunebaum's striking sketch three elements in particular, I believe, which contain possibly fruitful kernels of insight.

The first is his observation that, in Islam, the faith not only assisted in making the community at large indifferent, if not downright hostile, to science as an activity in its own right but rendered even the scientist himself almost defenseless in the face of attacks from the orthodox. If one ponders certain present-day reactions to extreme positions taken in the Islamic world (the Salman Rushdie affair is a case in point), one cannot help observing that, over large stretches of the history of this particular religion, extremists seem to have exerted a spell over their more moderate brethren's minds which is not, to a comparably persistent extent, to be encountered in other civilizations. True, von Grunebaum does not support his sweeping statement with any empirical evidence of half-hearted defenses of science, by Islamic scientists, against orthodox attacks. But if such cases may indeed be uncovered by historical research, it seems to me that we are then in possession of at least one important reason why Muslim civilization never produced a Scientific Revolution.

The second is the image called forth by von Grunebaum of the *awâil* sciences as a foreign intruder, never properly assimilated in the cultural tradition. Closely related to this is, third, his insistence on the failure of philosophy to be incorporated in the intellectual life of the Muslim community. However, von Grunebaum did little more than scratch the surface here. We turn now to Sayili, for whom these two points became the cornerstone of his entire argument.

6.2.3. *Sayili and the Failed Reconciliation between Science and Religion*

Unlike von Grunebaum, the historian of science of Turkish descent and American academic upbringing Aydin Sayili (a pupil of Sarton's at Harvard University in the late thirties; now a high science-policy official at Ankara) approached the issue at hand in a manner which betrays his thorough immersion in both Islamic and Western science and their respective histories.[40] By way of an appendix to his book *The Observatory in Islam and Its Place in the General History of the Observatory,* Sayili devoted a thoughtful essay to "The Causes of the Decline of Scientific Work in Islam." It came out in 1960.

Let me state at the outset that I like very much the unusual modesty with which Sayili presented his successive points. There is none of this lack of doubt with which so many theories and interpretations surveyed in the present book were originally put forward by their authors. Rather, Sayili displays a healthy awareness of how uncertain and, in the last analysis, speculative arguments in domains like the present one are doomed to remain. But alas, there is a less fortunate counterpart to Sayili's salutary restraint: his somewhat exasperating tendency to raise one point after another in fairly quick succession, without much effort being spent on integrating them in a unified argument. The reader is confronted with one eye-opening viewpoint after another of possibly great wisdom and fertility, yet few if any are allowed fully to run their own course, let alone to serve as the unmistakable core of the entire argument. In short, Sayili's points give much food for thought but in the end leave the reader even hungrier than before coming to dinner. Put overgrimly, Sayili's essay reminds one of that cruel KGB torture related by Solzhenitsyn which consisted of serving starved jail inmates with minuscule helpings of *haute cuisine* meals: tiny cutlets, two or three midget baked potatoes, thimblesful of real turtle soup. For the inmates, though, life itself was at stake, whereas in history writing slightly less vital issues come up. Sayili's essay just cries out for a somewhat more systematic successor expert to elaborate further such intriguing points as were left dangling by their originator.

Sayili's essay is divided in effect into two parts: one in which he gropes his way toward a suitable definition of the problem at hand, and one in which he makes a definite comparison between the attitudes toward science, philosophy, and religion in medieval Islam and in medieval Christendom. We take up first his search for a proper problem definition.

Groping our way toward the comparative approach.   Sayili begins by linking the issue of decline and the issue of why no Scientific Revolution took place in the context of Islam, in the following way. He defines the decline provisionally as a general decrease in scientific interest and vitality after they had peaked in the 9th and 10th centuries. Now one may reasonably suspect that, if the cultivation of science had persisted unabated through subsequent centuries, the contribution of Islamic science to the world would have been several orders of magnitude greater than it has been in fact. With evident approval Sayili quotes the late 19th-century translator of al-Bîrûnî, Sachau:

> The fourth century [A.H.; i.e., the tenth century A.D.] is the turning point in the history of the spirit of Islam. . . . But for Al Ash'arî and Al Ghazâli the Arabs might have been a nation of Galileos, Keplers, and Newtons.[41]

And indeed, Sayili goes on, it seems reasonable to assume that, but for the stagnation, science in Islam would have developed "in lines roughly equivalent to those in Europe."[42] However, what underlies discussions of this type is the hidden assumption that

> left to itself, science would progress more or less automatically and that its decline would have to be brought about by definite forces, would have to be imposed by outside factors. It is not by any means clear, however, that the progress of science in Europe, leading to the rise of the seventeenth century science, is less in need of explanation than is its failure to achieve a similar progress in Islam.[43]

After all, science became stagnant in other civilizations as well: in India, possibly also in China, and most certainly "among the Greeks themselves. Generally speaking, therefore, the stagnation of science should perhaps be considered at least as natural as its progress."[44] This point of view, so remarkably similar to Ben-David's ten years later, might well have become the cornerstone of the entire argument, in that, followed through to its logical conclusion, in the end it cuts loose the issue of decline from the question of why no 'Galileo' stage was ever reached in Islam.

However, Sayili leaves his point suspended in midair and passes on to a new, indeed quite basic consideration concerning the scientific heritage left by the Greeks. How ready was it for further advances? It is far from self-evident that Greek science, on the eve of its decay, still possessed an inherent capability for further growth. Rather, there are signs that Greek science had indeed reached the limits set to its natural progress:

> It has been pointed out that the transition from Greek science to the modern science of Galileos and Newtons involves a fundamental change of attitude; that although the Greeks had achieved mastery of the knowledge concerning the constant and the unvariable, they had never been successful in mastering the dynamic and the variable. There was therefore a question of opening up entirely new horizons for science, of embarking somehow upon the exploration of another fundamental aspect of reality.[45]

In retrospect we know, so Sayili continues, that the way to break through these natural limits set to Greek science was the experimental method. How could it be found? And here another original point of view is introduced, to wit, that the more the right method was being sought for, the greater the chance was that one day it would indeed be found:

> Such considerations give the impression that medieval science could transform itself into the seventeenth century science only by slow degrees and after a rather long period of gropings involving much work of a non-rewarding type. Under these circumstances, prospects of fundamental achievements could be much improved through the intensity of work. The greater the number of scientists and

the amount of scientific work, the more probable would the transitions in question become.[46]

This consideration suggests that, in those surely unwitting attempts to hit upon the experimental method, the role of Islamic science was, in effect, confined to "prepar[ing] the groundwork for the ultimate emergence of the new scientific era in Europe."[47] The image thus raised is that of an Islamic civilization both enriching the heritage of Greek science and enhancing its dignity, yet spending its energies in so doing, and thus being compelled to pass on the torch to another civilization ready to finish the work.

Somewhat unexpectedly in view of his preceding considerations of the naturalness of scientific stagnation, Sayili concludes with the following general statement of what it is that stands to be explained:

> While in the case of Europe one feels the need of explaining the emergence of the new era in science, in the case of Islam it is the failure of science to reach an equivalent level which requires explanation.[48]

And this is why, according to Sayili, we should seek explanations by means of *comparative* history. The purpose is twofold: to understand better both the decline of Islamic science and the rise of early modern science in Europe:

> In the study of the causes of stagnation and decline of science in Islam, therefore, comparisons with Europe would facilitate the intelligent choice and definition of problems and could also be helpful in checking the conclusions reached. Conversely, the study of the causes of the stagnation of science in Islam too should help improve our understanding of the developments in Europe which were pertinent in leading to, or ushering in, the seventeenth century science.[49]

Sayili, so we shall see, has remained to this day the only historian of science explicitly to combine two positions that otherwise appear only in separation or are denied or ignored altogether. He believes (and explains why he believes) that the comparative approach can heighten our understanding of the emergence of early modern science. He also holds that such comparisons may illuminate *both* sides of the same coin: not only the European side of the story— why the Scientific Revolution occurred where and when it did—but also its 'Oriental' counterpart as a topic of interest in its own right—in this particular case, why the Scientific Revolution failed to take place in Islam.

Having thus charted the goal of the inquiry, it remains to announce how to carry it out. Once again Sayili displays his undogmatic modesty. He confines himself to 'internal' causes, that is to say, to "cultural and intellectual" ones, which he regards as more directly "tractable" than "speculations on social, political, and economic factors." But he makes it clear that this is so only because he feels more at home with the former, and he grants at once that the latter, too, "may be very relevant to the subject in hand."[50]

Next, the decline we are going to find out about must be specified. The decline of Muslim civilization as such and of branches of knowledge other than philosophy and the sciences are not our subject matter, Sayili insists; only the "decrease of dynamism in science" is to be our topic.[51] This decline cannot, of course, be measured or located in time and space with great exactitude. For a general idea it suffices to say that such 9th- and 10th-century constellations of luminaries like Hunayn ibn Ishâq, Thâbit ibn Qurra, al-Râzî, al-Fârâbî, Ibn Sînâ, and al-Bîrûni were not repeated at later stages, despite the incidental appearance of individuals of comparable merit in later centuries.

We have already seen why, for Sayili, the total amount of science done in the two societies inheriting the Greek legacy is so important: The more scientific work was being carried out, the better the chances for hitting upon the experimental method, which, alone, could break through the barriers that had locked up the Greeks. This is why for Sayili decline sets in quite early: around 1000 the Golden Age of Islamic science is over. Combined with his reasons for pursuing the issue in a comparative sense, the above notion provides the general framework for his considerations in the second part of his essay. The historian's key task in the present domain turns out to consist in finding out in which of the two medieval societies, that of Islam and that of Christendom, science had the better chance of being cultivated with as much intensity as possible by as many people as possible. And since both societies were at bottom theocratic, the crucial challenge put to either was to find "a solution of the relations of faith and knowledge in a manner favorable to science and philosophy. This was, for medieval thinkers, a major intellectual task on which the fate of medieval science depended in great measure."[52] This, then, is the general program on the more detailed inquiry of which Sayili subsequently embarks.

The failed reconciliation. The Greek heritage, so Sayili argued, had to be reconciled with the reigning religion in order to make scientific progress possible. In Christian Europe such a reconciliation was indeed achieved, but not so in Islamic civilization. Clergymen, from their position in either society as the most learned, the most leisured, the most involved in education, in short, as the intellectual leaders of society, were the obvious candidates to digest Greek learning. So they did in Christendom; so they did not in Islam, where the *awâil* sciences were either ignored or actively opposed by the theologians. Whereas in western Europe philosophy and science became the handmaidens of religion, in Islam the distinction between the Arabic and the *awâil* sciences turned into a fundamental dichotomy overruling any other classification of learning and carrying with it the lesser dignity of the latter.

Islamic civilization was the first in world history to consider the acquisition of knowledge a thing necessary for every person; hence the Islamic origin of such institutions as the public library and the school for higher learning, or *madrasa*. However, only the Arabic sciences figured in the *madrasa* curriculum. In a certain sense the *madrasa* came too late. If set up during the Golden Age, it would surely have included in its curriculum the *awâil* sciences as well. "But Greek philosophical thought and scientific knowledge had already become suspect when the *madrasa* appeared in Islam."[53] As a result, "the transmission and the dissemination of the *awâil* sciences had to depend almost solely on private study and private instruction in Islam."[54] Clearly, the position of science and philosophy in the university in Europe (which, as an institution, may well, according to Sayili, have been modeled, at least partly, on the *madrasa*) was decisively different.

The odd paradox is that, at the beginning of Islamic expansion and throughout the Golden Age the *awâil* sciences, in particular Greek philosophy, had been received quite well. But when reaction set in, it was directed far more powerfully against philosophy. However, when philosophy was brought down, science was dragged down with it because of their indissoluble linkage.

What had at first appeared to be so attractive about the *awâil* sciences? Here we have to distinguish between the appeal, in the context of Islamic civilization, of science and of philosophy. Elements of Greek science had been the first to catch the attention of the conquering faithful in the decades after the death of the Prophet. "The major incentive for appropriating Greek learning seems to have come from medicine and astrology," and this is where the first systematic intellectual contact was made.[55] Later, the uses to which the sciences were put were expanded, yet what acceptance science was to gain in the society at large was always confined to its uses in a fairly restricted sense: Science was not appreciated, beyond the narrow circle of its practitioners, as an intellectual activity in its own right. In the first chapter of his book Sayili has much to say about the utility of Islamic science. Its usage went beyond the few functions enumerated by von Grunebaum—the *qibla*, the calendar, and medicine. Sayili mentions in addition the use of arithmetic in dividing legacies and of astronomical observation, primarily for astrological purposes but also for navigation and for traveling across the desert. The principle of judging the value of a certain branch of knowledge by its utility was generally regarded as quite legitimate.[56]

The chief point here is that, in Islam, apparently the accepted utility of science was not powerful enough a reason to sustain science as an acceptable activity down the centuries. In other words (and now we are moving toward comparative territory left unexploited by Sayili yet uncovered by his pertinent remarks), whereas links have been construed between the search for legitimation of early modern science and the utilitarian values fostered in Protestant circles, here we have a case of a religion which at least did not stand in the way of science being appreciated for its directly useful character, yet in the end was not prevented from turning against science for other reasons. But then those 'other reasons' must represent a key difference that serves to explain in part the different fate of science in western Europe (where the supposed utility of science helped legitimize scientific activity) and in Islam (where the actual uses to which science was put could not prevent its downfall). What, then, were these reasons?

To uncover them, we return to Sayili's argument and ask, with him, what theologians liked about the *awâil* sciences once these were introduced in the wake of medicine and astrology. Philosophy appeared of importance to early Islamic clergymen because it could help "in weaving the data of religion with the pattern of an appropriate body of rationalized thought."[57] Having freshly emerged from the desert, the upholders of the new faith still felt insecure when defending its tenets against the cultivated infidel they encountered after the conquests of the first caliphs. Here, Greek philosophy could help them out. However, as major differences between the Islamic faith and Greek thought began to dawn more clearly upon them, and as they began to feel more secure, "many theologians gradually began to frown upon the newly acquired knowledge from various standpoints, and these objections generally persisted through the centuries."[58]

Because philosophy and science were linked so closely in the class of the *awâil* sciences, the decline of philosophy under persistent theological attack could not leave science unperturbed. The harmful effect was exacerbated by "the scientists of Islam [being] less successful in formulating important new theories and fundamentally new conceptions in the field of science than they were in enriching the legacy of Greece by factual additions."[59] (We must note for further reference that a key element of the original *problem* is being invoked here as the *solution* to part of that very same problem!) Only in a case like astrology, which was popular both with the ruler and with the population at large, could such an *awâil* science "more than counterbalance theological censure."[60]

Thus a major thesis appears to be that, in Islam, the sciences suffered greatly from the divorce between philosophy and religion. However, Sayili continues, it has been claimed that western European nominalism, which aimed for exactly such a separation, in so doing helped to free science from religious fetters rather than damage it. Sayili accepts the anticipated objection but adds that there was nevertheless a crucial difference which explains the different outcome. In Europe, both science and philosophy had established a tradition of their own. But in Islam the radical dichotomy between the Arabic and the *awâil* sciences worked the other way:

> The Islamic dichotomy of knowledge too might have had a favorable effect upon the progress of science and the productivity of scientific work had it been accompanied by a well-established tradition of work in science and philosophy.
>
> In Islam the divorce of secular knowledge from religion was made too early from the standpoint of the progress of science.[61]

In Europe science and philosophy were first accepted and appreciated, and once the tradition had been established, their subsequent separation was accepted in its turn. In Islam, many scholars exerted their best efforts for a reconciliation between religion on the one hand and *falsafa* (philosophy) and the secular sciences on the other. Now why did these efforts ultimately come to naught? Sayili's answer brings him close to von Grunebaum:

> These attempts were detailed, conscientious, and patient, but the scholars of Islam were unable to convince themselves of the validity of the fundamental issues involved.[62]

Once again, neither in his essay nor elsewhere in his book is any evidence invoked to illustrate this important point. Rather, Sayili goes on to show that, remarkably, the arguments adduced by *Islamic* scientists in favor of a reconciliation of science, philosophy, and religion were to fall on much more fertile ground in *western Europe*. Why was this so? From the perspective of Islam, it looks as if the original appropriation of the Greek legacy had come too soon; had been overall too sudden and somewhat too rash. As compared to this, medieval Christendom had the great advantage that Christianity itself had grown up in a Hellenized environment.

Again, after raising this important point (elaborated much farther by Saunders, as we shall see), Sayili characteristically drops it in favor of a subsequent consideration. It is that, by necessity, in Islam attempts at reconciliation were undertaken by laymen, or at least by scholars in their lay capacities. Islam does not recognize any mediation between the believer and God, and, as a result, another route had to be taken than in Christianity, with its theological hierarchy and its councils, to get one's viewpoint elevated to the level of accepted doctrine. No other way than the tough one of a sustained outreach for consensus among the believers stood open for scholars who might wish to gain religious sanction for their scientific interests.

The fate of the awâil *sciences.*   What were the consequences of the failed reconciliation? In the first place, Sayili asserts,

> the Moslems hesitated to assume that natural processes were conducted according to certain fixed principles. This conception made their deductions as to the operations of the forces of nature take the form of a resigned doubt. For the Moslems there was a continuous interference of God, and of supernatural forces, in nature, or, at any rate, they did not feel they could deny that such was the case. One could not feel certain, consequently, that natural occurrences were the result of God's will to order and of the prevalence of immutable natural laws. Hence, a kind of intellectual stupor, at least in some fields of knowledge.[63]

Thus, in Islam, too much room remained for doubt, for superstition, and for a belief in the possibility of miracles, all of which worked altogether counter to advancing scientific laws. Suddenly changing the *tertium comparationis,* Sayili observes that "this was undoubtedly one of the widest gaps which separated Islam from the seventeenth century European science"[64] and goes on to find out at what point the gap had begun to reveal itself. In his view, a conception of nature as largely subject to miraculous, divine intervention had equally dominated Europe until roughly the 13th century. From then on, however, the belief that "God acted in accordance with certain natural laws," joined to a dwindling belief in miracles, became the rule.[65] Here above all the salutary effect of philosophy, in particular of Aristotelian philosophy, is to be recognized. And it is clear that, but for the reconciliation with religion meanwhile achieved in Europe but not in Muslim civilization, philosophy would not have been in a position to do this for science.

A similar, salutary effect of philosophy upon the sciences was that, in Europe, it put them in a systematic framework, whereas, in Islam, the body of scientific knowledge rather consisted "of isolated and independent sciences or items of information."[66]

Against these claims for the beneficial effects of philosophy upon science might be adduced the harm that came from an excessive submission to authority consequent upon the rapid appropriation, by two civilizations both originally deeply inferior to their Greek predecessor, of the latter's legacy. However, Sayili says, this theme of the unquestioning submission to the authority of the ancients has, for Europe, been much overplayed, and in the history of Islamic science, too, one finds relatively little authority worship and frequent enough expressions of a belief in the gradual progress of scientific thought.

Having thus come to the end of his sustained comparison,[67] Sayili introduces one more consideration of great, methodical interest. Altogether, he asserts, Muslim civilization was fairly stable. Therefore, the inferences made and the conclusions reached in his argument so far may be checked against the attitude displayed by Islam in later confrontations with the West. Attempts made in the 19th century by the Ottoman Empire to introduce Western educational principles into its school system, in particular, seem to confirm the impressions gained previously of "the Islamic tendencies of exaggerated specialization and antiphilosophical orientation."[68] More generally speaking, the willingness to learn from a higher-developed society that Europe had displayed toward Muslim society in the Middle Ages was not reciprocated by the latter once the former had gained ascendancy. Europe's greater receptivity, too,

> seems to have helped the development of scientific knowledge in Europe, and the failure of Islam to show an active and sustained interest in the scientific work going on in Europe was undoubtedly a factor of major significance.[69]

An excursion into the pitfalls of comparative history. Here Sayili's analysis ends. Before going further I wish at this point to make a few remarks about his method of comparative history. The way Sayili arrived at the method, and his defense of it, may speak for themselves. It is the way he carried out his comparisons between the Western and the Islamic Middle Ages that I find so very impressive because he manages, in a self-evident, unspoken sort of way to avoid just about all the manifold pitfalls that lie hidden in this highly disputed area. Many hindrances have blocked the road toward cross-cultural history of science in a comparative sense; many still do, and somehow Sayili seems completely untouched by the array of emotional explosives that lies ready in this particular minefield. What, then, are these explosives made up of?

The primary and original explosive is, of course, feelings of Western superiority. Although it is surely true that scholarly interest in cultures other than one's own has originated in the West, it is no less true that, right from the start, this interest has been tainted by a feeling that our own culture was the unique standard to set against all the others—the highest achievement of humanity so far and worthy of emulation by the backward members of any other given civilization. Here is not the place to go into the story (which belongs to the history of ethnography) of how this original attitude has over time evolved into almost its exact opposite: a complacent relativism that refuses to apply any standard of judgment at all. It is, however, relevant to point out here that the cross-cultural issue at stake in the present chapter tends to set in particularly sharp relief all those issues over which anthropologists have been fighting for many decades. For the simple fact is that the Scientific Revolution *did* occur in western Europe and nowhere else, and that a great deal of the process of Westernization that has gone on ever since all across the globe may rightly be seen as one ultimate consequence of that key event.

Discussion of the problem of how this came about, and why it was the West where early modern science was born, has been hampered from the start by the widely spread notion that there has never been any other science than Western science. This particular denial came from two different sources. It has been upheld in the tradition of Western chauvinism just discussed; as such it was castigated by Joseph Needham in words we quoted in section 6.1. But the same view has also been taken by considerable numbers of 'indigenous' scholars who, coming as they did from a purely humanist (often Confucian) education, ignored science or thought science a thing better not to have anyway.

Many ingredients have thus been lying ready to mar from the outset scholarly objectivity in answering the question of why no other civilization than the West produced early modern science. Put very bluntly, the temptation has always been there to answer, in effect (whether in writing or in private thought not failing to shine through the writing): 'Because those Chinese, or Arabs, or what-have-you, were too dumb, anyway'. Such a sentiment may receive a slightly more civilized guise in being phrased in such formulations as 'The Chinese, or the Arabs, do not think logically the way we do', or 'What those guys had was protoscience at best, not authentic science'. However, it is not rendered one whit more palatable that way for any civilized person—particularly for Chinese or Arabs with an interest in the past of their own civilization, of which science has over long stretches of time been one significant constituent among others.

Add to this that it *is* a hardly disputable fact, established beyond much doubt by anthropologists, that styles of thought do indeed vary considerably from one culture to another, and the stage seems now to have been fully set for endless confusion exacerbated by acrimonious disputes and, perhaps worst of all, for abstaining in despair from the debate altogether because it seems so pointless. This latter outcome, I am afraid, is precisely what appears to have happened in fact. Is not this one key reason why so very little attention has been paid by historians of the Scientific Revolution to this, in principle, fascinating domain of history where so many pet theories may receive further illumination?

Ways out of the knotty dilemmas involved have, after all, been sought—some of them more successfully so than others, to be sure.

One way out has been to deny that early modern science was a uniquely Western accomplishment and to argue that its key concepts may already be discerned in ideas originally developed in Islam or in China.

A more fruitful and realistic way out has been to develop standards for a truly *comparative* history. A history, that is, that takes the fact of Western priority in this particular domain of human achievement as a fact, not one to be unduly proud (or envious) of, but just as a remarkable fact that cries out for scholarly explanation through finding out how it is that the Scientific Revolution eluded other civilizations. If approached along such lines, the question is meant to throw light, not only upon Western civilization by bringing into sharper relief how it could produce early modern science, but also upon civilization p or q by focusing on features uniquely its own though less conducive overall to science in the post-17th-century sense.

Such an approach may be rejected nevertheless for good scholarly reasons because of the inherent danger that science in civilization p or q would not be seen as worthy of study in its own right. This, as far as I can see, is one of Sabra's principal reasons for opposing the comparative approach. Or the comparative approach may be rejected for purely methodological reasons, because it may be held that historians should keep their mouths shut about events that are merely conceivable rather than having taken place in reality. Here, as we shall see, Graham's principal objection appears to be situated.[70]

If, however, the comparative approach is accepted, it may still be carried out in several different ways.

One way is, for a Westerner, to lean over backward and demonstrate, wherever an occasion seems to present itself, how much earlier civilization p was than the West in developing such and such a concept, theory, tool, or method. This, as we shall find out, has been one overriding (even though generously conceded) flaw in Needham's otherwise so illuminating work.

Another way, for a Chinese or an Arab, might be to fall into that very same trap, though not of course out of a Needham-like feeling of guilt for the deeds of Western imperialism and a consequent desire to make up for them in retrospect but rather out of a wish to exploit those guilt feelings or out of cultural chauvinism or for whatever other reasons.

Having made all these preliminary remarks, I may now sum up my admiration for Sayili's achievement in this domain in just one sentence: He smoothly slides past all these mines. How objectively has Sayili analyzed pertinent features of his own civilization! There is no trace of chauvinism, or a vestige of self-hatred, to be discerned in his brand of comparative historiography—just a relentless interest in the past for its own sake governs his discussion throughout. He does not care whether the facts he brings up are 'favorable' or 'unfavorable' for Muslim civilization. He only wants to know how it is that science, in that particular environment, was arrested after so promising a start.

Sayili's flawless objectivity does *not* mean that none of the conclusions he draws from his exercise in comparative history could be disputed. But the point is that such a dispute may be held over nothing but the scholarly merits of one historical interpretation as set against another. Disputes along such lines presuppose no more than that most precious thing: a genuine *respect* for the wide diversity of the world's many civilizations of the past, not as being all equal, or all equally relative, but as worthy of a comparative inquiry into their respective characteristics, into what they had in common, and into what made them nevertheless characteristically different.

6.2.4. *Saunders and the Effects of Barbarian Destruction*

J.J. Saunders, in an article from 1963 entitled "The Problem of Islamic Decadence," shares certain key assumptions with Sayili. One is that Islamic science, despite heavy borrowings from Greek and other sources, nevertheless possessed many original features and, altogether, displayed a character uniquely its own. He also agrees that what stands to be explained is not the decline of Muslim civilization as such, but the decline of science, in particular. However, his criterion for the decay of science is somewhat more severe than Sayili's (whose work he does not appear to be familiar with). It is not only a failure to add to the treasure of knowledge but also the actual loss of portions of knowledge acquired previously. We shall see that, as a result, Saunders assigns a somewhat later date to the decay of Islamic science than Sayili did, a date that serves him well for his own explanation.

For Saunders, the starting point of the search for an explanation ought to be the different ways in which the Greek legacy was appropriated by both civilizations:

> Clearly any attempt to explain why Islam and the West followed such different paths must start with a comparison of the conditions in which each was born and reached maturity.[71]

The key thing about the rise of Christendom is that it came up within, and as part of, the Roman Empire, whereas Islam freshly emerged out of the desert. This difference in origin led to a deeply different imprinting in either case:

> The recovery of Greek in the West at the time of the Renaissance was hailed as the reconquest of a lost kingdom of the intellect which had always rightfully belonged to Christendom. . . .
>
> [As against this,] Islam started with no secular tradition behind it; its learned doctors, unlike the Jeromes and Augustines of the West, were untroubled by the distractions of a brilliant pagan culture and devoted their lives to the study of the holy book and the commentaries on it. It is this which has given Islam its imposing unity and theological exclusiveness. . . .
>
> [In sum,] classical Greece, with its rationalism and its curiosity about Nature, *belongs* to the West in a sense in which it does not and never did belong to Islam.[72]

Saunders hastens to add to this first conclusion of his that Islam, when it emerged out of the desert, appeared remarkably receptive and open-minded, and that, in von Grunebaum's words, "in the light of the crudeness of its origin, its achievement is extraordinary."[73] Nevertheless, as has always been true in the history of cultural transmission, its borrowings were highly selective. Hindu arithmetic was adopted, but not Hindu philosophy, which made no sense in light of the Muslim faith. Similarly, Greek drama was ignored, whereas Greek philosophy could be of help in defending the faith against the many Christians who lived in the freshly conquered lands. Thus faith was the core criterion throughout, and the fate of the *awâil* sciences depended on their acceptability to the faith's upholders.

So far we have learned little that had not already been either asserted or at least implied by both von Grunebaum and Sayili. But now comes the central turn in Saunders' argument:

> The first and most obvious reason for the 'freezing' of Muslim culture is therefore the predominance of theology and of an all-embracing religious Law (*shari'a*) and the absence of a rival secular tradition capable of challenging the dictatorship of the *'ulama*. This was Renan's explanation, but the further question is invited, why the defeat of philosophy and later of science occurred when it did, and why there was no recovery.[74]

To this latter question Saunders presents two answers. The first, relatively minor one is taken from the Isma'ilian heresy; the second from the barbarian invasions.

Saunders briefly describes the serious threat posed to orthodoxy during the 10th–12th centuries by the Isma'ilian sect. In the end, orthodoxy gained the upper hand, and this sealed the fate, not only of heresy, but of philosophy, too, since "the orthodox branded them as twin evils indissolubly linked."[75]

True, one is hard put to believe that, if the adherents of Isma'ilianism had gained the upper hand, science would have continued to interest them, for they were hardly freethinkers, even though it is also true that, as the Scientific Revolution suggests, the breakup of a monolithic belief system may greatly benefit science. Leaving speculation aside, the fact remains that the crushing of the heretics spelled nothing good to either science or philosophy.

An altogether more decisive event, however, was the barbarian invasions. It is their cumulative effect over several centuries that robbed Islamic civilization of its intellectual openness and of its creativity as far as secular issues were involved.

Civilization and the barbarian hordes.   Until fairly recent times, civilized life all across Eurasia was menaced every now and then by barbarian hordes breaking out of the steppes from West Africa to Mongolia. Western Europe was for many centuries subject to such raids, the Magyars (together with the Vikings) in the ninth century being the last to bring major destruction before settling down, whereas during that time Muslim civilization was scarcely affected. But just when western Europe began to recover, the tables turned, and from the eleventh century onward the Arab world had to face no less than four successive waves of invasions. The degree of destruction wrought by these differed, yet the cumulative impact was greater than any civilization could possibly absorb without serious, and often irreparable, damage. For example,

> Ibn Khaldun tells us that after three hundred years the havoc wrought by the Banu-Hilal in Barbary had not been repaired, and Iraq never recovered from the wanton destruction of its irrigation system by the Mongols.[76]

It is true that these ruthless conquerors were not entirely indifferent to intellectual pursuits—witness Hulagu Khan's willingness to allow Nasir al-Din to set up the Maragheh observatory—yet on the whole culture and education suffered dearly from such destruction.

What particularly tends to go underestimated by historians, Saunders continues, is the contrast that thus arises between the fate of the Muslim world and of western Europe. Too little is it appreciated how immensely lucky Europe was in the year 1241 when the Mongols, after having taken Silesia, were compelled by their own rules of succession to return in order to take part in the election of a new Great Khan:

> Had they pressed westwards to the Rhine and the Atlantic and overrun Germany, Italy and France, which they could probably have done with ease, there would have been no Renaissance, and the West, like Russia, would have taken centuries to reconstruct the shattered fabric of its civilization. Western Europe . . . emerged from the Dark Ages in the eleventh century, at the very time when the first barbarian blows were being struck at the world of Islam, and it was able from then onwards to build up a new civilization on the Atlantic fringe of the Eurasian

continent uninterrupted by the raids and devastations of Turks or Mongols or Bedouins. The rise to subsequent power and greatness of the West must be attributed, at least in part, to this circumstance.[77]

One consequence of the invasions was what Marxists regard as an independent cause of the decline of learning: economic retrogression. For centuries there had been a giant area of free commerce, stretching from Senegal to Canton, that was controlled by the Arabs. The invasions broke it up. The same happened to the social structure. There had never been autonomous cities such as began to emerge in western Europe. "A Muslim town had no corporate identity" and was filled with separate groups organized by ethnic origin and religion.[78]

> The contrast with the West could hardly have been greater. In the Muslim world there was no hereditary aristocracy, no clergy or monks, and no Third Estate. As commerce declined, the merchant element sank to the level of the artisans and craftsmen, and society resumed its old simple structure of landlords, officials and peasants.[79]

One more consequence of the invasions was that the monopoly of the Arabic language, which had been the unique carrier of both religious and secular knowledge in the Muslim world, was broken by Persian and, later, by the Turkish language, too.

The upshot of all this is that, in 1300, the world of Islam looked quite different from three centuries previously. "The free, tolerant, inquiring and 'open' society of Omayyad, Abbasid and Fatimid days had given place, under the impact of the devastating barbarian invasions and economic decline, to a narrow, rigid and 'closed' society in which the progress of secular knowledge was slowly stifled."[80] No more borrowings took place from the world outside Islam; Hellenistic philosophy now came to be seen as a danger to the faith. All this was codified in writing by al-Ghazzali, and Ibn Rushd's defense of the *awâil* sciences, one century later, failed to carry conviction.[81] A striking illustration of this is to be found in the works of the great historian Ibn Khaldun. With all his learning and erudition, he shows no respect whatsoever for philosophy, useless as it is for the believer:

> Ibn Khaldun has heard that the 'philosophical sciences' are now much cultivated among the Christians of Europe: his tone implies that this is what might be expected of unbelievers. Islam, being in possession of all truth, has no need to take anything from outside: the age of translation from the Greek is long over, and from the Latins the Muslims took nothing at all at any time.[82]

The net effect of the invasions on the life of the spirit and the mind in Islam was for men to turn inward. Facing one material disaster after another, only religion could give them consolation. And this is how it came about that

Islam, from the first an essentially religious culture, turned back to its origins; . . . profane science, which had always operated on the fringe and had never really cleared itself of the charge of impiety, was quietly abandoned as 'un-Muslim'. It has been said that 'Islam does not reach to the stars', it has none of the insatiable curiosity of the Greeks,

> To follow knowledge like a sinking star

> Beyond the utmost bound of human thought;

the Muslim's world is at rest, and he is at rest within it, and what strikes us as decadence, is to him repose in the bosom of eternal truth.[83]

Saunders concludes by means of a sketchy comparison with western Europe. In principle, so he asserts, what happened to science in Islam might equally well have happened to science in Christendom. However, there were key differences in both its inception and its subsequent development over time. As noted, Christendom was born and raised within the cradle of Hellenistic civilization. It suffered no more barbarian destruction *after* civilization had begun to flower. Its system of law was not as exclusive and comprehensive as the *shari'a;* Roman law and canonical law always kept each other in check. Islam knew of no separation between Church and State, which led to so much diversity in Christendom. In short, all sorts of elements that made for dynamism in western Europe were absent from the Muslim world in the shape it acquired after the barbarian invasions. There was no "stirring up the stagnant life of Islam, which experienced no Renaissance. Islam had discovered God: it felt no need to discover Man and the World."[84]

6.2.5. *Some Conclusions and Suggestions*

One general conclusion to be drawn from the preceding accounts seems entirely obvious. It is that the factual basis upon which viable explanations for the absence of early modern science in Muslim civilization may be erected is smaller than in any case discussed in previous chapters of the present book. For every single segment of the early history of science that has come up so far—be it science in ancient Greece, in the Western Middle Ages, or in the Renaissance—many more facts are now available than is true of Islamic science. One may even wonder whether, at the stage of scholarship reached by the three authors we have examined, and in the apparent absence of later comparative inquiries, it is not simply premature to risk such generalizations as presented to us—surely with a somewhat divergent expertise in the history of science—by von Grunebaum, Sayili, and Saunders.

The very starting point of their respective treatments—the decline of Islamic science—gives us reason to pause. When did it occur? Von Grunebaum mentions, vaguely, the 'later Middle Ages' as the time when decline set in. Sayili appeared to be more precise. His criterion is the nonrecurrence of clusters of great scientists, and he dates this right after the Golden Age; for him, the turning point is the 10th century. Saunders requires a loss of scientific knowledge in addition to its gradual leveling off and thus arrives at the 11th century as the onset of decline, with the year 1300 marking its completion. A decline thus defined happens to cover nicely the period during which Saunders' chief explanatory device raged through the Muslim world. In remarkable contrast, we have seen Sabra defining the decline as the lack of follow-up to promising research and, also, as the disappearance of virtually all scientific activity. These joint criteria lead him to date the decline around the year 1500.

From our sections on Greek stagnation we may recall a comparable spectrum of definitions of decline, along with somewhat divergent dates being assigned to the process, yet in the case of Islamic science the divergence is much more bewildering. As appears so often to be the case when historians set out to explain a given historical process, it is hard to make a clear separation between periodization and explanation. One cannot simply decide when a certain process got under way irrespective of how one intends to explain it. Yet besides this unavoidable predicament of the historian, we are also facing here the simple fact that, apparently, too little is known as yet about Islamic science to make even rough periodizations feasible.

Must we conclude from this that our three accounts are worthless, and that scholars had better forgo synthetic views in this area altogether until more raw facts have been assembled and agreed upon?

I do not think so. For one, those three accounts are simply too stimulating for such a trite conclusion. As I shall seek to show in what follows, they contain too many important ideas, too much that is highly suggestive, too much of that indispensable ingredient of historical research done in the truly grand manner: conceptualizations halfway between blank facts and empty abstractions. It also seems to me that, for the matter at hand—the definition and dating of the decline—there is an excellent way out. The same way out, in fact, available for the comparable case of Greek stagnation, namely, Ben-David's illuminating idea that, in traditional societies, the normal pattern is not the steady growth of science but rather intermittent growth alternated with periods of decline. We have seen Sayili, in particular, come close to such a view without pursuing it to its logical conclusion. The great advantage of applying Ben-David's point of view to the present case is that it enables one to untie a complex knot of really separate questions which, I think, our authors have been trying to handle as if they were just one. If we accept Ben-David's turning upside down of the issue of decline, we can begin to unravel the knot into at least four component strands:

1. Given that, in traditional societies, science was due to decay anyway, we may still ask when, in the case of Muslim civilization, it began to do so. The special thing about Islamic science then becomes, not that it displayed a pattern of flowering and decay, but rather that, and why, after 1500 it virtually died. When seen in this light, the issue of periodization is largely cut loose

from the issue of explanation, and much less ideological ballast, so to speak, hinges on the answers given.

2. This is so because finding out about the causes of decline of Islamic science thus loses much of its explanatory value for the causation of the Scientific Revolution. Such research certainly remains of great value for revealing the specific nature of this decline, just as the decline of Greek science, though conforming to the normal pattern in traditional society, nevertheless displayed characteristic features of its own. And many of those specific features of Islamic decline may serve as signposts to what it was that caused the normal pattern to be disrupted in western Europe for the first time. These signposts come best in view by asking the two questions that follow now.

3. We may ask why no Galileo or Kepler—but let us for brevity's sake just call this composite figure 'Galileo'—turned up in any civilization other than that of late 16th-century Europe. For such a man to appear a certain degree of scientific maturity is obviously required, and the problem at issue then becomes to define that degree in the absence of its actually having been reached. However, and this is crucial, there is no *inherent* reason why such a person could not have appeared in a scientific tradition due to decline later. For this historical possibility Archimedes, of course, is our crown witness.

4. We may ask next whether, once a 'Galileo' has turned up, this would have made the new type of science thus ushered in sufficiently palatable to the wider intellectual and social environment for an uninterrupted tradition of science-in-the-new-sense to have taken off at that point. In other words, were, in a given society, the intellectual status and the social acceptance of science such that 'Galileo' would not have remained an isolated freak but might rather have sparked off the type of unceasing, broadly cumulative scientific activity we have grown accustomed to in the West?

In brief, the two questions that, in my opinion, ought to be asked in neat separation are: *Could a 'Galileo' have emerged in a given civilization, and, if so, could he have prevailed?* I think that our three disquisitions on the decay of Islamic science can be clarified, and their fruitful elements sorted out, with these two core questions kept firmly in mind.

One way to show what we may gain in this way is to demonstrate what inconsistencies may follow when these questions are conflated (in addition to being mixed up with the issue of decline). The clearest example we have come across is to be found in Sayili's discussion, where the author began by treating on a par the question of decline and the problem of why no Scientific Revolution occurred in Islam. However, no more than twelve pages farther down he manages, in effect, to adduce the failure of Islamic scientists to advance substantially beyond the Greeks as one *cause* of why science never gained sufficient support in the Islamic community. Yet originally this failure was for him part of the *problem*. Along the lines of his treatment the circularity in the argument cannot be avoided. So, what we have to do is to take the questions involved apart, in the manner indicated, and analyze from there on what we may have learned now about the causes of the Scientific Revolution through the observation of its absence from Muslim civilization.

I perform this analysis in three successive stages. First I consider on what points the validity of the interpretations offered depends, partly or wholly, on specific conceptions of the Scientific Revolution (e.g., Sayili's point about the sheer amount of scientific work being done as the key to finding the experimental method). Next, I indicate what data about Islamic science may provide interesting test cases for certain explanations of the Scientific Revolution we have come across in previous chapters (e.g., what remains of Eisenstein's points about the printing revolution in view of the three dirhams paid by Avicenna for a fresh, routinely available manuscript copy of a commentary on the *Metaphysics*). Finally, I wish to enumerate those prominent features of Muslim civilization, as elucidated by our three authors, that may add most to our understanding of why Europe pulled off a Scientific Revolution that eluded the Islamic world (e.g., Europe's freedom from barbarian invasions from the 10th century onward).

How interpretations hinge on explanations. Sayili's discussion involves at least three elements of comparison whose validity hinges critically on his conceptions regarding the history of Western science. Those conceptions, although certainly not indefensible, have appeared in our previous discussions to be debatable issues in their own right.

The first of these is Sayili's idea that, for the flowering of science, it was necessary to achieve a thorough reconciliation of science and philosophy with religion before, at a later stage, they went their separate ways. A corollary of his view is that the nominalist separation between science and religion was good for science. I do not want to dispute this opinion here, but I do remind the reader of section 4.3 of the present book for an alternative conception of how science in the 14th century benefited, not from the separation from religion, but rather from (rationalist) philosophy. We also found discussed there the idea that nominalism was of no more than marginal significance for science, as the movement remained firmly encapsulated inside the Aristotelian framework and thrived mostly on clever imaginations.

Related to this is Sayili's praise for Aristotelianism as the chief means, in the West, to give science a systematic rigor it did not possess in Islam. In particular, I do not see why, in the West, the situation was so different in this respect from Muslim civilization. In *both* civilizations Aristotelian philosophy comprised substantial portions of natural philosophy—for example, problems of free fall, projectile motion, and the void—in essential separation from such other scientific topics as the reflection and refraction of light and the determination of specific gravity. One key achievement of the Scientific Revolution was to lift the topic of local motion out of its stifling environment, to put it on an almost entirely new footing, and to unite the new kinematics with loose topics in optics, hydrostatics, and so on under the overall heading of 'natural philosophy' in a new sense. On this specific score, it is hard to see why the challenge involved would have been easier to meet in the one civilization than in the other.

Finally, there is Sayili's core idea that the missing ingredient in the Greek heritage was 'the experimental method', and that just trying hard was the key to its discovery. I do not wish to repeat all over again what has taken the entire second chapter of the present book to set forth but leave it to the reader to decide whether he or she can accept, partly or wholly, what Sayili (a disciple of Sarton rather than of Koyré) here rather takes for granted. My purpose, after all, is not to enforce unanimity on the ultimately intractable problem of what the essence of the Scientific Revolution should be taken to be (such unanimity being as elusive as it would be undesirable). I only wish to show again and again the extent to which efforts to explain the Scientific Revolution really hinge on our core convictions as to what the event ultimately was about.

Islamic science as a testing ground.   Quite a few attempts at explanation of the Scientific Revolution that have passed muster in previous chapters may be put into sharper relief by means of considerations taken from the history of Islamic science. Several clues for intriguing testing procedures may be gleaned from what we have now learned about the latter. Altogether, I believe that Islamic science, precisely because it shares with the West the foundation on which it was built, is a veritable treasure-house for testing the validity of general assertions which, within the context of Western history alone, can never be made more than plausible at best without coming up for real proof or disproof. In essence, this was Zilsel's point, and I share it (although I also think that asking for 'proof' is more than you are to get in historical interpretations). In the present context I cannot do more than provide a brief checklist and express the hope that later research may take up these matters in the full measure of detail they deserve in my opinion.

1. *Technology and the arts and crafts.* Although, in Islam, science was cultivated chiefly because it was thought to be useful, the uses to which it was put do not appear to have been of a technological nature. Indeed, it is significant that David Landes, in the brilliantly comparative first chapter of his *The Unbound Prometheus,* where he considers why no *Industrial* Revolution took place in any Oriental society, confined himself to adopting von Grunebaum's and Saunders' explanations of why no *Scientific* Revolution occurred in Islam.[85] It might be worthwhile to examine the relations between scientists and artisans in Islam with a view to testing pertinent theories on that score. If fruitful and frequent contacts between the two groups were indeed established, this would tend to undermine Zilsel's and many others' contentions; however, none of our spokesmen cared to bring up this particular issue, and what little I have

read about city life in the medieval world of Islam makes it rather unlikely that such encounters took place with sufficient frequency. On the other hand, it is striking that, according to Sabra, an elaboration of the Archimedean approach took place in engineering circles rather than among scientists.

2. *The professionalization of science.* Another theme developed in previous chapters (mostly in connection with Ben-David's thesis on science in traditional society) is the inherent instability of the pursuit of science if dependent solely on support from the court. The fate of science in Islam seems fully to bear out the impression that, with little sustenance beyond the ruler's whims and fortunes, science was bound to suffer quite erratic bouts of intermittent flowering and decay. Now Ben-David also observed that in the European Middle Ages, for the first time, the establishment of university chairs in philosophy made it possible, though not indeed obligatory, for their incumbents to spend considerable portions of their official time on scientific matters if that was where their fancy took them. It is of interest to note that, at a comparable stage of Muslim civilization, there, too, an office arose—that of the *muwaqqit*—which its occupants were free to utilize for scientific purposes if they felt like it. We appear to have here an intermediate stage between the cultivation of science being entirely separated from permanent professional institutions and of science as practiced by men paid for doing just that. However, it is also clear that the availability of such halfway-institutionalized positions in a given society is not by itself sufficient to put science on a permanent footing; witness the fact that the *muwaqqit* proved unable to save Islamic science from its definitive decline.

3. *The uses and the legitimation of science.* The uses of science (whether real or confidently expected) provide an important point of debate over 17th-century science in Europe. In the case of Islamic science we appear to be confronted with a civilization where science, to the extent that it was valued at all, was valued in good part for its perceived utility. Still, in the end no enduring legitimation for the scientific enterprise was attained, as a result of overriding counterforces, among which the dominant values derived from religion were paramount. To my mind, this suggests two conclusions to be drawn for the debate about the Scientific Revolution. One is the wisdom of Ben-David's move to introduce certain 'intermediary concepts' between the utility of science on the one hand and its legitimation on the other: The fate of Islamic science seems to reinforce the impression that no direct causal link holds here. The other tends to confirm a tenet shared by Merton and many of his critics: that religious values *are* of paramount importance where the destiny of science in society is concerned. One may go further here and suggest that it is about time to step outside the narrow circle of the ongoing controversy over the Merton thesis and make it cross-culturally comparative. Little further insight stands to be gained in this domain of historical debate unless (more or less on the

example of Weber's comparative sociology of religion) the diverse values that went with the world's religions are introduced into the equation.

4. *From script to print.* On the face of it, the case of Islamic science fully bears out Eisenstein's core view that in a scribal culture no substantial advance in science can be expected beyond the preservation, as best one can, of the integrity of hand-copied ancient treatises. And quite conceivably, this may well be the ultimate outcome of a sustained test of Eisenstein's chief thesis against the conditions of scribal culture in Islam. Consider, on the other hand, what remains of her statement about the novelty of Copernicus' scrutinizing the entire *Almagest* at leisure in his hometown once we recall that, in Baghdad about 700 years earlier, a copy could be had in exchange for one donkey. Consider, somewhat more generally speaking, what Sabra has to say on the public library system throughout the Muslim world or on what the Muslim habit of writing on relatively cheap paper instead of on parchment or papyrus meant for the prices to be paid for a manuscript copy. In short, at issue is not why the invention of the printing press in the West did not affect Islamic science the way it affected Europe (for by then Islamic science had almost died out) but rather whether conditions of scribal culture in the Muslim world did not go far to alleviate restrictions set to the reasonably free distribution of reasonably standardized texts.

5. *The Voyages of Discovery.* Hooykaas called attention to the shock administered to humanist Portuguese literati when seamen began to report observations of phenomena—people actually living in the tropics, and so on—ruled out on *a priori* grounds by the ancients. One may wonder why no comparable shock was apparently felt by practitioners of the *awâil* sciences since, after all, they were part of a commercial free zone extending from West Africa to China and, hence, must have been aware of such phenomena at least five centuries earlier. Why did *they* not begin to doubt the reliability of Aristotle and Ptolemy? Why did *they* not conclude that, perhaps, the exalted but largely *a priori* reasoning of the ancients should be balanced more carefully by observation and experiment? Perhaps because, to them, the shock, if indeed there was one, must have come much more gradually?[86] Obviously, these are just questions, and maybe pointless ones. Nevertheless, they are perhaps worth asking.

6. *Greek stagnation.* We have found a variety of answers to the riddle of Greek stagnation, with the principal division being between those who thought that something fundamental was still wrong with science when the Greeks left it (too much reliance on the powers of pure reasoning; a conception of the cosmos as a living entity, etc.) and those who believed that but for some mostly external reasons (lack of appreciation of manual labor done by slaves; the Christian brain drain, etc.) a Scientific Revolution might have ensued at once. Now it is quite astonishing that none of these varied diagnoses has ever been applied to Islamic science, which surely represented the fullest embodiment of

Greek science before the European Renaissance. No one seems ever to have suggested that what was wrong with Islamic science—so wrong as to preclude a Scientific Revolution—was this regrettable lack of balance between reason and experience, or that Islamic society could not bring forth early modern science because it, too, held slaves, or that Islamic science got stuck because it continued to entertain the notion of a living cosmos.

The only scholar to begin to see what is at stake here was, once more, our perspicacious Sayili. He, however, in his characteristically frustrating way, dropped the point after one perceptive paragraph in which the fundamental shortcoming of the Greeks is identified as an inability to handle dynamic, as opposed to static, natural phenomena. Add to all this that the Archimedean tradition seems also to have been alive and well in Islam, so that once again the printing of Archimedes' work in 16th-century Europe is robbed of its unique impact, and we may be entitled to conclude that here, indeed, is a fruitful area for comparative research. The key question to be asked in this connection seems to me this: Given that Islamic civilization inherited Greek science as one coherent corpus and enriched it with both Hindu arithmetic and substantial finds of its own, how is it that all this was still not enough to break through the limits apparently set to Greek science in the first place?[87]

Western European science and the Islamic experience. Thus put, this question is but one variant of the general question we have come to phrase as follows: Could 'Galileo' have turned up in Muslim society? It is readily apparent, though, that virtually all that seems fruitful in the accounts of our three learned arabists pertains rather to the problem area delimited by our second question: If so, could he have prevailed?

We have come across one suggestion only—to be encountered in both von Grunebaum's and Sayili's essays—which seems to preclude the appearance of a 'Galileo' on the Arab scene. It is the idea that 'resigned doubt', a belief in miracles, and a lack of confidence in the possibility of establishing regularities among natural phenomena curbed much that might have become fertile in Islamic scientific pursuits. I admit that I find this general view a bit hard to reconcile with what I have learned about the proud achievements of such men as al-Bîrûnî, Ibn al-Haytham, and others. Sayili thinks that 'Galileo' might have turned up if the Golden Age of Islamic science had been allowed more time to develop further. From what I have learned about that Golden Age, I do not readily see why, *given the achievements and preoccupations of Islamic science,* in which both the problem of the planets and Aristotle's conceptions of motion were studied as they were in the West on the eve of the Scientific Revolution; given also that the Golden Age was of quite considerable duration by itself; given, finally, that no observer in Europe at the end of the 16th century could possibly have predicted as the next stage of advance the radical turnabout of our conception of nature that did in fact take place—given all this, I do not readily see why no 'Galileo' could possibly have appeared already in the course of the Golden Age of Islamic science, perhaps as its crowning achievement. For are we not, in denying such possibilities absolutely, indulging too much our historians' habit of stamping what is truly contingent with the seal of causally explicable inevitability? I am not advocating pointless speculation, but we should keep in mind the distinction between what happened indeed, what failed to happen but might have, and what could not possibly have happened.

Returning now to the issues handled by our three experts, I am much more impressed with four general features of Muslim civilization that help clear up the fate of science in Europe, in that they tend to show why, if indeed 'Galileo' had appeared among Allah's servants, he would in all likelihood not have prevailed among them.

The first is von Grunebaum's general point, analyzed in greater detail by Sayili, that philosophy—covering most of science as it did—after having served a fleeting need, consistently continued to be regarded, by the most vocal and literate men in society, as a foreign intruder. Von Grunebaum explains how this historical process is inextricably bound up with the fundamental aspirations of Islam; Sayili focuses rather on the failure to reconcile philosophy and science with religion on other grounds. Both may very well be right, and at any rate, here we seem to touch upon a seemingly trite yet fundamental condition for a Scientific Revolution to occur: that science not be perceived by the community at large as being directly opposed to its core values.

The second is an observation also made by both von Grunebaum and Sayili, namely, this remarkable tendency of Muslim scientists to identify at heart with the objections directed by, or on behalf of, the community against their activities. I have already made some comments on this point; let me confine myself here to adding that, whether or not this feature may be confirmed empirically, it does point the way toward a general, perhaps unsurprising conclusion that is nevertheless worth being spelled out in full: No Scientific Revolution is conceivable but for an at least semiprofessional cluster of scientists who exude a deep and abiding belief in the value and dignity of their own life's work.

Finally, Saunders has made us aware of two other conditions for a Scientific Revolution. One is the simple but, again, oh so fundamental idea that no Scientific Revolution is feasible, for centuries to come, in lands thoroughly destroyed by bands of foreign invaders. The other is that a society with some built-in checks and balances along the lines of the coexistence of Roman law with canonical law, of a neat separation between Church and State, of a vigorous, relatively autonomous city life, and of similar elements of budding pluralism is at a decisive advantage when it comes to preparing the ground for early modern science.

6.3. JOSEPH NEEDHAM AS A PIONEER IN CROSS-CULTURAL HISTORY OF SCIENCE

Whatever may be the vicissitudes of present-day politics in China, it remains nonetheless true to say that, in terms of civilization, we must educate ourselves in the world of the Chinese—if we do not succeed in mastering that grand lesson, there is no way for us to aspire to a complete, veritable humanity.

—Simon Leys, 1974[88]

In 1937, three Chinese graduate students at Cambridge University approached their biochemistry teacher—he could be expected to know something about the history of science, for he had recently written a vast volume on the history of his chosen specialty—asking him how it was that "modern science originated only in Europe."[89] Our Reader in biochemistry could not tell them the answer offhand; in fact, he had never thought about it. Under their spell, he took a crash course in the Chinese language. Only a few years later, in the middle of the Second World War, he grabbed a chance to go to China and plunge himself into the history of Chinese science. In 1944, on an isolated spot in Yunan Province, he began to entrust to paper his first thoughts on the subject, offering some preliminary, tentative answers to the question originally put before him by those inquisitive students. By then he had come to the conviction that the question was "one of the greatest problems in the history of civilisation."[90] He also thought that one book (already vaguely conceived in 1938 but to be written by him after the war ended) would answer it. Ten years later, in 1954, the first volume appeared of what had in the meantime been expanded into a projected seven-volume effort, organized so as to come up, in the end, with the definitive answer as far as the author was concerned. The plan for the overall content (left basically unaltered ever since) foresaw a division of these volumes into fifty sections altogether.[91] At first, the rate went at a volume a tome, each volume covering a number of sections; in the sixties volumes began to be split up into parts, and soon after, just one section succeeded in filling several tomes all by itself. Today, in 1992, fifteen lavishly produced tomes have appeared of *Science and Civilisation in China* (with abridged versions following in their trail), together with quite a few volumes of accompanying essays, many of which have also been devoted to answering the original question. The principal author, Noël Joseph Terence Montgomery Needham, is as old as the century, and it is uncertain whether the seventh and final volume of *Science and Civilisation in China* is to see the light under his authorship. Whether or not this is to happen indeed, it is safe to say that rarely in the history of 20th-century scholarship has such an innocent question yielded such impressive results.

Dimensions of the achievement. Indeed, few scholars who ever devoted themselves to appraising one or another portion of Joseph Needham's life's work have failed to express their profound respect for what he has accomplished. Not only has he disclosed to the world a huge, exciting area of scholarship whose existence had previously been either denied or ignored or, at best, studied only in small, highly specialized fragments,[92] not only has he himself taken the lead in delving up an enormous quantity of facts on the subject, but he has also gone ahead to synthesize these bare facts and to give them meaning as component parts fitting into a comprehensive pattern of historical understanding. Once *Science and Civilisation in China* is completed, we shall have for China what we do not have for Europe (or for any other civilization, for that matter)—a history of its science, its technology, and its medicine from the earliest times up to the impact of modernization, consistently set against the background of the society and the general culture in which science, technology, and medicine grew and manifested themselves. All this has been accomplished through generously acknowledged efforts of collaboration, yet the more and more independent contributions by collaborators have not, despite increasing specialization, led to the fragmentation of the original project. Even though he has insisted that "nothing would ever have been possible without the equal collaboration of Chinese friends," Needham has continued to put his own stamp on the entire effort throughout.[93] He has also continued over the years to accompany it with a huge array of essays intended to reorganize for specific purposes material meant for, or already covered in, *Science and Civilisation in China,* and to sum up vital parts for the benefit of those whose own professional concerns are too far removed from *Science and Civilisation in China* to read those thousands of pages in their entirety. Finally, and perhaps most impressively of all, Needham has consistently taken the grand view of history, in keeping with what Lynn White once wrote about him: "He is able to ask large questions because there is in him no trace of the vanity that quails at the prospect that someone may think his answers wrong."[94] While freely roaming over the continents and through the centuries, Needham has equally consistently kept himself distinct from those speculative historians who ignore the actual facts of history or use these exclusively to squeeze them into preordained patterns. Needham has consistently attempted to underpin his general pictures with facts and details of which he has done his utmost to gain firsthand knowledge.

Naturally, his interpretations, both on the grand scale and on the level of particulars, may come up for, occasionally harsh, criticism—his work has been highly controversial right from its inception. Yet I would not feel free to embark, in the following, upon a critical treatment of Needham's views on the Scientific Revolution and its absence in China without first summing up, as I have just done, what many before me have said in expressing their awed admiration for the overall achievement.

The nature and size of the 'Needham puzzle' and how to come to grips with it. Everything Needham does is done in the grand manner. His attempts to answer his Grand Question, in the varied guises it has over the years come to be clothed in, may be encountered at many places: in lectures-turned-essays, in scholarly papers, and in a number of passages strewn liberally over the available tomes of *Science and Civilisation in China.* Much of his writing has to do with the Grand Question only in a very remote sense, since *Science and Civilisation in China,* although set up to answer the question, has really turned into a comprehensive overview of whatever is noteworthy about ancient Chinese science, technology, and medicine in their entirety. For the purposes of the present book it would have been pointless to work my way through all this material, although, for acquiring a general overview, I have read three volumes of Colin Ronan's *The Shorter Science and Civilisation in China,* an authorized abridgment which covers volume 1 through most of 4 of the major series (by far the most important ones for our purpose). But my chief sources for Needham's evolving thoughts on why the Scientific Revolution eluded China have been the following three books of his:

—*The Grand Titration: Science and Society in East and West.* This book came out in 1969 and contains eight essays written from 1944 through 1964, all of which address the Grand Question in one way or another.

—*Clerks and Craftsmen in China and the West: Lectures and Addresses on the History of Science and Technology.* This collection appeared one year later; many of its nineteen essays are also of importance for coming to grips with Needham's approach to the Grand Question.

—*Science in Traditional China: A Comparative Perspective.* This collection of four essays came out in 1981. Only the preface and the introduction provide insights into Needham's ongoing thinking on our subject that had not already found expression in the earlier essays.

Since these hundreds of pages (he once called writing them "thinking aloud")[95] were never meant by their author to be more than preparatory to the final volume of the major work, Needham has never bothered to sum up, in one grand overview, where he stands on the Grand Question. As a result we face something of a jigsaw puzzle. There are scattered pieces everywhere, and there is also the near-certainty that these fit broadly into one grand pattern, yet that pattern still exists nowhere but in the head of the maker of the pieces. In a rather peculiar manner, Needham's essays are quite repetitive, yet complementary at the same time. He has an uncanny ability to unfold his many pet ideas at just about any length that suits his purpose of the moment. An argument that may take hundreds of pages in *Science and Civilisation in China* may appear, in abbreviated form, as the major theme in one essay, coming to the surface as a little subtheme elsewhere, or even being taken for granted in the argument of yet other essays. Furthermore, Needham's thought has changed and, in particular, has become more sophisticated over the past four decades, but, oddly enough, both more so than he himself seems ready to acknowledge and less so than some of even his most knowledgeable critics have given him credit for. By way of one, rather amusing example, Lynn White in 1984 hailed the 1981 collection *Science in Traditional China* as Needham's "most mellow" product to date, giving praise in particular to the "pluralistic and tentative tone" of its concluding passage. This passage, however, is identical with one written more than a quarter-century earlier.[96]

As a result of all this, Needham's readers appear to come away from a perusal of, in particular, *The Grand Titration* with oddly diverging notions of what his ultimate explanation is for why China did not undergo a Scientific Revolution. Thus I once heard a sinologist say that Needham thinks this was so because of the rival approaches of the Taoists and the Confucians, whereas one expert in comparative history thought that, in Needham's view, the lack of an autonomous merchant class is what inhibited a Scientific Revolution in China. I have also heard it said that, according to Needham, the retrenchment of Ming bureaucracy in the 15th century caused China's considerable edge in science and technology to go up in smoke. Given such a wide compass of purported explanations, it seems worthwhile to sort out all Needham's pertinent themes in some detail and to connect them, as far as possible, with the development of his thought over time. I have made an effort to draw a coherent picture from the books mentioned above (together with a couple of essays contributed by Needham to other volumes). With that preliminary result in hand, I went back to the pertinent volumes of *Science and Civilisation in China* to check it.[97] The procedure thus followed yielded one curious result—one final source of confusion over where Needham stands on the Grand Question. I found that some of the wilder claims he has made while 'thinking aloud' correspond to arguments in *Science and Civilisation in China* which are as a rule much more moderate and display much more good sense (e.g., the claim about translation which I reported in section 6.1 finds its counterpart in a quite balanced discussion of translation problems in *Science and Civilisation in China*).[98] Hence, at least where the Grand Question is concerned, the Needham of *The Grand Titration* and *Clerks and Craftsmen* appears as much more of a historiographical extremist than the Needham of large portions of *Science and Civilisation in China*. We find here one more instance of what has often been said about him: One may fiercely criticize his interpretations with the help of material he has made available himself. What all these contradictions (whether real or apparent) signify for our final assessment of the value of Needham's answer to his Grand Question is a matter that need not be settled here but to which we cannot avoid coming back at the end of section 6.5.3.

This, then, is how I have sought to come to grips with the 'Needham puzzle'. I can only hope that I have come up in the end with a sufficiently accurate and, as it were, prefigured outline of that elusive seventh, concluding volume of *Science and Civilisation in China*.

Some core features of Needham's thought before the Grand Question. Joseph Needham did not of course come to the comparative history of science with a blank mind. Four of his predilections and preoccupations in particular have greatly contributed to the peculiar tinge that has colored his work as a historian. These are his early experiences with toolmaking, his comprehensive attitude toward the sacred, his somewhat holistic conception of science, and his leftist political views. Together with this Westerner's 'Eastern' leanings, these features of Needham's thought and feeling have yielded a unique and, at times, startling blend, held together chiefly by what Needham calls his synthetic turn of mind, which makes him often see unity and congruence where others cannot but think in terms of opposites and incompatibilities.[99] We devote a few lines to each of these characteristics.

Although raised as a scientist, Needham counts among his formative experiences prolonged stints, during his high school days, "in the workshops, [acquiring] in this way a fundamental store of engineering knowledge gained among lathes and milling-machines and in the foundry."[100] This helps account for both his later insistence on the role of manual labor in the rise of early modern science and his facility in comparing technological feats in ancient China and in Europe. Needham is one of the very few historians of science to be equally well at home in the history of technology.[101]

Needham's religious sentiments also go back to his youth. "Fear and the irrational never became indissolubly connected in his mind with religion,"[102] and thus religion could become for him a liberating experience, providing him with a readily available channel to overstep the limitations of daily life. Although a devout member of the Church of England, he has always regarded its dogmas as just one, highly culture-bound, expression of the universally holy. Whether one regards the following passage (written in 1971 while meditating in a Japanese temple) as odd or moving or both, it is surely vintage Needham:

> The Buddhist faith is not my own, but the temple is numinous in the extreme, and to this a Christian cannot but respond, making his orisons in his own way. While the dean and canons chant the saving power of Shakyamuni and all Buddhas and Bodhisattvas, . . . , I praise the glory of the Holy and Undivided Trinity, hoping that this too may be acceptable to the eternal Tao, the Tao beyond speech, that cannot be spoken about.[103]

As an experimental biologist, spending his working hours over the intricacies of the living cell, Needham responded eagerly when exposed during his student days to Whitehead's 'philosophy of organism'. He cites with glee "Whitehead's immortal dictum, [according to which] physics is the study of the simpler organisms and biology is the study of the more complicated ones."[104] The conviction that there is much more to the subjects investigated by science than can be expressed mathematically has never left him and has deeply colored his idea of what the Scientific Revolution was about. Needham himself worked out this philosophy into a conception of what he called 'integrative levels'. Essentially, these are stages of ascending order and increasing organization which follow upon one another in time, from the atom and molecule up to the highest forms of organization we know of. In comparing, for example, the organization of an amoeba with a human, one finds that the latter's level of integration can be called higher in four distinct senses: a greater number and a greater complexity of the component parts; more effective control by the organism over its functions; greater independence of the organism from its environment; more effective methods for survival, reproduction, and control over the environment.

The key function Needham made these ideas serve was

the entire liquidation of all controversies of the 'vitalism–mechanism' type. It is useless to argue whether a science dealing with one integrative level can be 'reduced' to that dealing with another. Science at each level has to work with the concepts and tools and laws appropriate to that level. In genetics and embryology, for instance, we know of many regularities which cannot be affected by anything that biochemistry may discover. But on the other hand they can only attain their full meaning and significance in the light of what biochemistry will discover. . . . The higher or coarser levels cannot be explained in terms of the lower or finer ones, still less *vice versa;* but the universe remains quite enigmatic until we succeed, as from time to time we do, in finding the relations between various behaviours at two adjacent levels.[105]

In such a philosophy of science, evidently, developments in 20th-century science appear less as the perfection of Newtonian science than as the revelation of its mathematical one-sidedness, crying out for a more integral, ultimately organic, conception of nature.

Of nature, and of society, too. Needham has never believed that his integrative levels stop at the level of the human individual. Among his key formative influences was Herbert Spencer's extension of Darwinian evolution to the level of human society. For Needham, there is an "essential unity of cosmological, organic, and social evolution, in which the idea of human progress, with all due reservations, would find its place."[106] Despite occasional setbacks, humanity is inexorably set on its way toward "that world co-operative commonwealth which we see as the inevitable crown of social evolution."[107]

The progress thus written into history at just about every level manifests itself in many ways. Science, medicine, and technology are among the most obvious manifestations of human progress:

Here . . . humanity has always been marching in one column, and though the particular systems of natural philosophy may have been ethnically bound, . . . , the actual understanding and control over Nature has been passed down from mind to mind across all barriers to build up what the early Royal Society called the great edifice of 'true natural knowledge'.[108]

Not only science but the political organization of human society, too, is such a manifestation of progress, which finds its present-day embodiment in socialism. Needham's socialist creed, powerfully reinforced in 1931 by his exposure to Hessen's lecture, has ever since retained strong tinges of Marxism. However heterodox many of his views have become over the years, particularly on the class interpretation of history, Needham has never, to my knowledge, given up the idea that communist societies represent the pinnacle of human progress, in the sense of foreshadowing the future "world co-operative commonwealth." It is through such mental connections (finding their nadir in how he managed to weave the verbal contortions of dialectical materialism into his philosophy of integrative levels)[109] that Needham associated himself with some of the cruelest tyrannies the world has ever seen. His sinophilia and his socialism came together in his readiness during the Korean War to testify on behalf of the Chinese regime's accusations of bacterial warfare by the United States. Quite a few scholars in the West have never forgiven Needham for shoring up with his reputation these apparently groundless accusations, and the first volumes of *Science and Civilisation in China* were accorded a far less cordial reception than might otherwise have awaited them.[110]

And indeed, there is something of a mystery in how this open-minded, generous, and humane man could persist in his loyalty to such awful regimes. In 1971 he still saw fit to refer to the Great Terror in Soviet Russia as "the Stalinist 'illegalities'."[111] From whatever source such weasel words may spring, a large dose of genuine innocence and naiveté seems to be mixed through it. But whether or not he is politically extremely naive, it is certain that his political convictions have greatly colored the explanations he has come up with of why China never underwent a Scientific Revolution.[112]

The vicissitudes of cross-cultural history of science. In 1981 Needham observed, in a tone more resigned than bitter, that his pioneering effort had largely been kept separate from the two most adjacent academic departments: those of sinology and of history of science. Leaving aside the former, one cannot but agree on the latter. After all, it is not by chance that the present chapter is about topics not even touched by all those students of Western science we dealt with in earlier chapters. No integration whatsoever of findings about Oriental science into ongoing research on the Scientific Revolution has taken place. To cite just one example out of hundreds, Needham remarked in 1964 that A. Rupert Hall's article "Merton Revisited" (see section 5.2.5) might well have benefited from taking into account discussions about why no Scientific Revolution occurred outside western Europe (because addressing that issue would by itself have revealed the necessity of introducing 'external' points of view).[113] When, in 1983, Hall took up "The Problem of Cause" as one chapter of his rewritten textbook *The Revolution in Science,* he mentioned Needham in his preface as one of his four Cambridge mentors, along with Butterfield,

Koyré, and Singer, and went on to ignore both the man and his work throughout the chapter and, indeed, the whole book. Now why is this so?

Here is Needham's own answer:

> [Historians of science] are interested primarily in European post-Renaissance science, partly because they lack all access to primary sources in other languages. Sometimes they are interested in Greek science as well, very rarely in mediaeval or Arabic science. The last thing they want to hear about is non-European science, and that is partly because of their strongly Europocentric vision. The unspoken assumption is that because distinctively modern science originated only in Europe, the only interesting ancient and mediaeval sciences must also have been European. This glaring *non sequitur* still dominates the intellectual outlook in the West, in spite of all that has been done by enlightened comparative historians of technology, like Lynn White, who have shown over and over again the indebtedness of traditional Europe to discoveries and inventions made in the more easterly parts of the Old World.[114]

The core belief behind this mind-set, whether articulated or not, has been, according to Needham, that China (or any other Oriental civilization for that matter) never produced any authentic science. And this belief, in its turn, usually betrays a strong sense of European superiority. This is one of the few subjects on which Needham dwells with real vehemence, and he is curiously one-sided about it.

> Westerners, I reflect, still suffer seriously from the Four Evil Minds, greed, anger, foolish infatuation, and fear. Of greed, built into the structure of capitalist civilization, not much need be said. Anger or irritation at ways not their own is often to be found; and is not the idea that Western ways are the only universal culture a foolish infatuation? Fear, unreasoning fear above all, the fear of the alien and the unknown that expresses itself in phrases like 'hordes of gooks,' is an evil thing that Westerners must learn to overcome.[115]

There is quite something to all this, to be sure, and it is also clear that strong emotions lie behind such statements. Lynn White was only partly teasing Needham when he observed that "there are times when he seems to be doing personal penance for the Opium Wars."[116] But the other side of the coin is never even mentioned by Needham. On reading him, one would believe that 'greed' is bound up with just one particular form of social organization (and, indeed, much anticapitalist sentiment is rooted in the cheap equation of greed with capitalism, as if greed were less rampant in societies not built upon free enterprise).[117] One almost never hears him mention traditional Chinese fits of xenophobia, the immense sense of superiority (condescending rather than aggressive, but nonetheless real for that) with which the Middle Kingdom looked down upon the 'barbarians' beyond the Great Wall and the seas, and the remarkable lack of interest in other civilizations that marks just about every civilization other than the West (not by chance is anthropology, a full century after its inception, still by and large a Western science).

Thus, in one sense, Needham's insatiable curiosity about things Chinese testifies to a quite typically Western attitude. "Mankind is one great family," he has written, "and the scientific view of the world has clearly transcended all differences of race, color, and religious culture."[118] Is it not also clear that this lofty conception is, above all, Christian and Western-scientific in its original inspiration?

Yet this is not the only angle from which to look at the emotional roots of Needham's life's work, for which he abandoned without hesitation a highly promising career in biochemistry. Again, he has himself expressed the inherent interest of his pursuit in terms that cut it loose from whatever cultural roots it may sprout from:

> What happened in other civilizations is entirely worth studying for its own sake. Must the history of science be written solely in terms of one continuous thread of linked influences? *Is there not an ideal history of human thought and knowledge of nature, in which every effort can find its place, irrespective of what influences it received or handed on?* Modern universal science and the history and philosophy of universal science will embrace all in the end. [My italics][119]

It is with such a conception in mind, too, that Needham has tried to uncover not only the actual content of Chinese science but also the extent to which both Chinese and other Oriental scientific traditions have contributed to what he invariably calls 'universal science'. It is to these contributions that we turn now.

6.4. CONTRIBUTIONS OF NON-WESTERN SCIENCE TO THE SCIENTIFIC REVOLUTION

Needham's core view of the entire history of the world's science is set in determined opposition to a Spengler-like conception of world history as made up of carefully insulated, wholly incommensurable civilizations:

> While one can easily see that artistic styles and expressions, religious ceremonies and doctrines, or different kinds of music, have tended to be incommensurable; for mathematics, science and technology the case is altered—man has always lived in an environment essentially constant in its properties, and his knowledge of it, if true, must therefore tend towards a constant structure.[120]

There is one true, universal (or 'ecumenical') science; it has grown out of the early modern science that started in western Europe in the 17th century and has accumulated ever since; into the making of universal science went key elements of science as cultivated all across a great many traditional civilizations. Needham's favorite metaphor for the whole process is that of the many rivers of traditional science, all of which have flowed into the ocean of universal, modern science.[121]

The historically fixed yet contingent fact that early modern science arose only in western Europe has subsequently obscured how many foreign elements contributed to its origin. Francis Bacon started the process of retrospective appropriation, as we can see when we realize that the three 'mechanical discoveries' he put at the head of the new era of science because they had not been known to the ancients—the compass, the printing press, and gunpowder—are without exception Chinese inventions.

Nor are they the only ones. Needham has set up a comprehensive theory of how the transmission of inventions occurred in successive clusters, which, in their turn, may be connected with major political events facilitating, *casu quo* inhibiting, easy transmission. To mention just one such cluster by way of an example, in the late 13th and early 14th centuries gunpowder, a machine for reeling silk, the mechanical clock, and the segmental arch bridge turn up in Europe. All of these are, in Needham's opinion, Chinese inventions which may have found their way westward over the free commercial routes just established as a result of the Mongolian conquests.[122]

Yet we can understand how it is that the Oriental origin of this huge array of products of Eastern ingenuity was so quickly forgotten in the West—if noticed at all—once, with the Jesuits' arrival in Peking, the direction of transmission was reversed for the first time in history. One other reason for this besides retrospective appropriation is that the latent power of social transformation that these inventions and discoveries possessed came out in the open when they left home and were dropped into the society of medieval Europe. In China cast iron, gunpowder, papermaking, printing, and so on had not made much of an impact on society at large. Although Needham strongly opposes, as a conventional inanity, the notion that such inventions were only put to frivolous, silly usages, it is no less true for him that in western Europe they had the most drastic repercussions on society at large, different not only in number but in kind from the relatively quiet ripples they left at the surface of Chinese society. The social order in China remained generally stable under the limited impact of gunpowder, the compass, and so on, whereas "one cannot imagine the disappearance of feudalism in Europe without gunpowder, paper, printing, and the magnetic needle."[123]

Science and technology conflated and distinguished. At this point the reader may well wonder whether we are still engaged in a book about the *Scientific,* rather than about the *Industrial,* Revolution. Has not science been thoroughly mixed up with technology throughout our examples of transmission? It seems so, and indeed, we have arrived now at one of those highly confusing aspects of Needham's enterprise.

At first sight it may seem as if Needham makes no distinction whatsoever between science and technology. For example, it was entirely possible for him to write a paper dealing exclusively with technical invention under the title "Central Asia and the History of Science and Technology."[124] Here we see the two concepts really merge in his mind. At other occasions, which predominate in the great work but are quite rare in the accompanying essays, he keeps the two neatly distinct. At such occasions he recognizes that, in traditional societies, natural philosophies are largely ethnically bound structures of thought, liable to little transmission along the customary routes traveled by technical inventions.[125] Yet often, too (bridging the really incompatible notions in his head of actual merger on the one hand and radical separation on the other), Needham takes technology to be nothing but applied science: "One cannot separate science from technology, pure science from applied science—the two intertwine inextricably."[126]

Lynn White ascribed this (quite mistaken, in his view) equation of technology with applied science to the early imprinting Needham received at the hands of the only teacher in the history of science he ever had, the Cambridge historian of science, technology, and medicine Charles Singer.[127] I would like to suggest in addition that in China, as Needham has repeatedly emphasized, science was generally a much more practical activity than was the case in the Western tradition. This is precisely why many historians of science have denied that China ever had any science of its own: It was all technology, in their opinion. The volumes of *Science and Civilisation in China* testify that this is simply untrue, or at least not nearly true enough; yet it cannot be denied either that, if Needham had consistently kept the history of preindustrial technology out, as really belonging to a different cultural department, the series would, besides being of course seriously impoverished, also have turned out considerably smaller and thus, apparently in Needham's view, less convincing for making the point he has desired above all to make. Thus, given Needham's strong animosity to 'Westerners who do not believe in non-Western science', it has been ideologically important for him to keep science and technology in his treatment closely linked.

The results have, at times, been highly confusing. For example, although as a rule Needham defines his Grand Question as one of why the Scientific Revolution eluded China, in one essay he suddenly puts it as "why [the Chinese] did not succeed, as Europe's civilisation did, in giving rise to modern science and technology." And, a few lines further down, "Why was there no indigenous industrial revolution in China?"[128] No doubt the questions are related, since it is impossible to conceive of the Industrial Revolution without an antecedent Revolution in Science, yet overall quite different elements must be taken into account for answering the one or the other. (For example, the origin of the interconversion of alternating and rotary motion, however interesting if one considers the history of the steam engine, bears no relation whatsoever to the question of why the Scientific Revolution eluded China).

Also, throughout Needham's essays, conclusions inferred from technology only are often, through nothing but tacit extension, allowed to paint the picture for science as well. But the most damaging consequence of the effective conflation of science and technology in much of Needham's work lies, I think, in the problem area of the transmission of science, as if this were not already beset by sufficient pitfalls of its own. Lynn White even went so far as to conclude that, if one subtracts from Needham's manifold cases of premodern transmission all those that really concern technical inventions, nothing remains.[129] My own reading of Needham's work suggests that such is not quite the case. Still, quite divergent patterns of transmission can be culled from his work, with the net result that what one would like so much to possess—a grand historical survey of how much of Western science on the eve of the Scientific Revolution was due to non-Western sources—is still lacking. What bits and pieces of such a survey I have become aware of will now be presented very briefly in the following paragraphs, starting with an area to which Needham has contributed rather little.[130]

Islamic enrichments of Western science before the Scientific Revolution. The transmission of Islamic science to Europe may be seen as having occurred in three stages. The first stage consists of those Arabic works that were translated into Latin jointly with the Greek corpus in the 10th through 12th centuries. Works of authors such as Alhazen (Ibn al-Haytham), Avicenna (Ibn Sina), Averroes (Ibn Rushd), and many others became known in the West and considerably helped to shape natural philosophy in the age of scholasticism. Perhaps the greatest influence on the exact sciences in Europe was exerted by the algebraic techniques with which Islamic science (itself greatly enriched by Indian mathematics) had enriched the Greek legacy. The second stage is that quite limited number of elements from Islamic science that were passed on to Europe during subsequent centuries and that went into the making of European science on the eve of the Scientific Revolution. The third stage (which is not really one of transmission proper) began in the 19th century. It is the study, by Western historians of science, of the history of Islamic science. This has been undertaken by a relatively small number of scholars and is of course an ongoing concern, to which scholars in the Middle East have naturally contributed as well. Among the pioneers of cross-cultural research in this area may be mentioned Wiedemann, Sarton, Mieli, and Hartner. As we have seen, the great majority of the manuscript works that together make up Islamic science is 'still awaiting study'. In one sense, the third stage of 'transmission' has hardly yet begun. Our present interest, however, is focused rather on the second stage. The sole author I am aware of to seek a general perspective on the matter is Sayili, both in the conclusion to his book *The Observatory in Islam* and in an article from 1958, "Islam and the Rise of the Seventeenth Century Science."

Concerning the observatory Sayili's general conclusion is that "the observatory as an organized and specialized institution was born in Islam,"[131] where it underwent a considerable development of its own. The first of the two culmination points in its development was Maragheh observatory, which was founded with considerable support from the Mongol khan Hulagu and undoubtedly received Far Eastern influences. The second culmination point was Samarqand observatory in the fifteenth century. According to Sayili there are many indications that the fairly sudden advance in European mathematical astronomy made by Peurbach and Regiomontanus must be ascribed in part to the stimulus received from contemporary Islamic achievements centered in Samarqand.

Sayili is very careful not to infer transmission and influence from sheer preceding or simultaneous likeness of content. For example, although Ibn al-Nafîs' discovery of the circulation of the blood was translated into Latin in the sixteenth century, before fresh ideas on the topic appeared in Italy and England, Sayili does not consider the evidence for European borrowing conclusive. However, he does think that there is sufficient evidence for concluding that, besides the major feat of creating observatories in the first place, Islamic science contributed to European science on the eve of the Scientific Revolution in the following areas (all astronomical): (1) shifts in focus (both computational and observational) on certain parameters in Ptolemaean planetary theory that, in Europe, were to go into the making of the heliocentric theory; (2) instruments for astronomical observation (certain new instruments employed by Tycho may well go back to examples from the Muslim world); and (3) the tradition of *systematic,* rather than intermittent, observation of the heavenly bodies. All of which leads Sayili to conclude that

> though it is difficult to give conclusive proofs for each individual case, the cumulative impression gained leaves less room for doubt. . . . [Despite the decline of Islamic science] the claim can reasonably be made . . . that the positive contributions of Islam to the emergence of the seventeenth century science continued after the period of translations from Arabic had come to an end. There are reasons to believe, moreover, that this contribution of Islam was not trivial or negligible but was of some importance.[132]

Whatever the individual merits of the various claims making up such a modest conclusion, the net result that concerns us here appears to be that the common heritage in Greek science allowed a number of distinct influences from Islamic science to go more or less directly into the making of the Scientific Revolution in 16th and 17th century Europe.[133]

The Arabs between China and the West.   In a beautiful piece from 1948, "The Unity of Science: Asia's Indispensable Contribution," Needham put forward a most revealing thesis on the role of the Muslim world in the transmission of science from the West to the Far East and *vice versa.* We all know about the two large-scale efforts to translate treasures in science and philosophy: one—centered in Baghdad—from Greek into Arabic during the Golden Age of Islamic science and one from Arabic into Latin starting in the late 10th century in Spain. All this, so Needham asserts, can best be seen in terms of one, naturally heterogeneous yet in the end highly interdependent, system located around the Mediterranean, and the remarkable thing is that

> into this system . . . East Asian science was not incorporated. . . . It is not that there was no contact between Arabic civilisation and East Asian science; quite the contrary. But for some reason or other, when the translations were being made from Arabic into Latin, it was always the famous authors of Mediterranean antiquity who were chosen, and not the books of Islamic scholars concerning the science of India and China.[134]

He gives a number of examples, the most striking of which is al-Bîrûnî's grand survey of Indian science, written about 1012 and translated into a European language for the first time in 1888. There were also all kinds of personal contacts between Arabs and Chinese, for example, Ibn Battutah, who in the 14th century traveled to China and described what he saw there. Again, when in the middle of the 13th century Hulagu Khan allowed Nasir al-Din al-Tusi to erect an observatory at Maragheh, this led to a fairly intensive collaboration between Arab astronomers from all over the Muslim world, including Spain, and Chinese astronomers sent for by the khan. None of the results ever passed on to the West. Thus, Needham concludes,

> the examples which have been given show that there was no lack of contacts between Arabic and East Asian science, but it remains true that East Asian science did not filter through to the Franks and Latins, i.e. to precisely that part of the world where . . . modern science and technology were later to develop.[135]

What, then, did pass from China to the West despite the Arab filter?

Chinese enrichments of Western science before the Scientific Revolution. For starters, Needham maintains that, prior to the 17th century, China gave much more to the world than it took, both in technical invention and in scientific discovery. And this is not only true of possible influences from the West; it applies equally to contributions from India and the Arab world, to which China remained remarkably impervious. This is true even of Buddhist thought. Whereas the Buddhist religion penetrated deeply into the Chinese mind, the atomist speculations that emerge from its doctrine as one by-product left scarcely a trace in Chinese scientific thought, which, according to Needham, tends much more to Yin/Yang-like wave patterns. Despite many personal contacts between Arabs, Indians, and Chinese, Needham concludes that "evidence of a lasting influence either Indian or Western on Chinese science and medicine is hard to adduce."[136]

If we now reverse our viewpoint and look at what scientific discoveries passed from China to the West and thus went into the making of the Scientific Revolution, we must, first, cull the data from all over Needham's work and, next, filter out of the pertinent passages all that really bears on technology rather than on science. The net harvest appears to consist of five items: magnetism, alchemy, observational astronomy, cosmology, and the measurement of time. We take these items up in this order.[137]

Magnetism. This has been one of Needham's showpieces. The argument is typically made up of two parts: a demonstration that the Chinese were far in advance of the western Europeans in discovering key properties of magnetic substances and the subsequent claim that the latter owed at least their pre-Gilbert insights into magnetism to the former.

The point at issue, Needham maintains, is the directive properties of the lodestone, not its attractive capacity, of which the bare essentials were known in both civilizations.[138] The earliest ascertained mention of the magnetic compass in Europe is by Alexander Neckam c. 1190, whereas around 1080 Shen Kua clearly described two key properties of the magnetic needle: its constant direction southward and the existence of a slight declination to the east, subsequently measured at 15 degrees by K'ou Tsung-Shih in 1116. Earlier as well as subsequent texts refer to floating pieces of magnetized substance in the shape of a fish, which were made to serve as a magnetic compass. In short, "people in China were worrying about the declination before Europeans even knew of the polarity."[139] Now how did the Europeans find out about polarity and declination?

Since no contemporary reference to any magnetic phenomena except the sheer fact of attraction is known in the culture-areas of the Indian and the Muslim worlds, knowledge of the directive properties of magnetic substances, Needham concludes, was probably not passed on to Europe overseas (foreign mariners watching Chinese seafarers handling their compasses, for example) but along an overland route. Needham acknowledges that this seems odd, since, in order to avoid the Indians, the Byzantines, and the Muslim peoples, quite uncivilized northern tribes must be supposed to have passed on the knowledge. Nevertheless he accepts the conclusion, adding that much more research is needed on how, precisely, the Qara-K'itai may have accomplished this feat. He is even prepared to accept as an inescapable consequence that the marine application in Europe of the new knowledge about the magnet had taken place independently of the prior construction of a mariners' compass in China. It is striking that Needham, although apparently prepared to grant to the Europeans the ability to think of the application on board ship of a foreign discovery, does not appear to give a thought here to the one inference that, for the uncommitted reader, cries out to be made from all these data, namely, that no one ever passed on the Chinese knowledge of the directive properties of the magnet, but that these, too, were independently rediscovered in Europe. We shall see presently how it is that this commonsense solution is not entertained here by Needham.[140]

Let us grant transmission as a fact for the moment. Then "the priority in knowledge of polarity, induction, remanence, declination, etc.," on the side of the Chinese allows Needham thoroughly to exploit what this means for the non-Western contribution to the Scientific Revolution:

> Magnetical science was indeed an essential component of modern science. All the preparation for Peter of Maricourt, the greatest medieval student of the compass, and hence for the ideas of Gilbert and Kepler on the cosmic role of magnetism, had been Chinese. Gilbert thought that all heavenly motions were due to the magnetic powers of the heavenly bodies, and Kepler had the idea that gravitation must be something like magnetic attraction. The tendency of bodies to fall to the ground was explained by the idea that the earth was like an enormous magnet drawing things unto itself. The conception of a parallelism between gravity and magnetism was a vitally important part of the preparation for Isaac Newton. In the Newtonian synthesis gravitation was axiomatic, one might almost say, and spread throughout all space just as magnetic force would act across space with no obvious intermediation. Thus the ancient Chinese ideas of action at a distance were a very important part of the preparation for Newton through Gilbert and Kepler.[141]

Note how grandiosely the claim has now been expanded since its inception in Needham's usage of Shen Kua's few lines of text as his exciting starting point. Note also that what began as an argument on the *directive* properties of the magnet is carried to its ultimate conclusion through reference to those *attractive* properties that were studied earlier by Gilbert than by any Chinese scholar.

Alchemy. Very much has been made, in *Science and Civilisation in China,* of the Taoist origins of alchemy as a means to prolong life, the ultimate purpose of the alchemical work being to find the elixir of life. Here, Needham claims, lies the second "great extra of the scientific revolution":

> For chemists Paracelsus deserves to be thought of along with Galileo, and by his momentous statement that 'the business of alchemy is not to make gold but to prepare medicines' he was bringing into Europe the age-old Chinese belief in elixirs of life, ultimate source of all medical chemistry and chemotherapy.[142]

Regarding possible routes of transmission beyond the fact that the Arabs, too, indulged in a great deal of alchemy we find little but speculation. Also, in what precise sense Paracelsian iatrochemistry contributed to the Scientific Revolution is left wholly unexplained beyond the bold assertion that "the iatrochemical revolution of Paracelsus [was] hardly less important in the origins of modern science than the work of Galileo and Newton."[143]

Observational astronomy. Not only Paracelsus but also "Tycho Brahe . . .

was a patently Chinese figure."[144] What this odd phrase means is explained in another essay in a few more words:

> It was the father of modern observational astronomy, Tycho Brahe, who in the sixteenth century introduced both the Chinese practices, the equatorial mounting and the equatorial co-ordinates, into modern science, which has never since departed from them. His explicit reason was the greater instrumental accuracy, but he possessed Arabic astronomical books, and the Arabs knew well what the Chinese usage was.[145]

This, in its turn, is further explicated in *Science and Civilisation in China*, without, however, moving beyond the domain of more or less probable guesses.

Cosmology. On this topic Needham is also quite brief. As we shall see when discussing Chinese astronomy, the lack of a counterpart to mathematical planetary theory in the Ptolemaean tradition was made up for by the absence, in Chinese cosmology, of the notion of a fixed Earth at the center of a walled-in universe. The Earth, in the Chinese view, was taken to be flat; but Needham tends to play down this feature as much as he can. It does not, at any rate, prevent him from claiming in passing that "to the break-up of the naïve cosmology of medieval Christendom, seen still in Dante, Chinese influence indirectly contributed."[146] The conception of the universe shared by most established Chinese astronomers was one of infinite space, lighted by stars and planets of unknown substance floating through it. No fixed dates are set to the creation; no obsession with circular motion to be got rid of—small wonder, then, that

> their influence was a liberating one when Europeans were breaking forth from this prison. Whether any breath of it reached men such as Giordano Bruno and William Gilbert, who attacked the Ptolemaic–Aristotelian spheres before the end of the sixteenth century, we do not know, but it is quite sure that fifty years later European thinkers who were adopting Copernicanism and abandoning the spheres drew much encouragement from the knowledge that the wise astronomers of China (Europe's sinophile period was just beginning) had never had any use for them.[147]

Measurement of time. One potentially accurate way to measure time is to break up into countable units the equable outflow of liquid from a tank. The instrument built upon this principle is called the clepsydra. A quite sophisticated version was invented by Ktesibios around the second century B.C.; a much cruder version was employed by Galileo in measuring the time of descent in his experiments with bodies rolling over inclined planes. The mechanical clock that begins to appear all across Europe from the end of the 13th century onward is built upon other principles: by means of an intricate gearing mechanism and of an escapement controlled by, while also regulating, an oscillator, the downward motion of a suspended weight is made both uniform and countable. The appearance of the mechanical clock in the known source material is quite sudden, and no obvious forerunner operating along similar principles has been found.

Roughly such was the state of the historiography of time measurement when Joseph Needham appeared on its scene. In the late fifties, inspired by an idea of Derek de Solla Price, Needham discovered Chinese texts dating from the end of the 11th century with descriptions of the contemporary construction of an incredibly elaborate water clock that had been employed as the timepiece of a huge armillary sphere. Reconstruction in accordance with the extensive description confirmed that Su Sung's clock must have measured time much more accurately than was accomplished by any mechanical clock before Huygens' invention of the pendulum clock. Also, Needham and his collaborators were able to trace several earlier versions of Su Sung's clock down to the 1st century A.D. in the history of China.

This was obviously a major find in its own right, and it has accordingly been hailed as such by historians. However, for Needham more was at stake. A link had always been missing in the evolution from the clepsydra to the mechanical clock, he claimed, and Su Sung's clock now provided it. Here was an intermediate form of clock, based indeed upon the even outflow of water but also fitted out with an escapement, that constituted the missing link in the evolutionary line. "The Chinese hydro-mechanical clock thus bridges the gap between the clepsydra and the weight- or spring-driven clock."[148] What must have happened is that knowledge about Su Sung's clock was transmitted to Europe. In passing on either the construction of the escapement itself or at the very least "the knowledge that the problem of mechanical time-keeping had in principle been solved,"[149] Su Sung's clock set dynamical Europe on the road toward the final elaboration of the mechanical clock.

This, in the briefest of outlines, is the story enthusiastically set forth by Needham in *Heavenly Clockwork* of 1960 (written together with de Solla Price and Wang Ling) and frequently retold since with no more than slight adaptations that leave the claim fully intact.[150] Again, Needham has not failed to spell out its implications for the Scientific Revolution:

> The escapement was the first great achievement in the control of power. The mechanical clock, so familiar to all of us today, was truly at its birth a cardinal triumph of human ingenuity; it constituted perhaps the greatest tool of the Scientific Revolution of the seventeenth century, it trained the craftsmen who were needed for making the apparatus of modern experimental technique, and it furnished a philosophical model for the world picture which grew up on the basis of the 'analogy of mechanism'. . . .
>
> . . . When to all this one adds the simple fact that the measurement of time is one of the handful of absolutely indispensable tools of modern science, it can be seen that I-Hsing and Su Sung started something.[151]

The onus of proof. These, then, are Needham's five claims for significant contributions made by the Chinese to European science on the eve, or during the first stages, of the Scientific Revolution. No doubt the most glaring flaw in the argument has become clear by now. As always, Needham has himself generously laid the flaw bare without hedging about it: "The details of any transmission are still obscure," he freely grants in a passage to be found in *The Grand Titration.*[152] Only, Needham himself does not see the fact that no historical source has so far revealed any detail whatsoever of the ways of transmission as a flaw at all. Unlike Sayili, Needham thinks that in cases such as these *the burden of proof rests on those who wish to deny transmission.*

Affirmanti incumbit probatio, Roman law has it (upon him who makes an assertion it is incumbent to prove it). But Needham maintains that

> as in all other fields of science and technology the onus of proof lies upon those who wish to maintain fully independent invention, and the longer the period elapsing between the successive appearances of a discovery or invention in two or more cultures concerned, the heavier that onus generally is.[153]

In other words, cross-cultural transmission is supposed to be the 'natural' thing, and independent discovery or rediscovery the aberration that has to be proved. It is true that there are many cases of early Chinese discoveries and inventions which Needham feels sure have indeed been made, or remade, independently in Europe, for instance, the seismograph, the differential gear, and steel-making.[154] Only, it remains unclear what general criterion allows Needham to decide for or against transmission, since in many cases the absence of data on ways of transmission is not accepted by him as a criterion at all.

Although it is true that, if no transmission took place, there will necessarily be no source material to indicate transmission, the absence of source material need not necessarily *prove* independent rediscovery, even though it would seem to count against transmission rather than for it. Thus the two possibilities are not logically symmetrical. But since one can argue endlessly over whether the affirmation that stands in need of proof is transmission or independent rediscovery without ever getting anywhere, it would seem better to seek another criterion if one wishes to say anything at all about these really quite doubtful yet highly important matters. For the sole case of the mechanical clock David Landes, in his *Revolution in Time,* has employed such a rival criterion, namely, into what historical pattern the undisputed facts of the case fit best.[155]

The alternative pattern sketched by Landes comes down to the following. It is highly misleading, and indicative of the intellectual disasters wrought by looking at history in Needham-like 'evolutionary' terms, to regard Su Sung's timepiece as a 'missing link' between the clepsydra and the mechanical clock. Rather, it should be seen as the culminating effort of what was essentially "a magnificent dead end."[156] The clepsydra principle allowed of a relatively high degree of accuracy in measuring time, almost certainly higher than what the mechanical clock could achieve prior to Huygens' decisive invention. Su Sung's marvelous ingenuity managed to squeeze the absolute maximum out of what the clepsydra principle is inherently capable of. Owing to many destabilizing factors, though (e.g., corrosion and dirt, or the viscosity of the liquid changing with temperature), it cannot maintain its short-term precision in the long run. As against this, the mechanical clock that was invented during the European Middle Ages was inherently capable of far greater precision in both the short and the long run, even though this did not become manifest until centuries after Su Sung's clock had been erected—and, significantly, destroyed and forgotten. What we are facing here, then, is two different civilizations trying out with equal ingenuity two fundamentally different principles of time measurement, one of which happened to possess far greater developmental possibilities than the other.

Landes' brilliant argument naturally goes a great deal deeper into the technicalities of clock-making than serves our purposes here (the principal point of contention being whether Su Sung's 'escapement' has more in common with its European counterpart than an opportunity to apply the same term to both).[157] The important point for us is that his mode of arguing can be generalized far beyond the issue of the measurement of time and that, together with certain sustaining chains in his argument, it is to inform deeply the conclusion we are to reach at the end of the present chapter on what Needham has to say about the causes of the Scientific Revolution and why it eluded China.

But there is also a conclusion in store for our present issue of precisely what Eastern contributions went into the making of the Scientific Revolution.

A turnabout and a conclusion. We have found Needham's five cases of Chinese contributions to the emergence of early modern science heavily flawed on several counts, of which the most important are the absence of sources that even begin to point at transmission and the consistent aggrandizement of what Chinese finds, even if passed on indeed to western Europe, may have meant for early modern science in the best of cases. The Needham we have seen making these grandiose but generally ill-founded claims is the self-confessed 'preacher' of a message he feels strongly he has to teach the West:

> Everyone who undertakes a big inter-cultural job like this must naturally project his own system of beliefs in doing so—it is his opportunity to preach (and I use the word quite advisedly) to his own and later generations. If sometimes we have written like barristers pleading a case, or sometimes over-emphasize the Chinese contributions, it has been consciously to redress a balance which in the past tilted over much too far on the other side. We were out to redress a secular injustice and misunderstanding.[158]

What makes Needham so endlessly intriguing a thinker is that, besides the posture described here so frankly, there is also another Needham, one who knows very well that he has been wildly overstating his case. This other Needham, who takes the floor throughout large portions of the main work, could alternate his bolder claims with sober statements like the following:

> Western science . . . developed on the whole without the benefit of either Indian or Chinese contributions.
>
> [Also:] The mutual incomprehensibility of the ethnically-bound concept systems did severely restrict possible contacts and transmissions in the realm of scientific ideas. This is why technological elements spread widely through the length and breadth of the Old World, while scientific elements for the most part failed to do so.
>
> [And even more strongly elsewhere:] The sciences of the mediaeval world were in fact tied closely to their ethnic environment, and it was difficult if not impossible for people of those different environments to find any common basis of discourse. For example, if Chang Hêng had tried to talk to Vitruvius about the Yin and the Yang or the Five Elements, he would not have got very far, even if they could have understood each other at all.[159]

Is not this a much more plausible conception of how civilizations interact when truly sophisticated and intricate structures of thought are at issue? And does this not underscore the conclusion to which the present section inexorably seems to lead—a conclusion, curiously, not all that different from the one at which the 'other' Needham himself arrived at the end of volume 1 of *Science and Civilisation in China*? This conclusion is that, although much about the transmission of science in the premodern world is still obscure, it looks as if only the closely related scientific systems of the Muslim world (enriched in its turn by Hindu mathematics) and of western Europe achieved fruitful interaction: 'closely related', because of the common basis in the Greek legacy, and 'fruitful', in that scientific ideas emerging in the Muslim world indeed enriched to some extent the Greek legacy both in the course of and after its initial reception in the West. The problem of Greek stagnation cannot therefore be exhaustively defined as the problem of how it was that the eventually stagnant scientific corpus of ancient Greece gave rise to the Scientific Revolution right after it had passed to western Europe, for the fact is that important additions had come to be made to it in the meantime. Yet those additions seem to have come almost exclusively from the Muslim world, the West's elder stepbrother in science. To distant China's science the West does not seem to owe much: partly because of the 'translation filter', partly as a result of the incommensurability of the natural philosophies of China and the West. No better demonstration of how wide the gap was than the fact that a 12th-century translation into Chinese of fifteen chapters of Euclid's *Elements,* arranged by Nasir al-Din, is

reported by Needham to have rested on the shelves in the Imperial Library.[160]
It was resting there very quietly, with no one aware of its explosive potential—
one of those haunting time bombs of history that somehow never ignited.

6.5. WHY THE SCIENTIFIC REVOLUTION ELUDED CHINA

Having thus established that the comparison between Chinese and Western
science of premodern times really pits against one another two by and large
independent traditions, expressive of quite independent modes of coming to
grips with the world and our place in it, we carry out the comparison in three
stages. We start with an outline of parallels and contrasts between Chinese and
Western science as interpreted by Needham. Next we address the manifold
questions and answers that together constitute Needham's stance on his Grand
Question. Finally, we point at some alternative conceptions suggested by some
of Needham's many critics. A concluding section rounds off the discussion.

6.5.1. Chinese and Western Science: Parallels and Contrasts

One, perhaps the major, contrast between Chinese and Western science, so
Needham asserts, is the mostly practical nature of large portions of the former.
The Damascene scholar al-Jahiz already knew this well (around 830): "The
curious thing is that the Greeks are interested in theory but do not bother about
practice, whereas the Chinese are very interested in practice and do not bother
much about the theory."[161] This is not to say, Needham cautions, that China
lacked philosophical conceptions dealing with the universe at large, such as
the neo-Confucian synthesis of the eleventh and twelfth centuries. Yet the Chinese were not system-builders like the Greeks. "When they strive only to 'understand the high' without 'studying the low', how can their understanding of
the high be right?" Ch'eng Ming-Tao asked in the 11th century.[162] The remark
is directed against the Buddhists, but it serves equally well as a summing-up
of a significant contrast between the Chinese approach and the attitude characteristic of the West up to the time of the Scientific Revolution.

In the history of Chinese thought one finds a certain split that has no counterpart in the West. On the one hand, there was, in the Taoist tradition, much
empirical observation of natural phenomena; on the other hand, logical and
rational thought were largely confined to the problem of how to achieve and
maintain social order and cohesion—this being the preoccupation of the Confucians and the Legalists, neither of which schools had any interest in nature.
Such a gulf, Needham argues, "is not found to anything like the same extent
in European history."[163] The only school of Chinese thought to combine rationalized thought with an interest in nature, the Mohist school, dropped out of the
history of China at an early date. The members of this school were followers

of Mo Ti (5th century B.C.), a defector from early Confucianism who proclaimed a message of utilitarianism and universal love. The Mohist Canons and Expositions, which date from about 300 B.C., contain much of interest about causality, time and space, phenomena of light, and so on. The Mohists applied "logical processes . . . to zoological classification and to the elements of mechanics and optics."[164] According to Needham, these Mohist texts come closer than any other in the history of Chinese thought to the spirit of Western science, and it remains a key problem in the history of Chinese science to understand why the Mohist school never survived its original suppression, along with so many other schools of thought, when China was united for the first time under the Ch'in:

> The Mohists had no deductive geometry (though they might have developed one), and certainly no Galilean physics, but their statements often give a more modern impression than those of most of the Greeks. How it was that their school did not develop in later Chinese society is one of the great questions which only a sociology of science can answer.[165]

Chinese astronomy, mathematics, and physics. Passing on now from these general observations to those sciences which, in the West, were to constitute the backbone of the Scientific Revolution, we find that astronomy in China was put on a radically different footing from the Western variety—different indeed, but neither opposed to, nor incompatible with, one another. "While Greek astronomy had always been ecliptic, planetary, angular, true and annual, Chinese astronomy had always been polar, equatorial, horary, mean and diurnal."[166] On this basis a set of highly accurate scientific instruments was developed, partly for observation and partly for emulating in a model the heavenly motions. Among the latter armillary spheres stand out.

One of the remarkable things about Chinese astronomy is that it managed to achieve a measure of precision more than comparable with its Western counterpart in the absence of something like Euclidean geometry. Isolated geometrical rules, such as Pythagoras' theorem, had been known of old and had also been provided with ingenious proofs; yet no deductive geometry in the Western sense was built upon such groundwork. Chinese mathematics rather focused on arithmetical and algebraic procedures.[167]

In both astronomy and mathematics, then, one can say that the strengths and weaknesses of Chinese and Western science were complementary rather than opposite: "the attention of Chinese and Europeans had been concentrated upon different aspects of Nature."[168] The same is true of physics. Here the strong points of the Chinese were optics (in the Mohist tradition), acoustics, and magnetism (in connection with the invention of the compass needle, as we have seen), whereas scarcely any kinematics or dynamics is to be encountered in the Chinese tradition. Atomist conceptions were foreign to the Chinese, whose thinking in terms of the complementary polarities of Yin and Yang rather predisposed them, in Needham's view, toward field theories where action at a distance is entirely natural rather than in need of further explanation in terms of material particles.

We may conclude at this point with a remark that, as far as I have been able to establish, is nowhere made by Needham in quite so outspoken a fashion. It is very simply that the two key scientific issues around which the Scientific Revolution largely materialized, namely, the interconnected problems of the orbits of the planets and of free fall and projectile motion on Earth, thus could scarcely have arisen as significant problems in any hypothetical further development of the science of traditional China.[169]

The Chinese scientist in society. In a typology of inventors' positions in Chinese society which Needham presents on half a dozen pages of *The Grand Titration,* it is almost impossible to sort scientists out from pioneers in technical invention.[170] I present it here as put forth by Needham, although rearranged in descending order of social prominence.

Princes. Relatives of the emperor received a good education, yet they were not eligible for service in the bureaucracy. Wealthy, and with plenty of leisure, they were in a good position to foster science and technology if they wished. Needham adduces one prince, under the Han dynasty, who indeed served as a patron to astronomers, alchemists, and naturalists. He also mentions two later princes who made some notable inventions.

High officials. Most examples mentioned by Needham were really inventors rather than scientists, with two exceptions. The first is Chang Hêng, who in the 2nd century A.D. rose to the position of president of the Imperial Chancellery. "[He] was not only the inventor of the first seismograph in any civilization but the first to apply motive power to the rotation of astronomical instruments, one of the outstanding mathematicians of his time, and the father-figure in the design of armillary spheres."[171] The other is Shen Kua, "one of the greatest scientific minds in Chinese history,"[172] the man of the magnet, who served as an ambassador and as an elder statesman, and assembled around him a group of plebeian technicians of whom he appears to have been the patron.

Specialized bureaucrats. Under this category fall the astronomers, who were civil servants in a specialized bureau, often with premises in the imperial palace. It is well known that the role of the emperor as the pivotal point linking the favor of Heaven with the welfare of the people at large had everything to do with the principal drive toward accuracy in the description of the heavenly motions.

Minor officials. Many highly creative inventors, although qualified for service in the bureaucracy, never made it into the higher echelons of officialdom. Needham strongly suggests that this had much to do with the huge gap between them and "the sophisticated scholars nursed in the classical literary traditions." In the absence of a common language between the *literati* and the technically minded, the latter's status was bound to remain lowly.[173]

Commoners, master craftsmen, artisans. Here Needham locates the largest group of inventors; on this level, too, no mention is made of contributions to science.

Men of semiservile status. This category contains just one case, called indeed 'exceptional'. He is the 6th-century Hsintu Fang, who was a client to two successive aristocratic patrons, for one of whom (the owner of a large collection of scientific instruments) he is said to have written several scientific books.

Slaves. Here again just one example is given, of a man (Kêng Hsün) who was enslaved after rebellion. He appeared to be a gifted maker of armillary spheres and was finally freed and appointed to a low position in the bureaucracy.

Much of the evidence here assembled is anecdotal rather than satisfying minimal requirements of a true historical sociology of professions. To the extent that any conclusion seems at all warranted, it must be that, as far as science is concerned, its cultivation remained reserved for members of the higher mandarinate, with three significant exceptions: the astronomers, who formed a specialized agency inside the bureaucracy; the Taoist alchemists; and the makers of scientific instruments, who could be found in just about every social layer. This means in its turn that, unlike the situation in ancient Alexandria and in Muslim civilization, scientists were not dependents belonging to the entourage of a court but rather courtiers and officials themselves. In contrast, patron–client systems were relatively common where technical invention is concerned. A vast and rarely bridged social chasm appears to have separated officialdom from the lowly inventors.

An overall characteristic of the Chinese approach to science. The practical orientation of much Chinese science that we have discussed above "did not," Needham cautions, "imply an easily satisfied mind."[174] What did satisfy the minds of the Chinese in science did not follow the lines of Western thought to the extent that this was shaped by the uniquely Greek tradition of deductive geometry. Rather, their approach to science may be characterized by the following strong points:

Classification of natural phenomena. At a scientifically rather sterile level, classification might take the guise of the "pigeon-holing [of] novelty" in the symbolic filing system that over the centuries had grown out of the textbook of divination, the *I Ching* (Classic of Changes).[175] The doctrines of Yin/Yang and of the Five Elements (really processes or states of matter rather than substances) gave occasion to very intricate, associative classificatory systems which display key characteristics of Chinese thought (more on this in the next section). Particularly fruitful scientific categorization, however, took place in astronomy (star catalogues, parhelic phenomena) and in the organic sciences (pharmacopeias, classification of diseases, etc.).

Scientific instruments. Here traditional China, with its refined armillary spheres (to which the Chinese conventions in positional astronomy lent themselves much better than those of premodern Europe), observatory equipment, seismographs, and so on was far in advance over medieval Europe.

Observation. The accumulation of data on a wide variety of subjects and the subsequent recording of these were carried out "with a persistence hardly paralleled elsewhere."[176] For example, the body of Chinese observations on novae, comets, and meteors, sustained over centuries, still serves as a useful source for present-day radio astronomers. Under the category of painstaking, accurate observation fall also such feats as the discovery of the hexagonal shape of the snowflake some 1800 years before Kepler devoted his famous New Year's letter to the subject, and the observation of sunspots about 1700 years before this topic came up for discussion between Galileo and his Jesuit colleagues. Particularly remarkable is the degree of *precision* often reached in observations. One example is the early measurement of the declination of the magnet, originally in a geomantic, and later also in a navigational, context.

Experiments. It is true, Needham allows, that Chinese scientists "failed (like all medieval men, Europeans included) to apply hypotheses of modern type, [yet] they experimented century after century obtaining results which they could repeat at will."[177] Examples of this are trials on resonance in bells and strings and pharmaceutical experiments on animals carried out by alchemists. It is in domains like these that the Chinese came closest to the threshold separating traditional from early modern science. True, "none of [their experiments were] carried out on isolated and simplified objects," but neither were those of the Europeans before Galileo. Hence, Needham does not "propose to claim this honour for the medieval Chinese [even though they] came just as near it theoretically, and in practice often went beyond European achievements."[178]

If one considers the list of features of Chinese science just outlined, and focuses on the last-mentioned item, in particular, one may well wonder why no breakthrough on lines comparable to the Scientific Revolution was the next stage for a civilization that apparently had advanced thus far in science—farther, in one sense, than any other medieval civilization at least prior to the 15th century. And indeed, it was not until this surprising stage of advancement finally dawned upon Joseph Needham at some moment in the late forties or early fifties that he arrived at the definitive formulation of the Grand Question that had begun to exercise its spell on him more than ten years earlier.

6.5.2. Needham's Key Questions and How He Has Answered Them

The question put to Needham in 1937 was, it may be recalled, why modern science originated only in Europe. Needham has consistently transformed this question into one variant or another of "the problem of the origin of modern science in Semitic-Occidental (Euro-American) civilization on the one hand, and its failure to arise in Chinese civilization on the other" (1944) or (with two significant additions) "why modern science and technology developed in Europe and not in Asia" (1960).[179]

Once he had fully immersed himself in traditional Chinese science and technology and became more convinced every day of their richness, an additional question began to take shape in his mind, a question which further enhanced the paradoxical nature of the problem. "The more you know about Chinese civilization, the more odd it seems that modern science and technology did not develop there,"[180] and thus his second problem became: "Why was it that between the second century BC and the sixteenth century AD East Asian culture was much *more* efficient than the European West in applying human knowledge of Nature to useful purposes?"[181] How far did this superiority go? Needham surely thinks that, on the eve of the Scientific Revolution, China had a considerable edge over Europe in technology ('applied science'). More often than not he has flatly asserted that this was true in 'pure' science, too,[182] but on rare occasions he appears to think rather that the conceptual structures of science in China and in the West were of roughly equal complexity and depth, with the one stronger on this point and the other on that, but with no clear edge for either.

Leaving technology out of the picture once again as much as we can, we thus find ourselves addressing something like the following question: How is it that the science of traditional China, on a par, to say the least, in brilliance and depth with the West prior to the Scientific Revolution, nevertheless did not spontaneously produce early modern science? If I have counted well, Needham has rejected four possible answers to this question and has come up with six answers which he fully endorses. It would be a mistake, though, to believe that Needham came to his answers with a blank mind where the nature and causes of the Scientific Revolution itself are concerned. Needham's views on the Scientific Revolution that did occur in Europe have deeply shaped his answers to the question of why it did not occur in China. Therefore, we have to examine these prior conceptions first.

Needham's conception of what the Scientific Revolution was about. Needham does not appear to have made an independent study of the Scientific Revolution. His views represent a highly selective precipitate of assorted portions of the literature on the subject. Yet the unique Needham tinge is there to color it all, too.

The most extensive description of what the Scientific Revolution was about that Needham has given is the sharpest and most comprehensive definition I know of within the space of less than one page:

> When we say that modern science developed only in Western Europe at the time
> of Galileo in the late Renaissance, we mean surely that there and then alone there

developed the fundamental bases of the structure of the natural sciences as we have them today, namely the application of mathematical hypotheses to Nature, the full understanding and use of the experimental method, the distinction between primary and secondary qualities, the geometrisation of space, and the acceptance of the mechanical model of reality. *Hypotheses of primitive or medieval type distinguish themselves quite clearly from those of modern type. Their intrinsic and essential vagueness always made them incapable of proof or disproof,* and they were prone to combine in fanciful systems of gnostic correlation. In so far as numerical figures entered into them, numbers were manipulated in forms of 'numerology' or number-mysticism constructed *a priori,* not employed as the stuff of quantitative measurements compared *a posteriori.* We know the primitive and medieval Western scientific theories, the four Aristotelian elements, the four Galenic humours, the doctrines of pneumatic physiology and pathology, the sympathies and antipathies of Alexandrian proto-chemistry, the *tria prima* of the alchemists, and the natural philosophies of the Kabbala. We tend to know less well the corresponding theories of other civilizations, for instance the Chinese theory of the two fundamental forces Yin and Yang, or that of the five elements, or the elaborate system of the symbolic correlations. *In the West Leonardo da Vinci, with all his brilliant inventive genius, still inhabited this world; Galileo broke through its walls.* [My italics][183]

Two things seem to me particularly remarkable about this passage. One is its striking correspondence (especially in the italicized sentences) to Koyré's conception of the Scientific Revolution as the starting point of transition from the 'world of the more-or-less' to the 'universe of precision' (see section 2.3.3).[184] Needham's way to catch the opposition has been to say "that Chinese science and technology remained until late times essentially Vincian, and that the Galilean break-through occurred only in the West."[185] In this manner Koyré's distinction is expanded beyond premodern Europe alone to apply to traditional society generally—a move fraught with fertile possibilities, and to which we shall return.

Further, here as elsewhere Needham has consistently marked Galileo's life's work as the true beginning of the Scientific Revolution.[186] But this focus on Galileo by no means implies that the Scientific Revolution constituted much *more* than the point where the breakthrough to modern science occurred. Needham phrased the matter in his own, inimitable way:

For anyone who is a mathematician and a physicist, perhaps a Cartesian, this may not be welcome; but I myself am professionally a biologist and a chemist, more than half a Baconian, and I therefore do *not* think that what constituted the spearhead of the Galilean break-through constitutes the whole of science. What happened to crystallize the mathematization of experimental hypotheses when the social conditions were favourable does not exhaust the essence.[187]

445

The Scientific Revolution as 'the spearhead, but not the whole, of science': such a description, in its turn, does not exhaust Needham's conception of the event. In his mind, the Scientific Revolution is inextricably bound up with the Renaissance and with the Reformation, these three events forming together the peak landmarks in the underlying process of the dissolution of feudalism and the rise of capitalism from the 15th through the 18th centuries. This entire historical process must be seen, he has insisted, as "a kind of organic whole, a packet of change, the analysis of which [so he wrote in 1963] has hardly begun."[188] This view of things is very close to that of his great friend and colleague Bernal (section 3.6.2) and was probably worked out in close collaboration with him. Inevitably, it has far-reaching consequences for the *explanation* of the Scientific Revolution.

Needham's explanations of the Scientific Revolution.   Needham appears to believe that we still know very little of how this complex of changes—the rise of capitalism with its three principal landmarks—occurred. But this has not prevented him from coming up with explanations of the one landmark that holds his abiding interest. These explanations go back to those of Zilsel, although more than a tinge of Hessen is to be detected here and there, too (discussed in the present book in sections 5.2.2 and 5.2.4). Needham readily granted (in 1964) that the 'hypotheses' in Hessen's paper had been "relatively crude," yet (so he disarmingly added) "that was surely no reason why they should not have been refined" rather than condemned.[189] Such refinement, then, he found in Zilsel's articles. Here he discovered what he thought necessary above all: explanations which

> in no way neglect the importance of a multitude of factors in the realm of ideas—language and logic, religion and philosophy, theology, music, humanitarianism, attitudes to time and change—but [which] will be most deeply concerned with the analysis of the society in question, its patterns, its urges, its needs, its transformations.[190]

One key element in Needham's explanation of the Scientific Revolution is the support science received from the early merchant-adventurers and "from the princes who supported them, and based their power on them."[191] Here we meet our familiar (section 5.2.4) alleged connection of merchants' interest in a quantitative economics with the rise of a quantified natural science. The other key element in the explanation is the social distance between the scholar and the artisan, which, Needham maintains, can be found all through history, not to be overcome until the democratic undercurrent of early capitalism put an end to it: "The scholar, from being the support of the king, becomes the comrade of the artisan."[192] What capitalism did, was to bring forth both modern science and democracy, and the one cannot thrive without the other, since science itself is a deeply democratic enterprise.

These are extremely crude notions indeed—bleak colorings within a facile schema of simplified-Marxist origins. They were put on paper in 1944, and I am surely not claiming that Needham has failed to refine his pertinent views considerably in later years. However, they serve the historian of Needham's thought in that they reveal the mind-set with which he began to approach his Grand Question—a mind-set that has become much more sophisticated over the years but is in essence still there. For the crux of the matter is that Needham's first answers to his question of why China did not undergo a Scientific Revolution took, without exception, the form of mirror images of his views on why Europe did undergo one.

Mirror images of the fate of science in European and in Chinese society. The process of mirror-imaging in Needham's mind is palpable in the following quotation from 1947, which sums it up in a nutshell:

> To ask why modern science and technology developed in our society and not in China is the same thing as to ask why capitalism did not arise in China, why was there no Renaissance, no Reformation, none of those epoch-making phenomena of that great transition period of the fifteenth to the eighteenth centuries.[193]

But this means that we seem now to have worked our way into a terrible quandary. In our historiographical survey we have come to the question of why the Scientific Revolution eluded China because we were less than fully satisfied with the sum total of explanations historians have put forward of how early modern science arose in western Europe (section 5.4). Joseph Needham's aim has pointed in the exactly opposite direction: Feeling fully secure about what caused the Scientific Revolution, he wanted to know why it had not taken place in China (although he once said that "for us the Chinese experience through the ages is a control experiment for Europe").[194] Thus it may be asked how, given what we are aiming at, we can learn anything at all from the solutions to the Grand Question Needham has come up with.

Well, to the extent that these remained sheer mirror images we cannot indeed learn anything useful from them for our purposes. But Needham did not remain stuck in that stage of his thinking forever. Therefore the full answer to our quandary is that comparative history, whether one wants it to or not, works both ways. Needham's struggle with his Grand Question, so I can already assure the patient reader, has yielded, as one mostly unintended by-product, some gems of insight into what may have caused the Scientific Revolution to take place in Europe.

I begin with taking stock of Needham's early mirror images.

Democracy. Having established to his own satisfaction that early modern science could only emerge in a democratic environment and that early capitalism provided it, Needham gave himself a fairly easy time (we are still in 1944) in showing that the Chinese social order, which he calls 'feudal bureaucracy', precluded a comparably democratic development. True, the mandarinate contained some democratic elements in that office was not hereditary and that the road to officialdom stood open (in principle) for every talent irrespective of social origin, if only one made it through the examinations. Nevertheless, "that particular sort of democracy associated with the rise of the merchants to power, that revolutionary democracy associated with the consciousness of technological change, that Christian, individualistic and representative democracy with all its agitating activity . . .—that China never knew until our own day."[195]

Confucianism versus Taoism.   As everybody knows, Confucianism is above all a social philosophy, espousing principally that social order and cohesion are to be maintained, not so much by law imposed from above, but rather by a 'natural' social ethics stamped by *li* in the sense of proper ritual. Neither Confucius himself nor his later adherents displayed an interest in the workings of nature. In a paper dating originally from 1947,[196] Needham pitted antinaturalist Confucianism, associated by him with social philosophy, with rationalism, with textual scholarship, with feudal bureaucratism, with the established official, and with maleness, against Taoism, with its favorable impulses for science and technology, with its receptivity and lack of prejudice, with its focus on the arts and crafts, with its pre- and antifeudal collectivism, with its revolutionary political impulses, and with its worship of the feminine. Taoism is painted here as the perennial opposition movement in China, forever trying to undermine the supremacy of Confucianist bureaucracy: "For two thousand years the Taoists fought a collectivist holding action, only to be justified by the coming of socialism in our own time."[197] In this early paper Needham does little more than illustrate the science-mindedness of Taoism with a few bits and pieces from the great Taoist authors (I have indicated at the beginning of the present chapter how dubious these seem to be from the sinologist's point of view), projecting upon them the 'progressive' stance of scientists of 17th-century Europe. The Taoists, he asserts, unlike the Confucians, truly cared for nature, and that is why Galileo's adversaries, who refused to look through his telescope, displayed "a very Confucian attitude [whereas] Galileo corresponded rather to the Taoists, who had an attitude of humility towards Nature and were anxious to observe without pre-conceptions."[198] At this stage of Needham's thought the picture is straightforward: Taoism stands for everything that Needham considers progressive, both in science and in politics, whereas Confucianism represents everything that works against progress. In the feudal-bureaucratic structure of Chinese society, however, unlike Europe where feudalism gave way to capitalism, Confucianism prevailed, so that the route toward progress as embodied in Taoism was blocked. Therefore the answer to the Grand Question is, ultimately, to be found once more in the different social structures of China and Europe.

Clearly, Taoism is scarcely being examined here as a movement of thought in its own right. Rather it functions as an entity projected from Europe upon China with some local veneer added. Over the years, however, Needham has become much more knowledgeable about the School of the Way and what its adherents did for science.[199] Taoism is surely the philosophy that appeals most to him personally; he even entitled an autobiographical piece "The Making of an Honorary Taoist." What he likes best about Taoism is its receptivity toward phenomena; its attitude of "going with the grain of nature"; its refusal to coerce man or nature; its doctrine of *wu wei,* or nonaction (or, in Needham's preferred translation, "no action contrary to Nature"),[200] in the sense of not trying to force the course of events but flowing with them; its streak of cheerful fatalism. Taoism appeals both to his religious sentiments and to his scientist's mind, and thus he could quote with evident approval the Chinese philosopher Fêng Yu-Lan, who stated that "Taoist philosophy is the only system of mysticism which the world has ever seen which is not fundamentally anti-scientific."[201]

Yet Needham has become aware in the meantime that, although "the Tao, the cosmic order in all things, [does] work according to measure and rule, [nevertheless] the tendency of the Taoists was to regard it as, for the theoretical intellect, inscrutable."[202] The consequences of seeking a *tao,* or natural order, that cannot be found were twofold. Science was indeed fostered in the School of the Way, but only science with a strongly empirical, observational bent. Moreover, an early Taoist belief in the possibility of achieving immortality led to the development of a very elaborate alchemy aiming at the discovery of the elixir of life. Taoism surely embodied a great deal of superstition and magic, but this, too, worked both ways: It hampered a scientific conception of things, but at the same time it fostered the activist stance toward nature that often goes with magic.

Early on in his career as a Chinese scholar, as we have seen, Needham thought that, given a more favorable socioeconomic setting, the Taoist school might have produced a Galileo.[203] It seems unlikely that he still believes this (although he has never repudiated his early papers), and altogether his answer to the Grand Question that is embodied in the opposition between Confucianism and Taoism has quietly receded into the background.

No early modern science without merchants.   In a paper dating from 1953 Needham focused upon another aspect of China's social structure: the position, and the mentality, of the merchant.[204]

We have already come across the term 'feudal bureaucracy' a couple of times. Its usage by Needham goes back to a conceptual apparatus forged by the sinologist Karl August Wittfogel in his *Wirtschaft und Gesellschaft Chinas* (1931; China's Economy and Society).[205] The key idea is that it is highly misleading to equate China's socioeconomic and political organization with western European feudalism. Rather, China's one-man rule, with a huge bureaucracy lording it over an impotent peasantry permanently available for *corvée* labor to build the irrigation works and keep them going, ought to be given a characterization of its own, and 'feudal bureaucracy' (later generalized and expanded by Wittfogel into 'Oriental despotism' coupled with 'the Asian mode of production') is the concept Wittfogel came up with. Needham found it substantially endorsed by Chinese scholars during his visits to China. One recurrent theme in the history of China, so Needham goes on, is that the bureaucrats invariably succeeded in preventing the merchants from becoming a social group with an autonomous power base in society. Although there was much commerce and, hence, great wealth all across the realm (as everybody knows who has ever glanced at Marco Polo's travel account), the bureaucracy took care to keep the trade in such vital products as salt and iron in its own hands; to levy very heavy taxes on commercial gains, alternated with arbitrary expropriations to prevent the ongoing accumulation of wealth; and, last but not least, to despise the merchant as a socially inferior person. The merchants even internalized this sentiment, as they themselves believed that there was no higher career for their sons to aspire to than to become an official rather than the successor to their fathers' business.[206] As a result there was no counterpart in China to the 'liberating city air', or the 'citizen's legal security', of the townships of medieval Europe. Each city was governed without any restraint in the name of the emperor by one civil and one military prefect representing him.

All this meant that the merchant class never succeeded in "imposing its mentality on the surrounding society."[207] And now we are ready for the crucial link to be made: Without such a merchant mentality, early modern science is impossible. Needham is no longer quite certain why this is so: "What the exact connection was between early modern science and the merchants is of course a point not yet fully elucidated."[208] Yet the link itself is clear, for the reasons already indicated: Without the merchant mentality, the gulf between the scholar and the artisan cannot be overcome, and it is the merchant who fosters physical science through his need of precise measurements. Therefore, when we consider that physics was precisely the scientific discipline in which traditional China was weakest, this rounds off the picture neatly.

It only remains—so Needham ends this third and final account of his in which China serves as little but a mirror image of Europe—to explain how, if all this is true, it fits in with the fate of *Arabic* science. Perhaps, he suggests, here lies a confirmation after all, as the decline of Islamic science may be explained through the fact that what began as a typical merchant society (consider the Prophet's early profession!) was later bureaucratized under Persian influence.

From projection toward genuine comparison. What we have seen so far is little more than the projection of Needham's pet ideas about Western science and society upon the *tabula rasa* of their Chinese counterparts, with the invocation of Wittfogel's fruitful concept as almost the only element based fully upon phenomena belonging to China proper. But as Needham began to find out more and more details about Chinese science, and began to find them increasingly interesting in their own right, the European picture began to fade into the background, and the Chinese picture began to shine through somewhat thinner filters of conceptualizations derived exclusively from the European experience. All this happened in the fifties and early sixties: the most flexible period, so it seems to me, in Needham's enduring preoccupations as a student of science and civilization in China.

Also, Needham has become somewhat less sure about the causes of the Scientific Revolution in the West. He still believes Zilsel was overall right in connecting the rise of early modern science with the role of the 'higher artisanate', yet he knows less and less well why this was so, as he wrote in 1964: "It is *the intimate connections* between the social and economic changes on the one hand and the success of the 'new, or experimental' science on the other which are the most difficult to pin down" [my italics].[209]

As a result of these shifts in his thought, projection began to give way to authentic comparison. We shall try to find out now what fresh limits his mind has set to these acts of comparison.

Bureaucrats and science in traditional China.   In a paper from 1964 Needham distinguishes between two aspects of the effect feudal bureaucracy had upon the fate of science in traditional society.[210] On the one hand, it stimulated science; on the other, it acted as a brake. We discuss these two opposite roles in succession, although Needham does not claim that they marked successive stages of Chinese history.

Naturally, the bureaucracy had an interest in the prediction of events with large political and societal consequences, such as earthquakes and floods that threatened both the population and the elaborate irrigation works. The connection with early Chinese achievements in seismography, with the construction of rain gauges, etc., seems obvious. Also, the State had a direct interest in certain sciences for other reasons. Thus, astronomy was part of the civil service; vast encyclopaedias were prepared under State auspices; expeditions were carried out by the State with scientific purposes among others (e.g., the survey of large stretches of a particular meridian or the charting of constellations visible only in the Southern Hemisphere).[211]

In another, less direct sense, too, Needham claims that science (and its applications, in particular) benefited from the political order. Bureaucratic rule, although monolithic in the sense of not tolerating the emergence of rival sources of power, was nevertheless aimed at noninterference as much as possible, particularly on the village level:

> All through Chinese history, the best magistrate was he who intervened least in society's affairs. . . . It might . . . be said that this non-interventionist conception

of human activity was ... propitious for the development of the natural sciences.[212]

The connection construed here seems particularly far-fetched: The preference of the bureaucracy for action at a distance corresponds with 'action at a distance' as a physical principle that tends to foster "early wave-theory, the discovery of the nature of the tides, the knowledge of relations between mineral bodies and plants as in geo-botanical prospecting, or again in the science of magnetism."[213]

Thus we see that China's overall superiority over the West prior to the Scientific Revolution, of which Needham has now become convinced, is mostly technological in nature. As we might expect, it is primarily 'applied science' we are dealing with. And, however tenuous the connections are that Needham has come up with, it is, in his view, China's feudal bureaucracy that is to a considerable extent responsible for the edge thus acquired. ("By contrast science in Europe was generally a private enterprise. Therefore it hung back for many centuries.")[214] What, then, did the bureaucracy have to do with Chinese science being nevertheless overtaken by that of western Europe?

On this score Needham's views have scarcely undergone any change. Of the fatal inhibition of a possible rise of early modern science in China because of the absence of an autonomous merchant class he is as convinced as ever.[215] This implies that a Scientific Revolution in China was hampered by the bureaucracy indirectly, in strangling the merchant class. But Needham has now discovered that, in certain respects, the bureaucracy also hampered science *directly*. I have found two examples. One of these follows neatly the lines of Zilsel's pet argument of the gulf between the scholar and the artisan, extended as it now is to China. It concerns a text from the third century A.D., in which the memory of one particularly inventive engineer is honored:

> Fu Hsüan describes how Ma Chün was quite incapable of arguing with the sophisticated scholars nursed in the classical literary traditions, and in spite of all the efforts of his admirers, could never attain any position of importance in the service of the State, or even the means to prove by practical test the value of the inventions which he made. [Needham proceeds at once to expand this summing-up of the text in question into the following conclusion:] No document throws more light than this upon the inhibitory factors affecting science and technology which arose from the feudal-bureaucratic tradition of the scholar-gentry.[216]

Note, incidentally, the confusion of categories with which this passage ends. It illustrates with particular cogency one more hardy perennial among Needham's Marxist beliefs: his consistent tendency to look at the bureaucracy as a *social group* rather than as the embodiment of what Wittfogel, in *Oriental Despotism,* came to call 'a State stronger than society'. It was the State monopoly on power that strangled merchants; it was a scholar who failed to communicate with an inventor.

The other indication of bureaucracy preventing Chinese science from crossing the threshold to early modern science is that the principle of nonintervention, even though conducive to science in many respects, nevertheless stood in the way of its further development:

> So long as 'bureaucratic feudalism' remained unchanged, mathematics could not come together with empirical Nature-observation and experiment to produce something fundamentally new. The suggestion is that experiment demanded too much active intervention, and while this had always been accepted in the arts and trades, indeed more so than in Europe, it was perhaps more difficult in China to make it philosophically respectable.[217]

This brings to an end Needham's explanation of how China's sociopolitical order, while originally fostering science, in the end thwarted a possible Scientific Revolution. For those of my readers who may find that, on our search for an oasis of authentic insights, we are still crossing the desert of Needhamian mirror images I have good news in store: The explanations that remain to be discussed are much more rewarding, in that they display less the stamp of preordained patterns and move on a rather high level of sophistication supported by lovingly collected detail. The first of these is Needham's answer to the Grand Question on the lines of the absence, in the thought of traditional China, of the conception of laws of nature.

The Laws of Nature. Once again, the original idea goes back to Zilsel.[218] In his search for evidence for the gulf between the scholar and the artisan prior to the Scientific Revolution Zilsel had come across the usage, in the works of Agricola (the 16th-century author of a celebrated book on mining, *De re metallica*), of the concept of natural law in the sense that was to become common in 17th-century parlance but which had been absent in previous usage. Before the late Renaissance, Zilsel argued, one can find examples all across history, from the Babylonians onward, of the idea of a deity giving laws that the things of nature must obey, from the stars down to the minerals. The idea is present both in the Judaeo-Christian tradition and in its Graeco-Roman counterpart, and Zilsel goes on to show how it develops further in close connection with the Roman distinction between positive law and 'natural law' (in the sense of law laid down from above). He then finds that the transition to the *scientific* concept of laws of nature, in the sense of a fixed regularity to be observed in natural phenomena, occurred, as mentioned, in engineering circles and passed on speedily from there to such regular users as Kepler, Descartes, Boyle, and Newton. Zilsel goes on to ascribe the emergence of the idea of laws of nature to the concomitant rise of the absolutist state. Barring the latter speculation, this paper was surely Zilsel's most perceptive contribution to the history of science. Joseph Needham was more than ready to use it as a starting point for a widely cast inquiry into the question of whether any concept employed in Chinese science and philosophy ever came close to natural law in the Western sense, and, if not, why not.

His argument is rich and detailed—two features which inevitably drop out of a brief summing-up, so that, in the outline to be presented now, the argument may well look more like its rather skeletal predecessors than is warranted by the full text.[219]

Needham adopts the conclusion that in the West two distinct conceptions of natural law had by the 17th century come to be fully differentiated: law as valid for all human beings and laws obeyed by everything in nature that is not human. He goes on by passing in review all possible Chinese philosophies and finds that none of their central concepts comes sufficiently close to such a differentiation. One important case in point is Legalism. This was a philosophy of strong and coercive law (*fa*), to be applied by the ruler in order to keep society stable and bend it to his will. Not by chance, Legalism became the state philosophy of the totalitarian-minded Ch'in, who united China for the first time in 221 B.C. Not by chance either, the speedy downfall of the dynasty spelled the end of the Legalist school as well, and the idea of positive law never regained respect in Chinese thought. The Legalists had no interest whatsoever in the workings of nature, yet the bad odor of the concept of *fa* had enduring, and highly negative, consequences for the emergence of a concept of natural law in China.

Confucianism, equally uninterested in nature, did not believe in law but rather in *li,* proper ritual, as the warrant for social order. Taoists, on the contrary, were concerned about nature, but (as noted before) the *tao* is ultimately inscrutable. So is the central concept of the neo-Confucians of the 11th and 12th centuries, which comes down to order and pattern in nature but not to law.

This is a key distinction in Needham's view. All across Chinese thought, so he sets out to demonstrate in a number of strikingly beautiful passages, one finds that order is surely being taken note of, only it is not taken to come from above: "We do not see Heaven command the four seasons, and yet they do not swerve from their course," as one commentary on the *I Ching* has it.[220] The key conception here is, in Needham's eloquent summing-up:

> Universal harmony comes about not by the celestial fiat of some King of Kings, but by the spontaneous co-operation of all beings in the universe brought about by their following the internal necessities of their own natures.[221]

To illustrate further the incompatibility of such a conception with the notion of a celestial Lawgiver, Needham adduces a quotation from the 3rd century B.C., which he finds "truly sublime" and which I cannot, indeed, resist passing on to the reader:

> "The operations of Heaven are profoundly mysterious. It has water-levels for levelling, but it does not use them; it has plumb-lines for setting things upright, but it does not use them. It works in deep stillness. . . .

Thus it is said, Heaven has no form and yet the myriad things are brought to perfection. It is like the most impalpable of featureless essences, and yet the myriad changes are all brought about by it. So also the sage is busied about nothing, and yet the thousand executives of State are effective in the highest degree.

This may be called the untaught teaching, and the wordless edict."[222]

Needham goes on to show that such a conception of how regularity in the universe comes about has a great deal to do with the absence, in Chinese thought, of a personal God. The Highest Being of the Chinese was never thought of as a Creator, and insofar as this Being had received some personal traits very early on in Chinese civilization, it was rapidly depersonalized. Hence, no idea emerged that man might recognize rational order in nature through the decipherment of a divine code:

It was not that there was no order in Nature for the Chinese, but rather that it was not an order ordained by a rational personal being, and hence there was no guarantee that other rational personal beings would be able to spell out in their own earthly languages the pre-existing divine code of laws which he had previously formulated. There was no confidence that the code of Nature's laws could be unveiled and read, because there was no assurance that a divine being, even more rational than ourselves, had ever formulated such a code capable of being read. One feels, indeed, that the Taoists, for example, would have scorned such an idea as being too naïve to be adequate to the subtlety and complexity of the universe as they intuited it.[223]

We shall discover soon that Needham tends to agree, on this latter point, with his scornful Taoists.

The explanations weighed in the balance. We have now come to an end of our survey of answers given by Needham over time to his Grand Question insofar as this is concerned with the Scientific Revolution taken as a phenomenon all by itself. And the question comes up naturally what he, himself, thinks of their relative worth. We have collected five explanations. Is each of them exclusively true? Are they all equally true? Or what? Where does Needham strike the balance between them?

We have by now become sufficiently familiar with the workings of Needham's mind to say with some confidence that, to him, the explanations complete one another nicely—nothing for him to grant most-favored-explanation status to just one among them or to create a neat hierarchy between them. However, if one looks at the chronological order, not only of the papers in which his explanations were originally set forth but also of the respective summing-up passages in which most of these continue to live on in later papers, one can see that the 'democracy' explanation has receded very much to the background, and 'the merchants', 'bureaucracy', and 'laws of nature' figure much more prominently, with the 'Taoism' explanation taking on a more moderate position among the chains in the arguments set up in the three others.

Still, one great divide is being made among the explanations. Needham has over the years opened his mind considerably to a search for causes wherever he can find them: in the ways of the mind as well as of society, and indeed all across civilization at large. Yet he has never ceased believing that some of the 'inhibitory factors' were more decisive than others. In the shortest formulation possible: China, so Joseph Needham thinks, did not undergo a Scientific Revolution because the socioeconomic circumstances were not right.

At the start of the entire inquiry, in 1944, this sentiment was expressed very crudely and naively:

> Had the environmental conditions been reversed as between Euro-America and China, all else would have been reversed too—all the great names in the heroic age of science, Galileo, Malpighi, Vesalius, Harvey, Boyle, would have been Chinese and not Western names; and in order to enter today fully into the heritage of science, Westerners would have to learn the ideographic script just as the Chinese now have to learn alphabetic languages because the bulk of modern scientific literature is written in them.[224]

Yet at the end of his much more subtle analysis of such a typically 'mental' difference between East and West as the presence or absence of the idea of laws of nature, essentially the same conclusion is drawn:

> Always in the background stood the concrete forces of the social and economic life of Chinese society, out of which arose the transition from feudalism to bureaucratism, and which could not but condition at every step the science and philosophy of the Chinese people. Had these conditions been basically favourable to science, the inhibiting factors considered in this lecture would perhaps have been overcome.[225]

There are two sources, in Needham's mind, of his persistent conviction that the 'socioeconomic circumstances' (in particular, the strangling of the merchants' autonomy by the bureaucracy) were decisive in the end.[226] One source is wholly obvious and has been played up very highly by many of his early critics: the unredeemed Marxist's conviction that the economic substructure ultimately 'determines' the intellectual superstructure (section 3.6.2). But another, perhaps even more tenacious and emotionally very powerful root of this belief is Needham's desire to *exonerate* the Chinese, as it were, from their 'failure':

> When full account is taken of the environmental conditions, it is not so much to the credit of Europeans that they developed modern science and technology, nor yet so much of a reproach to our Chinese friends that they failed to do so. The abilities were everywhere, but the favourable conditions were not.[227]

In other, cruder, words: The Chinese should by no means be thought to have been too dumb to achieve a Scientific Revolution![228] Which brings us to those possible answers to his Grand Question that have been explicitly *rejected* by Needham.

Rejected explanations.   Joseph Needham has rejected four possible responses to his Grand Question. These come under the following headings: personal features, climate, the role of language, and the conception of time. We take them up in this order.

Personal features of the Chinese people. Needham has wasted few words on explanations along 'physical-anthropological' and 'racial-spiritual' lines. He remarks quite simply that his dealings over many decades with Chinese friends and colleagues have shown him that they are entirely (quoting an Italian traveler in 1515) *di nostra qualità.*[229] It is all right to study human races in a scientific, comparative sense, but that is something quite different from an explanation of the Scientific Revolution through European superiority over all other peoples.[230]

Climate. One reason why comparing Western and Chinese science is so instructive is that "the complicating factor of climatic conditions does not enter in. Broadly speaking, the climate of the Chinese culture-area is similar to that of the European,"[231] and therefore differences in climate cannot even begin to explain the absence of a Scientific Revolution in China, as they conceivably might if the Grand Question were to be asked about India.

Language and script. The idea that Chinese science might have been influenced adversely by difficulties inherent in the language, and in ideographic script in particular, is considered by Needham in somewhat greater detail than the two preceding cases since the possibility has often been suggested. However, he rejects it fully, chiefly because the Chinese language and script have proven themselves quite capable of assimilating the concepts and technical terms of modern science. Also, the classical Chinese language offered excellent opportunities for the pithy formulation of deep insights into nature.[232]

Conceptions of time. This is the only rejected explanation that Needham has found worthy of consideration in depth. In fact, one of his finest essays, "Time and Eastern Man," culminates in this very rejection.[233] Central to the argument is the idea (widespread among theologians, in particular) that one unique feature of Western civilization has been its conception of linear, unidirectional time, as opposed to the cyclical conceptions of time to be encountered in other civilizations, including the Graeco-Roman world. If life acquires its full meaning only during the brief time span marked by Crucifixion and the Last Judgment, this makes for a much more activist stance than the fatalism fostered by the notion that, in due time, everything is to repeat itself anyway. Needham first expands this overall idea further in the direction of science, showing in particular that linear time lends itself far better to being mathema-

457

tized than recurrent time, and goes on to inquire whether the Chinese conception of time was predominantly linear or cyclical.

However tempting, I have to pass by all the manifold considerations leading Needham to his ultimate conclusion. Since Needham tends to accept the premise of the entire argument, he agrees that, "in so far as the traditional natural philosophy was committed to thinking of time in separate compartments or boxes perhaps it was more difficult for a Galileo to arise who should uniformize time into an abstract geometrical co-ordinate, a continuous dimension amenable to mathematical handling."[234] Therefore Needham grants that this may have been one obstacle for the further mathematization of nature in ancient Greece. However, because Needham finally concludes that among the various conceptions of time in ancient China linear time was overwhelmingly predominant over recurrent time, no inhibitory factor is to be encountered in this particular area of human thought.

> The conclusion springs to the mind. If Chinese civilization did not spontaneously develop modern natural science as Western Europe did (though much more advanced in the fifteen pre-Renaissance centuries) it had nothing to do with China's attitude towards time. Other ideological factors, of course, remain for scrutiny, apart from the concrete geographical, social, and economic conditions and structures which may yet surface to bear the main burden of the explanation.[235]

This brings to an end our catalogue of explanations rejected by Needham. One more response to the Grand Question has been going around that has often been ascribed to him, but which, as I shall show, he has avoided at just about all costs, and for most revealing reasons. The idea is simply this. If it is true that the Chinese up to the 15th and 16th centuries had a considerable edge in matters scientific, then it may seem rather obvious to seek the cause of the disappearance of the edge *not only* in the upheaval of the Scientific Revolution that took place in the West *but also* in a possible decline occurring in Chinese science during the Ming dynasty, which ruled from 1368 to 1644.

Retrenchment under the Ming as an explanation not considered by Needham. As ever, excellent grounds for pursuing such a course of reasoning are provided by Needham himself. Take the celebrated case of the Chinese Voyages of Discovery under the admiral Chêng Ho, which came to an end around 1450. Needham has devoted an extensive paper to these voyages,[236] which took the Chinese as far as the east coast of Africa just half a century before Vasco da Gama was to sail into the same harbors. Needham pays much attention to what was, in one sense, a spectacular non-event: this missed opportunity for the East to discover the West rather than the other way round. He waxes eloquent, on page after page, in contrasting the greedy, monotheistic aggression of the Portuguese with the patronizing yet tolerant benevolence displayed by Chêng Ho and his crew toward their hosts. However, although Needham is far too honest to hide from the reader what finished the voyages off, the reason is tucked away in a subordinate clause, and some hard searching is required to find it.[237] The reason was, quite simply, that the imperial bureaucracy decreed an end to the whole enterprise as part of a movement of overall retrenchment in the face of threats from barbarians outside the Wall. Later authors have turned this general retreat into the cornerstone of a larger argument in which the decision to call Chêng Ho home for good figures as one signpost among others.[238] But why is Needham so reluctant to face these facts squarely?

It is not of course up to me to say whether this turning inward of the empire under the Ming affected the subsequent fate of science in China as well. Needham himself, in any case, feels assured that it did. When, in *Science and Civilisation in China,* he is dealing at his leisure with mathematics or with astronomy, he is quite outspoken about a certain stagnation that took place in these sciences during the Ming period, and at a few scattered places one even comes across a generalized statement to the same effect.[239] However, when it comes to drawing the lines together and seeking causes for the nonemergence of early modern science in China, Needham bows out in silence, apparently refusing to take this seemingly obvious course toward a possible answer to the Grand Question. Why? Naturally, one can only speculate, yet I feel reasonably assured that to follow up his own observation of a period of stagnation is incompatible with his overriding idea that all through the centuries Chinese science and technology have displayed unilinear, steady, uninterrupted progress.

Needham's grand picture of the overall contrast between the developmental lines of Western and Chinese science is as follows. China and the West (ancient Greece, that is) started on a par, with perhaps some advantage for the West because of its advanced geometry; then the West passed through the Dark Ages, which lack a counterpart in China, for China's science continued to grow at a steady pace. This went on and on, giving China a substantial edge over Europe, until Europe, in the magnificent passing maneuver of the Scientific Revolution, overtook China and ushered in the age of cumulative, universal science, to which China has since adapted itself (see Fig. 1). This picture, then, is Needham's answer to "the cliché of stagnation, born of Western misunderstanding."[240] Obviously, it is incompatible with assigning even one period of decline to the development of China's science. It is even less compatible with ascribing to it alternating periods of flourishing and decline. There is, in other words, no room at all left in Needham's broad conception of things Chinese for Ben-David's picture of intermittent growth alternated with losses and a general decay in scientific activity as the normal pattern for science in traditional society.

There are nonetheless reasons for assuming that science in China conformed in reality to Ben-David's, rather than Needham's, pattern (some of these reasons making their appearance, to be sure, in the fine print of *Science and Civilisation in China).*[241] Not bringing up Ben-David's name or work, David Landes has shown for the case of Su Sung's water clock that there is ample

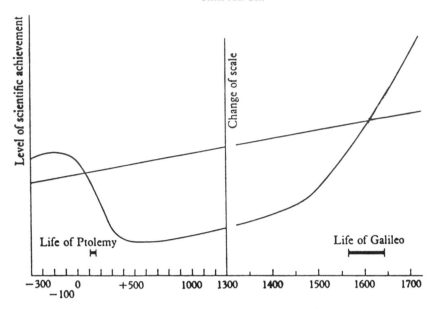

Figure 1. Left-hand portion of a diagram by Needham. The straight line represents the development of science in China; the curved line, in the West. From J. Needham, *Clerks and Craftsmen,* p. 414. © Cambridge University Press 1970. Reprinted with the permission of Cambridge University Press.

reason to think in such terms rather than in terms of continuity. "Over a span of fifteen hundred years only a half-dozen, perhaps only four, astronomer-clockmakers kept the great tradition alive for China or, more accurately, revived it at intervals," writes Landes, adopting Needham's figures.[242] Needham sells this as an example of the amazing continuity of Chinese scientific engineering over time. For Landes it rather betokens "the repeated lapse of knowledge over long periods, so that each great clockmaker had to search in old records for the forgotten secrets of earlier reigns,"[243] a conclusion strongly reinforced by the fate of Su Sung's wonder clock, which was used for about a generation but then, after its removal and subsequent failed reconstruction, was so thoroughly forgotten that (despite some faint later efforts) it took Joseph Needham himself, in 1960, to bring it to light again—a 'magnificent dead end' if ever there was one.

Needham's essay on the discovery of magnetic declination in China can be seen to follow quite the same pattern. Once having squeezed all he could out of Shen Kua's pioneering document, Needham set out at once to find predecessors. His terminology is nothing but revealing. Having determined the names and works of two previous thinkers about the magnet, he asks: "How now can we fill the gap between +80 (the date of Wang Ch'ung's book) and +980 (the approximate date of the birth of Wang Chi)?"[244]

'Filling the gap' so as to conform to a pattern that *must* display continuity on sheer *a priori* grounds and in the face of contrary evidence advanced by Needham, himself! It might be worthwhile for future historians of Chinese science tentatively to adopt Ben-David's perspective and to consider whether the facts of the history of Chinese science do not rather fit into that particular pattern. As one by-product it might further be inquired to what extent the Grand Question could be, at least partly, resolved by pointing to one particular period of decline in the science of traditional China owing to the movement of retrenchment under the Ming bureaucracy.

One truly comprehensive answer. So far we have been examining Needham's answers to the question of why the Scientific Revolution, considered so far as one unique and relatively isolated event, had no counterpart in China. We came across five answers given, four answers rejected, and one answer avoided. What these have in common is that they are answers to a question asked about the Scientific Revolution as such. Let us recall now that, in Needham's view, the birth of early modern science was really bound up with the Renaissance and the Reformation into that one grand process of the dissolution of feudalism and the rise of capitalism in late medieval and early modern Europe. Let us also recall an earlier observation of Needham's, namely, that so many indigenous inventions which left China relatively unperturbed had quite drastic consequences, when transferred to Europe, for social order there. These two viewpoints have been taken together by Needham in the most comprehensive and the most enlightening answer that he has, in my opinion, given yet to his Grand Question. It is not, to be sure, an answer about any specifics of the Scientific Revolution; we already went through all those. It is an answer to the question of what (if I may be allowed to strip Needham's formulation of its Marxist rhetoric about the dissolution of feudalism and all that) made European society fit for bringing forth the Scientific Revolution—an answer gained through a sustained comparison with its Chinese counterpart. It has long been customary to consider the West's peculiar features in terms of the *Abendland* and its 'Faustian spirit'; here, then, is Needham's attempt to pin down to more distinct historical phenomena that spirit that supposedly animated Europe and made, among many other things, early modern science possible. It is, in short, Needham's personal contribution to the everlasting riddle of what it was that set Europe apart, with Needham as the one contributor to the debate so far to realize clearly that this riddle involves the birth of early modern science, too, and not in the last place either.

How, then, to characterize the fundamental difference between the developmental principles of the two societies, China and western Europe?

> There was a certain spontaneous homoeostasis about Chinese society [whereas] Europe had a built-in quality of instability. . . .
>
> Cathay had been self-regulating, like a living organism in slowly changing

equilibrium, or a thermostat—indeed the concepts of cybernetics could well be applied to a civilization that had held a steady course through every weather, as if equipped with an automatic pilot, a set of feedback mechanisms, restoring the *status quo* after all perturbations, even those produced by fundamental discoveries and inventions. Struck off continually like sparks from a whirling grindstone, they ignited the tinder of the West while the stone continued on its bearings unshaken and unconsumed.[245]

This homoeostasis, Needham continues, is quite a natural state for a society to be in, and Europe was the exception. "To what then was the instability of Europe due? Some have referred it to the aspirations of the never-satisfied Faustian soul,"[246] but such an explanation cannot satisfy Needham:

I would prefer to think in terms of the geography of what was in effect an archipelago, the perennial tradition of independent city-states based on maritime commerce and jostling military aristocrats ruling small areas of land, the exceptional poverty of Europe in the precious metals, the continual desire of Western peoples for commodities which they themselves could not produce (one thinks especially of silk, cotton, spices, tea, porcelain, and lacquer), and the inherently divisive tendencies of alphabetic script, which permitted the growth of numerous warring nations with centrifugal dialects or barbarian languages. By contrast China was a coherent agrarian land-mass, a unified empire since the third century BC with an administrative tradition unmatched elsewhere till modern times, endowed with vast riches both mineral, vegetable, and animal, and cemented into one by an infrangible system of ideographic script admirably adapted to her fundamentally monosyllabic language. Europe, a culture of rovers, was always uneasy within her boundaries, nervously sending out probes in all directions to see what could be got—Alexander to Bactria, the Vikings to Vineland, Portugal to the Indian Ocean. The greater population of China was self-sufficient, needing little or nothing from outside until the nineteenth century . . . , and generally content with only occasional exploration, essentially incurious about those far parts of the world which had not received the teachings of the Sage. Europeans suffered from a schizophrenia of the soul, oscillating for ever unhappily between the heavenly host on one side and the 'atoms and the void' on the other; while the Chinese, wise before their time, worked out an organic theory of the universe which included Nature and man, church and state, and all things past, present, and to come. It may well be that here, at this point of tension, lies some of the secret of the specific European creativeness when the time was ripe.[247]

This truly inspired passage, born as it seems to me from an intimate familiarity with the respective characters of these two widely divergent civilizations, teaches us more about the environment in which the Scientific Revolution became possible than any other I have ever come across. It also betrays in a very interesting manner Needham's personal oscillations between the two ways of life so eloquently compared here. He belongs to the West and is aware, better than most of us, of its unique features, among which he acknowledges modern science as its most admirable fruit. But he has also come to believe that the ways of China were altogether more natural, more balanced, more *organic* in short, than those of Europe. He pinpoints a certain 'schizophrenia' in the European soul; he abhors Vasco da Gama's (and his successors') dealings with the natives compared to Chêng Ho's; he finds the entire idea of laws of nature "rather naïve."[248] Let us recall at this point that, to Needham, the Scientific Revolution represented 'the spearhead, not the whole of science', since key elements of fully universal science were still left out of the Newtonian synthesis. And Needham has speculated repeatedly about the historical possibility of China's producing a Scientific Revolution after all—not the 'mechanical' Scientific Revolution of the West, to be sure, but rather an 'organic' Scientific Revolution of its own. We must not leave unexplored this revolution that never materialized.

An alternative, organic Scientific Revolution?   We have met in the course of the present book Galileo, intuitive empiricist; Galileo, mathematical scientist; Galileo, experimenter; Galileo, impetus theorist; Galileo, Aristotelian methodologist; Galileo, technical inventor. Are we now finally facing that most unexpected of all Galileos: 'Galileo, organic scientist'?

Needham has ever wavered over the issue that is encapsulated in this (perhaps a little too playful) formulation of mine. "The *philosophia perennis* of China was an organic materialism,"[249] and this, Needham feels sure, is the very element that needs to be fused with the mechanical science of the Scientific Revolution so as to relieve the latter of its one-sided, mathematical reductionism. Now is such a 'philosophy of organism' a stage not to be reached by science until it has gone through its Newtonian phase, or might the entire transition from traditional to universal science have taken an independently 'organical' course instead of, or alongside, the 'mechanical' route science has taken in fact? "[Could] the recognition of statistical regularities and their mathematical expression have been reached by any other road than that which Western science actually travelled?"[250]

At times Needham does not think so. The Taoists, he has said (and the premise of the argument, though not its conclusion, has been repeated since by scientists considerably less knowledgeable than Needham about Chinese science and philosophy),

> were groping for what we may call a world-picture of the kind Einstein was later to draw in the West ... without laying the right foundations for a Newtonian one. By that path science could not develop.[251]

But on other occasions he is less sure: "Who shall say that the Newtonian phase was not an essential one?"[252]

Now *if*—so Needham has asserted—it would have been possible in principle for science to take another route than the Newtonian, and *if* socioeconomic conditions in China had been more favorable, *then* a Chinese Scientific Revolution might have taken place, of which one thing is certain: the "science of Nature which would have been developed . . . would have been profoundly organic and non-mechanical."[253] Let us go ahead and find out a few things about this organic, nonmechanical science.

One of the acknowledged triumphs of Needham's scholarship has been his success in making sense of a form of natural philosophy that flowered under the Sung dynasty and is commonly named 'neo-Confucian', even though it was actually a synthesis of Confucian, Taoist, Buddhist, and quite original notions. No one had been able to make much of the texts of the five leading thinkers who, in the 11th and 12th centuries, together formulated a philosophy based on the concepts of *li* (quite another word than its Confucian homonym meaning 'ritual') and *chi*. Needham recognized in the teaching of the five leading neo-Confucian thinkers something intimately familiar to himself: a philosophy of organic development more than faintly reminiscent of his own conception of integrative levels. Taking *li* as 'universal pattern', or 'principle of organization', and *chi* as 'matter-energy', Needham managed to convey a great deal of meaning to an otherwise rather incomprehensible conception of the universe. To what extent Needham has read his own views into these ideas is not up to the outsider to decide; what is certain is that his pertinent chapter in volume 2 of *Science and Civilisation in China* makes breathtaking reading.[254]

The analysis of neo-Confucianism is not made for its own sake only (Needham's analyses rarely are); he regards this philosophy as the pinnacle of an organic approach to nature that has pervaded the conception of nature through all the centuries of traditional China. This overall conception runs equally across the theory of the Five Elements; the theory of the Two Forces, Yin and Yang; the world of ideas embodied in the *I Ching;* many elements of Taoism; and so on and so forth. Needham has often called the mode of thought expressed by it "correlative thinking."[255]

"In primitive thought anything can be the cause of anything else: everything is credible; nothing is impossible or absurd."[256] Two principal routes lead out of primitive thought: one is the Western road of 'mechanical causation' first trodden by the Greeks; the other is the Chinese road. In the Chinese conception,

> nothing was uncaused, but nothing was caused mechanically. . . . The regularity of natural processes is conceived of, not as a government of law, but of mutual adaptations to community life. . . . [The way of the Chinese was] to systematise the universe of things and events into a structural pattern which conditioned all the mutual influences of its different parts.[257]

Also (but now taking 'cause' in the sense of 'mechanical cause'),

> for the ancient Chinese, things were connected rather than caused. . . . The universe is a vast organism, with now one component, now another, taking the lead

at any one time, with all the parts co-operating in a mutual service which is perfect freedom.

In such a system as this, causality is not like a chain of events, but rather like what the modern biologist calls the 'endocrine orchestra' of mammals where, though all the endocrine glands work, it is not easy to find which element is taking the lead at any one time. And we should be clear about it; modern science needs concepts like this when considering questions like the higher nervous centres of mammals and even Man himself. But leaving modern science aside, *it is clear that the concept of causality where the idea of succession was subordinated to that of interdependence dominated Chinese thinking.* [My italics][258]

We need not follow Needham on his subsequent attempts to show that this system of ordered association was the intellectual counterpart to the administrative filing systems of the bureaucracy, nor on his argument meant to demonstrate that, through the venue of Leibniz' correspondence with the Jesuits in Peking, this mode of thought has considerably influenced Western philosophical thought ever since. Enough has been said to see that, for Needham, the philosophy of organism is the key solution to the problem of the mathematical reductionism to which modern science, with all the understanding of nature it has given us ever since the Scientific Revolution got under way, has remained subject—a reductionism that should now be overcome because it has reached its limits. We may recall that this entire conception is one that Needham had worked out already before he embarked on his Chinese adventure (see section 6.3). What has happened since is that he found it back in the deepest recesses of ancient Chinese philosophy—hence its enduring fascination for Needham. Now why, so Needham has wanted to know, did this organic science not give rise to a Scientific Revolution in a manner similar to how the mechanical thought of the Greeks ultimately brought forth (though not without some admixture with products of Eastern ingenuity) early modern science as we know it?

As noted, there are only two possible answers to this question, both considered by Needham. The one is that this route cannot be traveled, because the Newtonian stage is indispensable. Reluctantly, Needham has sometimes adopted this answer (which, of course, is the natural one for those who do not agree at all that a philosophy of organism is what present-day science stands in need of).[259]

His alternative, and more frequently given, answer is that the socioeconomic conditions were not right.[260] But this is an odd answer indeed. Recall that those socioeconomic conditions of China were supposed to be unfavorable to *mechanical* science in the Western sense, chiefly because the merchant mentality could not arise in China and because this very merchant mentality, according to Zilsel and to Needham, was indispensable for *mechanical* science to emerge. One fails to see, however, how even on Needham's own terms the absence of an autonomous merchant class could inhibit the coming about of an *organic* Scientific Revolution in China. What remains to be concluded, then, is that either a possible further post-Sung development of neo-Confucianism fell victim to movements of retreat under the Yüan and the Ming, or, indeed, that along the road of organicism no cumulative science can come into existence. I am afraid that the latter inference is by far the most plausible.

6.5.3. *Further Perspectives Offered by Needham's Critics*

Everything Needham does is done in the grand manner. It is not by chance, therefore, that, ever since *Science and Civilisation in China* began to appear, his work has received a greater amount of critical attention than has been paid, I believe, to any other historian of science. Again, I could not think of working my way through all this criticism, entire volumes of which have seen the light. I have confined myself to those criticisms and comments that appear to have yielded the most fruitful alternative approaches to *our* core problem: what we can learn from science in traditional China in order to understand better the what and why of the Scientific Revolution.

As far as I know, Joseph Needham is fully aware of these perspectives of a younger generation. He has defended his own views against some of them (notably Nathan Sivin's) and has gone ahead in his own work happily to ignore them.[261] The topics in question may be categorized as follows: the science of the Jesuits as a test case, the Needham legacy, the nature of comparative research, and the methodological status of the Grand Question.

After the Scientific Revolution came to China. It may be recalled from our discussion of Islamic science that Sayili, on reaching the end of his analysis, announced that one possible way to test his conclusions would be to find out what happened when Islam took up the confrontation with early modern science or parts thereof. In the Muslim world, this confrontation started quite late—centuries after the Scientific Revolution, in fact. In China, however, the confrontation began while the Scientific Revolution was just under way. Many of its early products were employed for purposes of conversion to the Christian faith by those Jesuit missionaries who succeeded in gaining entrance to the imperial palace in Peking. The celebrated Matteo Ricci, who was allowed to settle in Peking in 1601 and died there in 1610, was the first of a whole array of learned Jesuits, all of whom were well informed about the state of science in their homelands and, as part of their missionary strategy, freely passed on products of Western science to any Confucian *literatus* who cared to hear about it. What, then, did the *literati* make of Western science?

It has been pointed out by the sinologist and historian of science Nathan Sivin in particular that, in a sense, China *did* undergo a Scientific Revolution.[262]

By this he means the exposure of members of the Chinese elite to Western science as transmitted by the Jesuits, and he has insisted that there is a great deal to learn from how they reacted. After all, the episode was

> the last major face-to-face encounter of non-Western and European science in world history. By the eighteenth century modern science was crossing national boundaries on the coattails of Empire, and competition . . . on the basis of . . . abstract merits had become a thing of the past.[263]

If, then, the Chinese picked up the principles and the results of the new science readily once they were exposed to them, it would serve as a strong indication that, through their own indigenous scientific development, they had already come close to their own revolution. If not, it is more than worthwhile to find out why precisely they failed to accept and utilize Western science at that time, thus providing the historian with a ready-made yardstick to gauge the distance between two largely autonomous structures of thought about nature.

I have been able to locate precisely one instance of such usage made by Needham of post-Jesuit science and its reception by Chinese scholars. It concerns his claim about the lack of a concept of natural law in Chinese thought, which he goes on to underpin by observing that "the introduction of the idea of Laws of Nature to China after the time of the Jesuit Mission met with little response."[264] The fundamental inability of the Chinese to grasp the concept is illustrated by quoting from an exciting 18th-century report on the findings of one missionary:

> "When we teach [the Chinese atheists] that God, who created the universe out of nothing, governs it by general Laws, worthy of his infinite Wisdom, and to which all creatures conform with a wonderful regularity, they say, that these are high-sounding words to which they can affix no idea, and which do not at all enlighten their understanding. As for what we call laws, answer they, we comprehend an Order established by a Legislator, who has the power to enjoin them to creatures capable of executing these laws, and consequently capable of knowing and understanding them. If you say that God has established Laws, to be executed by Beings capable of knowing them, it follows that animals, plants, and in general all bodies which act conformable to these Universal Laws, have a knowledge of them, and consequently that they are endowed with understanding, which is absurd."[265]

All this is truly marvelous stuff, of which one would like to see a great deal more. Why has Needham not given it?

The answer is to be found in just about the oddest paper he has ever written in my opinion. Entitled "The Roles of Europe and China in the Evolution of Oecumenical Science," it was written in 1966 with the explicit idea in mind to make it foreshadow general conclusions to be reached in the still-to-be-written final volume of *Science and Civilisation in China*.[266]

Recall Needham's view of the patterns of progress in Chinese and Western science: steady, unilinear growth in the one as opposed to recovery from the Dark Ages and a little help from the East, followed by the sudden outburst of the Scientific Revolution, in the other. From this point in the 17th century onward science in the West begins to grow exponentially, and two questions come up about what subsequently happened to science in China. The first is: At what distinct points in time were the various Chinese sciences *overtaken* by their Western counterparts? And the other: When did a particular branch of Chinese science *merge* with its Western counterpart so as jointly to turn into truly universal, or 'oecumenical', science? The first point in time is called by Needham the 'transcurrent point'; the second one the 'fusion point'.

Now the upshot of the paper is to arrive at what Needham grandiosely dubs the 'law of oecumenogenesis'. It asserts that the more organic a particular branch of science is, or rather, the higher the integrative level of the phenomena studied by that branch of science, the wider the temporal gap between the transcurrent point and the fusion point. This may sound a little esoteric, and will now be duly illustrated. The transcurrent point in mathematics, astronomy, and the physical sciences coincided, of course, with the Scientific Revolution, and the corresponding Chinese sciences fused with the science of the Scientific Revolution very rapidly; Needham locates the fusion point at the middle of the 17th century. In botany, Western science overtook its Chinese counterpart around 1780, with the work of Adanson, and fusion took place at the end of the 19th century. Western medicine, so Needham decides after some deep soul-searching, did not acquire a noteworthy edge over Chinese medicine until the turn of the 20th century, with the fusion point still lying far ahead in the future.

Thus we see indeed that the distance in time between the transcurrent points and the fusion points grows with the degree of organicity of the branch of science involved. Next chemistry is called in by Needham to serve as a fully independent test case, and, lo and behold, with its transcurrent point located around 1800 with Lavoisier and Dalton, and 1880 being appointed the moment of fusion, "a figure intermediate between those for the physical sciences and for botany" duly turns up.[267] (Fig. 2 displays in the format of a diagram the whole picture as Needham conceives it.)

The interest of this incredible piece for us lies, of course, in the determination of the fusion point in the mathematical sciences. Here Needham says flatly that

> the mathematics, astronomy and physics of West and East united very quickly after they first came together. By 1644, the end of the Ming dynasty, there was no longer any perceptible difference between the mathematics, astronomy and physics of China and Europe; they had completely fused, they had coalesced.[268]

As evidence for these assertions we find (beside the curious reiteration of fundamental *differences* between the respective approaches and conceptual

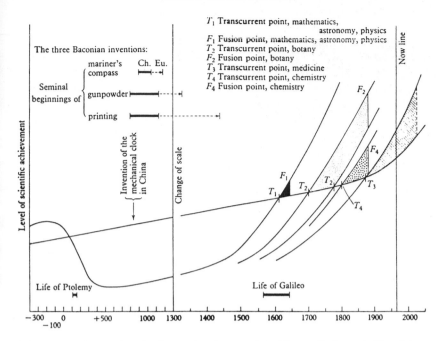

Figure 2.   Diagram by Needham, already reproduced in part in figure 1. The right-hand portion represents Needham's "law of oecumenogenesis" explained in the text. From J. Needham, *Clerks and Craftsmen*, p. 414. © Cambridge University Press 1970. Reprinted with the permission of Cambridge University Press.

frames) little more than Father Verbiest equipping Peking observatory with astronomical tools constructed in accordance with Western principles; the adoption by European astronomers of equatorial coordinates such as the Chinese had employed for ages; an 18th-century astronomical screen in Korea which combined elements from both traditions; and, finally, the possibly independent construction of telescopes by Po Yü in 1635.

This is all, and the entire conceptual framework in which such bits and pieces have been set by Needham evidently leaves no millimeter of room for an in-depth inquiry into the reception of 17th-century Western science by contemporary Chinese. Yet there appears to be ample reason for such an inquiry. The American sinologist Jonathan Spence observed in 1984 that to him the sum total of these early exchanges suggests "how magisterially the attempt at interchange failed. . . . One can argue with some force that China did not enter the world of universally valid modern science in any significant way until the twentieth century."[269]

Take astronomy as one example. It is particularly instructive because what little Needham has to say in *Science and Civilisation in China* on the episode of the Jesuits concerns mostly the astronomical domain, with much of his attention focused on their underestimation of Chinese scientific achievement and on their prevarications over the Copernican hypothesis (held under suspicion of heresy in Rome from 1616 onward—fifteen years after Ricci's arrival in Peking, that is).[270] Once again, if one screens the fine print, so to speak, sufficient facts peep through to evoke a picture altogether different from what the law of oecumenogenesis prescribes. That picture is not drawn by Needham, but Sivin outlines it. In his paper, "Why the Scientific Revolution Did Not Take Place in China—Or Didn't It?" he has shown that, in this area of vital concern to the Chinese, Western superiority was indeed accepted by a number of experts. However, these men confined their conceptual reforms to the procedures of observational astronomy; nothing like the Galilean breakthrough ensued.[271]

Or take the measurement of time. David Landes rounded off his idea of the 'magnificent dead end' into which the water clock tradition had run as opposed to the developmental possibilities of the Western mechanical clock by reciting in fascinating detail what happened when the Jesuits took such clocks with them to Peking and demonstrated them at court.[272] Generally speaking, the very notion that those men from the West might have made superior clocks ran up against the condescending contempt in which barbarians were generally held. One early Ching emperor indeed used his new mechanical clocks, to the dismay of his less punctual, because clockless, officials ("Why are the memorials late?" the emperor demanded to know after due consultation of his watch). Courtiers and concubines took their clocks as nothing but funny toys, and none of all these timepieces (both the Western-made ones and the Chinese copies modeled after them) ever left the imperial palace. So much for the rapidity of cultural transmission, let alone fusion, as far as the introduction of the Western, mechanical clock is concerned.

In a similar vein, the Dutch sinologist P. M. Engelfriet is at present examining the translation into Chinese of the first six books of Euclid that was prepared by the Confucian scholar Hsü Kuang-ch'i under the aegis of Father Ricci.[273] Although no final conclusions have come forward yet, it is crystal clear that this second Euclid translation did not leave many more ripples on Chinese intellectual (let alone social) life than its predecessor had achieved in the twelfth century. The translator was elated; a few later mathematicians tried to combine Euclid's method of proof with methods they found back in the indigenous tradition they helped restore, but that was about all; the remainder of the *Elements* had to wait until 1857 to be translated into Chinese in its turn.

What follows from all this is not of course that the 17th-century Chinese were too dull to grasp these matters. The point is rather that cross-cultural interaction is a very intricate affair. Many culture-bound attitudes of either acceptance or rejection may stand in the way of a smooth transmission, and even where there is a ready willingness to learn the symbolic representations by means of which various civilizations express themselves, these may be harder to grasp by foreigners than Joseph Needham, in his enthusiastic proselytizing moods, has often fancied. I am certainly not claiming that it is impossible, but I would claim that very painstaking in-depth research is needed in order to arrive at a full understanding of any system of symbolic representations coming from a foreign culture, including its science. Which leads us inevitably to the tough question of the extent to which Joseph Needham's life's work can be conceived as the embodiment of such a full understanding.

Has the time come for answering the Grand Question? I would not so easily bring up the question just put, which I am in the end not qualified to answer, if it had not in effect been asked by some of Needham's most knowledgeable critics. Although they have not said so in quite so many words, it has been suggested by Nakayama and by Sivin in particular (and much in our preceding account tends to support such a suggestion) that just about everything Needham has ever written to answer his Grand Question should be regarded as *one giant projection of Joseph Needham's collected preconceptions upon China's society and world of thought.*[274] Let us consider very carefully to what extent such a conclusion is warranted.

The first thing to do is to determine which Needham we are talking about. Do we mean Needham the proselytizer; the historiographical extremist; the man who went to such enormous, wholly counterintuitive lengths to vindicate for the Chinese a steady tradition in advanced science, superiority over Europe until the advent of early modern science, and a rapid catch-up thereafter? Or do we have in mind Needham the patient, surely one-sided yet eminently reasonable scholar who seems to have anticipated just about every single objection that has occurred to us to make while working our way through the essays? Which of these men are we talking about? Will the real Needham please stand up?

I have little doubt that, if thus summoned, *both* would arise; so often Needham deems to be complementary what to other minds presents itself as wholly incompatible. There is little point, therefore, in setting up an argument over who the true Needham is. The important thing to recognize is that, notwithstanding the reasonableness and the many qualifying statements of long stretches of *Science and Civilisation in China, the simplified, radicalized picture that arises from most of the essays has never ceased to govern Needham's mind whenever he directs it toward the Grand Question.* He has surely elevated into an art form his knack for arranging his chapters and papers in such a way as to make his reader leave them with a sense of 'it all fits'. But the counterpart to such a sense of fit is, naturally, the suspicion dawning upon the reader that the fit reflects the arrangement of the argument rather than an order inherent in the material adduced to underpin it. In addition to all we have come across already, I may point here to Needham's habit of mentioning from time to time that a given topic still awaits examination in far greater detail, yet he already feels sure that, 'when all information is in', we shall find the answer to be such-and-such. All of which enhances our suspicion that the reader's sense of 'it all fits together' may be no more than an artifact of the coherence of Needham's preconceptions rather than reflecting a genuine pattern of Chinese science and society.

We have meanwhile become aware where his principal preconceptions stem from. There are essentially two. One is his Marxism—fairly rigid in his early days as a Chinese scholar but becoming more fluid ever after, although we have also found some definite limits set to flexibility. These limits have appeared generally there where his second basic preconception comes in: his urgent wish to uphold the precise opposite of the clichés about China adopted by oh so many unthinking, or unwitting, 'Westerners' in Needham's and previous generations. And the question arises of whether Needham has not unwittingly fallen into the very trap he has wished above all to avoid: to have Europe uppermost in mind when ostensibly writing about China.

If I have understood well what Sivin (a severe but no less respectful critic) means to convey, it is indeed his opinion that, in one sense, the study of science and civilization in China, particularly where the issue of the Scientific Revolution is concerned, still has to begin. "Although every culture must experience much the same physical world, each breaks it up into manageable segments in very distinct ways," Sivin has written,[275] and we must therefore find out exactly how the Chinese broke up their world before we can even think of comparing the results with those reached for cultures that do it quite differently. And apparently Sivin does not believe—as Sabra, too, does not believe—that the historian is free to study a non-Western civilization in its own right and for its own sake if he approaches it right from the start with the question in mind of why it *failed* to culminate in an event characteristic only of the West. Not until we know much more about Chinese science in this deeper sense, and most urgently about the *unity* and the *social values* of science in China, so Sivin has argued, may it perhaps make sense to bring up again the Grand Question—if in that distant future anybody still cares to ask it.[276]

Sivin has been joined in bringing up the appropriateness of the Grand Question by his close colleague, A. C. Graham, a British sinologist. Graham's objections to the Grand Question are rather of a methodological nature; they are also more radical in their negative outcome. His argument has been set forth in a deeply probing article "China, Europe, and the Origins of Modern Science: Needham's *The Grand Titration*." Together with two papers by Sivin, this is among the most stimulating contributions to the underlying problems of comparative history of science that I have ever seen. Graham's subtle argument comes ultimately down to the simple point that one cannot answer questions about what failed to happen. His principal reason for thinking so appears to be that, when a range of necessary conditions for an event to happen presents itself, one must show that their effect could not have been reached by other means as well, and this can only be done if the event for which the condition was necessary did in fact take place. His counsel, therefore, is that we should ask the question only in the 'positive' variant: not 'why did the Scientific Revolution elude China?' but rather 'why did the Scientific Revolution happen in Europe?' This advice neatly closes the circle of our entire inquiry, as Graham (not a professional historian of science) seems blissfully to assume what the two preceding chapters of the present book have taught us to call into serious question: the confident expectation that students of the Scientific Revolution are in a position to come up with definite causation patterns not at all in need of independent testing against the experience of those civilizations where the Scientific Revolution *failed* to occur.

True, the appearance of such a logical circle had been predicted all along by Koyré (section 5.2.1: "all explanations, however plausible they may be, ultimately turn around in a circle"). Nevertheless, with all my doubts about the tenability of much of Needham's work on his Grand Question, I remain persuaded that, along with the personal stimulus it gave him to embark on the entire work in the first place, it has also yielded some truly valuable insights into that very Scientific Revolution about which Graham believes that one cannot really become wiser by comparing it with what did not happen elsewhere.

But it is also true that Needham's methods of comparison deserve some harsh critical treatment. He has himself called his method one of 'titration': "the determination of the quantity of a given chemical compound in a solution by observing the amount of a solution of another compound at known strength required to convert the first completely into a third, the end-point being ascertained by a change of colour or other means."[277] This means *in casu* that he and his collaborators "are always trying to fix dates" so as to establish priorities and to give credit where credit is due, in this way " 'titrat[ing]' the great civilizations against one another."[278] Whatever its merits in the thirties, professional standards in the writing of the history of science have meanwhile made such 'titration' obsolete. On reading Needham's answers to the Grand Question, far too rarely one feels that an entire corpus of thought has been studied 'from within', as one consistent whole, with no other purpose in mind for the time being than just to understand what it has to say. Another side to really the same coin is Needham's belief that humanity's march toward truth is inevitable and inexorable, and that science provides it. One need hardly go along all the way with present-day relativism to see that quite a great deal more room for the contingent and the unpredictable is to be desired in coming to grips with the what and why of the Scientific Revolution, both in the West and in other civilizations.

And therefore again: Are we yet ripe for an answer? Would it not be better to take *Science and Civilisation in China* as one giant mine of factual information (much of it quite reliable, especially in the later volumes, but other portions less so) but discard its interpretation entirely? Luckily, Needham's best critics have been much too sensible for suggesting such an easy way out. The answer to partly mistaken syntheses is not to give up the business of making syntheses, but rather to prepare better ones. Besides, grand syntheses are *always* mistaken in certain respects. There are *always* discrepancies between a conceptual skeleton and the historical facts that are adduced to clothe it with flesh, although it is also true that, in Needham's case, the balance has shifted more than one would wish in the direction of his preconceptions at the cost of the facts. There is this multiplier in his head which, as soon as he has come across a piece of text that calls up one of his pet convictions, compels him to squeeze out of the text inferences that go way beyond what the evidence can bear. Yet, on the other hand, the equally unusual excitement that one often experiences on reading Needham occurs precisely because one finds him pondering over minute details of some arcane, ancient Chinese philosophy and then the next minute he may be telling us all about the Faustian spirit—from what distinguishing features of European life this so-called spirit really stems, and how the Scientific Revolution cannot possibly be understood without due consideration of those features. Such grandness of scope is not to be lost in a cry for more up-to-date scholarship.

All this seems to me the more true because Needham's work has yielded some indispensable insights into the Grand Question and has also revealed fruitful areas of further research inconceivable without his preceding wrestling with the angel (as Lynn White has in all seriousness called Needham's persistent effort to find answers to his Grand Question).[279] We shall list those results and those promising venues for further research in the next section, thus bringing the China part of the present chapter to a close.

6.5.4. *Some Conclusions and Suggestions*

We consider first whether a conclusion about the possibility of a Scientific Revolution in China follows from Needham's researches that is altogether different from the one he has been willing to allow himself. We inquire next what little in the history of Chinese science seems to furnish an opportunity for testing explanations of the Scientific Revolution based solely upon the Western experience. Finally, we list those key ideas of Needham and his critics which appear to be fruitful ones for enhancing our understanding of the nature and causes of the Scientific Revolution.

Organic materialism reconsidered.   Among the most inspired passages in *Science and Civilisation in China* are those which Needham devoted to expressions of what he called the perennial Chinese philosophy of organic materialism. 'Going with the grain of nature'; *wu wei* as the highest form of action; the spontaneous flow of things; their organic wholeness as opposed to their mechanical divisibility; 'correlative thinking'; causation taken as interdependence rather than as successive 'if/then' relations: all such and many similar notions have been presented by Needham as a route out of primitive thought alternative to the one taken by the Western approach to nature. At that very point of his argument a conclusion seems to impose itself which Needham has ever refused to draw. But if we discard the hopes he has set upon such an organic philosophy as the required complement to modern 'billiard ball physics', and if we also discard his insistence that such a principal difference in respective approaches to nature has no independent interpretive validity but must in the end be reduced to overriding differences in Chinese and European society;[280] if we discard all these *a priori* convictions held over a lifetime, is there not every reason to arrive at a conclusion which generalizes what Landes had to say about the fate of timekeeping in China and the West? The conclusion would be that there appear indeed to be at least two roads which lead out of primitive thought. Both of these were fully worth pursuing; neither possessed an apparent in-built superiority. Just as Su Sung's water clock was probably a more accurate timekeeper than its mechanical counterpart in medieval Europe, just so one may well, if one likes, call conceptions of nature in traditional China more advanced in several respects than natural philosophy in Europe prior to the Scientific Revolution. Only, the one approach happened to possess greater developmental possibilities than the other. Organic materialism ultimately ran into a 'magnificent dead end', whereas somehow out of Aristotelianism and some other, scattered (and meanwhile enriched) remnants of the Greek legacy a kind of science could be forged that went on to conquer the world.

Such a view of things European and Chinese, which Needham has steadfastly evaded, cannot be proposed with any assurance, entirely dependent as it is upon data and interpretations advanced by Needham himself. If it does make any sense, though, it puts into a quite different perspective the historiographical status of Needham's Grand Question. The short answer to it becomes something like: China had no Scientific Revolution because such an outcome was not contained in the developmental possibilities of an organic approach to nature in the 'correlative' mode of the Chinese. This answer seems to throw us back upon our original point of departure: historians' efforts to understand the birth of early modern science from causes to be found in the thought and mode of life and outer events of western Europe itself. But this does not imply in its turn that our Eastern detour has been in vain, and that there is nothing at all to be learned for our purposes from the Chinese experience. We go on to list what these lessons are, starting with those few domains where China appears to offer a suitable testing ground for 'Western' explanations of the birth of early modern science.

Chinese science as a testing ground.   As a result of both Needham's working methods and the overall 'foreignness' of Chinese, compared with Islamic, science, relatively little material for testing purposes seems to present itself. We have already seen how for Needham's own explanations the Jesuits' efforts at disseminating Western science may serve as a fruitful testing ground; scholarly attention appears to be focusing increasingly in this direction. It might further seem as if Zilsel's idea of the social barrier between the scholar and the artisan and how its ultimate breakdown ushered in the Scientific Revolution may profitably be checked against the Chinese experience. However, Zilsel's explanation has been so compulsively projected by Needham upon China that an unprejudiced inquiry into the relations in traditional China between the scholar and the artisan (necessarily to be coupled with a thoughtful distinction between scientific and technological activities) remains to be done. When one recalls Zilsel's belief in the possibility of independently testing his thesis against the Chinese experience (section 5.2.4), the outcome is surely an ironic demonstration of the unexpected ways in which 'independent testing' in the social and the human sciences may go astray.

We seem to be left, then, with the case of the printing press. This, after all, was a Chinese, rather than a Western, invention (a statement that remains true, irrespective of whether Gutenberg independently reinvented the invention or rather had picked up some rumors about Chinese practice). No doubt Eisenstein's suspicion is right that the Chinese experience with paper and printing provides a suitable testing ground for her explanation of the Scientific Revolution. On this topic Needham has made some remarks which might well be employed as the starting point for such a testing procedure. His remarks strike one as particularly open-minded; in this domain nothing of ideological contention is at stake for him. Here are a few consecutive passages to whet the appetite:

> The sinologist is much less dependent upon unreliable manuscripts of fitful and sporadic location than students of other civilisations. . . . Since the making of paper began about + 100 and printing about + 700, practically everything in Chinese is either printed or lost. . . . The losses of Chinese literature have been incalculably great, partly because the flimsy replication medium of paper was invented so long before the mass replication technique of printing, but partly also because the great inter-dynastic upheavals and foreign conquests destroyed even whole editions of printed books. There are tens of thousands of books of which we know only the titles. Without printing there would have been a million.[281]

All of which suggests that (1) printing is a necessary but hardly a sufficient condition for a smooth transmission of the written word over successive generations; (2) not so much the invention of printing itself as, rather, how it went on to function in society is the crucial variable (an inference hinted at by Eisenstein, too, but not elaborated by her further; see the end of section 5.2.9).

European science and the Chinese experience.   One always wants more than one gets. Kepler turned the whole science of planetary motion upside down— yet we want him to have discovered the law of universal gravitation as well.

Mersenne, amid thousands of plodding pages through which he established a novel science of musical acoustics, made half a dozen truly brilliant, but entirely disconnected, remarks stretching far beyond his own science of music— we withstand with some difficulty the temptation to castigate him for failing to see that those loose ideas, if duly connected, were all that was required to forestall Sauveur by half a century. Needham has been the first to embark on comparative research in the history of science on the grand scale—we easily take the original act for granted and go on to harp on how many traps he has fallen into. But we should not cling to those, just as it would be unfair, and unwise, to conclude that his entire enterprise has been mistaken right from its inception.

At least three conceptions of Needham's in particular—conceptions which could only have arisen from his comparative point of view—are fraught with the most fruitful kernels of insight into the nature and causes of the Scientific Revolution. These are (1) the challenge posed by the study of science in non-Western civilizations to all historians of science to combine internal and external perspectives in enhancing our understanding of the birth of early modern science; (2) the expansion of the idea of the 'world of the more-or-less' to non-Western civilizations; and, finally, (3) the conception of European civilization as 'a culture of rovers'.

1. *The challenge to the discipline of the history of science.* Joseph Needham has maintained that the study of the history of science in other civilizations than the West poses a unique challenge to the profession. He has couched this conviction in rousing prose, which should not mask the kernel of precious truth that lies hidden there:

> If you reject the validity or even the relevance of sociological accounts of the 'scientific revolution' of the late Renaissance, which brought modern science into being, if you renounce them as too revolutionary for that revolution, and if at the same time you wish to explain why Europeans were able to do what Chinese and Indians were not, then you are driven back upon an inescapable dilemma. One of its horns is called pure chance, the other is racialism however disguised.[282]

The case is characteristically overstated, to be sure, since it is by no means clear why no independent explanations could be found in terms of the specific forms scientific thought has taken in any given civilization all by itself, propelled by its own, internal logic. On the other hand, one would have to be fully content with self-imposed blinders if one attempted to come to grips with the inescapable question Needham has put before us without considering the specific guises political and socioeconomic life in other civilizations has attained or the extent to which the dominant values in a society have fostered science. True, the thing that, with Needham, passes for 'sociology' has more often than not come down to mere obsolete, Marxist rhetoric, but that is surely no reason to forgo the search for a better sociology. I have done my best to show in the present book that the conceptual groundwork for such a sociology has been with us for quite some time. Also, one must be more careful than Needham has been in seeking precise links between the internal and the external factors, which in my view is done best by always starting with the former, not because they are necessarily primary but because they yield the best questions. Here Lynn White's wise words are surely to be heeded:

> A complex society like that of traditional China or that of medieval Europe . . . is composed of a vast web of activities, ideas, and emotions. Innovation can start anywhere in this network and then travel in the most unexpected ways and directions, producing results that could scarcely have been guessed before the fact.[283]

I just called Needham's Grand Question 'inescapable', and indeed I think that here lies one great message that Needham has passed on to the discipline of the history of science. One may dismiss the Grand Question from a positivist assurance that it is 'meaningless'; one may spell out in detail why it cannot be answered, just as any astronomer in the late 16th century could tell Kepler that his questions were meaningless and could not be answered—the question is not going to go away for all that. Refusing to ask the question in one formulation or another comes perilously close to explaining the Scientific Revolution by chance and, here I fully agree with Needham, "to attribute the origin of modern science entirely to chance is to declare the bankruptcy of history as a form of enlightenment of the human mind."[284] With the as yet unfinished monument of *Science and Civilisation in China* and its accompanying volumes before us, we cannot go back and ignore the principal lessons the history of science in the East has in store for students of the Scientific Revolution: lessons in the sense of bringing up fresh ideas; lessons to be gained from testing procedures; lessons stretching all across the worlds of both the intellect and its immediate and farther-flung surroundings.

2. *The 'Vincian' world prior to the Scientific Revolution.* We have seen how Needham expanded Koyré's conception of the 'world of the more-or-less' so as to encompass those civilizations which, in Koyré's here somewhat narrow-minded opinion, had lacked science altogether. Before the advent of 'the universe of precision' in 17th-century Europe, not only science in medieval Europe but advanced science in every other traditional civilization still partook of the world of the more-or-less. This was the world in which "no one ever sought to go beyond the practical usage of number, weight, and measure in the imprecision of daily life," to cite Koyré, or the world of "[the] intrinsic and essential vagueness [of] hypotheses of primitive or medieval type [making] them always incapable of proof or disproof," to cite Needham.

I regard this conception as a splendid definition-in-a-nutshell of the full import of the Scientific Revolution. It marks in the most succinct manner what fundamental change the birth of early modern science has entailed. Koyré's powerful idea has remained very much underexploited, and so has Needham's equally powerful expansion of it. Beyond using it to set off ethnically bound structures of thought all across the Old World against the universality of the type of science which was born in Europe, not even Needham has done much with the idea.[285] One reason for this neglect in their own work may have been the close relation of the idea to Euclidean geometry: Both Koyré and Needham tended to confine their notion of 'precision' to geometrical precision. This brings the idea needlessly close to the kind of mathematical reductionism that has been one consequence of the Scientific Revolution. Both Koyré and Needham have regretted this consequence. Neither thinks that post-17th-century science has been particularly beneficial on balance, chiefly because both think that the mathematical approach to natural phenomena has tended to rob with increasing vehemence and power the 'universe of precision' of the very features that make life livable. Perhaps this is why neither has pursued further the characterization of the Scientific Revolution as the transition from the 'Vincian' to the 'Galilean' world.

It so happens that Needham's one passage on this concept has inspired his critic Graham to expand the idea of the universe of precision further still, in such a way as further to weaken its restrictive link with Euclidean geometry. Graham, in the very essay which starts by doubting the appropriateness of asking Needham's Grand Question, goes on to take up for critical scrutiny some of Needham's answers and ends by giving a by and large original answer of his own making (inspired, to be sure, both by Needham's passage and by an idea of his close colleague Sivin). Graham begins this answer by observing that the original strangeness of Chinese texts disappears once one realizes that its authors reason just as we do in our daily lives—by weighing advantages and disadvantages, by invoking analogy, by citing precedent, and so on. However,

> what we do miss, as Nathan Sivin observes, is 'the notion of rigorous demonstration, of proof'. This concept of proof, it may be necessary to insist, is narrower than any vague idea of 'Reason' that could be supposed to characterize Western thought in general. Even in the West it requires quite a special temperament to appreciate the full value of the geometrical proofs we learn as schoolchildren, demonstrations that are far in excess of the ordinary demands of common sense. . . .
>
> It is therefore hardly profitable to make the vague accusation that Chinese thinkers lack our respect for reason. . . . What matters is that most Chinese thinkers (like ourselves, in most of our thinking outside the exact sciences) exchange arguments of varying and indefinite weight without seeing any point in putting premises and conclusion in the same form, filling in all steps however obvious, and pressing every line of thought to its logical end.[286]

The notion of the universe of precision inhabited by the exact sciences has been expanded here beyond its origin in Euclidean geometry, thus making it much more widely applicable. In a seemingly simple manner a very clear-cut distinction has been reached between the 'Vincian' world from before the Scientific Revolution and the 'Galilean' world ushered in by that revolution. One may go on to assay what precise measures of mathematization, of experiment, of empirical observation, etc., went into the making of early modern science—here, I think, is one irreducible core of the transformation wrought by the Scientific Revolution. And it remains, as it always has, but now perhaps with reinforced cogency, to inquire into the conditions that made the appearance of this new world of rigorous proof and disproof possible.

3. *What set Europe apart.* The comparative search for the sources of Europe's uniqueness has gone on for quite some time already. Historians have taken part in the search, and so have sociologists, economists, philosophers, theologians, and so on and so forth. With two exceptions only, historians of science have kept outside the debate. This is most unfortunate, because that centerpiece of their preoccupations, the Scientific Revolution of the 17th century, is not only one of the primary manifestations of Europe's uniqueness but was also (as I have argued in the introductory chapter) instrumental in solidifying Europe's peculiar ways further. Therefore the debate in question can be ignored by historians of science only at their own intellectual loss—but also at the loss of all the others. No debate over the why of Europe's unique performance as the one traditional society to usher in modernity can be complete without full consideration of the Scientific Revolution in the dual role just indicated.

The two exceptional historians of science to have contributed to this debate, and to have conceived their own life's work as partaking in it, are Lynn White and Joseph Needham. In the profession of historians of science, both may be called honored outsiders or even mavericks: White in the sense that he was primarily a historian of technology with a rather odd perspective on early modern science and its medieval predecessor,[287] and Needham in a sense that has amply been discussed in the preceding pages.

Needham's conception of what distinguished Europe most from China has found expression on two distinct levels. On the one hand, there is this brilliant shortlist of features: Europe's spiritual dichotomies; its explorative urges, owing, partly, to a perennially adverse balance of trade with the East; its pluralistic divisions, due, in part, to the heterogeneity built into its alphabetical script. Such generalizations are the very stuff an enhanced understanding of how this peculiar civilization could bring forth the Scientific Revolution is to be made of. It is most unfortunate that Needham has never more than just touched upon such features in passing, preoccupied as he has remained for a lifetime with, in his opinion, larger bait. This larger bait is the overall social structure of China as compared to Europe. But here Needham has fallen very far short of what appears most needed: a subtle, nonreductionist historical-sociological approach to the comparison of the societies of China and the West. If we wish to gain a glimpse into basic features of such a comparison we must direct our steps beyond the Needham track and see what is to be learned there.

Beyond the Needham track.   Two commentators—Benjamin Nelson and Nathan Sivin—have called attention to how odd it is that Needham has never even mentioned (let alone made use of) Weber's extended comparison between Chinese and European society intended to explain why modern capitalism arose in the latter but not in the former.[288] True, this has never been Needham's problem (even though in his own mind the birth of capitalism has always remained indissolubly linked with the birth of early modern science). True, too, some important components of Weber's comparison (which dates from 1913/1920) have been shown by sinologists and sociologists alike to be seriously defective in light of decades of subsequent research. True, again, nowhere are these deficiencies more glaring than in the very domain of the natural sciences, of which Weber denied the very existence in the history of indigenous Chinese thought.[289] Despite all this, Weber's essay in comparative historical sociology has one great advantage over Needham's, which is that it is built upon a very carefully constructed basis of very subtle concepts which (notwithstanding their frequent origin in Western phenomena) are well capable of being applied to the analysis of other civilizations.[290] Weber's analysis of patterns of domination, of 'elective affinities' between distinct social layers and their respective religious inclinations, and so on puts us in a scholarly world infinitely more sophisticated than that of the hardy perennials in Needhamian sociology: the merchants and the bureaucrats and the antagonism between them. If comparative research into the sciences of China and of Europe is to gain further ground, it may perhaps fruitfully be done by linking Needham's best results on features of Chinese science with a social analysis that preserves what strengths remain in Weber's approach.

   Still, there is one overriding issue which requires tools from the sociologist's conceptual apparatus not furnished by Weber, who contributed nothing to a sociology of science. This is the question of what may be called the broad pattern of Chinese science. Was it governed by steady growth or rather by alternating fits of advance and stagnation? Against Needham's favorite picture of unbroken advance, the expatriate Chinese physicist Wen-yuan Qian has proposed in 1985 an image of what he has called 'the great inertia'.[291] For him, an unsolvable paradox resides in the ongoing and, to him personally, shameful inability of China to modernize as set against the supposed glory of the scientific achievement of traditional China proclaimed by Needham. Granting a small number of achievements, Wen-yuan Qian concludes nevertheless on both historical and methodological grounds that as a whole traditional China represents a case of 'scientific stagnation'—the very picture so much resented by

Needham. Whatever the merits of Wen-yuan Qian's argument, it would surely require support from considerations which only a historical sociology of science can provide. As argued before, the conceptual apparatus developed and applied by Joseph Ben-David would appear to provide a splendid starting point for finding out more about the developmental line of the science of traditional China. One further step in the same direction has recently been taken by the historian of science Harold Dorn. In his book of 1991, *The Geography of Science,* he set forth a rich, thought-provoking thesis on two distinct landscapes for science, obeying a 'hydraulic' and a 'rainfall' pattern, respectively. On the one hand, there are the 'hydraulic' societies in history. Dorn adopts a moderate version of Wittfogel's theory of 'Oriental despotism'—where climate requires for a sustained agriculture to become possible a good deal of water manipulation (irrigation, drainage, flood control), there is a strong tendency for government to turn into a centralized bureaucracy. Such bureaucracies, Dorn argues, have often displayed throughout history a strong interest in scientific research of a utilitarian kind. Unlike the Greek variety, science in hydraulic societies was carried out in institutions, it was most often done anonymously, and it was almost wholly directed toward knowledge useful for society. This particular pattern keeps returning, whether we consider science in Egypt, Babylonia, India, China, or Mayan civilization. In an analytical framework such as this— where big waterworks, a centralized bureaucracy, and the useful application of science, as opposed to the 'pure' variety that was born only in Greece, all go together—ample room seems available to understand how science in China represents one specimen among many others of a particular sociological type and, as such, could not but take a different route from science as cultivated in the manner of the Greeks.[292]

Such a conception certainly has an aura of plausibility about it. Still, Needham's researches have made it clear that science in traditional China was a great deal more than a purely bureaucratic affair, even though his impressions of the place of science in Chinese society leave much to be desired. Again, Ben-David's approach furnishes opportunities for taking up in less cursory fashion than Needham has done the sociology of Chinese science: in particular, the joint issues of *who* cultivated science as distinct from technology, and in what, if any, layers of society these people found support. Furthermore, how was science regarded by both its cultivators and the wider public: as an activity valuable in its own right? as inappropriate for whatever reason? as an innocent pastime? It is key questions such as these that must be answered about the place of Chinese science in society before a full-fledged comparison between science in China, the Muslim world, and the West becomes at all feasible.

The need for such an overall comparison has been urged upon us by no one quite so imperatively as by the versatile scholar whose name keeps recurring in these pages—Joseph Needham.

6.6. THE HARVEST OF THE COMPARATIVE ROUTE

"A full-dress comparison," so Needham wrote in 1961, "will eventually have
to be made between the effects upon science and technology of the Confucian–
Taoist view of the world as compared with those of Christendom and Islam."[293]
Here, in his view, lies the true dichotomy:

> In spite of the schoolboy ideas which we of the West acquired about the great
> gulf between Crusaders and Saracens, one feels, on better acquaintance, very
> reluctant to regard Arabic civilisation as 'oriental' at all. The culture of Islam,
> whatever its desert origins, was really much more closely allied to the European
> culture of mercantile city-states than to the Chinese culture of agrarian bureau-
> cratism.[294]

One need not adopt the reasoning hook, line, and sinker to find the conclusion
illuminating up to a point, at least where the history of science is concerned.
Now that we have reached the end of our discussion of the Scientific Revolution of western Europe and of such reasons as have been advanced to explain
why no comparable event took place in either the Muslim world or China, we
have at long last put ourselves in a position to compare with one another the
fates of science in these three civilizations.

What conclusions we reach are very much to be qualified by perhaps the
greatest lesson to be drawn from the entire preceding account: the fact that the
independent study of the exact sciences of China and the Muslim world, of
the unity of science or the lack thereof, and of the values of science in society
still has a very long way to go. Keeping this proviso in mind all the time, we
proceed now to connect those valuable insights we have encountered along the
way. In order to do so, we take up, as a heuristic guiding thread, our two questions of (1) the possibility of our composite 'Galileo' turning up anywhere
outside Europe and (2) whether his hypothesized appearance might have led
to a sustained tradition of scientific research such as has been ushered in by
the Scientific Revolution. Is a neat dichotomy along the lines of China on the
one hand and Islam and Christendom on the other, such as suggested by Needham, what we eventually come up with?

An Arabic or a Chinese 'Galileo'?   Taken literally, these questions surely
yield nothing but empty speculation. What they are meant to get across is a
sense of the structural features of science as cultivated in one or another civilization. Taken in this sense, our conclusion has been that there are no obvious
structural reasons for why no 'Galileo' could have manifested himself during
the Golden Age of Islamic science. We have noted that Islamic science had
come to encompass both the full corpus of Greek science (now appearing as a
unified whole for the first time) and considerable enrichments from both India

and China. Medieval Europe, on adopting portions of the Greek legacy and of its Islamic enrichments, inherited somewhat less than this. As Needham has demonstrated, the Chinese contributions were filtered out of the translation effort, and it is also fairly unlikely that Europe adopted directly any scientific findings (as distinguished from technical inventions) from China later. In a sense, therefore, the Western-medieval heritage was poorer than the full corpus of Islamic science had been. Must we now conclude that the few additions that accrued since to the Western heritage prior to the Scientific Revolution made the key difference? Note that the Archimedean heritage was not truly part of those additions. It had rested as an isolated little pocket in the late Hellenistic corpus and in the Arabic scientific corpus, as well as in that of the European Middle Ages, not to gain prominence until mathematical humanists seized upon the legacy subsequently to be exploited to the full by Galileo. The study of planetary motion, too, was carried out in basically the same manner in late Hellenistic science, in Islamic science, and in European science prior to Kepler. From the amount of available knowledge and the quality of the available procedures in the key sciences where the breakthrough was eventually to take place, then, no decisive reason appears to come forth for why only science in the medieval West underwent a further development to such a point as to make the work of 'Galileo' possible as the very next step, however radical. It is not clear why such a further development should not have been in reach of Muslim civilization, too. Therefore I have concluded that there is nothing to rule out as principally inconceivable the appearance of 'Galileo' during the Golden Age of Islamic science.

With China, the situation has appeared profoundly different. Despite the great ingenuity of the Chinese, despite much experimentation, much accurate observation, much wonderful classification, and despite many ingenious scientific instruments, no Scientific Revolution was anywhere near in sight along lines at all comparable to the one in Europe. We have also seen how very implausible speculations are in the end on the possibility of an 'organic' Scientific Revolution in China. Translated into the terminology adopted here, all this means that, in traditional China, no 'Galileo' could possibly have come forward. Which, of course, at once robs of any possible meaning our second question of whether 'Galileo' could have prevailed in China. One more reason why this question, as far as China is concerned, cannot be answered is that so little is yet known about the place of science in Chinese society. As opposed to this, we have found, as the net result of our survey of explanations of the decline of Islamic science, a few very good reasons for why no 'Galileo' could have prevailed in the Muslim world: because of the ongoing invasions, and because the value system inherent in the Islamic faith robbed scientific activity of indispensable support both from the scientists themselves and from the community at large as represented by its most vocal and influential intellectual leaders.

The overriding opposition. The considerations just advanced imply that I cannot go along all the way with Needham's dichotomy between the world of the Far East and the world around the Mediterranean. True, Islamic science and Western science stem from the same root, and, from the point of view of their respective contents, the way they were cultivated naturally had much in common. But the environments in which the two functioned were drastically different—so different, in fact, as to lead in the one case to decay and extinction and in the other to the Scientific Revolution. Over and above Needham's dichotomy, therefore, stands the opposition between the range of traditional societies and the one traditional society that broke through its limits and went on to create, to its own astonishment, the modern world. 'The differentiation of the West from the Rest', to a deepened understanding of which Needham has greatly contributed, still stands before us as the ultimate problem to be faced. And we end this chapter by introducing from the literature a wider perspective on the Scientific Revolution—one that may serve to bring into sharper focus the process of differentiation from the point of view of natural science.

Approaches to nature in East and West and their religious backgrounds. In his book of 1968, *Science and Civilization in Islam,* the Iranian historian of science (also a Sufi philosopher and an American-trained physicist) Seyyed Hossein Nasr argued that, toward the end of the Middle Ages, the civilization of Islam quite consciously declined to take the necessary steps toward a Scientific Revolution. Its scientists might have pursued their researches to their logical conclusion, but they refused to do so because they sensed that no good could come in the end from *scientia* devoid of *sapientia.* The Scientific Revolution as the triumph of science without wisdom—from such an anticipated image late Islamic scientists recoiled, finding a timely way back to an introspective contemplation of divine unity as an end in itself, to which science, in the grand scheme of things, ought to remain subordinate. Leaving aside the question of whether what we are wont to call the 'decline' of Islamic science resulted from a conscious decision,[295] it is clear that the interpretation given here implies a certain idea about the place of nature in the world-pictures of Christendom and of Islam, respectively. Widening further his point of view in another book, *Man and Nature* (also of 1968), Nasr wrote:

> The main reason why modern science never arose in China or Islam is precisely because of the presence of metaphysical doctrine and a traditional religious structure which refused to make a profane thing of nature. Neither the 'Oriental bureaucratism' of Needham nor any other social and economic explanation suffices to explain why the scientific revolution as seen in the West did not develop elsewhere. The most basic reason is that neither in Islam, nor India nor the Far East was the substance and stuff of nature so depleted of a sacramental and spiritual character, nor was the intellectual dimension of these traditions so enfeebled

as to enable a purely secular science of nature and a secular philosophy to develop outside the matrix of the traditional intellectual orthodoxy. Islam, which resembled Christianity in so many ways, is a perfect example of this truth, and the fact that modern science did not develop in its bosom is not the sign of decadence as some have claimed but of the refusal of Islam to consider any form of knowledge as purely secular and divorced from what it considers as the ultimate goal of human existence.[296]

Here the West is being offset against the Rest because it alone among the world's civilizations robbed nature of its sacredness and turned it into a profane thing. It need not surprise us that Joseph Needham, for whom the Scientific Revolution betokens the ushering in of universally valid science rather than a profound trespassing and distortion of the natural order of things, would have none of such a view. In one of the 'Author's Notes' in *Science and Civilisation in China* he has taken issue with it, arguing that Nasr aims for a return to a medieval world-view, to which there can be no going back, and that, in a striking phrase, "the scientist must work *as if* Nature was 'profane.' "[297]

This observation invites a further question, which Needham has not, to my knowledge, gone on to consider. What peculiar conception of nature provides such liberty to treat it 'as if it were profane'? Adamant as Needham is about the ultimately socioeconomic explanation of the different scientific experiences of the world's civilizations, he has never made a detailed comparison between attitudes toward nature in different civilizations and how these may go back to differences in attitudes toward the sacred and the profane. In other words, what we do not find with Needham is anything like a counterpart to Weber's comparative researches, in which, not the 'economic ethos', but rather the 'conception of nature' in the great religions of the world is the principal domain of inquiry. There is one exception, though. Needham once noted in passing that the Buddhist contribution to Chinese thought was, and could not but be, scientifically sterile from the outset, because

> one of the pre-conditions absolutely necessary for the development of science is
> an acceptance of Nature, not a turning away from her. If the scientist passes the
> beauty [of Nature] by, it is only because he is entranced by the mechanism. But
> other-worldly rejection of this world seems to be formally and psychologically
> incompatible with the development of science.[298]

Considerations such as these lend themselves well to being expanded further. Weber, in his comparative studies, went to great lengths to distinguish a variety of attitudes toward the world among different religions, extending from the world-rejection of the Indian religions to the harmonious acceptance of the world-as-it-is that is taken to characterize Chinese thought and life. Not only did Weber seek to connect these distinct attitudes to the socioeconomic positions of a variety of social layers to which the most vocal bearers of the world-views expressed by these various religions belonged, he also examined their attitudes toward the world to find out (this being the ultimate objective of his pertinent researches) whether these attitudes could give rise to the kind of methodical, self-disciplined, rationalized orientation of life which he regarded as indispensable for the spirit of modern capitalism to arise.

Ample room seems available to pursue such in-depth comparative studies with a somewhat different objective in mind: to determine what attitudes toward the world are fitted best to that peculiar treatment of nature that manifested itself in the Scientific Revolution. A possibly helpful pointer in that direction has been given by the Dutch historian and philosopher of culture P.F.H. Lauxtermann.[299] He thinks that modern natural science could not emerge in such an otherworldly-directed civilization as that of India, in which man sought to renounce nature altogether and to turn rather into himself. Nor could it emerge in the this-worldly civilization of China, where life was oriented toward harmony with nature. Early modern science could arise only where we find the typically Judaeo-Christian awareness of man's tragic duality: taking part in nature while transcending it at the same time. Wandering thus on the narrow ridge between the renunciation of nature and the absorption into it, people in the West have adopted a dual attitude which alone made possible the very type of involved detachment toward nature that enables the man or woman of science to turn it into an object. Once taken for an object, nature can be subjected to mathematical analysis of the phenomena it presents to us, and to that characteristically Western determination to bend it to human will.

Promising as such notions are in my opinion, they run an inherent risk of explaining too much (as their propounder is well aware). Even if the Judaeo-Christian conception of nature is to be considered a necessary condition of the Scientific Revolution, it is certainly not a sufficient one, if only because the conception had been around for many centuries prior to the birth of early modern science. Hence, within the broad framework of the Judaeo-Christian view of things more specific determinants must still be sought.[300] This search has gone on for quite some time already. It is that particular historiographical tradition to which we keep returning in which those features of Europe are identified that made the region break with traditional modes of life and thought. Among these, such specific features have been advanced as the relative autonomy of the medieval city, the discipline of Benedictine monastic life, administration by law (Weber), the so-called 'military revolution' (Roberts), the mechanical clock (Landes), and a unique technological drive (White). More such features are likely to be yielded by future research; a deeper insight into their inner connectedness (and this is where Needham's as well as Saunders' and Sayili's special contributions to the issue come in) is equally likely to be gained in the future. None of these and similar features can serve to explain that the Scientific Revolution was bound to occur. Taken together, all may go quite some way toward explaining why, *if* such an event was in the offing, Europe could not but be the place where it materialized.

A very long trail of thought has moved us through the present book from the European Middle Ages conceived as a specimen *par excellence* of stagnancy and scientific sterility toward the very same period being regarded as the seedbed of much that set Europe apart, not least where its unique product of early modern science is concerned. Having at long last completed our round-trip through assorted bits and pieces of the history of the world's science, we will now draw the lines together in considering where fifty years in the life of the concept of the Scientific Revolution have meanwhile led us.

PART THREE

SUMMARY AND CONCLUSIONS: 'THE BANQUET OF TRUTH'

To exclude such inquiries [using bold, causal hypotheses] would be to secure ourselves from the poison of error by abstaining from the banquet of truth.

William Whewell

❧ SEVEN ❧

THE SCIENTIFIC REVOLUTION: FIFTY YEARS IN THE LIFE OF A CONCEPT

7.1. INTRODUCTION TO PART III

In each case, after exploring the movement in its own terms, we shall give it the most constructive appraisal we can—asking how it might fulfill its promise more completely than it already has and at what points the limits set by its present presuppositions need to be left behind.

—E. A. Burtt, 1965[1]

WHAT KIND OF CONCLUSION would best fit the preceding argument? The answer depends on how one characterizes the argument that has preceded.

The history of the historiography of a given topic of the past may be written in a variety of ways, with two extreme possibilities presenting themselves. One way would have been for me throughout this book simply to report summaries of historians' conceptions of the birth of early modern science, listing these chronologically or topically as the case required. Large stretches in the preceding argument do come down to precisely this. Such an act of historiographical bookkeeping has a certain value in its own right, I think. Reducing, as it were, a small library to the confines of one book provides the reader with handy access to significant portions of the literature, and also perhaps with a sense of what has been accomplished in the field and where fresh researches may profitably be anchored. Still, more could be made of the exercise, or so at least I thought when I embarked upon it.

That 'more' might have consisted—and this is our other extreme—in a procedure like the following: to start with a full-fledged conception of the Scientific Revolution and its causes fixed unmoveably in my head and, from there on, to produce a survey of the existent literature on these subjects by way of taking every contributor's measure as set against this gauge of my own prior making. If this had been my purpose, the conclusion to the argument would almost write itself. It would in bald, authoritative sentences spell out the Truth about the Scientific Revolution, implying if not expressly stating that the reader had better subscribe to it. Conclusions like this, in which one aspect of a given domain of the world is proclaimed the one and only fitting key to that domain, with the entire preceding argument organized to provide evidence for its over-all validity, frequently make their appearance in the world of scholarship (not so often in the genre of the history of historiography, to be sure).[2] Their fate may be to fall by the wayside or to catch on and become fashionable for a while. In either case they will eventually be realized by the world of scholar-ship to be no more than just one contribution (whether more or less enlight-ening) among many, destined to live on, if at all, in that more modest and more fitting guise.

But it has not been my purpose to sell to the reader any particular concep-tion of my own of what the birth of early modern science really and truly was about. Wishing to proceed neither deductively (measuring everything along the way against a fixed standard) nor inductively (starting with a blank mind), I have been seeking a middle way between these two possible extremes. On the one hand, I thought no harm was involved in the absence, in my mind, of a definite, clear-cut picture of the Scientific Revolution when I set out on what has, all along, been for me much of a voyage of discovery, with more and more elements of what I came across beginning to fall into place as I went along. For, on the other hand, I could hardly fail to come to this study armed with certain broad, less or more explicated ideas of my own. These have served me throughout as guidelines—guidelines to organize the account; to find hidden treasures in the literature I surveyed; to connect the thoughts of one historian with those of others; to discover inconsistencies and trace these to their, often enlightening, roots; to seek, in short, common denominators.

To seek and, very gently and cautiously, to establish common denomina-tors seems to me a vital task to perform for the historian engaged in writing the history of the writing of history. The benefits, I venture to hope, may extend beyond the world of historical scholarship. There are many scholars outside the historical discipline who would like to make use of historians' findings for their concerns in, for example, social science or philosophy. They occasionally complain that the historical products they encounter and would like to utilize for their own, mostly more generalizing purposes are insufficiently homoge-neous for straightforward usage. They demand (as one sociologist of technol-ogy once inimitably phrased it) a 'homogeneous data base' from the historian.[3] Historians are wont to laugh uproariously when confronted with such an im-possible demand, and go on to ignore it entirely. The ensuing frustration on the part of scholars in search of data fit to be generalized may lead them either to homogenize pertinent historical writings themselves, to the enduring wrath of the historians, or to get exasperated and quit the past altogether, to their own intellectual loss. It is here, in these unhappy border regions where mutual irrita-tion and incomprehension reign, that the historian of historiography has an important, almost wholly neglected task to fulfill—one that in fact extends to his fellow historians as well. He naturally refuses to squeeze historians' interpretations and explanations into one homogeneous data base, for life is much too rich for such a procedure—far too often, in history, the exceptional, the unique, the unexpected, and indeed 'unexpectable' have determined what happened. But on the other hand, where tens of thousands of pages have been devoted over many decades to the understanding of one particular episode of the past by historians who, being human like all of us, are different in what they want to achieve, in how they go about their inquiries, in what kind of facts catches their attention, in how they marshal facts into conceptions of varying scope and depth—where one is confronted with such a bewildering diversity it may be useful to have a guide ready to map these products of the historical imagination at its best, to sort them out, and to bring a measure of coherence to what has in the meantime come to strike both the outsider and the professional as a dizzying chaos. If this sorting out is not done with the requisite respect for its variety, which is the variety of life itself, the exercise has been in vain. Throughout the previous chapters, therefore, my aim has been to strike a proper balance between ungraspable chaos and unpalatable homogeneity, by means of a procedure which it might be good to term 'gentle homogenization'.

Those broad guidelines which assisted me from the outset have been subjected to a process of ongoing cross-fertilization with historians' ideas that formed my proper subject. Thus sharpening along the way, they yielded in the end a somewhat coherent pattern. Or rather two patterns. Half a century of historiographical effort began to acquire a shape of its own, and so did the phenomenon at which the effort has throughout been directed—the Scientific Revolution itself.

Hence, our concluding Part III is made up of two chapters. The present chapter is meant to outline the historiographical picture as it arises from the preceding survey. In most of the book I felt compelled by my material to present it thematically rather than in chronological order. The summing-up now takes the form of a stage-by-stage account of fifty years in the life of the concept of the Scientific Revolution. I take the concept from its 'prehistory', through its coinage by Koyré, to its later adventures close to the present day (that is to say, to the early nineties). The issue of the causes of the Scientific Revolution is woven through the account, which is rounded off with a perspective on what I think would be wise, and what would be less wise, directions to take in future scholarly dealings with this versatile, yet indispensable concept. I hope that what would inevitably have appeared as a bloodless skeleton if presented right at the start now takes on life through the flesh that has been examined throughout the main body of this book.

In the next and final chapter I set forth in highly condensed form my present understanding of the Scientific Revolution as it has gradually taken shape in my mind through writing the present book. And I wish to caution very expressly that I do *not* regard my own conception of the Scientific Revolution as definitive in any way, nor has it been my purpose in this book to arrive at such a conception. My primary aim has been to present to the reader a world of historical writing of great beauty and depth on a topic of immense importance, irrespective, in one sense, of what my own opinions on the subject are. Over and above all else I have striven to be *fair* to the historians whose ideas I have discussed, and whose imaginative and penetrating minds I have come to admire more and more as I went along. *These* men and women who devoted their scholarly lives to no less worthy a subject than the birth of early modern science have been my chief topic. In writing the present book I have not so much aimed to convince the reader of any viewpoint of my own, to the extent that I have one, but rather to induce in him or her, through an exposure to the combined ideas of these eminent scholars, trains of thought he or she might not otherwise have set in motion.

7.2. THE RISE AND DECLINE OF THE CONCEPT OF THE SCIENTIFIC REVOLUTION

Even though Enlightenment thought took a great deal of inspiration from the illumination of the human mind brought about by 17th-century science and also distilled out of its perceived course over time a conception of science as developing in a revolutionary, convulsive sort of way, thinkers in the 18th century did not make a *historical* category out of the episode. Only Kant went ahead to categorize it; inevitably his category was of a philosophical rather than of a historical nature.

In the 19th century two budding historical categorizations of the birth of early modern science emerged, soon to vanish almost without trace. Whewell came to a deep understanding of the transformation in the sciences during the period of their most drastic revolution; yet he did not distill a concept out of the event, owing to the overall organization of his inquiry but chiefly because of his overriding, philosophical interest in revolutions rather than in one Revolution. Rather than making up for this, the subsequent rise of positivism precluded a fertile follow-up to such a promising beginning. Still, another kernel of historical categorization came from those very quarters. Comte saw what Whewell failed to see: the structural unity of the history of science. But he lacked completely what Whewell possessed in abundance: a sense of history. Positivist preconceptions continued to dominate views on 17th-century science all through the 19th and early 20th centuries, with Mach providing the most influential example of how this philosophy stood squarely in the way of such historical conceptualization as was increasingly marking other branches of the study of history at the time.

This situation did not change until one philosopher of science with strong positivist leanings, Duhem, became interested in the history of science (for reasons residing more in his unique set of political and religious beliefs than in his philosophy) and turned his sudden discovery of medieval science into a full-fledged theory about the *true* birth of early modern science as having taken place in the 14th, rather than in the 17th, century. This theory—as inherently unlikely as it was presented persuasively—brings us to the edge of the point where the concept of the Scientific Revolution could arise.

Another strand, one that went directly into the making of that concept, was the insight of the American philosopher Burtt, arrived at in large measure on the strength of his own thought, that a study of 17th-century science could help explain the mathematical reductionism of science he had come to admire almost as much as he detested the threat it implied to the autonomy of the human spirit. Out of this exercise came a book that knit events from Kepler and Galileo through Newton (with a Copernican prelude) much closer together than had ever been done before.

It is from these two strands, in particular—Duhem's outrageous theory and Burtt's coherent interpretation—that the mathematical philosopher Koyré drew inspiration to turn his antipositivist creed to new, productive usage. It so happened that the concept of the Scientific Revolution was born when Koyré turned his attention away from the history of mysticism and religion to the history of early modern science, just at the time when Burtt was exchanging the latter for the former. Thus Burtt's *Metaphysical Foundations* served as a bridge for both men, for Koyré on the route toward the history of early modern science and for Burtt on the route away from it—in one sense this book forms the pivot in the story of how the concept came about.

The concept was forged in *Études Galiléennes,* published a little more than half a century ago now. In the guise originally given to the concept by Koyré, it was as narrow in temporal scope as it was sharp in its definition. The Scientific Revolution, now emerging as a full-blown *historical* concept for the first time, was seen as a mutation—no less sudden for having been long prepared—in human thought. The mutation was taken to have occurred during a few decades before and after 1600 and to consist of the replacement, by Galileo and Descartes, of the closed, purposive, qualitative Cosmos of the Greeks and of medieval Europe with the conception of the infinite space of Euclidean geometry in which minute particles wander aimlessly to and fro, to be studied by a human mind inexplicably present there as well. The concept of the 'Scientific Revolution' denoted, in its original guise, not a historical period so much as an event, or rather a highly interconnected range of events. Its coinage by Koyré brings to a close the story of the 'prehistorical' Kantian and Enlightenment conceptions, the varied intermediate labors, the birth, and the first cries of this new, historical concept. We go on to outline what happened during youth, adolescence, and early maturity.

The concept widens and turns inward.   In the late forties the concept began to catch on. It caught on in America, where the way Koyré had arrived at his concept became the model for how the history of science could be turned into exciting intellectual history. It caught on, in an adverse sense, with another mathematical thinker who from studying Duhem had come to rather similar notions as Koyré had, even though Dijksterhuis, because of his professed belief in the continuity of the history of science, rejected the idea of a *revolution* in science. Butterfield and Hall—the two scholars who applied the concept in the most productive manner in Britain—appeared more flexible where terminology was concerned. Although they, too, had little time for Koyré's sharp discontinuity, they saw that the concept could be employed in the service of a continuist stance on the birth of early modern science, provided the concept was widened in time and made less sharp in definition. Both Butterfield and Hall used the term for a very large time span in the history of science, even though both also recognized some sort of 'inner core' Scientific Revolution, stretching roughly from Galileo to Newton, inside the full, 'outer' Scientific Revolution. It is primarily in such an enlarged and rather indefinite guise that the concept spread all across the budding profession. Koyré went along with the widening of its temporal scope, so that his Scientific Revolution eventually covered the entire period from Copernicus through Newton. He did his best to keep his definition as sharp as he had originally made it, although he never sought to reconcile his original definition with the widened period to which it was now meant to refer.

One important (possibly decisive) motive for the process of widening was that, in Koyré's original conception, the Scientific Revolution had been confined almost exclusively to the fields of mathematical physics and astronomy. Nonmathematical optics, electricity, magnetism, chemistry, the life sciences, etc., had no clear function in it. Koyré continued practically to ignore these fields (quite in keeping with his original conception, but less compatible with the wider time scale on which he later projected the Scientific Revolution), whereas Butterfield and, in particular, Hall juxtaposed all scientific developments from the 15th through the 18th century in a story left by and large unstructured.

In Koyré's original conception the Scientific Revolution had been exclusively an adventure of the *mind*. Thus it did not seem to fuse readily with Olschki's attempt, undertaken one to two decades earlier, to pinpoint the background of early modern science in the world of technology and the arts and crafts. During the thirties Olschki's effort was made to merge with less subtle, largely (though not exclusively) Marxist-inspired attempts to account for the birth of early modern science. Such attempts, however, failed to catch the very issue that stood to be explained if Koyré's Scientific Revolution made sense at all: *why* the intellectual mutation as defined by him had taken place the way it did. With adherents to 'external' explanations failing to make that link sufficiently precise, and with Hall in particular proclaiming the purely intellectual character of the event as sufficient reason for seeking its explanation in the sole realm of the intellect, the historiography of science in the externalist tradition and the historiography of science organized around the new concept of the Scientific Revolution went their by and large separate ways.

As noted, the original definition of the concept in its original sharpness had already come under great strain through the widening of the time period to which it was taken to refer. The exposure of Koyré's overstatements where the historical reality of Galileo's experiments was concerned finished it off. Meanwhile the process of dilution went on relentlessly under its own steam. Hall confined himself in the end to one sole defining characteristic of his long-drawn-out Scientific Revolution: the transition from the rule of the evil sisters of mysticism, magic, and superstition to the triumph of rational thought. For the rest, both the 'Koyréan' and all other, nonmathematical sciences were just described by him in their ongoing development over time, with some of these punctuated by more drastic reversals than others. That is to say, what had been born as a sharply defined concept was now being turned into the overall framework of a story. This may well have been the most fateful turn in the as yet short history of the concept of the Scientific Revolution.

The fact is that, when taken as little more than a fitting label for one or two centuries in the ongoing history of science, the Scientific Revolution covers a bewildering variety of events and currents of thought—varied in content, varied in geographical scope, varied over time. How to give the concept, thus widened, a new unity after the original coherence had been tacitly discarded? This problem has plagued historians of the Scientific Revolution to the present day.

Seeking to structure afresh a diluted concept. In seeking to meet the challenge, both Kuhn and Westfall made an effort to salvage a measure of conceptual coherence for such a diluted notion. Both sought to give structural meaning to key events in just about all of 17th-century science (but leaving medicine out). The major advantage of Kuhn's attempt, besides the intended and naturally ensuing incorporation of the 'Baconian sciences', was that he showed how his distinction between two strands in the Scientific Revolution dovetailed nicely with current attempts at *explanation* that had so far got stuck in the all-encompassing nature of the Scientific Revolution in a geographical and disciplinary sense. The chief drawback was that Kuhn got entangled in confusion over how the corpuscularian view of things fitted into the picture. The major advantage of Westfall's arrangement, besides the intended incorporation of chemistry and the life sciences as well as the equally problematical corpuscularian tradition, was that, as a perhaps unintended by-product, he turned the hitherto *static* concept of the Scientific Revolution into a *dynamical* one. He showed that the continuation over time of the Scientific Revolution had not been a foregone conclusion right at its inception, but that the new scientific movement underwent productive tensions of its own which, in culminating in the Newtonian synthesis, brought the affair to a happy end not otherwise guaranteed. The main drawback of the account was that Westfall ran into trouble over the role of experiment outside the mathematical tradition, conflating along the way a variety of meanings of the term 'mechanical'.

Add to these remaining, unresolved problems that no one had found a sufficiently fitting place for Descartes in the unfolding developments of the Scientific Revolution—old-fashioned in his search for an all-encompassing doctrine of the world and man, largely misconceived in his imagined, corpuscularian mechanisms, yet somehow indispensable in providing the new scientific currents with a suitable, metaphysical substructure.

Add further that there was also an unresolved problem of *dating* the Scientific Revolution. Not even everyone agreed on the provisional outcome in Newton's work (Butterfield and Hall, in particular, extending the outcome farther up the centuries), but there was much more disagreement over when the Scientific Revolution should be taken to have started: with Copernicus, or earlier still, or with Galileo and Kepler and possibly Descartes? In this matter much depended on historians' ideas on continuity and discontinuity, which, in their turn, were linked with ideas about *causes*.

For explaining the event had meanwhile remained a tough job all along. The search for precise connections between events in the outer world and such key *explananda* as the intellectual deeds of the primary heroes proved mostly elusive. The mathematization of nature, in particular, has remained too hard a nut to crack for those few efforts that, in a very odd and opaque selection process, were filtered out and made to serve henceforth as standard explanations.[4] This was true of the Merton thesis, or at any rate of the various misunderstandings and distortions by means of which commentators turned Merton's original thesis into an explanation of the Scientific Revolution out of utilitarian motives deriving from features of Protestantism. It appeared no less true when, in the sixties and seventies, the Hermeticist explanation made its appearance. It was found to explain quite a bit if linked to the domain of the Baconian sciences, whereas it appeared virtually powerless to handle the process of mathematization of nature pioneered by Kepler and Galileo, in particular.

One way out of the difficulty has been to seek to explain away the fundamental novelty of the Scientific Revolution by interpreting the episode as just the next step following almost seamlessly upon previous thought in the various traditions of medieval and Renaissance natural philosophies. Invariably the novelty thus explained away (mostly by an appeal to continuities in the formulation of scientific method) proved to have a way of regaining a hearing through the backdoor of the explanation.

The net result has been that even those standard explanations which were meant at the outset to address the event in its entirety could be seen to apply to no more than one or another of its constituent parts at best, be these parts defined geographically, temporally, or by scientific domain.

Such, then, are some of our difficulties, our despairs, our dilemmas, in putting to productive analytical usage the concept of the Scientific Revolution. What is one to do with such an intractable concept? Rob it of its last vestiges of meaning? Discard it? Or reanimate it? The first cries of the newborn are now fifty years behind us—what should we expect, and what should we encourage? Death at middle age through collision with too many, too formidable obstacles? Or resuscitation?

The quandary deepens. Some fifty years after its original coinage, the concept of the Scientific Revolution has come to serve a number of functions strikingly different from those intended by its pioneer. Efforts to examine one or another aspect of what has proved to be a very slippery concept have certainly not ceased. Still, over the past ten to fifteen years we find the term increasingly being used, too, for pet episodes in 17th-century science about which an author wishes to make a point that is thought to appear in a light more lofty and worthy of our attention if adorned with the label 'Scientific Revolution'. Anglocentric approaches, in particular, are liable to slip into such usage. The tendency to identify the Scientific Revolution with events in England has been reinforced as the center of the professional cultivation of the history of science moved to the Anglo-Saxon world and as the knowledge of languages other than just English began to wither.

For the rest, the notion of the Scientific Revolution has taken on a quiet life of its own as an innocent label covering, for example, university courses on science in the 16th and 17th centuries. Nor does anyone frown if the label 'Scientific Revolution' is pasted on a conference devoted to an assorted array of topics extending from 15th-century perspective to early 19th-century medicine.[5] Such bleak, disjointed usage of the term can be noticed all across the discipline, and one predictable result has been the proposal to abolish the concept altogether. For example, in 1988 the historian of science Jan Golinski put on the agenda "the question as to whether the notion of a coherent, European-wide, Scientific Revolution can survive continued historiographical scrutiny."[6] One leading sentiment behind his and similar suggestions appears to be a sense of exasperation that what at bottom we would desire the concept to achieve—sharp definition applicable over the full geographical and temporal scope of all of 16th- and 17th-century science in Europe—has proved elusive over all fifty years of the concept's existence.

One final, powerful drive toward this very outcome appears to reside in the thrust given to large portions of the most serious work that has over the past decades been devoted to aspects of the Scientific Revolution. This thrust has been directed at showing that what appeared to an earlier generation of historians as a smooth, rapid, and wholesale victory for the new science had, in the reality of the past, followed a more crooked path, with remnants of nonscientific modes of thought preserving a vitality of their own all through the period of the Scientific Revolution. Such an approach, while providing a needed corrective, contributed its share to blurring what sharp edges of the original concept still remained, as can best be seen by taking up a recent collection of papers which embody the approach, entitled *Reappraisals of the Scientific Revolution* (1990). The net effect its perusal has on the reader is to make him or her wonder (not wholly at variance with the editors' intentions, I take it) whether such a thing as the 'Scientific Revolution' mentioned in the title ever took place at all.[7]

Is the 'Scientific Revolution' going the way of all historical concepts? Perhaps—so I hear by now some readers mutter—perhaps such ongoing dilution of a once sharp concept is just what one should expect. Perhaps we have all along been considering here nothing else but one more case of a general truth historical concepts must obey. The idea of the Renaissance, which was once given by Burckhardt a clear-cut, unambiguous conceptual meaning, has largely dissolved since, chiefly because the features he deemed uniquely characteristic of the episode denoted by the concept proved to apply to other periods as well.[8] Similarly, we need not wonder that the original, sharp meaning Koyré gave to his concept of the Scientific Revolution has been blurred by the results of later historical research. The concept has by now fulfilled its once useful services; the time has come to discard it. After all, historical concepts are nothing but metaphors, which one should beware to reify; they may help focus the historical imagination for a while, but we should never forget that they are no more than lenses placed between our vision and the ultimately unknowable reality of a past irrevocably behind us.[9]

I can go along to some extent with such arguments out of the tool-kit of the particular philosophy of history called 'narrativism'. I readily grant that what we have been chronicling here displays a well-known cycle in the lives of historical concepts, and that the outcome—apparent dissolution—need not therefore surprise us. It is also true that it would make little sense to urge a 'return' to the original meaning of Koyré's concept. Fifty years of historical research hugely stimulated by that very concept have made us lose forever the 'innocence' with which it could once be put forward.

Still, in reducing the case to a universal pattern one easily loses sight of the peculiar, highly contingent manner in which the cycle manifested itself in our particular case. Also, I would dispute that the concept of the Scientific Revolution is *nothing but* a metaphor. It refers directly to a past event, or episode, of undisputable historical reality: the emergence of early modern science. Around 1700 something was manifestly there which had been absent from Europe (or anywhere else in the world) no more than one or two centuries earlier: the Newtonian synthesis and a host of concomitant approaches, ideas, and intimations of a novel, immensely successful way to pursue the secrets of nature. The task to give coherent meaning to the range of events that yielded this particular outcome remains in any case. There is no better way to fulfill it than to let the search be guided (not dominated) by a unifying (yet not a monolithic) concept. The challenge we face is not, therefore, to regret, or passively to acquiesce in, the inevitable dissipation of the original concept. It is, rather, to retain the concept and fill it with new meaning in keeping with the best results historical research has meanwhile put at our disposal.

What if the concept is allowed to wither away for good? Despite the apparently intractable quandary to which the concept of the Scientific Revolution seems to have led us, its effective abolishment would constitute both a premature capitulation and the thoughtless giving-up of an indispensable historical concept. One reason for finding it indispensable is that a number of very fertile, mostly underexploited explanations of the Scientific Revolution would thereby be left suspended in midair. Or do we truly wish those explanations to lack henceforth an *explanandum* because the concept of the Scientific Revolution has stealthily been abolished behind their backs?

But the principal reason why I hold the concept to be indispensable is that it denotes that unique moment in history when the West succeeded in acquiring an intellectual and operative mastery of nature. To let go of the concept therefore comes down to shrinking back before the challenging task of coming to grips with one of the key problems of world history: the sources, the manifestations, and the consequences of Europe's breaking out of the ways of traditional society. Efforts to treat early modern science as if it were just one intellectual current among countless others the world has seen and will be seeing; efforts to uncover such subjective elements in early modern science as cannot fail to beset every human enterprise—efforts in this vein, while often worthwhile in themselves, should not blind us to the elementary fact that science since the onset of the Scientific Revolution has at the same time been much more than all that—that the Scientific Revolution is the embodiment of humanity's beginning to gain a consolidated, coherent grasp of nature. That grasp has been, and always will be, incomplete; its advance over time is riddled with setbacks; it is not even clear at what level of reality natural phenomena are being apprehended; and yet, grasp it is, on a scale unlike anything yielded by any other mode of approaching nature tried out before and/or elsewhere. The question of how this breakthrough occurred, and how it could occur, is what has continued to occupy historians of science, and what has therefore occupied us throughout the present book. But the whole search for historical understanding on this score derives its true significance from an awareness of what science since the Scientific Revolution, as distinct from any other human activity, ultimately is about. And it is the very special event of its emergence in the 16th and 17th centuries that has set the West apart—the decisive event that sealed for good Europe's already ongoing drift away from traditional modes of acting and thinking, as Herbert Butterfield, but also Joseph Needham and J. D. Bernal, in their own curious manner, well saw.

Why has so central a concept as that of the Scientific Revolution however defined played so small a role in the writing of history at large—outside the narrowly circumscribed world of the historiography of science, that is? Lynn White warned the profession back in 1966

> that although the discipline of the history of science is today winning increasing recognition among philosophers, it is becoming less exciting to general historians, economists, social psychologists, sociologists, and engineers, than it was twenty-five years ago when historians of science were heatedly discussing not only science as a self-contained intellectual activity, but also the nature and extent of science's connections with other human concerns. I fear for the future health of the history of science if it abides in its present faith that all scientific conception is immaculate.[10]

Perhaps because of the sly dig at the end, White's wise warning has been to little avail. It is true that what in the thirties and forties passed for social history was mostly of a rather crude and prejudiced kind. Despite the hidden vision of greater things ensconced at its core, its inborn reductionism formed an impediment rather than a boost to such a goal as the integration of history of science into general history. It is also true that the historian of scientific ideas has a tough job to perform anyway. Understanding scientific texts of the past is a painstaking enterprise in its own right. It must obey high professional standards and requires, besides much erudition and factual knowledge of a fairly disparate character, an uncommon combination of uncommonly divergent qualities. It is no less true that 'general' historians have paid more than due homage to the goddess of the Two Cultures in keeping their awed distance from what went on in the history of science. All this is true, and yet—the Scientific Revolution is one episode central to the history of humankind; why have we made it so peripheral?

I believe that we, students of the Scientific Revolution however defined, owe it to the worth of our shared subject to retain the concept; to redefine it again and again, until it has acquired the joint features of coherence, factual accuracy, explanatory power, and relevance in the framework of history at large. I do not think that such an aim is within easy, or speedy, grasp. But I do not think either that prospects for the hoped-for outcome are so dim at all. For consider what treasures of unexploited or underexploited ideas about, and approaches to, the problem of the nature and causes of the Scientific Revolution we have delved up in previous chapters.

7.3. IDEAS FOR FUTURE CONCEPTIONS OF THE SCIENTIFIC REVOLUTION

In passing in review now a range of particularly promising ideas, I am not going to list here those suggestions for fresh research topics that have already been made along the way (the reader can find these assembled in the Index under "Suggestions"). Nor is it my intention to exhaust, or to argue all over again, the full collection of historians' ideas hitherto treated. Rather, I wish to recall a sample of certain broad, clustering ideas which might profitably be pursued in future efforts to come to grips with the Scientific Revolution.

There is, in the very first place, the idea of relative discontinuity. It expresses the notion that certain things may change quite suddenly in history, even though it remains true that, on examination, such drastic changes, or 'mutations', can invariably be seen to have been prepared for in earlier times. A reverse formulation of the same principle would say that, even though a search for continuities in history is never in vain, this does not obliterate the distinction between gradual shifts over time and relatively sudden changes, which one might go on therefore to call with good conscience 'revolutions'.

There is, further, the idea that a comprehensive way to catch the kind of novelty that came to prevail in the course of the Scientific Revolution resides in a transition away from the world of the more-or-less, inhabited by all traditional societies including Europe, on to the universe of precision inaugurated by the mathematization of nature pioneered in the decades around 1600.

There is the idea that this 'inner Scientific Revolution' was somehow encapsulated in an 'outer Scientific Revolution' that had already been around for at least a century.

There is the idea that the Scientific Revolution can profitably be seen to have flowed onward in two currents: one mathematical/experimental, the other characterized by some mixture of heuristic experiments, corpuscularian models of the material world, and analogies taken from the domain of the machine.

There is the idea that these two currents interacted in various ways, with elements of harmony and elements of mutual incompatibility providing much of the dynamics that kept the movement going.

There is the idea of the overall contingency of both the event itself and its mode of spreading throughout its intellectual and its social environment.

There is the idea of the search for a lost unity of knowledge, undertaken by those who could not reconcile themselves to the fact that the new science was no longer part and parcel of a comprehensive philosophy—a search taking the guise of the construction of new, all-encompassing philosophies built in part upon new, scientific insights.

There is the idea of the search for a lost unity in the cosmos, undertaken by those who could not reconcile themselves to the fact that the new science appeared to rob the world of its ultimate meaning—a search taking the guise of the reaffirmation of mystical and magical modes of thought meant to safeguard, in one mode or another, the autonomy of the human spirit.

There is the idea that past science should, where possible, be put into the context of its intellectual and its social environment. To this is coupled the idea that such a 'new contextualism' works best if applied undogmatically, shrinking away in particular from even the appearance of reducing (as distinct from relating) the ways of the creative mind to the ways of society. The approach might best be anchored, not so much in esoteric forms of relativism, but rather in social history broadly conceived. In such a manner it might (by way of one important by-product) regain for the Scientific Revolution the deserved interest of an audience outside the confines of the discipline of the history of science.

There is the idea that the swift acceptance of the new science all across Europe was not due solely to its intellectual superiority but equally constitutes a major riddle in its own right. A social history of science might be called in to help resolve that riddle by focusing on such ideas as (1) the pervasive system of patronage as the key means of support for science at the time, (2) an all-European crisis of authority, (3) a sceptical crisis, and (4) the search for legitimation of the new science being fostered alike by the utilitarian promise of the new science, by values stemming from a variety of Christian creeds, and by the concomitant rise of a scientistic movement.

There is the idea that it is about time to break out of the narrow circle of 'standard' explanations of the Scientific Revolution and to seek to incorporate lesser-known causal connections in our accounts as well. In so doing, there is no need to confine such causal connections *a priori* to either 'internal' or 'external' ones; the point is rather to be clear about what the *explanans* is (from whatever domain one takes it), and how precisely it connects with the *explanandum*.

There is the idea that the problem of how to connect the process of the mathematization of nature with whatever causal agent seems to present itself (in particular, if taken from the surrounding society) has proved almost wholly intractable so far, whereas a variety of such causal agents could be connected to the 'Baconian' sciences in an enlightening manner.

There is the idea that, on seeking causes of the Scientific Revolution among the many currents of thought which immediately preceded the event, the key issue never to lose sight of is the extent to which the protagonists of the new science were to *transcend* those earlier patterns of thought they had been reared in.

There is the idea that minimal requirements for a Scientific Revolution to run its course are (1) a self-confident intellectual atmosphere of pride felt in the worth of science by its practitioners, (2) a community living by core values with which the values associated with science are perceived to be compatible, and (3) a region left free of large-scale, ongoing devastation.

There is the idea that a society animated by budding elements of pluralism provides an important, perhaps indispensable, nurturing ground for early modern science to arise and prosper.

There is, finally, the idea that any range of causes of the Scientific Revolution we may come up with ultimately finds its proper framework in the overriding issue of 'the differentiation of the West from the Rest'.

The task that remains.   The task ahead does not lend itself very well to a one-man show. What remains to be done is the biggest part of all: to incorporate these and possibly many other ideas into a partly old, partly novel conception of the Scientific Revolution. The present book is the embodiment of my belief that, in order to gain a chance to look ahead in the most promising direction, the first thing for us to do was to look backward so as to receive guidance from the men and women who shaped the field, and thereby to sharpen our gaze. From this effort at looking back I distilled such clustering ideas as have been recalled and listed above. Clearly, even if one were to adopt these ideas in their entirety, one may still employ and combine them in many different ways. My ultimate aim with this book has been to challenge professional historians of science (aided where possible by historians otherwise specialized) to go ahead and draw afresh such comprehensive pictures of the Scientific Revolution as may yield a deepened understanding of the event. And it is only with this particular reservation in mind that I move on now to presenting, by way of just one specimen among many possible others, a sketch of my present understanding of the episode. It consists of two parts: a structured outline, put almost in catchwords, of the event as I see it and a survey of the causal links that seem to fit best the event thus interpreted.

❧ EIGHT ☙

THE STRUCTURE OF THE
SCIENTIFIC REVOLUTION

8.1. AN OUTLINE OF THE EVENT

IN THE SKETCH THAT FOLLOWS the origin of the ideas that underlie it is
taken for granted, since these ideas have been treated at length in Parts I
and II. The sketch, if taken to stand by itself, may give the impression that
it is possible to treat of past events in an apodictic manner lacking all proper
nuance and qualification. But every single sentence that follows should be read
as if prefixed with a 'possibly' or a 'but also' or an 'as far as we think we know
now'. After all, the sketch is no more than an afterthought to the whole of the
argument that has preceded, and the most suitable test for it would be to have
it written out in book form to see whether it can encapsulate the succession of
historical facts that together make up the birth of early modern science. But
that would fill another book, which, indeed, I hope to write next. For now, what
follows is no more than just that—a highly provisional sketch.

The broad landscape of science. In traditional society, the investigation of
natural phenomena is one among many activities that may be undertaken to
fulfill the need of human beings to make sense of the world around them.
Civilizations may well prosper in the absence of such investigations; their inherent worth is not so great as to make them indispensable. In societies that do
provide a basis for pursuits which we tend in retrospect to acknowledge as
'scientific', such a basis is invariably small and unstable. Science, if pursued
at all, is pursued as a rule in the framework of a larger conception of how
things in the universe cohere. If pursued as an activity in its own right it is
most unlikely to survive indefinitely. A positive valuation of science varies
widely across cultures and during successive periods; in general, science
stands or falls with fits of such positive valuation.

In two early civilizations the investigation of natural phenomena condensed spontaneously into a more or less coherent, more or less systematic
body of ideas. One of these civilizations was China, where the condensation
took the overall guise of an organic materialism—a splendid system of thought
capable of yielding important insights, yet ultimately running into a dead end.
The other was Greece. Here loose computing rules as developed in Egypt and

506

Mesopotamia were transformed into an abstract system of deductive geometry (with only one small segment of the empirical world revealing a capability for being subjected to mathematical rule: the domain of musical intervals). The formation of the system was paralleled by, but not integrated in, a proliferation of broad conceptions of the cosmos, with the one devised by Aristotle offering a relatively great deal of room for scientific findings based on empirical research. With one very important exception, geometry played no part in these broad philosophies. The exception was Plato's philosophy, for which mathematics provided the epistemic model. When, under the spell of this philosophy, an effort got under way to find geometrical Ideal Forms in natural phenomena other than those of musical consonance, three more domains underwent a process of incipient mathematization: statics, optics, and what became 'mathematical' (as distinct from 'physical') astronomy. At that point the development (which stretched out over centuries) came more or less to a halt. The net legacy of Greek science thus consisted of (1) a highly unified body of deductive geometry, (2) a range of comprehensive philosophies in which some elements of observational science found a place, and (3) a few small pockets of mathematized science.

Three major, enduring civilizations were to adopt this corpus. One—the Byzantine Empire—inherited in a physical sense the surviving manuscripts which materially embodied the legacy. Beyond preservation it added hardly a thing to their contents. Islamic civilization, once come into being, proved more inquisitive. On having appropriated the legacy, it went on to enrich it and also to provide it with a coloring distinctly its own. One principal enrichment from outside consisted of a system of reckoning by position adopted from India (whereas a range of borrowings from China later failed to cross the language barrier). Some important enrichments from within concerned the improvement of parameters in mathematical astronomy and speculations on local motion inside the framework of Aristotle's natural philosophy; also the elaboration of the Archimedean tradition in solid and fluid statics. In the intellectual environment offered by Islam, the combined corpus of science and philosophy was for a time considered a valued instrument to combat by reasoned argument the unbelievers and also to perform a number of other services dictated by the Islamic faith (e.g., determining the *qibla* and the calendar, dividing legacies, restoring the health of the faithful, and predicting through the planetary constellations the fate of their rulers). Despite such local coloring acquired by the scientific enterprise in Islam, in the end the pursuit of science-in-philosophy came to a halt.

In the meantime in Europe another civilization had arisen, which had materially inherited a small trickle of Greek science. It went on in the 12th and 13th centuries to appropriate, through the channel of translations from Arabic and Greek, a larger portion of the original heritage, together with most of the enrichments that had meanwhile accrued to it.

The reception in Europe of part of the Greek legacy. Out of the many strands of which the heritage consisted the comprehensive, empirically minded synthesis offered by Aristotle was selected to satisfy the intellectual needs of the secular and regular clergy teaching in the newly founded universities all over Europe. With the Archimedean tradition and Plato's mathematical idealism falling to the wayside, Aristotelianism gained supremacy, whether cultivated according to its letter or to its overall spirit. Inside the nominalist current of Aristotelianism, in particular, a measure of intellectual daring was regained, which was employed for exploring some alternative ideas on local motion among many other subjects. But inside the overall framework topics like free fall and projectile motion occupied a very marginal position, so that their hesitant reform remained wholly inconsequential for the overall integrity of the system. What was done at most was to fill perceived gaps and to repair noted inconsistencies in accordance with the presuppositions of the system at large. No reform in the margins could imperil the huge measure of plausibility attributed to Aristotle's system in view of its joint features of logical coherence and of empirical (in the sense of human daily experience) grounding. Not until the phenomena of free fall and projectile motion gained a more central position in the whole complex of natural phenomena demanding attention could the edifice of Aristotelianism be turned upside down. They acquired such a position as one unintended consequence of Copernicus' proposed reform in that particular domain of science defined of old by the problem of how to account in a mathematical model for the observed complexities in the motions of the planets.

The reform was due to the third and final stage in the European reception of the Greek corpus, whereas the productive elaboration of its unintended consequences owed much to the particular local coloring scientific activity acquired in the whole of European civilization at the time.

The reception in Europe of the whole Greek legacy. The fall of Byzantium gave Europe a second, more comprehensive chance to appropriate the whole of the Greek corpus; some seized upon it as a means to escape from the dead end into which Aristotelianism appeared to them to have run. The primary aim of restoring to the letter the words of the Greeks gave way almost imperceptibly to elaboration and improvement. It was in this context that the Archimedean tradition was revived and hesitatingly directed toward novel purposes. It was in this context, too, that Copernicus decided to alter two of Ptolemy's presuppositions and to find out how far this decision could get him in restoring to its proper harmony the 'monster' into which the Ptolemaean body had meanwhile turned in his view. One of his alterations—the idea that the Earth is really and truly a planet—indeed yielded a somewhat more harmonious arrangement of the heavens. Yet the advantages gained from this feature and from somewhat improved calculations seemed so small in comparison with the outrageous price at which they had to be bought if taken to represent reality that the obvious solution for most later astronomers was either to stick to the Ptolemaean setup altogether or to treat the Copernican hypothesis as fictional in the accepted sense of the term 'hypothesis' (or to seek a compromise solution). Thus the realist core of the hypothesis went mostly ignored, with the surrounding machinery of orbital parameters being adopted in the framework into which it seemed best to fit anyway: that of traditional, mathematical astronomy.

There the reform might well have rested forever but for a particular coloring present around the edges of the movement occupied with recovering the Greek corpus.

The European Coloring. Just as the Greek corpus, once appropriated in the world of Islam, was not only extended by its students but also given a coloring of its own reflecting specific features of this particular civilization, so did the civilization of Europe tinge the corpus with features specifically its own. The Greek corpus, on being transplanted to western Europe, acquired around it an intellectual environment made up of many different strands. No coherent pattern kept them together; none of them was wholly original *qua* intellectual content; still, something novel colored it all. Five characteristic kernels can be identified. One was the urge for a very accurate observation of given phenomena of nature. This expressed itself primarily in geographical, botanical, and anatomical description, but it came also to the fore as an enrichment of traditional astronomy (notably in Tycho's program). Another kernel was the application of mathematics to concerns of Renaissance artists. Third, there was an uncommonly positive valuation of manual labor, expressing itself in (1) some guidance gained in the world of empirical technology toward the solution of specific theoretical problems, (2) hopes that in the domain of the artisan a reform of science might take its origin, and (3) a new boost for alchemical theory and practice. A fourth kernel was formed by the elevation of magical notions from a conjuring business to a lofty plane of abstract insight into three layers of ultimate reality, and of how these might be put to use for human ends. A fifth, closely related kernel resided in the rise of iatrochemistry.

The sheer proliferation of competing philosophies, ideas, and points of view put forward in the course of the adoption of the Greek legacy and of the formation of its European Coloring led some thinkers to an epistemic scepticism adopted from its Greek predecessor, but fitted out now with a somewhat more outspoken anti-Aristotelian thrust.

In structural terms, we have here the same constellation of science as had obtained in the Muslim world: the adoption of the Greek legacy, its modest enrichment inside an overall framework left intact, and the aggregate thus produced being complemented by some civilization-specific pursuits of its own making. There is no inherent reason why the European constellation should have escaped the destiny of all science at all previous occasions: to come to a standstill at some point. All previous experience might have led the detached observer of the scene of European science in the second half of the 16th century to expect the Golden Age of European Science soon to come to an end as its Hellenistic and its Islamic predecessors had. But then something wholly unprecedented happened.

The onset of the mathematization of nature.   The wholly novel thing is that the Copernican hypothesis caught on. That is to say, its realist core and its underlying mathematical harmony exerted such a spell over the minds of a very few thinkers that they were inspired to overcome all objections arising from common sense, from everyday experience, from God's word taken literally, and from elementary Aristotelian doctrines on local motion. The combined efforts of Kepler and of Galileo, in particular, to turn all possible objections into so many cornerstones of a new conception of the world ushered in a new period for science, marked by an intellectual and (as a promise actually to be fulfilled later) operative mastery of nature. Kepler, with the indispensable help of Tycho's accurate observations, mathematized nature in the sense of creating a mathematical physics of the heavens, utterly distinct from the by and large fictional model on which many of astronomers' best efforts had been spent from Ptolemy up to and including Copernicus himself. Galileo, that perceptive Archimedean who discovered the impossibility of 'Archimedeanizing' the best Aristotelian account of motion available, saw that free fall and projectile motion were the key phenomena which (although to all appearances their thrust was directed against an acceptance of Copernicanism) could be turned into the cornerstones of a new science of motion fit to support the heliocentric doctrine taken as physically true. Bringing Archimedean techniques to bear on these phenomena, in particular, he mathematized nature in the sense of subjecting an as yet modest range of terrestrial phenomena to mathematical treatment.

What the two men, Galileo and Kepler, thus ushered in jointly was an almost wholly new 'form of life'—the universe of precision. Which is not to say that the world of the more-or-less previously inhabited by humanity all over the globe was left behind at one stroke. The Scientific Revolution was not over yet—it had hardly begun.

The dynamics of the revolution.   The dynamics of the extraordinarily complex chain of events that ensued may in broad terms be thought of as follows. Almost from the day of its effective emergence at the hands of Kepler and Galileo, the universe of precision has proved to possess an unheard-of transformative power, which nothing in the world can escape once touched by it (which is what justifies its being called a 'universe'). Still, that very same budding universe of precision, although inherently capable of spreading like an oil stain, was not yet quite able independently to hold its own in the world into which it had just been born. It needed (besides a social environment about which more is to be said further on) to be fed with ideas and practices deriving from the conglomerate of mental attitudes we have called the 'European Coloring'. But this conglomerate was itself among the first to be reached, and transformed, by the expanding universe of precision. Two domains in which this overall dynamics manifested itself with particular cogency were those of experiment and of the corpuscular interpretation of the material world.

The universe of precision begins to expand. Nature as mathematized by Galileo, in particular, finds itself represented at a different level of abstraction than nature apprehended by daily experience. Being aware of the difference, the problem arose for Galileo of how to ensure that the mathematically expressed laws he found, whose validity could of necessity be vindicated solely on that abstract, idealized level of his, were valid in some way at the level of experience, too. The required means to establish this were discovered by Galileo to reside in experiment. He had already used experiment heuristically in his search for the mathematical regularities underlying free fall and projectile motion; experiment also appeared to him to provide the means by which to bridge the gap between phenomena represented at an idealized and at an empirical level. For this usage he could find inspiration in ancient predecessors (the monochord; a variety of tools which obey the law of the lever) but also, to some extent, in the general atmosphere of the European Coloring—in particular, in a positive valuation of manual labor and in the domain of craft practice considered as a resource of empirical phenomena artificially produced. The definition and the justification of the epistemic mode that goes with the invocation of artificially produced bits of nature called in to bridge the domains of experience and of abstract mathematics were to occupy many thinkers dedicated, from Galileo onward, to the further expansion of the universe of precision. That epistemic mode proved to be one of tentative, partial knowledge. Such a form of knowledge—at first grasped and intuitively applied rather than nailed down in an explicit doctrine of scientific method—came very gradually to be seen as acquired through chains of ongoing, hypothetico-deductive reasoning (as it was to be called much later) and as seeking its ultimate vindication in its pragmatic adequacy rather than in a need to be anchored in an all-encompassing philosophy of supposedly perennial validity.

Meanwhile—independently of the contemporaneous formation of a budding universe of precision—a number of thought-constructions belonging to either the Greek legacy or to its surrounding European Coloring strove to be turned into cornerstones of precisely such an all-encompassing philosophy. Ready candidates from the latter domain were the chemical and the Hermeticist views of the world; from the former, the corpuscularian conception inherited from Democritean atomism. In the end corpuscularianism proved victorious. It was the only doctrine that appeared to be compatible with the budding universe of precision, which here already demonstrated its capacity for influencing and, in the act of influencing, of transforming ever more segments of its environment. The corpuscularian doctrine was put forward almost simultaneously in a variety of intellectual atmospheres, with Beeckman being the unsystematic explorer and Gassendi and Descartes the men to seize upon his idea as a suitable means to counter the sceptical onslaught. Their somewhat distinct brands of corpuscularianism were not simply taken over from the ancients but were given a novel twist inspired by the universe of precision. Material particles were conceived as endowed with the defining, geometrical properties of size, shape, and location, and this is how the corpuscularian program came to consist of the explanation of the material world through the motions of tiny particles marked by an—in principle—infinite variety of shapes and sizes.

Taken by itself, the corpuscularian view of things proved almost wholly sterile; only very rarely did the sheer description of phenomena in corpuscular terms add to a deepened understanding. Its sterility manifested itself right at the start in Gassendi's work as well as in how the urge to corpuscularize everything dead or alive tended to smother creative impulses in the thought of both Beeckman and Descartes themselves. At a later stage, efforts to corpuscularize such phenomena as chemical reactions, the workings of the heart and the blood vessels, or the reproduction of living beings proved equally sterile. Still, the corpuscularian world-view had a potential to fertilize other approaches if combined with them. In particular, the combination of a mathematical science and a corpuscularian conception of the world proved quite fertile in drawing novel branches of science other than those explored by Galileo into the domain where precision reigns.

Sciences mathematized with the aid of corpuscularianism. The mathematization of nature pioneered by Galileo in the decisive area of local motion had already made some headway under its own steam, in domains like solid statics (independently carried beyond its Greek confines by Stevin and Mersenne), fluid statics (transformed by Galileo and Torricelli and axiomatized by Pascal), optics (expanded and reformed by Kepler), and musical arithmetic (transformed into a branch of mathematical physics by Galileo, Beeckman, and Mersenne). In the next generation the approach was, on the Continent, put under the joint aegis of Galileo and Descartes, yielding in the work of its principal practitioner, Huygens, the mathematization of such domains as the kinematics of impact, of centrifugal motion, and of the propagation of light. However fertile the combination was for a time, in the end it proved to run up against too many obstacles, as neither the mathematical exactitude of Kepler's laws of planetary motion nor the quantitative rigor of Galileo's laws of falling bodies proved derivable by means of the explanatory tools accepted in the corpuscularian world-view. Wherever the idea, inherent in that world-view, of *the force of a body's motion* came into play, conceptual confusion at once became rampant. It was not clear, however, in what direction a reform of the concept of force could be undertaken without giving up ground just gained by the new scientific movement.

The emergence of the Baconian sciences. Meanwhile, the fertilizing role of the European Coloring did not remain confined to the two areas already mentioned (those of experiment and of accurate astronomical observation). Around the turn of the century two novel thought-constructions crystallized out of it.

In one of these the whole conglomerate of the European Coloring, with all its incoherence and its intrinsic vagueness, was provided with an ideology of its own by Francis Bacon. In keeping with the magical-Hermeticist inspiration, in particular, Bacon's ideology of science was marked above all by an activist utilitarianism.

In the other, a number of inquiries was undertaken with the aid of tools adopted from, and in an intellectual atmosphere determined by, the conglomerate with its overtones of manual labor, empirical concern, accurate observation, belief in the harnessing of occult yet powerful forces, and a broadly 'spiritual' conception of things. These inquiries touched in particular phenomena arising in chemical reactions and in magnetic and electrical action, thus giving rise to the onset of the 'Baconian' sciences. As the century moved on, these two latest products of the European Coloring—the Baconian sciences and the Baconian ideology—were rapidly shorn of the 'spiritual' atmosphere out of which they had arisen in the first place; here again the influence of the universe of precision made itself felt. Whereas the reign of mathematical/experimental science—whether or not combined with a corpuscularian view of things—stretched over the Continent, the Baconian complex marked by and large what was going on in England.

The preparation of the Baconian Brew, and what came out of it. Around the middle of the century, this Baconian complex—the sciences plus the ideology—was further enriched when (through the agency of Mersenne, but also of Hobbes, Charleton, and other English thinkers with contacts in France) the corpuscularian world-view spread across the Channel and, in England, went on to merge with the Baconian complex. Thus came into being a most curious brew, made up principally of two hard-to-combine ingredients: heuristic experiment and a corpuscularian conception of the material world. Boyle and Hooke went on to become its most prominent *cuisiniers*. Although held together by the Baconian ideology, the Baconian Brew lacked all intellectual rigor, since corpuscularian mechanisms could only be imagined, not subjected to experimental testing. Still, the Brew was a potent one, partly on its own merits but chiefly in that in the end this direct offspring of the European Coloring provided the universe of precision with one final, fertilizing contribution. Out of the Brew (in particular, out of its alchemical condiment) Isaac Newton distilled a by and large novel conception of force as an active principle. Wholly novel in this was his becoming aware that here was the very conception required to overcome the dead end into which the combination of mathematical science joined to a corpuscularian substructure had meanwhile been running on the Continent.

Newton subjected his idea of force to the two principal requirements he drew from the universe of precision, which he, almost alone in England, inhabited: susceptibility to both mathematical rigor and experimental testing. For the time being Newton's magnificent effort to construct a new account of the sum total of nature's phenomena out of these few constitutive elements—the actions of diverse forces upon particles in motion, handled mathematically and checked by experiment—proved an aim set too high. But the domain where the approach did succeed, which was that of a new dynamics employed to construct nothing less than 'the system of the world', was so definitive in its firm grasp on a large array of natural phenomena and so promising in its overall approach that at the very moment when the *Principia* saw the light this wholly unprecedented movement of thought that we call 'early modern science' may be considered to have come into the world for good. The mode of its triumph was hard to predict yet, dependent as it was on many outside factors; the fact of its triumph was by now as foreordained as anything in our fragile world can be.

A look backward.   With this utterly tentative outline of the Scientific Revolution and its prehistory now behind us, two things remain to be done. We must throw a glance backward and consider which domains of human endeavor had, in the course of the whole episode of the Scientific Revolution, been touched by the budding universe of precision, and how and to what extent these domains had been transformed thereby. That done, we must examine the array of causal agents that determined the whole movement, the diverse explanatory ranges of these diverse agents, and the inner connections between them.

In looking back from the vantage point of the Scientific Revolution provisionally completed in the Newtonian synthesis, we consider first the various constituent parts of the Greek heritage. The tiny pocket of mathematized science at one edge of the legacy had undergone a huge transformation and an equally huge expansion. On its way to covering both a rational mechanics and an equally mathematical dynamics of the solar system, it had come to make up the universe of precision, ready to transform in its turn its intellectual and its social environment and to go on doing so up to the present day. Among its first, most obvious victims was natural philosophy as cultivated in the Aristotelian framework. During the Renaissance, Aristotelianism had risen to the challenge emerging from a plurality of rival philosophical viewpoints and had even managed to regain its vitality in the process. But it was not, in the end, up to the onslaught coming from the universe of precision. However unfair some of the attacks directed against it, however tenacious its hold on Europe's universities, however anxious the search for mixed forms—in the end Aristotelianism was powerless to withstand the new universe. University curricula very slowly began to be adapted to the new science, while those corporate, yet flexible adherents of Aristotelianism—the Jesuits—wound up seeking refuge in the Baconian sciences as a metaphysically neutral platform for their urge to demonstrate afresh the Society's intellectual superiority. Nor did the search for a unifying, all-embracing philosophy die out with the gradual demise of Aristotelianism. Descartes, Hobbes, Spinoza, Leibniz, and many lesser men sought to forge component parts of the universe of precision into elements of such a comprehensive philosophy. As more mundane practitioners of the new, mathematical/experimental science became better aware of the partial, tentative, hypothetical (in a new sense of the term) character of their own methods, the incompatibility of this new mode of knowledge with the old-fashioned search for an all-encompassing philosophy became ever more evident. Still, the aptitude of the human mind to seek truths beyond the tentative and the partial and to locate them in comprehensive thought-constructions is such that the tension between these two modes has remained with us to this day.

Something similar is true of that other large-scale effect of man's entering the universe of precision: the attendant loss of meaning of the cosmos at large. Religion, once the very agent to give coherent meaning to the world, appeared to face tough choices: While all kinds of solutions were being tried out at the time, in the end its adherents seemed compelled either to resign and end up in unbelief or to turn against science or to proclaim the religious domain separate from the domain of science or to enter into a perennial race to give its dogmas ever new interpretations in accordance with ever-changing scientific insights. The same goes for spiritual values not connected to religious dogma, which in the context of the European Coloring meant above all a Hermeticist conception of the world. The defenders of 'spirit' at large either had to adapt (as the Cambridge Platonists did) or to seek a vindication of at least a measure of autonomy for 'spirit' outside the reign of the universe of precision. This particular search, too, has never since ceased.

What about the other components of the European Coloring besides Hermeticism? Out of selected elements of alchemy and iatrochemistry (adopting practices rather than tenets) the universe of precision oversaw the forging of a more rigorous method for dealing with chemical reactions, which nonetheless remained waiting for a proper mode of mathematization. The construction of scientific instruments, too, underwent drastic changes as soon as the domain was touched by the universe of precision (which happened almost at once). For centuries, the making of clocks, backstaffs, astrolabes, etc., had been an empirical affair, practiced at the outskirts of a variety of craft guilds. What ensued in the course of the century was the rise of a novel kind of craft dedicated to quite novel, theory-absorbed standards of rigor and accuracy. At the same time, a research context was gradually built around the scientific instrument so as to engage it in ongoing cross-fertilization with the universe of precision. Many novel abilities had also to be acquired by the builders of the instruments in the course of a gradual learning process. More than that, the whole domain of empirical technology underwent a process of transformation because it had to obey the new standards of rigorous accuracy. A new kind of engineer (the Harrison/Smeaton/Watt type) still had to arise before the fruit could be harvested that was far too hastily thought to lie just around the corner: a science-based technology.

Meanwhile the organization of the sciences themselves was transformed as well. On the intellectual level, the system of the liberal arts found itself in disarray, having lost its rationale in the novel universe of precision. Subjects like local motion, the nature of light, the explanation of color, etc., were moved out of their original framework of natural philosophy and, after much to-and-fro searching, found themselves partly or wholly reassembled in the end as components of a new entity: mathematical/experimental science. On the social level, practitioners of the new mode of doing science began to organize their communications by means of informal meetings and exchanges of letters, yielding in the end formalized societies and regular journals. The societies, in particular, provided the means by which something of a scientific community came into being and by which the confident *mores* of that increasingly self-confident community were gradually worked out in practice.

What shape had the universe of precision itself acquired at the provisional, open-ended, final stage of the Scientific Revolution? It was no longer mathematical science pure and simple, such as it had emerged at the hands of its pioneers, Kepler and Galileo. It had meanwhile absorbed one remaining constituent part of the European Coloring: accurate observation. Although manifest at many levels of scientific endeavor, accurate observation came to the fore with spectacular results in telescopic and in microscopic research, in particular. Some of these could be incorporated in mathematical theory—most could not, and for a long time to come the mathematical sciences and the Baconian sciences were to live on together, inside the universe of precision, by way of an often uneasy compromise. Still they held something in common. Measurement where possible, and as accurately as possible, in the framework of a mathematical theory where possible; rigorous causal thinking guided where possible by the feedback provided by experiment—such methodical rules became the hallmark of the sciences. This is how the outsider came to perceive the scientific as distinguished from all other enterprises; this is what held together the elements of that particular entity that, by the end of the 17th century, had come to make up the universe of precision.

8.2. SORTING OUT POSSIBLE CAUSES

One overriding question now imposes itself: How could this universe of precision come into being?

No simple answer is likely to satisfy us. If there is some measure of truth in the—highly simplifying—outline just sketched, the complexity of the event cannot but be matched by a certain complexity in the pattern (or patterns) that caused the event. In fact, our question of one general 'why?' dissolves in almost as many 'why' questions as there are assertions to be found in our outline. In other words, our outline itself yields the range of questions that must be asked if we wish to explain the Scientific Revolution. It is not surprising that answers will come forward in the guise of those very explanations our survey in Part II has revealed to be the most fruitful ones, neatly distributed so as to match the various component parts of our outline.

The general type of explanation to be sought must be one that is proper to the historian's mode of understanding. In our Part II we have found virtual consensus over at least one tenet. It holds that historical causation is unlike causation in the natural sciences in that not if/then relations but less clear-cut patterns stand to be invoked. The following formulation, due to Lynn White, seems to me by and large to capture that consensus (at least if coupled to the idea that the historical *fact* remains the ultimate arbiter):

> Historical explanation . . . is seldom a matter of one billiard ball striking another, of 'causes' in the narrow sense. It is much more often a process of gradual illumination of the fact to be explained by gathering around it other facts that, like lamps, seem to throw light on it. At last the historian arrives at a sense that the central fact on which he is focusing has become intelligible.
>
> [This is] the sort of explanation, necessarily common among historians dealing with large phenomena, that can neither be proved nor disproved with any rigor but that must be accepted or rejected on grounds of general coherence or incoherence.[1]

Under such a broad umbrella formulation (counterparts to which can be found in the methodological reflections of many 'general' historians, too)[2] a host of distinct *types* of cause may still be invoked, just as we have found these being invoked or at least effectively employed all over our Part II: internal and external causes; necessary and sufficient causes; contributing and precipitating causes; direct and indirect causes; favorable and inhibiting causes; intended and unintended causes; sharply defined causes of 'key junctures' and broadly operative causes of large historical processes. All such kinds of causes may be used wherever serviceable to throw light on whatever appears obscure, with two extremes only to be avoided at all costs: indiscriminately to invoke the whole of what preceded (which amounts to a radical determinism perhaps defensible as a philosophical position but not as a means to gain an understanding of history) or to ascribe anything at all to chance without giving reasons for doing so.

Explaining the prehistory of the Scientific Revolution. With our tools in place, we start asking questions. In our outline the Scientific Revolution has been taken as an event inherent as a developmental possibility in the Greek corpus of science. That is to say, the Greek corpus was inherently capable, as its Chinese counterpart was not, of yielding in the end something more or less akin to early modern science as we have come to know it. Whether or not this potential outcome was ever to be realized—that was the issue at stake throughout.

A great deal militated against such an outcome, to be sure. Inhibiting causes are to be found in everything in the human mind and in human society that goes against the kind of knowledge embodied in early modern science— against knowledge, that is, that allows itself to be corrected systematically by empirical feedback; that is partial and tentative; and that stops short of no possible subject of inquiry, whether held sacred or taboo in the society at large or not. Little more can here be said about such inhibiting causes, since a cognitive-social psychology needed to study these in detail and in greater depth is hardly available yet. However, we have witnessed the actual workings of at least a few such inhibitory causes in the case of Muslim civilization, where values perceived to be central to the faith were in the end felt to be incompatible with those of science.

This is not to say that the kind of breakthrough that did take place in Europe around 1600 could not possibly have occurred in Islam; it only means that no *follow-up* would have occurred in the latter. Given its overall environment, no further expansion of mathematized science would have taken place; mathematical science would not have entered the universe of precision.

It is at this point in our preceding analysis that one crucial contribution has come to the fore which was absent both in Hellenistic and in Muslim society but which was characteristic of western Europe only. The 'Islamic Coloring' hardly entered into productive interaction with the Greek corpus of science; the 'European Coloring' did. Why was this so? Why was the Islamic Coloring so much poorer, *at least where science was concerned,* than its European counterpart? One may of course respond that this particular civilization sought its strength in other pursuits, but this answer only moves the problem around without coming to grips with it. One way to answer it resides in a comparative, historical sociology of science-and-religion, which is hardly existent yet, but to which our three authors on the decline of Islamic science have already contributed some important insights.

The other way to seek to explain the coming into being, in Europe but not in Islam, of a 'Coloring' fit to engage in productive cross-fertilization with the Greek legacy appropriated at much the same time is to investigate the uniquely

European background of its specific Coloring. Here we must carefully distinguish between two levels of causal analysis. On the one hand, we may identify specific elements in the European Coloring as products of such features as mark specifically late medieval and Renaissance Europe but no other civilization at the time. This appears to apply to the following two elements, in particular:

1. Respect for work done with one's hands. Scholarly consideration of where this uniquely European feature finds its origin has almost without exception pointed at a typically Christian conception of labor as a form of prayer, which sanctified the domain of the menial and thus made it respectable.

2. An urge for accurate observation. At least three direct sources of this urge present themselves: (*a*) a respect for empirical facts as God-given rather than as explainable-away by man's superior intellect; (*b*) a respect for empirical facts over and above their rash rationalization, arising equally from prior experience in the Voyages of Discovery; (*c*) the example of the great Renaissance artists in their readiness to leave stereotyped forms of expression behind.

On an analytical level distinct from these relatively precise and straightforward causal attributions, significant portions of the overall landscape of 15th and 16th century Europe may be invoked as the indispensable background to the rise of the European Coloring. One thinks here of the kind of explorative urge that led to the Voyages of Discovery; of how Latin, but not Byzantine, Christianity acquired such activist traits; of the fragmentation of power that increasingly came to mark European politics; and of many other manifestations of specifically Western modes of thought and action. What we touch here, in short, is, once again, that time-honored enigma of the West and the Rest—the huge problem of what exactly made the West break out of the ways of traditional society. At the present stage of historical analysis this problem is rather a mess, with some authors invoking secular trends of more or less questionable validity, and others pointing at one or another specific feature of Europe without much thought being spent on how these interacted and how they may be thought to hang together. Although our understanding has surely moved beyond the 'Faustian spirit' invoked of old, nothing sufficiently coherent has taken its place. Lynn White pointed at the origin in Latin Christianity of attitudes toward labor and nature that gave rise to the unique technological dynamism of medieval Europe. Needham pointed at origins of Europe's fragmentation in its script; he ascribed its 'roving' qualities—its lack of 'homeostasis'—in part to its geography-fostered, maritime orientation and to its perennially negative trade balance with the East. Weber pointed, among other things, at the corporate autonomy of the medieval city; at the disciplining effect of the Benedictine monastic rule, and at the rise of administration by law. Landes pointed at the mechanical clock. Kernels of brilliant insight often lie hidden behind a listing of such truly or supposedly unique Western traits. Still, all these traits and secular trends are urgently in need of being assembled in their turn, of being sorted out, critically compared, 'gently homogenized', and where possible considered in terms of common denominators. But that, as the saying goes, is another story (which I hope to be able to take up in the foreseeable future).

Explaining the breakthrough.   Back now to the causes of the Scientific Revolution. In the interpretive part of our sketch I argued that the overall constellation of an enriched legacy in Greek science joined to a specific Coloring around it, although necessary for the great breakthrough to ensue, was not by itself sufficient to provoke it.

The breakthrough that, around 1600, did ensue, appeared to us to result from roughly the following sequence: Copernicus' reform of Ptolemy's solution to the problem of the planets out of a felt need to restore its lost harmony; the acceptance by Kepler and Galileo of its realist core; the elaboration of that realist core into a mathematical physics of the heavens; the mathematization of free fall and projectile motion as a ready means to remove otherwise insuperable objections to the realist core of Copernicanism; the coming about, thereby, of a new, inertial conception of motion in an overall setting of abstract, idealized nature linked to the domain of the empirical by the artificial means of experiment.

A range of causal agents of various dimensions can be seen to be involved in this sequence:

1. The perception of Ptolemaic planetary theory as a 'monster'. For such a perception at least two things were required. One was the sheer passage of time (time spent by astronomers over the centuries on tinkering in the margin of the system); the other, a belief in the significance of 'harmony'. The former cause helps explain why no 'Copernicus' succeeded Ptolemy at once, in the first or second century A.D.; it leaves open, however, the possibility that a heliocentric hypothesis might have been worked out in the framework of Islamic civilization. The latter cause moves us toward a peculiar niche in the European Coloring still scarcely explored for its explanatory potentialities for the Scientific Revolution: that particular thought-complex where musical arithmetic, conceptions of universal harmony, and a 'spiritual' view of the cosmos intersected in the 15th and 16th centuries. Here another contributing (certainly not a necessary) cause may well prove to reside.

2. Kepler's acceptance of the realist core of the Copernican hypothesis, coupled to his willingness to let empirical facts gain the upper hand over his theoretical construction (the celebrated '8 minutes of arc' divergence). Here Kepler's Christian-inspired realism has been offered as an explanation, which at this key juncture directed him away from the ordinary course of Platonic/ Pythagorean-inspired thinking. (After all, harmony and mathematics had been all over the place; what was unique was Kepler's painstaking loyalty to the hard facts whenever his fancy carried him too far away). Equally crucial was the

availability of a store of very accurate observational data, of which the causal background (as noted already) has likewise been claimed to reside in a partly Christian-inspired respect for the factually given.

3. Galileo's performance in employing Archimedean techniques for the mathematization of free fall and of projectile motion. At this (to the minds of many historians of the Scientific Revolution) most critical juncture of all, the search for causes has proved singularly frustrating. The point at issue here is, once again, not what Galileo inherited, along with many other thinkers, from the Greek corpus, but rather what he accomplished in transcending that heritage and in transforming portions of it almost beyond recognition. What made him see that, and how, the Archimedean approach could be applied to such hitherto intractable natural phenomena as free fall and projectile motion? Our discussions throughout Part II have yielded two specific causes, both situated in the midst of the European Coloring—more precisely, in the 'arts, crafts, and empirical technology' complex.

One possible, precipitating cause has appeared to reside in the mechanical clock—not so much in the instrument itself as, rather, in its wide distribution all over Europe; in an ensuing habituation to a modicum of precision in daily life; and, especially, in a consequent awareness that the passage of time may in one way or another be subjected to mathematical treatment.

The other cause—the arts-and-crafts background—operated on two distinct levels. On one level, it sharpened Galileo's eye for the kind of questions that might fruitfully be asked in the domain of motion (with his Archimedean upbringing telling him how to move the resulting insights at once to a more abstract plane of analysis). On the other level, it contributed to his discovery of artificial experiment, employed both heuristically (in the wake of his father's experiments with the monochord)[3] and for providing the required bridge between an ideal and an empirical level of analysis of nature's phenomena.

So much for Galileo. If we survey now the explanations given of various key junctures in the sequence of events that ushered in the universe of precision, we can readily see that (especially where Kepler's and Galileo's personal accomplishments are concerned) our collection of causes, although certainly not insignificant, hardly exhausts the *explanandum*. At least two crucial elements have been left wholly unexplained so far. One is that a smaller or larger residue of *creative genius* has still been left out of the account altogether. The other is that, over and beyond these key events, a much more general movement of thought had been sweeping over Europe. Let us consider both explanatory gaps in succession.

Long ago, in Carlyle's innocent times, we were still free to exalt 'great men' without blushing. We have become wiser since; we have learned to see how men and women held by posterity to be 'great' were no less rooted for that in their own times. We have even learned to explain their greatness in terms of their times. But, in so doing, we have become at times a little bit too wise. Once we have exhausted every single respect in which we can ground a creative thinker (or doer) in his or her own times, a residue is still left. That residue is not perhaps wholly inexplicable by itself—it is that property called of old 'genius', and cognitive psychology is not powerless in analyzing a range of component parts of genius. But when all is said and done a residue still remains. One may rightly point at Kepler's Christian humility in face of the hard, stubborn fact, but no one would take this to exhaust the riddle of Kepler's mind. One may reconstruct from history that, owing to Copernicus' hypothesis, a vulnerable spot in Aristotle's system had acquired a new significance, and that this was more than the system could bear once the situation was realized with crystal clarity by a highly unorthodox thinker at Pisa who had just tried in vain to put impetus accounts of free fall and projectile motion on a mathematical footing. We can, if we are lucky, point at ideas and practices in the world at large that assisted Galileo in perceiving that this was so. But those ideas and practices were there for all of his contemporaries to pick up. If Galileo had not so fully grasped, and exploited, the opportunities that lay ready here, there is no guarantee whatsoever that somebody else might have done it. History is filled with opportunities going unnoticed and therefore not seized upon, and it requires a genius to notice them and seize them. It is rather naive to 'explain' Galileo and Kepler as just geniuses transforming traditional conceptions in natural philosophy from scratch. It is no less naive to explain their genius away by reducing it all to the ripeness of a situation that was really ripe only by the historian's hindsight—and by the foresight of those very geniuses who saw what lesser men might never have noticed.

This is the more true because the appearance of these two geniuses on the European scene did not wholly stand by itself. All over Europe something was 'in the air'. Independently of one another, both greater and lesser thinkers began to 'see' things that had gone unnoticed for ages. Tycho Brahe recognized a nova in the sky where no scholar for centuries had seen one because such phenomena were ruled out *a priori* by the reigning philosophy. A telescope was suddenly constructed out of parts and by means of tools which had been lying ready for centuries.[4] Marin Mersenne discovered, without employing any tools or tricks, the presence of harmonics never distinguished before in your everyday musical note. All across Europe such central categories of Aristotelian thought as 'substantial form' and the like, which for centuries had made perfect sense, began to lose their meaning for critical thinkers. The idea of a wholesale application of mathematics to physics was in the minds of many men (such as Beeckman and Descartes) who had little idea of how to do it. Why did all this happen in the course of those few decades around 1600? Bits and pieces have causally been linked to significant *explanantia* here and there. We may talk of 'thinking-caps', we may formalize these into 'gestalt-switches', we may seek to reduce all this to the assiduous reading of printed books, yet ultimately we are facing here that residual riddle of history: Sensibilities shift, and no one knows quite where they come from; no one knows quite where they go.

Explaining how the universe of precision caught on.   In our outline, we have seen how the Scientific Revolution moved onward—how the universe of precision caught on. We must now distinguish between three motor drives that were operative here: (1) the inherent transformative power of the universe of precision; (2) the ongoing process of cross-fertilization with elements from both the Greek legacy and its European Coloring; (3) the acquisition of social support through a hit-and-miss process of legitimation of the scientific enterprise.

The first of these motor drives—the inherent power of the movement—is by and large self-explanatory. The temptation to ascribe in good part the capacity of the universe of precision to catch on to the fact that properties of nature objectively there were now being systematically apprehended for the first time is hard to resist. Whether the mathematization of nature is so irresistible a process because ultimately our world *is* mathematical, or whether other ultimate constituents of reality are responsible for this, remains a matter of speculation. What counts for us is the plain, experiential fact of its irresistibility.

Which is not meant to say that this in-built motor operated wholly in isolation. In its early stages, the universe of precision might well have collapsed. After Kepler and Galileo the movement might still have run down, possibly to be picked up again at other times, in other civilizations. What was still decisively required was the construction of a solid building out of a host of dispersed elements; what was equally needed was social support to keep the movement going.

The construction of that solid building, which is the story of how the Scientific Revolution moved from Kepler and Galileo to Newton, also leaves little to be explained beyond the dynamics outlined in the previous section—with one important exception. The whole movement from Galileo and Kepler to Newton, with so much cross-fertilization and transformation going on inside so disparate a range of currents of thought, can hardly be imagined if not running on reasonably effective communications, which only the printed word could furnish. One can still imagine the performance of Kepler and of Galileo to have taken place in a scribal culture. One may also be fairly sure that in an intellectual environment by necessity devoted in large measure to the *preservation* of cultural goods, their effort at renewal would fairly quickly have petered out.

By contrast, the story of how the new movement gained social support is virtually riddled with occasions to adduce pertinent causes. After all, in the face of such a radically novel mode of understanding, countervailing forces may be expected to have sought to contain the flood. And indeed, there was a strong urge to dam it. The loss of an all-encompassing picture of the world was far too worrying for many to bear. What made elites at various places all over Europe so eager nonetheless to support the new science?

One answer is that they had been supporting science already for quite some time. In Italy especially, science had become part and parcel of the system of patronage, adding to the glory of the patron as his support of the arts did. But when Galileo laid the groundwork for a new science, and strenuously sought support for it beyond the narrow basis he had gained for his own person, his effort failed. In the end the religious issue and the fashion-bound cultural interests of so socially rigid a group as the Italian aristocracy put an end to most organized support for science in Italy. Nor did the kind of inherently unstable sustenance Kepler and many lesser thinkers all over southern Europe enjoyed lend itself to becoming an enduring source of support.

To be sure, in an increasingly war-ridden Europe, where many sources of traditional authority had gone up in smoke, there was a great demand among reigning elites for anything that might help restore stability. Could the new science cater to that demand? Its Janus face of disruption of vested ideas and interests on the one hand and of utilitarian promise linked to a metaphysically neutral mode of acquiring knowledge on the other made the outcome—support, yes or no?—very much a matter of regional, even local contingency. At best, one may point here at greater chances for support in circles which were better able to meet the doctrinal challenge (hence, Protestant regions emerged as overall somewhat more science-minded than Catholic regions), and which were engaged in a search for social ascendancy (hence, socially fluid northern Europe emerged as somewhat more science-minded than the frozen aristocracies of the south). In the end, one specific complex of ideas—the Baconian ideology that at the middle of the century had come to form the protective shell around the Baconian Brew—provided the most coherent kernel around which a highly disparate range of social aspirations and projections could crystallize. The utilitarian sentiments of English Protestantism coincided neatly with the spirit of Baconian experiment and its high promise, with the new science offering at the same time a neutral meeting ground for all possible persuasions. Here the scientific enterprise found the kind of enduring legitimation without which it cannot survive. No such ideology had arisen in either Hellenistic or Muslim society. When the utilitarian promise was finally fulfilled, a less period-bound source of legitimation could take over: the image of science as the goose that lays the golden eggs. For the time being, the scientistic ideology that grew out of Baconianism served as an effective source of legitimation required to see the new science through the stage of its first, provisional consolidation in the Newtonian synthesis.

A final consideration.   In seeking to capture the involved nexus of causes that together released that extraordinary event, the emergence of early modern science in 17th-century Europe, we have been moving on a high level of explanatory discourse. But we must equally keep in mind that, if in 1241 the Great Khan Ogodai had not died, and if traditional Mongolian rules of succession had not prescribed the presence in the flesh at the election meeting in Kara-korum of the leaders of armies on the verge of overrunning and pillaging all of Europe as they already had Russia and large parts of the Muslim world, no Scientific Revolution is likely to have taken place in a region so stricken as Europe would have been for a very long time to come—possibly forever. In the end it is not either causal chains *or* coincidence, but rather an unpredictable mixture of the two, that reigns in history.

✎ NOTES ✐

ONE

1. From John Dryden's *Of Dramatick Poesie: An Essay;* vol. 1, p. 12, in M. Summers' 1931 edition of *Dryden: The Dramatic Works.* With further help kindly given by Geoffrey Cantor, I found this quotation through A.C. Crombie, "Historians and the Scientific Revolution," p. 12 (where 1693 is mistakenly given as the date of publication of the essay).

2. There is one excellent monograph in which de Kadt's evolving political views are set forth, interspersed with illuminating comments: R. Havenaar, *De tocht naar het onbekende: Het politieke denken van Jacques de Kadt* (Journey into the Unknown: Jacques de Kadt's Political Thought) (Amsterdam: van Oorschot, 1990). The abstract at the end of this book is probably all that has ever been written about de Kadt in English, his acquaintance with such men as Sidney Hook and Franz Borkenau notwithstanding.

3. J. de Kadt, *Het fascisme en de nieuwe vrijheid* (Fascism and the New Liberty) (Amsterdam: Querido, 1939).

4. De Kadt derived the phrase from Max Eastman's *Marx and Lenin: The Science of Revolution* (London, 1926).

5. D.S. Landes, *Revolution in Time,* p. 25.

6. One indisputable feature of Watt's steam engine—the chief technological motor of industrialization—is that the machine is entirely dependent on a prior awareness of the void and of atmospheric pressure, which are insights gained in the course of the Scientific Revolution (as is argued with particular cogency in D.S.L. Cardwell, *Turning Points in Western Technology*). Also, as D.S. Landes has shown in *The Unbound Prometheus* (pp. 21–33), the cost-benefit rationality underlying early industrial society was marked throughout by scientific modes of thought that stem equally from the 17th century.

7. H. Butterfield, *The Origins of Modern Science, 1300–1800;* the quotation is on p. vii of the revised edition of 1957.

8. R.S. Westfall, "The Scientific Revolution," pp. 7–8.

9. H.F. Cohen, *Quantifying Music: the Science of Music at the First Stage of the Scientific Revolution, 1580–1650.*

10. H.F. Cohen, "Over aard en oorzaken van de 17e eeuwse wetenschapsrevolutie: Eerste ontwerp voor een historiografisch onderzoek" (On the Nature and Causes of the 17th-Century Scientific Revolution: Preliminary Draft of a Historiographical Research Project) (Amsterdam: van Oorschot, 1983), inaugural address at Twente University of Technology, 9 June 1983. I later gave two "work-in-progress" talks about my ongoing research at meetings of the History of Science Society in 1986 and 1989.

11. I.B. Cohen, *Revolution in Science,* particularly ch. 26, "The Historians Speak."

12. H. Guerlac, *Essays and Papers in the History of Modern Science;* in particular the first 65 pages of the book. T.S. Kuhn, *The Essential Tension;* in particular the preface and chs. 1, 3, 5, and 6. H. Kragh, *An Introduction to the Historiography of Science.* J.R.R. Christie, "The Development of the Historiography of Science." A. Thackray, "History of Science," in P. Durbin (ed.), *A Guide to the Culture of Science, Technology, and Medicine* (New York: Free Press, 1980), pp. 3–69.

13. Stanley L. Jaki's *The Origin of Science and the Science of Its Origins,* although in a formal sense the only book that aims to cover the same ground as mine, is hardly a useful contribution to the topic. As more than one critic has observed, it is really apologetics (of a rather malicious kind, too), not history writing. To Jaki the history of the historiography of the Scientific Revolution seems divided into three stages: All those who lived before Pierre Duhem and failed to anticipate him; Pierre Duhem, who was RIGHT; and, finally, all later historians, who without exception failed to perceive that this is so. The urge to ascribe the birth of modern science to Roman Catholic thinkers like Buridan and Oresme overrides every other consideration Jaki brings to bear on the subject.

14. Ignoring this fact of historical life is what makes Jaki's wholesale vindication of Pierre Duhem's conception of the birth of modern science such an oddity.

15. I.B. Cohen, *The Newtonian Revolution: With Illustrations of the Transformation of Scientific Ideas.*

16. N.M. Swerdlow, "The History of the Exact Sciences" (paper circulated at a History of Science Society meeting at Madison, 30 October–3 November 1991; kindly forwarded by the author).

17. Interesting data on this topic are spread throughout A. Koyré, *De la mystique à la science.*

18. For example, Roy Porter, in his article "The Scientific Revolution: A Spoke in the Wheel?" has tried to lay bare the ideological underpinnings in the Cold War of what I have called here 'the Great Tradition'. Much of what Porter has to say makes good sense to me, yet in the end one finds that the unique achievement of those bold thinkers who shaped the tradition has almost entirely disappeared from view.

19. E.A. Burtt, *In Search of Philosophic Understanding,* p. 93 of the second edition.

20. From a journal piece of 1928, quoted in J. Huizinga's *Briefwisseling* (Correspondence) (Utrecht: Veen/Tjeenk Willink, 1990), vol. 2, p. 218.

21. One exception is studies, such as Landes' *Unbound Prometheus,* of what the Scientific Revolution meant for technology in particular and for industrialization at large. Or take Jean-François Revel's two-volume *Histoire de la philosophie occidentale,* which is organized around the Scientific Revolution as a key break in the history of Western philosophy as well. (I cannot refrain from mentioning here an article I wrote in Dutch on how the Scientific Revolution stands at the origin of the split in our culture best known as 'the two cultures' ["De Wetenschapsrevolutie van de 17e eeuw en de eenheid van het wetenschappelijk denken" [The Scientific Revolution of the 17th Century and the Unity of Scientific Thought], in W. Mijnhardt and B. Theunissen [eds.], *De*

twee culturen: De eenheid van de wetenschap en haar teloorgang [Amsterdam: Rodopi, 1988], pp. 3–14.

No doubt there are many more such studies. Yet one is struck most often by authors who assign to modern science a great transformative influence exerted ever since the 17th century upon their chosen topic but who nonetheless appear entirely unaware of the historical literature on the Scientific Revolution. For example, the historical remarks scattered across Theodore Roszak's fascinating *The Cult of Information* (New York: Pantheon, 1986) might have benefited from a modicum of awareness of the literature on the Scientific Revolution. The same is true of another captivating study: Christopher Small, *Music, Society, Education,* 2nd rev. ed. (London: Calder, 1980). In ch. 4, "The Scientific World View," Small sets up an argument to demonstrate that "it is no accident that tonal functional harmony began to take over western music precisely at the start of the seventeenth century: the daylit rational world that this music was to inhabit almost exclusively for the next three centuries, despite the efforts of the Romantics to break out of it, is the same world as that of scientific rationalism" (p. 81). His argument is powerful and persuasive, yet one wishes that the author had examined more deeply the 17th-century emergence of 'scientific rationalism'. There must be countless similar examples.

22. H. Butterfield, *The Origins of Modern Science, 1300–1800.* E.J. Dijksterhuis, *The Mechanization of the World Picture* (1950 in Dutch; 1961 in English translation). A.R. Hall, *The Revolution in Science, 1500–1750.* R.S. Westfall, *The Construction of Modern Science: Mechanisms and Mechanics.* R.R. Palmer and J. Colton, *A History of the Modern World* (New York: Knopf, 1950; many revisions since).

TWO

1. I.B. Cohen, *Revolution in Science,* pp. 391–400.

2. For example, in *Dictionary of the History of Science* A.R. Hall writes on p. 379, s.v. "Scientific Revolution," that "the phrase has been current since the late 17th century," right after having defined the term as "the transformation of thought ... about Nature through which the Graeco-Islamic tradition was replaced by modern science." Or see T.L. Hankins, *Science and the Enlightenment,* p. 1: "The expression 'Scientific Revolution' had been coined by mathematicians like d'Alembert."

3. See for an overview I.B. Cohen, *Revolution in Science,* ch. 6.

4. I raised this question on p. 374 of my article "Music as a Test-Case."

5. I.B. Cohen, *Revolution in Science,* chs. 1, 4, 5, 12, and 13.

6. Jean Le Rond d'Alembert, s.v. "Expérimental," *Encyclopédie,* vol. 6, p. 299: "la renaissance proprement dite de la Philosophie."

7. Ibidem: "les siècles d'ignorance"; "ces tems ténébreux"; "cette méthode vague & obscure de philosopher."

8. Ibidem: "il ne faut pas espérer que l'esprit se délivre si promptement de tous ses préjugés. Newton paru."

9. Ibidem: "Newton paru, & montra le premier ce que ses prédécesseurs n'avoient fait qu'entrevoir, l'art d'introduire la Géométrie dans la Physique, & de former, en réu-

nissant l'expérience au calcul, une science exacte, profonde, lumineuse, & nouvelle.... la lumière a enfin prévalu."

10. Ibidem: "plusieurs autres vérités qui sont aujourd'hui la base & comme les élémens de la physique moderne."

11. Ibidem: "quand les fondemens d'une révolution sont une fois jetés, c'est presque toujours dans la génération suivante que la révolution s'achève."

12. J.F. Montucla, *Histoire des mathématiques,* vol. 1, pp. i, viii. I thank N.M. Swerdlow for calling my attention to Montucla by means of his paper "The History of the Exact Sciences" (see ch. 1, note 16). On perusing Montucla's four volumes I have come to share Swerdlow's enthusiasm, even though I do not think that Montucla advanced very far a *conceptual* understanding of the birth of early modern science. But I share Swerdlow's view that in a sense we are still following in Montucla's footsteps.

13. J.F. Montucla, *Histoire des mathématiques,* vol. 1, p. 559: "cette heureuse révolution dans les esprits."

14. Voltaire, *Essai sur les moeurs et l'esprit des nations* (1745–1785) (modern ed. by R. Pomeau [Paris: Classiques Garnier, 1990], 2 vols.).

15. The edition of Kant's *Kritik der reinen Vernunft* I have used was originally edited by Raymundt Schmidt in 1926, and published in Hamburg (Felix Meiner) in 1976 as vol. 37a of the Philosophische Bibliothek series. The argument in Kant's preface to the second edition discussed in the text is only a few pages long and is found right at the beginning. Since there are so many editions, not only in German but also in English translations, I do not give page numbers. The English of the quoted passages is a blending of phrases found in Meiklejohn's and Kemp Smith's translations of the *Critique* with phrases of my own.

16. "der sichere Gang einer Wissenschaft."

17. "ein blosses Herumtappen."

18. "Vielmehr glaube ich, dass es lange mit ihr (vornehmlich noch unter den Ägyptern) beim Herumtappen geblieben ist, und diese Umänderung einer Revolution zuzuschreiben sei, die der glückliche Einfall eines einzigen Mannes in einem Versuche zustande brachte, von welchem an die Bahn, die man nehmen musste, nicht mehr zu verfehlen war, und der sichere Gang einer Wissenschaft für alle Zeiten und in unendliche Weiten eingeschlagen und vorgezeichnet war. Die Geschichte dieser Revolution der Denkart ... , und des Glücklichen, der sie zustande brachte, ist uns nicht aufbehalten."

19. "den Heeresweg der Wissenschaft."

20. I.B. Cohen, *Revolution in Science,* ch. 15, lists a huge array of authors who adopted the phrase 'Kant's Copernican revolution' without bothering to look up Kant's preface, where the phrase is alleged to be found.

21. See for a similar distinction d'Alembert's entry "Expérimental," p. 298. The difference is that for d'Alembert this is a distinction among others, whereas for Kant it is the key point.

22. "Ich folge hier nicht genau dem Faden der Geschichte der Experimentalmethode, deren erste Anfänge auch nicht wohl bekannt sind."

23. I have used the following edition of Whewell's works: G. Buchdahl and L.L. Laudan (eds.), *The Historical and Philosophical Works of William Whewell,* 10 vols.

(London: Frank Cass, 1967). Vols. 2 through 6 of this edition are photomechanical reproductions of W. Whewell, *History of the Inductive Sciences, from the Earliest to the Present Time,* 3rd ed. 3 vols. (London: Parker, 1857), and of idem, *The Philosophy of the Inductive Sciences, Founded upon Their History,* 2nd ed. 2 vols. (London: Parker, 1847).

There is a considerable literature on Whewell, most of which is focused on his philosophical views, with much less concern for their historical underpinnings or for how the two domains hang together in Whewell's work. One exception is an illuminating article by Yehuda Elkana, "William Whewell, Historian." Large portions of this article were used by Elkana for his Editor's Introduction to *William Whewell: Selected Writings on the History of Science.* Here one may find further references to the literature on Whewell. Another exception is G.N. Cantor, "Between Rationalism and Romanticism: Whewell's Historiography of the Inductive Sciences." Here the context of Whewell's effort is set forth in great detail (e.g., with what surprising speed Whewell acquired mastery over a subject virtually ignored by him until the mid-1830s). This paper by Cantor is part of an interesting, recent collection edited by M. Fisch and S. Schaffer, *William Whewell: A Composite Portrait.* Fisch also published *William Whewell Philosopher of Science.* Although Fisch focuses here on a reconstruction of Whewell's philosophy, he also offers some interesting comments on the relation between Whewell's *History* and his *Philosophy.*

I.B. Cohen, *Revolution in Science,* pp. 528–532, lists examples of Whewell's usage of the term 'revolution'. A more extended study of the same topic—although confined to a philosophical analysis—is F. Schipper, "William Whewell's Conception of Scientific Revolutions."

24. W. Whewell, *Philosophy* I, p. iii of the second edition.

25. Ibidem, pp. 8–9.

26. A quite different assessment of Whewell's sense of history is on p. 56 of Henry Guerlac's article "A Backward View," reprinted in his *Essays and Papers,* pp. 54–65. One example of Whewell's sensitivity as a historian, in addition to all that is said in the main text of the present section, is that, scattered across the *Philosophy,* one encounters a great number of remarks which show how well Whewell was aware that the science of music (ignored by most historians of science up to the present day) is a legitimate object of study for the historian and the philosopher of science.

27. W. Whewell, *History* I, pp. 318, 326.

28. W. Whewell, *Philosophy* II, p. 40.

29. W. Whewell, *History* II, p. 99.

30. W. Whewell, *History* I, p. 9.

31. W. Whewell, *History* II, p. 137.

32. W. Whewell, *History* I, p. 8.

33. W. Whewell, *Philosophy* II, p. 208.

34. Ibidem, p. 230.

35. Ibidem.

36. W. Whewell, *History* II, p. 41. F. Schipper, "William Whewell's Conception of Scientific Revolutions," pp. 44–45, found that Whewell employed the term 'revolution' in science with the greatest frequency when a revolution in the philosophy of

science was simultaneously at stake, so that "mostly the term is used to characterize the development of the new way of doing science during and after the Renaissance." This confirms my observation that what we now call the Scientific Revolution was recognized by Whewell as a particularly revolutionary era in the history of science.

37. W. Whewell, *Philosophy* II, p. 208.

38. W. Whewell, *Philosophy* I, pp. 162–163.

39. Y. Elkana, Editor's Introduction, p. xxvi. The text of Whewell's essay is reproduced by Elkana on pp. 385–392 of the *Selected Writings* mentioned in note 23.

40. I have used a modern reedition of Comte's *Cours:* A. Comte, *Philosophie première: Cours de philosophie positive, Leçons 1 à 45*, ed. M. Serres, F. Dagognet, and A. Sinaceur. The quotation is on p. 39 of this edition: "Cette révolution générale de l'esprit humain est aujourd'hui presque entièrement accomplie: il ne reste plus, comme je l'ai expliqué, qu'à compléter la philosophie positive en y comprenant l'étude des phénomènes sociaux, et ensuite à la résumer en un seul corps de doctrine homogène. Quand ce double travail sera suffisamment avancé, le triomphe définitif de la philosophie positive aura lieu spontanément, et rétablira l'ordre dans la société."

41. Ibidem, p. 27: "Il est impossible d'assigner l'origine précise de cette révolution; car on n'en peut dire avec exactitude, comme de tous les autres grands événéments humains, qu'elle s'est accomplie constamment et de plus en plus.... Cependant, vu qu'il convient de fixer une époque pour empêcher la divagation des idées, j'indiquerai celle du grand mouvement imprimé à l'esprit humain, il y a deux siècles, par l'action combinée des préceptes de Bacon, des conceptions de Descartes, et des découvertes de Galilée, comme le moment où l'esprit de la philosophie positive a commencé à se prononcer dans le monde en opposition évidente avec l'esprit théologique et métaphysique. C'est alors, en effet, que les conceptions positives se sont dégagées nettement de l'alliage superstitieux et scolastique qui déguisait plus ou moins le véritable caractère de tous les travaux antérieurs. Depuis cette mémorable époque, le mouvement d'ascension de la philosophie positive, et le mouvement de décadence de la philosophie théologique et métaphysique, ont été extrêmement marqués."

42. W. Whewell, *Philosophy* II, p. 329.

43. Ibidem.

44. Ibidem, p. 379.

45. W. Whewell, *Philosophy* I, p. x (preface to the second edition).

46. W. Whewell, *Philosophy* II, p. 324.

47. Ibidem, p. 325. Whewell's handling of Comte's treatment of Kepler is both accurate and fair. Whewell refers to Tome I, p. 705, of Comte's *Cours*. In the 1975 edition cited in note 40 this corresponds to pp. 287–288.

48. W. Whewell, *History* I, p. 323.

49. H. Guerlac, in an article "Some Historical Assumptions of the History of Science" (reproduced in his *Essays and Papers*), regarded Comte as "the first to conceive of, and to baptize, the Scientific Revolution" (p. 33), his chief evidence being such passages as quoted above in notes 40 and 41. But, as far as conceptualizing the Scientific Revolution goes, Comte does no more than Kant had done: He demarcates wrong science from right science and dates the demarcation to the 17th century. Guerlac, while acknowledging the "narrow rigidity of [Comte's] thought" (p. 32), was much impressed

with Comte's awareness of the "underlying unity of the sciences" and with the importance Comte therefore professed to attach to their history (see for examples pp. 52–53 and pp. 463–464 of the 1975 reedition). In "A Backward View" Guerlac shows indeed that in the 19th century a thin line connects Comte's advocacy of an "histoire générale des sciences," through a Parisian university chair, with Paul Tannery's pioneering work at the turn of the century in the historiography of science.

50. The book to be discussed in the present section is E. Mach, *Die Mechanik in ihrer Entwickelung historisch-kritisch dargestellt* (Leipzig: Brockhaus, 1883). It was translated by T.J. McCormack as *The Science of Mechanics: A Critical and Historical Account of Its Development* (Chicago: Open Court, 1893). The translation was reedited twice to bring it up to date with current German reeditions. Here and there I have changed somewhat the wording of McCormack's generally excellent, authorized translation. In the following notes my page numbering conforms to the 6th English edition (1960), which includes all alterations made by Mach up to 1912. The German originals quoted in the notes are taken from the 4th edition, which came out in 1901. Rather than page numbers I list the numbers of the sections in which they appear.

51. E.N. Hiebert, s.v. "Ernst Mach," *Dictionary of Scientific Biography,* vol. 8, pp. 595–607; see in particular pp. 600–603.

52. E. Mach, *Science of Mechanics,* p. 330: "Für von vornherein einleuchtend kann man gewiss einen Satz nicht halten, welcher erst seit so kurzer Zeit allgemein anerkannt ist" (II, 10.3).

53. D. Shapere, *Galileo: A Philosophical Study,* pp. 3–8. These few pages gave me an excellent introduction to my examination of Mach's views on the birth of early modern science.

54. E. Mach, *Science of Mechanics,* p. 151. (The clause "Dynamics was founded by Galileo" is no longer present in the 3rd English edition. In the 2nd edition it is to be found on p. 128.) "Wir gehen nun an die Besprechung der Grundlagen der Dynamik. Dieselbe ist eine ganz moderne Wissenschaft.... Gegründet wurde die Dynamik erst durch Galilei. Dass diese Behauptung richtig sei, erkennen wir leicht, wenn wir nur einige Sätze der Aristoteliker der Galilei'schen Zeit betrachten" (II, 1.1).

55. Ibidem, p. 159: "Wir müssen zuvor bemerken, dass damals alle die Kenntnisse und Begriffe, die uns jetzt geläufig sind, nicht vorhanden waren, sondern dass Galilei dieselben erst für uns entwickeln musste" (II, 1.4).

56. Ibidem, p. 174: "Ein ganz neuer Begriff, auf den Galilei geführt wurde, war der Begriff *Beschleunigung*" (II, 1.13).

57. Ibidem, p. 167: "Wir sehen nun ... , dass Galilei nicht etwa eine *Theorie* der Fallbewegung gegeben, sondern vielmehr das *Thatsächliche* der Fallbewegung vorurtheilslos untersucht und constatirt hat" (II, 1.8).

58. Ibidem, p. 168: "Es gibt kein Verfahren, welches sicherer zur *einfachsten,* mit dem geringsten Gemüths- und Verstandesaufwand zu erzielenden Auffassung aller Naturvorgänge führen würde" (II, 1.8).

59. Ibidem, p. 141: "Wollen wir Galilei's Gedankengang ganz verstehen, so müssen wir bedenken, dass er schon im Besitz von instinctiven Erfahrungen ist, bevor er an das Experiment geht.... Für den wissenschaftlichen Gebrauch muss aber die gedankliche Nachbildung der sinnlichen Erlebnisse noch *begrifflich* geformt werden. Nur so

können sie benutzt werden, um zu einer durch eine begriffliche *Maassreaction* charakterisirten Eigenschaft durch eine begriffliche *Rechnungsconstruction* die davon *abhängige* Eigenschaft der Thatsache zu finden, die theilweise gegebene zu ergänzen. Dieses Formen geschieht durch Heraushebeen des für wichtig Gehaltenen, durch Absehen von Nebensächlichem, durch *Abstraction, Idealisirung.* Das Experiment entscheidet, ob die Formung genügt. Ohne irgend eine vorgefasste Ansicht ist ein Experiment überhaupt unmöglich, indem letzteres durch erstere seine Form erhält. Denn wie und was sollte man versuchen, wenn man nicht schon eine Vermuthung hätte?" (II, 1.4).

60. Ibidem, p. 130 in the 2nd edition: "Der moderne Geist, den Galilei bekundet, äussert sich gleich darin, dass er nicht fragt: *warum* fallen die schweren Körper, sonder dass er sich die Frage stellt, *wie* fallen die schweren Körper, nach welchem *Gesetze* bewegt sich ein frei fallender Körper?" (II, 1.2). Remarkably, in the 3rd edition (p. 155) the sentence quoted in the text is changed into: "In the riper and more fruitful time of his residence in Padua, Galileo dropped the question as to the 'why' and inquired the 'how' of the many motions which can be observed." Why Mach altered his original line is explained at the end of the present section.

61. D. Shapere, *Galileo,* passim.

62. E. Mach, *Science of Mechanics,* p. 170: "Es wäre ein Anachronismus und gänzlich unhistorisch, wollte man die gleichförmig beschleunigte Fallbewegung, wie dies mitunter geschieht, aus der constanten Wirkung der Schwerkraft ableiten. 'Die Schwere ist eine constante Kraft, *folglich* erzeugt sie in jedem gleichen Zeitelement den gleichen Geschwindigkeitszuwachs, und die Bewegung wird eine gleichförmig beschleunigte.' Eine solche Darstellung wäre deshalb unhistorisch, und würde die ganze Entdeckung in ein falsches Licht stellen, weil durch Galilei erst der heutige Kraftbegriff geschaffen worden ist" (II, 1.9).

63. E.J. Dijksterhuis, *Val en worp,* p. 271; E. Mach, *Science of Mechanics,* pp. 168–169.

64. E. Mach, *Science of Mechanics,* p. 182: "Auch der schiefe Wurf konnte ihm keine wesentlichen Schwierigkeiten mehr bereiten" (II, 1.17).

65. Ibidem, p. 169: "dass die Vorgänger und Zeitgenossen Galilei's, ja Galilei selbst nur *sehr allmählich,* von den aristotelischen Vorstellungen sich befreiend, zur Erkenntniss des Beharrungsgesetzes gelangt sind" (II, 1.8).

66. Ibidem: "Wohlwill's Untersuchung ist sehr *dankenswerth* und zeigt, dass Galilei in seinen eigenen bahnbrechenden Gedanken schwer die volle Klarheit erreichte und häufigen Rückfällen in ältere Anschauungen ausgesetzt war, was von vornherein sehr wahrscheinlich ist" (II, 1.8).

67. "Lorsque nous voyons la science d'un Galilée triompher du Péripatétisme buté d'un Cremonini, nous croyons, mal informés de l'histoire de la pensée humaine, que nous assistons à la victoire de la jeune Science moderne sur la Philosophie médiévale, obstinée dans son psittacisme; en vérité, nous contemplons le triomphe, longuement préparé, de la science qui est née à Paris au XIVe siècle sur les doctrines d'Aristote et d'Averroès, remises en honneur par la Renaissance italienne" (P. Duhem, *Études sur Léonard de Vinci: Ceux qu'il a lus et ceux qui l'ont lu,* vol. III, *Les précurseurs parisiens de Galilée,* p. vi). I have used the second impression—Paris: De Nobele, 1955 (first edition—Paris: Hermann, 1906, 1909, and 1913, respectively). In section 4.3 I discuss

some features of Duhem's ten companion volumes *Le système du monde: Histoire des doctrines cosmologiques de Platon à Copernic*.

There is one biography of Duhem: S.L. Jaki, *Uneasy Genius: The Life and Work of Pierre Duhem*. Despite the wealth of material collected in this book, its heavy partisanship makes it unreliable as an interpretation. One article on topics taken up in the present section is R.N.D. Martin, "The Genesis of a Mediaeval Historian: Pierre Duhem and the Origins of Statics."

68. P. Duhem, *Études* III, p. xi: "Ainsi, de Guillaume d'Ockam à Dominique Soto, voyons-nous les physiciens de l'École parisienne poser tous les fondements de la Mécanique que développeront Galilée, ses contemporains et ses disciples."

69. Ibidem, pp. ix-x: "Si l'on voulait, par une ligne précise, séparer le règne de la Science antique du règne de la Science moderne, il la faudrait tracer, croyons-nous, à l'instant où Jean Buridan a conçu cette théorie, à l'instant où l'on a cessé de regarder les astres comme mus par des êtres divins, où l'on a admis que les mouvements célestes et les mouvements sublunaires dépendaient d'une même Mécanique."

70. R.N.D. Martin, in "The Genesis of a Mediaeval Historian," found that Duhem's interest in tracing the origins of Leonardo's ideas goes back to his surprise discovery—made in the fall of 1903, when the first installments of *Les origines de la statique* were already in press—of manuscripts from the school of Jordanus Nemorarius. Martin uses this to argue that Duhem's thesis was not due to his extrascientific beliefs. I agree but, as I show further on, those beliefs *did* lead Duhem to aggrandizing his views.

71. In vol. III of the *Études* Leonardo is still counted as one of Galileo's 'Parisian' forerunners (see p. xiii).

72. P. Duhem, *Études* I, pp. 1–2: "L'histoire des sciences est faussée par deux préjugés, si semblables qu'on pourrait les confondre en un seul; on pense couramment que le progrès scientifique se fait par une suite de découvertes soudaines et imprévues; il est, croît-on, l'oeuvre d'hommes de génie qui n'ont point de précurseurs.

"C'est faire utile besogne que de marquer avec insistance à quel point ces idées sont erronées, à quel point l'histoire du développement scientifique est soumise à la loi de continuité. Les grandes découvertes sont presque toujours le fruit d'une préparation, lente et compliquée, poursuivie au cours des siècles. Les doctrines professées par les plus puissants penseurs résultent d'une multitude d'efforts, accumulés par une foule de travailleurs obscurs. Ceux-là même qu'il est de mode d'appeler créateurs, les Galilée, les Descartes, les Newton, n'ont formulé aucune doctrine qui ne se rattache par des liens innombrables aux enseignements de ceux qui les ont précédés. Une histoire trop simpliste nous fait admirer en eux des colosses nés d'une génération spontanée, incompréhensibles et monstrueux dans leur isolement; une histoire mieux informée nous retrace la longue filiation dont ils sont issus."

73. Ibidem, p. 156: "Non plus que la Nature, la Science ne fait point de saut brusque."

74. For example, P. Duhem, *Études* I, pp. 127–128, and III, pp. 33–34 (where Duhem calls Kepler and his contemporaries "les initiateurs de la Science moderne"), 51, 54, 258–259, 263–264.

75. P. Duhem, *Études* III, p. xiii.

76. The sole exception is Jaki.

77. P. Duhem, *Études* I, p. 162: "un esprit vraiment scientifique."

78. P. Duhem, *Études* III, p. 55: the history of the impetus theory "nous la montrerait ensuite se dépouillant de sa forme purement qualitative pour revêtir une forme quantitative plus précise."

79. p. 56: "désormais, elle y trouvera des adeptes que la lecture des anciens a formés aux habiles procédés de la Géométrie, qui la traduiront en langage mathématique, qui expliciteront ainsi les vérités qu'elle contenait en puissance et la détermineront à produire la Science moderne." The passage goes on as follows: "Dans les écrits de Léonard de Vinci, nous saisissons cette science parisienne au moment même où elle passe de l'esprit médiéval à l'esprit moderne." Note that these are two distinct 'spirits' after all!

80. Ibidem, p. 227.

81. In *Études* III, pp. 258–259, Duhem takes a somewhat more favorable view of Galileo's "geometrical genius."

82. D.G. Miller, "Duhem, Pierre-Maurice-Marie," *Dictionary of Scientific Biography,* vol. 4, pp. 225–233; a discussion of Duhem's religious and political extremism is on p. 232. S.L. Jaki, in *The Origin of Science,* p. 138, note 10, denies Miller's charges but gives no other grounds for his denial than the countercharge that Miller wrote out of "malice or miscomprehension."

83. P. Duhem, *Études* II, p. 412, quotes the proposition in question: "Quod prima causa non posset plures mundos facere."

84. Ibidem: "S'il nous fallait assigner une date à la naissance de la Science moderne, nous choisirions sans doute cette année 1277 où l'Évêque de Paris proclama solennellement qu'il pouvait exister plusieurs Mondes, et que l'ensemble des sphères célestes pouvait, sans contradiction, être animé d'un mouvement rectiligne."

85. P. Duhem, *Études* III, p. 227: "C'est à la Logique, à la Physique des Parisiens qu'en Italie, les initiateurs de la Science moderne empruntent des armes pour combattre les enseignements surannés du Philosophe et du Commentateur; ceux qui s'efforcent de secouer le joug de la tyrannique routine ont les yeux fixés sur Paris, dont la Scolastique nominaliste est, depuis des siècles, en possession de la liberté intellectuelle." Another statement to the same effect is on p. 320 of P. Duhem, *Système du monde,* vol. 4: "Voilà pourquoi nous ne comprendrions rien à l'avènement des idées qui devaient placer la Terre au rang des planètes si nous ignorions comment l'Église catholique a lutté contre les Métaphysiques et les Théologies léguées à l'Islam par l'Antiquité hellénique."

86. P. Duhem, *Études* III, pp. xiii-xiv: "Cette substitution de la Physique moderne à la Physique d'Aristote a résulté d'un effort de longue durée et d'extra-ordinaire puissance.

"Cet effort, il a pris appui sur la plus ancienne et la plus resplendissante des Universités médiévales, sur l'Université de Paris. Comment un parisien n'en serait-il pas fier?

"Ses promoteurs les plus éminents ont été le picard Jean Buridan et le normand Nicole Oresme. Comment un français n'en éprouverait-il pas un légitime orgueil?

"Il a résulté de la lutte opiniâtre que l'Université de Paris, véritable gardienne, en ce temps-là, de l'orthodoxie catholique, mena contre le paganisme péripatéticien et néoplatonicien. Comment un chrétien n'en rendrait-il pas grâce à Dieu?"

87. P. Duhem, *Études* I, p. 123: "la chaîne de la tradition scientifique."

88. Ibidem: "Léonard ne nous apparaît donc plus comme un génie isolé dans le temps, sans lien avec le passé comme avec l'avenir, sans ancêtres intellectuels comme sans postérité scientifique; nous voyons sa pensée se nourrir des sucs de la science des siècles précédents pour féconder à son tour la science des siècles futurs; maillon admirablement solide et brillant, il reprend sa place dans la chaîne de la tradition scientifique."

89. She addressed the narrower question in 1967 ("Galilei und die scholastische Impetustheorie," in *Ausgehendes Mittelalter,* vol. II, pp. 465–490). Maier collected her essays in seven volumes, to which one more was added after her death. All eight were published in Rome by Edizioni di "Storia e Letteratura." The first five have the common subtitle *Studien zur Naturphilosophie der Spätscholastik.* Their titles are: vol. I, *Die Vorläufer Galileis im 14. Jahrhundert* (1949); vol. II, *Zwei Grundprobleme der scholastischen Naturphilosophie* (1951); vol. III, *An der Grenze von Scholastik und Naturwissenschaft* (1952; revised version of an earlier edition of 1943); vol. IV, *Metaphysische Hintergründe der spätscholastischen Naturphilosophie* (1955); vol. V, *Zwischen Philosophie und Mechanik* (1958). The three later volumes have the common title *Ausgehendes Mittelalter: Gesammelte Aufsätze zur Geistesgeschichte des 14. Jahrhunderts.* Vol. I appeared in 1964, vol. II in 1967, and vol. III (edited by A. Paravicini Bagliani) in 1977. In 1982 S.D. Sargent edited and translated a selection of her essays under the title *On the Threshold of Exact Science: Selected Writings of Anneliese Maier on Late Medieval Natural Philosophy.*

90. Maier to Dijksterhuis, 28 June 1948: "als es anfing unerträglich zu werden" (Dijksterhuis Archival Collection, no. 4, on which more in note 103).

91. Ibidem. See for other biographical and bibliographical details Sargent's introduction to the collection of essays by Maier mentioned in note 89.

92. In each of these essays the argument comes to virtually the same conclusion; also, a variety of other topics that came to the fore as a result of Duhem's work were discussed by her with a similar thrust.

93. A. Maier, *Vorläufer,* p. 148: "der impetus wird durch dieselben Widerstände, die zum Zustandekommen einer Bewegung unerlässlich sind, allmählich zerstört.... Es gibt in der Inertialbewegung ein Moment, von dem man schlechterdings nicht abstrahieren kann, und das ist die Trägheit des bewegten Massenpunktes. Man kann von äusseren Hindernissen und Kräften absehen, aber nicht von der Masse des proiectum. Und der Unterschied zwischen scholastischer und klassischer Mechanik ist eben der, dass diese Trägheit von der letzteren als der eigentliche bewegungserhaltende Faktor aufgefasst wird, von der ersteren dagegen als Widerstand gegen diese Bewegung und den sie verursachenden impetus."

94. Ibidem, p. 149: "Eine analoge Anwendung auf die irdische Mechanik zu machen ist ihm nicht eingefallen und konnte ihm nicht einfallen."

95. Ibidem, p. 154: "Darum noch einmal: der Uebergang von der Impetustheorie zum Prinzip der Inertialbewegung hat nichts zu tun mit irgendwelchen weltanschaulichen Wandlungen, es wird einfach eine physikalische Theorie durch eine andere ersetzt, die den zu erklärenden Phänomenen besser gerecht wird. Und dieser Uebergang vollzieht sich in ganz analoger Weise wie der entsprechende im 14. Jahrhundert: hier und dort hat man zunächst durchaus die Absicht, bei der traditionellen Erklärung zu

bleiben und will diese nur in Einzelheiten modifizieren, um gewisse apparentia zu retten. Aber diese Einzelheiten, so stellt sich nachträglich heraus, sind von so fundamentaler Bedeutung, dass sie das ganze alte Gebäude zum Einsturz bringen. Im 14. Jahrhundert bezog sich die Korrektur auf das 'subiectum' der bewegenden Kraft, im 17. auf die Rolle der vis inertiae—und in beiden Fällen wurde mit diesen anscheinend geringfügigen Modifikationen eine neue Epoche in der Geschichte des physikalischen Denkens eingeleitet."

96. Ibidem, p. 1: "Grundsätzlich hat Duhem sicher recht, wenn er in der Naturauffassung des 14. Jahrhunderts eine Vorstufe und Vorbereitung der klassischen Physik sehen will."

97. Ibidem: "nur hat er im einzelnen die scholastischen Lehren oft in zu modernem Sinn interpretiert und zu viel aus ihnen herausgelesen."

98. Ibidem, p. 6: "wir müssen versuchen, uns von den physikalischen Theorien des 14. Jahrhunderts und dem weltanschaulichen Rahmen, in den sie hineingehören, ein objektives und möglichst exaktes Bild zu machen, müssen feststellen 'wie es eigentlich gewesen ist.'"

99. Ibidem: "Das ist seit Duhem die strittige Frage."

100. Ibidem, pp. 1–2: "Im Ganzen genommen ist die Geschichte der exakten Naturwissenschaft im christlichen Abendland, von ihren Anfängen im 13. Jahrhundert bis in das 18. hinein, eine Geschichte der allmählichen Ueberwindung des Aristotelismus. Diese Ueberwindung ist nicht in einer einzigen grossen Revolution erfolgt, wie man es lange Zeit angesehen hat, und andererseits auch nicht in einem stetig verlaufenden Emanzipations prozess, der sich gleichmässig über die Jahrhunderte erstreckt, sondern in einer Entwicklung, die sich in zwei grossen Phasen vollzieht, von denen die erste ihren Kulminationspunkt im 14., die zweite im 17. Jahrhundert hat."

101. Ibidem, p. 5: "eben im Sinn einer ersten Phase in einem grossen Entwicklungsprozess, dessen zweite und entscheidende Phase ins 17. Jahrhundert fällt."

102. One exception is Edward Grant (see section 4.3.).

103. In addition to the books and papers by Dijksterhuis referred to in the notes that follow I have also consulted the Dijksterhuis Archival Collection, which is kept at the Museum Boerhaave, Leiden, Netherlands; registered under "Arch 471." D.J. Struik wrote a brief sketch on Dijksterhuis in a new foreword to the 1986 reedition of *The Mechanization of the World Picture* (by Princeton University Press). K. van Berkel is currently preparing a monograph (in Dutch) on Dijksterhuis' life and works.

104. In a letter dated 13 October 1955, I.B. Cohen expressed the hope that *Val en worp*, together with *De mechanisering van het wereldbeeld* and *Simon Stevin* (1943), would be translated into English. Dijksterhuis answered on 21 October that translating *Val en worp* "would require a thorough remodelling for a new edition and I am too busy to undertake this" (Dijksterhuis Archival Collection, no. 2). In the end, *De mechanisering* was published in German and English translations, whereas a condensed version of the *Stevin* book in English appeared posthumously.

105. E.J. Dijksterhuis, *Val en worp: Een bijdrage tot de geschiedenis der mechanica van Aristoteles tot Newton*, p. v.

106. E.J. Dijksterhuis, "Doel en methode van de geschiedenis der exacte wetenschappen," pp. 10 sqq. Maier subscribed to the same ideal: *Vorläufer*, p. 6.

107. E.J. Dijksterhuis, *Val en worp,* pp. 302–303.

108. Ibidem, p. 167.

109. Ibidem, p. 170.

110. Ibidem.

111. Ibidem, p. 170–171.

112. Ibidem, p. 171.

113. Ibidem, pp. 234, 264; cf. E.J. Dijksterhuis, *The Mechanization of the World Picture,* IV:92–97 (references to this book will be made by section numbers rather than by page numbers).

114. E.J. Dijksterhuis, *Val en worp,* p. 433.

115. Ibidem, p. 433–434.

116. Ibidem, p. 285, note 4.

117. Ibidem, pp. 261, 284.

118. E.J. Dijksterhuis, *Mechanization,* I:1. From surviving correspondence with his regular translator, C. Dikshoorn, it appears that she frequently consulted him throughout the process, so that the translation of *Mechanization* may be regarded as broadly authorized (Dijksterhuis Archival Collection, nos. 37, 38, 174). Despite this procedure the translation makes somewhat awkward reading and is not always very accurate. I have therefore altered the wording of the translation wherever I thought fit.

119. Ibidem, I:2.

120. Ibidem.

121. Ibidem, V:7. In the English version the text of this paragraph differs substantially from the original Dutch, without apparent authorization, so I have retranslated the entire passage here.

122. Ibidem, V:8.

123. Ibidem, V:9. (The terms for scientific periods are in conformity to usage generally adopted by scientists since Bohr.)

124. Ibidem.

125. Ibidem.

126. Ibidem, IV:2.

127. For Kepler, see ibidem, IV:56; for Galileo, IV:77–78, 83, 103.

128. This impression is strongly reinforced by Dijksterhuis' lecture "Ad quanta intelligenda condita," pp. 118–122.

129. E.J. Dijksterhuis, *Mechanization,* V:5.

130. K. van Berkel informs me that Dijksterhuis' principal reason for sticking to the term 'mechanization' had little to do with the merits of the term in the context of the history of science (it rather resided in ongoing debates over the reform of the secondary school system in the Netherlands, in which he was very much involved).

131. Descartes to Mersenne, 11 October 1638: "il a basti sans fondement" (*Oeuvres de Descartes,* 12 vols., ed. Adam and Tannery [Paris: Vrin, 1897–1913], 2:380). (Edition consulted—2nd impression of 2nd ed., in 11 vols., 1971–74.)

132. E.J. Dijksterhuis, *Val en worp,* p. 343.

133. Ibidem, p. 353.

134. For a detailed exposition of Dijksterhuis' mathematical hypothetico-deductivism, see his "Ad quanta," passim.

135. E.J. Dijksterhuis, *Mechanization,* I:1.

136. Ibidem, IV:98.

137. See also E.J. Dijksterhuis, "Ad quanta," pp. 119–120. One curious, fleeting passage in Duhem's *Études* (III, p. v) suggests a similar view of the role played by the recovery of Archimedes' work in bringing about the emergence of early modern science.

138. Two interesting collections have been devoted to Koyré. One is a book edited by P. Redondi: A. Koyré, *De la mystique à la science: Cours, conférences et documents, 1922–1962.* The other, also edited by Redondi, is a special issue of *History and Technology* 4, 1–4, 1987, under the title "Science: The Renaissance of a History. Proceedings of the International Conference 'Alexandre Koyré,' Paris, Collège de France, 10–14 June 1986." For the biographical data mentioned in the text here and further on see C.C. Gillispie, s.v. "Koyré," *Dictionary of Scientific Biography,* vol. 7, pp. 482–490; Redondi's collections; and I.B. Cohen's article "Alexandre Koyré in America: Some Personal Reminiscences."

139. Koyré's *From the Closed World to the Infinite Universe* is a bit of an exception in this respect; so is *Études d'histoire de la pensée philosophique*—twelve essays collected by Koyré three years before he died.

140. A. Koyré, "Galileo and Plato," in idem, *Metaphysics and Measurement,* p. 22.

141. A. Koyré, *Études Galiléennes,* 2nd ed. (Paris: Hermann, 1966) (the first edition, dated 1939, came out in 1940). J. Mepham published a translation under the title *Galileo Studies* (Atlantic Highlands, N.J.: Humanities Press, 1978). For the passages quoted in the text I have generally preferred to make my own translations. The passage quoted here is as follows (*Études Galiléennes,* p. 12): "profonde transformation intellectuelle dont la physique moderne, ou plus exactement classique, fut à la fois l'expression et le fruit" (*Galileo Studies,* p. 1).

142. Ibidem, p. 74: "Nous voyons bien: le mouvement s'émancipe; le Cosmos se disloque; l'espace se géométrise" (*Galileo Studies,* p. 35).

143. Ibidem, p. 156: "La nature ne répond qu'aux questions posées en langage mathématique, parce que la nature est le règne de la mesure et de l'ordre" (*Galileo Studies,* p. 108).

144. Ibidem, pp. 206–207; *Galileo Studies,* p. 155.

145. It is this very point which Shapere misses in the interesting argument set up in ch. 5 of his *Galileo* to show that Koyré had overlooked the fundamental agreement between Plato and Aristotle in this respect (see p. 138 in particular). It is true that Koyré himself, in his subsequent rhetoric on "Galileo and Plato," contributed much to obscuring the point.

146. A. Koyré, *Études Galiléennes,* pp. 282–283: "Et l'objection galiléenne implique, bien au contraire, que le réel et le géométrique ne sont nullement hétérogènes et que la forme géométrique peut être réalisée par la matière. Bien plus: qu'elle l'est

toujours.... La forme géométrique est homogène à la matière: voilà pourquoi les lois géométriques ont une valeur réelle, et dominent la physique" (*Galileo Studies*, p. 204).

147. Ibidem, p. 290: "Le mouvement obéit à une loi mathématique. Le temps et l'espace sont liés par la loi du nombre. La découverte galiléenne transforme l'échec du platonisme en victoire. Sa science est une revanche de Platon" (*Galileo Studies*, p. 208).

148. Ibidem, p. 291: "Mais jamais victoire n'a été plus chèrement payée" (*Galileo Studies*, p. 209).

149. A. Koyré, *Newtonian Studies*, pp. 6–7. I left the wording as it stands, because it is apparently Koyré's own.

150. A. Koyré, *La révolution astronomique: Copernic—Kepler—Borelli*, pp. 15–17.

151. A. Koyré, *Newtonian Studies*, p. 5. In the preface to *From the Closed World to the Infinite Universe* Koyré takes the 140 years between the publication of Copernicus' *De revolutionibus* and Newton's *Principia* as the time span of the Scientific Revolution, during which the road from the closed world to the infinite universe was traveled in its entirety. Compared to this, so Koyré now stated, the topic taken up in *Études Galiléennes* had merely constituted the "prehistory" to that "great revolution."

152. A. Koyré, *Études Galiléennes*, p. 131: "La géométrisation à outrance—ce pêché originel de la pensée cartésienne—aboutit à l'intemporel: elle garde l'espace, elle élimine le temps. Elle dissout l'être réel dans le géométrique. Mais le réel se venge" (*Galileo Studies*, pp. 91–92).

153. Ibidem, p. 13 (*Galileo Studies*, p. 2).

154. Ibidem, p. 227: "la bonne physique se fait a priori" (*Galileo Studies*, p. 166).

155. Ibidem, p. 79 (*Galileo Studies*, p. 37).

156. Ibidem, p. 181: "On reste confondu devant la hardiesse, et le radicalisme, de la pensée de Bruno, qui opère une transformation—révolution véritable—de l'image traditionelle du monde et de la réalité physique" (*Galileo Studies*, p. 141).

157. A. Koyré, *Révolution astronomique*, pp. 9–10 (cf. p. 221).

158. A. Koyré, "An Experiment in Measurement," in idem, *Metaphysics and Measurement*, p. 91.

159. A. Koyré, "Gassendi et son temps," in idem, *Études d'histoire de la pensée scientifique*. This piece appeared originally in *Tricentenaire de Pierre Gassendi, 1655–1955, Actes du Congrès* (Paris: Presses Universitaires de France, 1957). It derives from a talk first given at another occasion in 1953.

160. A. Koyré, *Newtonian Studies*, p. 12; cf. idem, *Metaphysics and Measurement*, p. 130.

161. A. Koyré, *Newtonian Studies*, p. 6.

162. A. Koyré, "De l'influence des conceptions philosophiques sur l'évolution des théories scientifiques," in idem, *Études d'histoire de la pensée philosophique*, pp. 253–269. This is an extended version of a paper originally presented in English as a lecture for the American Association for the Advancement of Science. In February 1955 it appeared in *The Scientific Monthly* under the title "Influence of Philosophic Trends on the Formulation of Scientific Theories," pp. 107–111; here the phrase 'mathematical realism' appears on p. 111 (in the French version it is on p. 267). The paper also appeared in P.G. Frank (ed.), *The Validation of Scientific Theories* (Boston, 1956).

163. A. Koyré, "Du monde de l'à-peu-près' à l'univers de la précision," in idem, *Études d'histoire de la pensée philosophique,* p. 359: "une historiographie infecté par le virus de l'épistémologie empiriciste et positiviste qui a fait, et qui fait encore, tant de ravages parmi les historiens de la pensée scientifique." (This piece appeared originally in *Critique* 28, 1948.)

164. A. Koyré, "Les origines de la science moderne," in idem, *Études d'histoire de la pensée scientifique,* pp. 48–72. This piece appeared originally in *Diogène* 16, 1956, pp. 14–42; an English version is in *Diogenes* 16, Winter 1956, pp. 1–22. The phrase quoted in the text is on p. 18 of *Diogenes;* in the French collection it is on p. 67. (Margaret Osler thinks [private communication] that Koyré is conflating here instrumentalist and positivist conceptions of science.)

165. Ibidem, pp. 18 and 22 of the English version, from which the translation is taken; pp. 67–68 and 72 in the French collection.

166. A. Koyré, "De l'influence des conceptions philosophiques." I have compounded the quotation from the French version, p. 267 ("la solution improfitable, ou impossible, de problèmes déclarés dénués de sens") and its English counterpart, p. 111, where it appears garbled into "the allegedly unprofitable, impossible, or meaningless task."

167. A. Koyré, "Monde de l' 'à-peu-près,'" p. 343: "jamais elle n'a voulu admettre que l'exactitude puisse être de ce monde."

168. Ibidem, p. 347: "caractère approximatif."

169. Ibidem, p. 349: "personne n'a jamais cherché à dépasser l'usage pratique du nombre, du poids, de la mesure dans l'imprécision de la vie quotidienne—compter les mois et les bêtes, mesurer les distances et les champs, peser l'or et le blé—pour en faire un élement du savoir précis."

170. Ibidem, p. 350: "la structure générale du 'monde de l'à-peu-prés.'"

171. R.S. Westfall has noted throughout his *Never at Rest: A Biography of Isaac Newton* that Newton was the one alchemist to bring precision and rigor to the field. See also idem, "Making a World of Precision: Newton and the Construction of a Quantitative Physics'.

172. A. Koyré, "Monde de l' 'à-peu-près,'" p. 350: "Ce n'est pas le thermomètre qui lui manque, c'est l'idée que la chaleur soit susceptible d'une mesure exacte."

173. A. Koyré, *Newtonian Studies,* pp. 24–25.

174. E.A. Burtt, *The Metaphysical Foundations of Modern Physical Science: A Historical and Critical Essay* (London: Kegan Paul, Trench, Trubner, 1924). In 1925 the New York publishing house Harcourt, Brace marketed the book in the United States. Part of their issue was given a separate cover, which reads: "*The Metaphysics of Sir Isaac Newton: An Essay on The Metaphysical Foundations of Modern Science,* by Edwin Arthur Burtt, AB, STM, Assistant Professor of Philosophy in the University of Chicago. Submitted in partial fulfillment of the requirements for the degree of Doctor of Philosophy, in the Faculty of Philosophy, Columbia University." Apparently this title was used only for the dissertation version. In 1932 Routledge & Kegan Paul published a revised edition, which has been reprinted since without further alterations (but for the unfortunate cover given to some reprints, omitting the word 'Physical' from the title,

which has led to some bibliographical confusion). I have used the 1972 reprint, to which all subsequent page numbers refer.

Two recent articles devoted in large measure to the *Metaphysical Foundations* are by Lorraine Daston, "History of Science in an Elegiac Mode: E.A. Burtt's *Metaphysical Foundations of Modern Physical Science* Revisited," and by Gary Hatfield, "Metaphysics and the New Science."

175. This statement does not of course imply that the *Metaphysical Foundations* lacks all intellectual ancestry. The mathematical interpretation of 17th-century science was pervasive at the time, and Burtt mentions Whitehead, Cassirer, and Broad as men who worked in the same direction (p. 15). The point is that of all these and many other thinkers, only Burtt, Dijksterhuis, and Koyré were to elaborate such views into detailed examinations of the birth of early modern science.

176. Just two examples are what A.R. Hall, *The Revolution in Science, 1500–1750,* and E.L. Eisenstein, *The Printing Press as an Agent of Change,* have to say on Burtt. For R.S. Westfall, on the other hand, Burtt's book has been an ongoing source of inspiration (private communication).

177. E.A. Burtt, *Metaphysical Foundations,* pp. 4–5 and 10–11 (italics removed).

178. Ibidem, p. 11.

179. Ibidem, pp. 13–14.

180. Ibidem, pp. 15–17.

181. Ibidem, p. 21.

182. Ibidem, pp. 298–299. Note that Burtt was apparently not aware of such radically alternative currents as represented by Goethe or by previous Hermetic thought.

183. See the final pages of the book, where Burtt, while talking about 'spirit', adds: "Mayhap we must wait for the complete extinction of theological superstition before these things can be said without misunderstanding."

184. Ibidem, p. 22.

185. Ibidem, p. 26.

186. Ibidem, p. 53.

187. Ibidem, pp. 54–55.

188. Ibidem, pp. 57–58.

189. Ibidem, p. 68.

190. Ibidem, pp. 70–71.

191. Ibidem, p. 73.

192. Ibidem, p. 79 (italics removed).

193. Ibidem.

194. Ibidem, p. 80.

195. Ibidem, p. 156.

196. Ibidem, p. 114.

197. Ibidem, p. 168.

198. Ibidem, pp. 209–210.

199. Ibidem, p. 203.

NOTES TO CHAPTER TWO

200. In the conclusion of his book, Burtt briefly made some points about these issues.

201. Examples are provided, to varying extents, by the works of Hooykaas, Metzger, Olschki, Pagel, Sarton, Singer, Thorndike, and others. See for Olschki and Hooykaas chapter 5 of the present book.

202. E.A. Burtt, *Metaphysical Foundations,* p. 203.

203. A. Koyré, *Études Galiléennes,* p. 11: "une véritable 'mutation' de l'intellect humain"; p. 12: "Une telle mutation—une des plus importantes, si ce n'est pas la plus importante depuis l'invention du Cosmos par la pensée grecque" (*Galileo Studies,* p. 1).

204. G. Sarton, *The History of Science and the New Humanism* (New York: Holt, 1931), passim. See also R.K. Merton and A. Thackray, s.v. "George Sarton," *Dictionary of Scientific Biography,* vol. 12, pp. 107–114; and the March 1984 issue of *Isis* (vol. 75, no. 276), devoted largely to Sarton, An interesting discussion of Sarton, which I do not always agree with, is provided by L. Pyenson, "What Is the Good of History of Science?"

205. C.C. Gillispie, s.v. "Koyré," *Dictionary of Scientific Biography,* vol. 7, p. 486.

206. H. Guerlac, "A Backward View" (ch. 5 of his *Essays and Papers*), p. 63. (I.B. Cohen tells me that this may have been an overstatement due to Guerlac's 'Cornellism'.)

207. A. Koyré, *Études Galiléennes,* p. 213.

208. A. Koyré, "De l'influence des conceptions philosophiques," p. 255.

209. Koyré's reference to Burtt in the closing lines of "The Newtonian Synthesis" (*Newtonian Studies,* p. 24) displays more understanding of Burtt's intentions.

210. E.W. Strong, *Procedures and Metaphysics,* pp. 10 and 8, respectively.

211. A. Koyré, "De l'influence des conceptions philosophiques," pp. 254–255.

212. E.J. Dijksterhuis, review of Maier's *Vorläufer Galileis,* p. 208.

213. Here is a selection of notes I made from the Dijksterhuis–Maier correspondence in the Dijksterhuis Archival Collection, no. 4 (see above, note 103):

D to M, 24 February 1948 (minute): In *Val en Worp* D has largely adopted Duhem's views, since there was almost no occasion for independent checking (excepting only Buridan's 12th Question). From M's writings he now gets the impression that much in Duhem stands to be corrected. He needs this for a new book he is in the process of writing. [This was, of course, to become the *Mechanization.*]

M to D, 30 June 1949: Thanks D for his review in Isis. Mentions "... den naturphilosophischen Problemen ... , die doch eigentlich das interessanteste Gebiet der Scholastik darstellen. [...] Und immer wieder findet man etwas, das man nicht erwartet hat und das in irgend einer Form in der neuzeitlichen Naturwissenchaft weiterlebt."

M to D, 14 November 1950: Emphasizes that she is neither a physicist nor a historian of science, but rather a philosopher.

M to D, 17 March 1956: "... da ich Kongresse hasse ..."

M to D, 11 February 1960: "Dass wir in einigen grundsätzlichen Punkten verschiedener Ansicht sind und wohl immer bleiben werden, lässt sich nun einmal nicht ändern.... Wenn Sie mir vorhalten, ich selber hätte mehrmals 'Vorwegnahmen' von künftigen Erkenntnissen konstatiert, so haben Sie zweifellos recht; oft habe ich es immerhin nicht getan—und jedesmal sehr ungern. Aber als ich anfing, mich mit diesen Problemen zu beschäftigen, war diese Art der Fragestellung seit einem Menschenalter üblich, und da bleibt dem Einzelnen ja schliesslich nicht viel anderes übrig, als sich—wenn auch unter gelegentlichem Protest—zu fügen."

214. In *Zwischen Philosophie und Mechanik,* p. 382, note 60, Maier came back once more to the issue (1958). And now, at long last, Dijksterhuis acknowledged defeat: In the final sentence in his article "Over een recente bijdrage tot de laat-scholastieke natuurfilosofie" (in *Dancwerc: Opstellen aangeboden aan D.Th. Enklaar* [Groningen: Noordhoff, 1959], pp. 222–234) he granted that Maier's objections to the connection upheld by him between scholastic impetus and Newton's *vis inertiae* were justified (p. 234).

215. A. Koyré, review of Maier's *Vorläufer Galileis,* p. 781: "j'ai été vivement étonné."

216. See note 213.

217. Ibidem.

218. Cf. A. Maier, *Zwischen Philosophie und Mechanik,* pp. 373–375: "Immer wieder kehrt die Diskussion mit einer seltsamen Hartnäckigkeit zu der Frage zurück, ob man sagen könne, dass die scholastische Naturphilosophie gewisse Erkenntnisse der späteren exakten Naturwissenschaft schon 'vorweggenommen' hat.... In allen diesen Fällen handelt es sich ja nicht um wirkliche Abhängigkeitsbeziehungen und um greifbare historische Kontinuitäten, sondern um ungefähre Entsprechungen, die sich lediglich einem abstrakten, von allen geschichtlichen Realitäten absehenden Vergleich erschliessen."

219. A. Maier, *Ausgehendes Mittelalter* I, pp. 456–457: "die scholastischen Philosophen haben nicht nur praktisch keine Messungen durchgeführt und keine Theorie des Messens aufgestellt, sie haben darüber hinaus aus prinzipiellen Ueberlegungen ein wirklich exactes Messen für unmöglich erklärt. Der Grund für diese Entscheidung ist in letzter Analyse weltanschaulicher Natur.... für Gott ist die ganze Welt in allen ihren Einzelheiten gezählt und gemessen, aber nicht für den Menschen: ihm ist nicht die Möglichkeit gegeben, diese Masse exact zu erkennen und sozusagen nachzuprüfen." (A translation into English of this essay is on pp. 143–170 of Sargent's selection. The passage quoted in the text is on p. 169; I preferred to make my own translation.)

220. The eight volumes lack indices for nonmedieval authors, so I devoted one additional perusal to the sole purpose of spotting Koyré's name. I failed, but others may be luckier.

221. E.J. Dijksterhuis, "The Origins of Classical Mechanics from Aristotle to Newton," p. 176.

222. Ibidem, p. 175.

223. See for a summing-up of Crombie's thesis his "The Significance of Medieval Discussions of Scientific Method for the Scientific Revolution."

224. A. Koyré, *Études d'histoire de la pensée philosophique,* p. 37: "Assertion curieuse, qui met à l'origine de la science moderne la proclamation par l'évêque de Paris de deux absurdités." An interesting article has been devoted to this topic by M. Clavelin: "Le débat Koyré–Duhem, hier et aujourd'hui."

225. A. Koyré, "Origines," p. 63; the quotation is taken from the English version in *Diogenes* 16, Winter 1956, p. 15.

226. L. White, Jr., *Medieval Religion and Technology,* pp. xi-xii. (N.M. Swerdlow has shown, in his paper cited in ch. 1, note 16, that in the early historiography of the mathematical sciences the Middle Ages had not been wholly neglected.)

227. Drake has brought together his researches on the topic in a book no less controversial than the research papers on which his reconstruction is founded: *Galileo: Pioneer Scientist* (Toronto: University of Toronto Press, 1990).

228. Shapere's *Galileo* provides one striking example of how the identification of the two issues was taken for granted at the time; see pp. 86, 129–130, and many others.

229. A recent summing-up of this line of inquiry is A.G. Debus, "The Chemical Philosophy and the Scientific Revolution."

230. That admirer is Joseph Agassi (personal communication from his one-time pupil N.J. Nersessian).

231. I.B. Cohen, *Revolution in Science,* p. 390.

232. Cf. ibidem, p. 398. Throughout ch. 26 Cohen charts earlier usages of the term.

233. H. Butterfield, *The Origins of Modern Science, 1300–1800* (London: Bell, 1949). A new edition came out in 1957, of which I have used the 1968 reprint. The quotation is on p. 8.

234. Ibidem, p. 1.

235. The expression has given rise to a great deal of amazement and confusion, discussed by I.B. Cohen in *Revolution in Science,* pp. 390–391.

236. H. Butterfield, *Origins,* p. 37 (first page of ch. 3).

237. Ibidem, p. 1.

238. Ibidem, p. 118 (second page of ch. 7).

239. A.R. Hall, *The Scientific Revolution, 1500–1800,* p. xiv.

240. Ibidem, pp. 6, 20–21.

241. Ibidem, p. 365.

242. Ibidem, p. 307.

243. Ibidem, p. 366.

244. Ibidem, p. xiii.

245. Ibidem, p. xi.

246. In the General Introduction to the two volumes.

247. M. Boas, *The Scientific Renaissance, 1450–1630* (I have used the Fontana paperback edition of 1970). The quotation is on p. 156.

248. Ibidem, p. 323.

249. A.R. Hall, *From Galileo to Newton, 1630–1720* (I have used the Dover edition of 1981). The quotation is on p. 30.

250. Ibidem, p. 18.

251. Ibidem, p. 34.

252. Ibidem.

253. Ibidem, p. 104.

254. Ibidem, pp. 216–221.

255. Ibidem, p. 38.

256. A.R. Hall, *The Revolution in Science, 1500–1750,* p. 115 (note 13).

257. Ibidem, p. 144.

258. Ibidem, p. 2.

259. Ibidem, p. 9.

260. Ibidem, p. 12.

261. Ibidem, p. 13.

262. Ibidem, p. vii.

263. Ibidem, p. 149.

264. Ibidem, p. 45.

265. Ibidem, p. 148.

266. Ibidem, p. 344.

267. Ibidem, p. 73.

268. Ibidem.

269. Ibidem, pp. 73–74.

270. In my opinion, neither Butterfield nor Hall ever fully grasped Koyré's analytical usage of his concept of the Scientific Revolution. Hall, in particular, tends to project his own broad usage of the term upon Koyré's core conception. I think that almost all Hall's objections to Koyré's handling of the Scientific Revolution (made in his 1987 article "Alexandre Koyré and the Scientific Revolution") can be traced back to this misunderstanding. It culminates in the following sentence on p. 490: "I make this contention on the assumption (which I believe Koyré shared) that there is an historical phenomenon which we call the scientific revolution that is larger and more diffuse than the development of classical mechanics from Galileo to Newton and beyond." It is true that Koyré himself gave occasion to the misunderstanding by widening his original concept of the Scientific Revolution in time only, without considering possible conceptual alterations to follow suit; this is the very job subsequently taken up by Kuhn and Westfall. (See further this book, section 7.2.)

271. The literature on Kuhn is virtually boundless. Still, I am not aware of a discussion of the topic brought up in the present section. *The Structure of Scientific Revolutions* came out in 1962 with the University of Chicago Press. I have used the second, enlarged edition of 1970. "Mathematical versus Experimental Traditions in the Development of Physical Science" appeared first in *Journal of Interdisciplinary History* 7, 1976, pp. 1–31. I have used its reprint as ch. 3 of *Essential Tension,* pp. 31–65.

272. In *Essential Tension,* p. 108, Kuhn identifies Dijksterhuis, Maier, and Koyré as the historians responsible for this coming-of-age of the discipline.

273. T.S. Kuhn, *Structure,* p. 3 (the reference to Koyré is on p. 2).

274. T.S. Kuhn, *Essential Tension,* p. 21.

275. H.F. Cohen, "Music as a Test-Case," in particular pp. 372 sqq.

276. T.S. Kuhn, *Essential Tension,* p. 34.

277. Ibidem, p. 35.

278. Ibidem, pp. 35–36.

279. Ibidem, p. 36.

280. Ibidem, p. 38.

281. Ibidem, p. 39.

282. Ibidem, p. 41.

283. Ibidem.

284. Ibidem, p. 42.

285. Ibidem, p. 43.

286. Ibidem, p. 44.

287. Ibidem, p. 45.

288. Ibidem, pp. 45–46.

289. Ibidem, p. 46.

290. Ibidem.

291. Ibidem, p. 47.

292. Ibidem.

293. Ibidem, p. 49.

294. Ibidem.

295. Ibidem, p. 50.

296. H.F. Cohen, *Quantifying Music,* passim. An excellent state-of-the-art survey *cum* anthology is in P. Gozza (ed.), *La Musica nella Rivoluzione Scientifica del Seicento* (Bologna: Il Mulino, 1989). In the 18th and 19th centuries the topic had not yet dropped out of history of science writing; witness in particular the vast surveys by J.F. Montucla and W. Whewell, both of whom paid considerable attention to the science of music through the ages.

297. T.S. Kuhn, *Essential Tension,* p. 37, note 5.

298. J.G. Yoder's recent, illuminating study *Unrolling Time: Christiaan Huygens and the Mathematization of Nature* strongly suggest that the mathematical element dominated entirely.

I agree with Hall, in his article "Koyré and the Scientific Revolution," that "the Koyré–Kuhn distinction seems to force itself between the investigations of different scientists, and even between the investigations of one individual" (p. 494); also with his conclusion: "Whatever the difficulty of defining and expressing the change, we should endeavour to see it as possessing historical consistency and homogeneity, and not as it were make schizophrenics of our subjects " (p. 495).

299. This is set forth at greater length in the final chapter of C. Hakfoort, *Optica in de eeuw van Euler* (Amsterdam: Rodopi, 1986) (to appear with Cambridge University Press as *Optics in the Age of Euler*).

300. See in particular ch. 2 of Kuhn's *Structure.*

301. Examples are chapters 1, 5, and 6 of *Essential Tension.*

302. Written in 1968 and revised in 1976; pp. 3–20 of *Essential Tension.*

303. R.S. Westfall, *The Construction of Modern Science: Mechanisms and Mechanics* 2nd ed. (Cambridge: Cambridge University Press, 1977). The first edition came out in 1971 (New York: Wiley); many unaltered reprints have since followed. The passage quoted in the text is on p. ix.

304. R.S. Westfall, *Force in Newton's Physics: The Science of Dynamics in the Seventeenth Century.*

305. R.S. Westfall, *Construction of Modern Science,* p. 1.

306. Ibidem, p. 3: "Whether the book [*De revolutionibus*] would initiate a revolution had yet to be determined." Westfall nonetheless takes the Scientific Revolution to have started with Copernicus (personal communication).

307. Ibidem, p. 31.

308. Ibidem, p. 38.

309. Ibidem, p. 41.

310. Ibidem, pp. 41–42.

311. Ibidem, p. 43.

312. Ibidem, p. 45.

313. Ibidem, p. 50.

314. Ibidem.

315. Ibidem, p. 64.

316. Ibidem, pp. 50 and 49–50.

317. Ibidem, p. 81.

318. Ibidem, p. 82.

319. Ibidem, p. 86.

320. Ibidem, p. 104.

321. Ibidem, p. 123.

322. Ibidem, pp. 132 and 134.

323. Ibidem, p. 138.

324. Ibidem, pp. 142–43.

325. Ibidem, p. 159.

326. E.J. Aiton, "The Concept of Force" (essay review of Westfall's *Force in Newton's Physics*), *History of Science* 10, 1971, pp. 88–102; see p. 94.

327. In *Construction of Modern Science* Newton's alchemical pursuits are not yet mentioned in this connection; the theme was taken up shortly after 1971 by both Westfall and B.J.T. Dobbs. More on this in sections 3.3.2–3.

328. R.S. Westfall, *Never at Rest: A Biography of Isaac Newton.* I have used the paperback edition of 1983. The passage quoted is on pp. 16–17.

329. See on the nature of Mersenne's 'mechanicism' J.A. Schuster's unpublished thesis "Descartes and the Scientific Revolution, 1618–1634," pp. 419–422. R. Lenoble, the author of what before P. Dear's *Mersenne and the Learning of the Schools* (Ithaca, N.Y.: Cornell University Press, 1988) came out was the only serious monograph on the life and works of this important 17th-century scientist, entitled his book *Mersenne ou la naissance du mécanisme* (Mersenne or the Birth of Mechanicism). This unfortunate title has led to much confusion. See also my *Quantifying Music,* pp. 99–115 and 198–199.

330. Pascal in the *Pensées* (fragment 79 in Brunschvicg's numbering): "*Descartes.*—Il faut dire en gros: 'Cela se fait par figure et mouvement'; car cela est vrai. Mais de dire quels, et composer la machine, cela est ridicule; car cela est inutile et incertain et pénible." See for an extensive discussion of these matters R. Hooykaas, "Pascal: His Science and His Religion," esp. p. 127.

331. R.S. Westfall, *Never at Rest,* p. 17.

332. In class, where I have been using the *Construction of Modern Science* for many years with great satisfaction, I invariably find myself in trouble when explaining the role of experiments from this particular chapter.

333. A related idea is the 'Experimentally-orientated Corpuscular-Mechanical natural philosophy' distinguished by J. Schuster in his "The Scientific Revolution," pp. 238–241.

334. K. van Berkel, *Isaac Beeckman (1588–1637) en de mechanisering van het wereldbeeld* (Isaac Beeckman and the Mechanization of the World Picture), ch. 8.

335. See note 299.

336. In *Quantifying Music* I faced the challenge of coming to grips with a science not previously handled as part and parcel of the Scientific Revolution. My division into mathematical, experimental, and mechanicist approaches helped me to find significant connections between different theories. It worked quite well for the mathematical tradition into which I put the musical theories of Kepler, Stevin, and young Descartes, and for the mechanicist, that is, corpuscularian tradition which I used as the proper peg for Beeckman and mature Descartes. Placing Mersenne in the category 'experimental' also worked fine to the extent that it brought out the differences between his and Beeckman's respective approaches to a number of problems in the science of music examined by both men. However, the categorization itself continued to strike me as rather artificial, and although it did work reasonably well for the other supposed experimentalist in my account, Galileo, I never felt quite at ease with putting him there, happily reunited with his father who fitted so beautifully. One other limitation of my attempt is that I dealt only with those first-rate scientists who contributed in a major way to this one science during no more than the first stage of the Scientific Revolution, rather than facing the full task of finding an organizing principle that does justice to every important scientist during the entire period.

337. E.J. Dijksterhuis, "Renaissance en natuurwetenschap," p. 194.

338. In *Revolution in Science,* I.B. Cohen has some interesting remarks on the increasing time span allowed to 'revolutions' in the 20th-century usage of the term.

339. H. Butterfield, *Origins,* p. 180.

340. E.J. Dijksterhuis, *Val en worp,* p. 302.

THREE

1. Examples are Crombie (discussed in section 2.4.1), Randall and Wallace (discussed in section 4.4.3), and maybe also McMullin (discussed in the next section)— witness his (in my opinion, misconceived) opening statement that "those who argue for a 'revolution' in science around 1600 nearly always base their case on an alleged change in the methods of science at that time" (p. 331 of E. McMullin, "Empiricism and the Scientific Revolution").

2. McMullin deals with a number of such pitfalls. See, for example, the following passages: "The separation between explicit and implicit methodology, between what is said about method and the method actually practiced, is more troublesome in Newton's

work than in that of almost any other scientific writer" (ibidem, p. 358). "[Newton] wanted to insist on the empirical character of what he was doing and had no other way to describe this than the traditional language of deduction and induction" (ibidem, p. 362).

3. Overlooking this state of affairs constitutes, in my view, the central misconception in certain continuist claims (discussed in the present book in section 4.4.3) which are based on the family resemblance between Aristotelian and Galilean pronouncements on what constitutes scientific knowledge.

4. E.A. Burtt, *Metaphysical Foundations*, p. 203.

5. One example is M.A. Finocchiaro's book *Galileo and the Art of Reasoning*.

6. E. McMullin, "Empiricism," p. 333.

7. Ibidem, p. 334.

8. Ibidem, p. 333.

9. Ibidem.

10. In the case of Copernicus, I think, McMullin is not quite so careful. He ascribes to Copernicus a much more unambiguously 'realist' conception of the astronomer's job than is warranted by Books II through VI of *De revolutionibus*. See section IV of McMullin's essay; cf. E.J. Dijksterhuis, *Mechanization*, sections IV:8–16.

11. E. McMullin, "Empiricism," p. 357.

12. Ibidem, p. 363.

13. Ibidem, pp. 366–367.

14. E. McMullin, "Conceptions of Science in the Scientific Revolution."

15. From Burton Watson (trans.), *The Complete Works of Chuang Tzu* (New York: Columbia University Press, 1968), ch. 22 ("Knowledge Wandered North"), p. 241.

16. Dante Alighieri, *Divina Commedia,* Inferno IV, 131: "il maestro di color che sanno."

17. R. Hooykaas, in his *G.J. Rheticus' Treatise on Holy Scripture and the Motion of the Earth,* p. 142, beautifully sums up the tension inherent in so much Renaissance thought: "It is as if [Rheticus] had been swept away by his colourful dream and then finds himself awakening in grey daylight. It is true that an alternation of almost absolute certainty and total skepsis was not uncommon with scholars in the period of upheaval that we term the Renaissance. Old certainties were undermined; scholasticism was attacked from all sides, and other systems—platonism, hermeticism, etc.—were recommended. But there were so many 'true' systems; and who could without doubt establish which was the right one? Many Renaissance scholars loudly and boisterously overshouted their own doubts and then suddenly recognized that their thirst for a full understanding could not be quenched."

18. R.F. Jones, *Ancients and Moderns*, pp. 145–146.

19. Ibidem, p. 268.

20. Ibidem, pp. 146–147.

21. "Réponse de Blaise Pascal au très bon Révérend Père Noël" (29 October 1647): "Je sais que vous pouvez dire que vous n'avez pas fait tout seul cette matière, et que quantité de physiciens y avaient déjà travaillé; mais sur les sujets de cette matière, nous ne faisons aucun fondement sur les autorités: quand nous citons les auteurs, nous

citons leur démonstrations, et non pas leurs noms" (p. 374 in Chevalier's Pléiade edition).

22. R.F. Jones, *Ancients and Moderns,* pp. 221, 222.

23. P. Dear, "Totius in Verba," p. 149.

24. Quoted in R.F. Jones, *Ancients and Moderns,* p. 122 (I had no chance to check the quotation against the original).

25. R.S. Westfall, *Construction of Modern Science,* p. 116.

26. L. Olschki, *Geschichte der neusprachlichen wissenschaftlichen Literatur,* vol. III (see note 29 of ch. 5 for full bibliographical information), pp. 121–122: "Galilei zwang seine Zeitgenossen zur Erkenntnis, dass die wissenschaftliche Wahrheit kein Besitz der Menschheit, sodern ihr fernes Ziel war; er vertauschte die beruhigende Ueberzeugung eines bereits abgeschlossenen Wissens über die Natur mit der aufregenden Vorstellung einer Wahrheit, die das Ergebnis der Zeit, des Zweifels, des Fortschritts, einer unabsehbar fernen Zukunft sein sollte. Er bot an Stelle einer Totalität des Natursystems blosse Ansätze, um es zu formen und eine unübersehbare Fülle von Problemen und Rätseln, deren Lösung er der Zukunft überliess."

27. G.J. Whitrow's *Time in History* provides an exhaustive summing-up of all the varied issues that come up under the heading of 'Time in History', from the construction of calendars and the measurement of time to conceptions of history. It is the more significant that, in ch. 8, "Time and History in the Renaissance and the Scientific Revolution," nothing at all is to be found on the topic we address in the present section.

28. F.A. Yates, *Giordano Bruno and the Hermetic Tradition,* p. 1.

29. Edizione Nazionale of Galileo's works, vol. 8, p. 190 (opening passage of the Third Day): "Haec ita esse, et alia non pauca nec minus scitu digna, a me demonstrabuntur, et, quod pluris faciendum censeo, aditus et accessus ad amplissimam praestantissimamque scientiam, cuius hi nostri labores erunt elementa, recludetur, in qua ingenia meo perspicaciora abditiores recessus penetrabunt."

30. As he explains in the *Discours de la méthode* in particular, after the deduction, with mathematical certainty, of the principles of nature, nothing remains to be done except to fill in the details. Although the principles are indubitable, God might have established second-order phenomena in one way or another, and at that point experience comes in to decide exactly how He did establish these in fact. Dijksterhuis (*Mechanization,* IV:201) notes that "one may regard his method as an application in the grand manner of the *metodo risolutivo* and the *metodo compositivo,* or of analysis and synthesis, combined—an application marked by the peculiarity that the analytical part is being curtailed so drastically as to become virtually unrecognizable, since it no longer calls for deliberate observation; natural experience suffices to furnish the axioms needed for the synthesis." See also the passages on Descartes in the articles by McMullin discussed in sections 3.1.1–2.

31. E.J. Dijksterhuis, *Simon Stevin: Science in the Netherlands around 1600,* pp. 128–129.

32. J.E. McGuire and P.M. Rattansi, "Newton and the Pipes of Pan."

33. Isaac Beeckman to Marin Mersenne, 1 October 1629. In de Waard's edition of Beeckman's Diary the quotation is in vol. 4, p. 163: "Magna certe et magnificia, imo vere philosophica sunt, quae in illis libris *de Musica* promittis, quos si quidem, ut decet,

absolvas, nae tu nobis omnibus de rebus philosophicis posthac meditandi omnem an-
sam praeripueris."

34. A.R. Hall has one isolated remark on the topic in his *Revolution in Science,*
p. 74: "One could find everywhere confidence in past achievement as well as optimistic
expectation for further intellectual conquests in the future.... Some imagined that the
human quest for understanding of the ultimate mathematical and natural truths might
be completed in two or three generations if enthusiastically undertaken." R.F. Jones, in
Ancients and Moderns, has a few, much less precise, remarks on the future-orientedness
of the new science; e.g., pp. 119, 146.

One other example is in a letter by van Leeuwenhoek to the Royal Society (*Phil.
Trans.* 1703, vol. 23, no. 287, p. 1473): "A little while before the last Sickness and
Death of the famous Christian Hugens, Lord of Zuilichem, being together in his Study,
he told me, that we were arriv'd to the utmost degree of knowledge in our observations
of Heavenly Bodies, and consequently that there remain'd nothing more to be seen or
said concerning them: I might likewise say, that we have penetrated so deeply into the
great secrets of the Seeds both of Animals and Plants, that we seem to be at the end of
our Discoveries; but however I may be mistaken in those Sentiments." I found this
passage (quoted in the original Dutch) in K. van Berkel, *In het voetspoor van Stevin,*
p. 68.

35. In a somewhat different context some of these topics come up in Paolo Rossi's
The Dark Abyss of Time.

36. O. Gingerich quotes this passage in his Kepler entry in *Dictionary of Scientific
Biography,* vol. 7, p. 304, with a reference to Kepler's *Gesammelte Werke* 3, p. 408. The
Latin there reads as follows: "siquidem Deo placuerit justum humano generi spacium
temporis in hoc mundo indulgere, ad residua ista perdiscenda."

37. I found this quotation on p. 159 of Maurice Daumas, *Les instruments scientifiques aux XVIIe et XVIIIe siècles* (p. 119 of the English version): "Si un savant se
propose une question nouvelle, il l'attaque avec les forces réunies de tous ceux qui l'ont
précédé." This is cited as part of Condorcet's eulogy for Vaucanson (who died in 1782)
before the Académie des Sciences, but with no footnote appended.

38. T.S. Kuhn, *Structure,* especially chs. 2 and 11; see also a number of passages
in *The Essential Tension,* in particular pp. 228 sqq. Remarkably, after Kuhn's *Structure*
became famous, social scientists began proudly to call their disciplines 'polyparadig-
matic', thus borrowing prestige from Kuhn's sanctified terminology while subverting
entirely one of his basic points.

39. For Kuhn, as we have seen in section 2.4.4, the attainment of paradigmatic
status happened to different disciplines at different times. This is not to say, as we have
also seen, that he is fully consistent as to when these transitions took place, or even
what units of scientific endeavor are taken to be affected (problem areas, disciplines,
clusters of disciplines). It is nonetheless certain that neither from his *Structure* nor from
his "Traditions" (see note 271 in ch. 2) can the reader possibly emerge with the impression that at one definite point in time natural science as such became cumulative. In
fact, Kuhn's overall conception of science rules out such a possibility from the start:
Even though his 'normal science' is indeed cumulative, revolutions cut right through
this, and the gestalt-switches of which his revolutions are made up exclude cumulativity
as a constant feature of any science in its development over time.

40. H. Guerlac, *Essays and Papers,* p. 43.

41. Examples are the fate of the laws of falling bodies after Galileo (studied in detail in R.S. Westfall, *Force in Newton's Physics*) and the reduction of Kepler's laws to statements inside the 'saving the appearances' framework still common to most mathematical astronomers in the generation after Kepler's death.

42. In the Adam and Tannery edition of the *Oeuvres de Descartes,* the *Discours de la méthode* is in vol. 6; the passages in question are on pp. 62–63 and 68–74.

43. D. Gjertsen, *Science and Philosophy: Past and Present,* ch. 1, "When Did Philosophy and Science Diverge?"

44. Descartes to Mersenne, 11 October 1638: "que [Galilée] ne les [i.e., his topics] a point examinées par ordre, et que, sans avoir considéré les premières causes de la nature, il a seulement cherché les raisons de quelques effets particuliers, et ainsi qu'il a bâti sans fondement."

45. J.-F. Revel, *Histoire de la philosophie occidentale.* I remain greatly indebted to ch. 6 of vol. 2, in particular, which has forever colored my ideas about Descartes, even though I now find that Revel treats him too much as if the new science were already fully present when he appeared on the scene.

46. S. Shapin and S. Schaffer, *Leviathan and the Air-Pump: Hobbes, Boyle, and the Experimental Life,* esp. ch. 4. One particularly cogent formulation is on p. 115: "According to Hobbes, it was impossible to understand the air-pump experiments 'unless the nature of the air is known first.' "

47. C. Hakfoort is presently examining the thought of Wilhelm Ostwald as one exemplar for a future history of scientism (taken in the broad sense of a world-picture inspired by science).

48. This domain of historical inquiry is at the center of much attention at the Department of History of the University of Twente. Not only Hakfoort works on it (see previous note) but also J.C. Boudri, who is examining the relation between physics and metaphysics in conceptions of force in the 18th century; P.F.H. Lauxtermann, who is investigating Schopenhauer's theory of color in the context of his philosophical work; and P. Várdy, who is studying foundational principles of the physical sciences through the centuries.

49. The works by Yates that I have chiefly used here as well as in ch. 4 are the following: *Giordano Bruno and the Hermetic Tradition* (Chicago: University of Chicago Press, 1964); *The Rosicrucian Enlightenment* (London: Routledge & Kegan Paul, 1972) (consulted in the Paladin paperback version of 1975); and "The Hermetic Tradition in Renaissance Science," in Ch. S. Singleton (ed.), *Art, Science, and History in the Renaissance* (Baltimore: Johns Hopkins University Press, 1967), pp. 255–274 (consulted in the reprinted version on pp. 227–246 of the third and final volume of Yates' *Collected Essays* [London: Routledge & Kegan Paul, 1982–84]; the same volume contains an autobiographical sketch and a complete bibliography).

50. F.A. Yates, *Giordano Bruno,* ch. 21.

51. F.A. Yates, *Rosicrucian Enlightenment,* pp. 270, 271, 266.

52. A.R. Hall, *Revolution in Science,* p. 2.

53. A. Debus, review of Yates, *Giordano Bruno,* in *Isis* 55, 1964, pp. 389–391; quoted passage on p. 391.

54. B.J. Teeter Dobbs, *The Foundations of Newton's Alchemy, or "The Hunting of the Greene Lyon."* See for her rebuttal of Yates' claim, p. 19; for her statement of the thesis that Newton's concept of force was alchemical in origin, pp. 210–213. Westfall published a number of papers on Newton's alchemy, with his final views on the topic to be found in his *Never at Rest* of 1980.

55. The principal collections are M.L. Righini Bonelli and W.R. Shea (eds.), *Reason, Experiment, and Mysticism in the Scientific Revolution,* and B. Vickers (ed.), *Occult and Scientific Mentalities in the Renaissance.* An overview of the debate is given by P. Curry in an essay review of the latter volume: "Revisions of Science and Magic." See for a recent survey B.P. Copenhaver, "Natural Magic, Hermetism, and Occultism in Early Modern Science."

56. P. Curry, "Revisions of Science and Magic," p. 300.

57. F.A. Yates, "Hermetic Tradition," p. 242.

58. The phrase is of course an adaptation from Balthasar Bekker's *Betooverde Wereld* (1691; The Enchanted World). To translate *Entzauberung* by 'disenchantment' became (perhaps unfortunately) standard usage decades ago.

59. M. Weber, "Wissenschaft als Beruf," p. 536: "Die zunehmende Intellektualisierung und Rationalisierung bedeutet also nicht eine zunehmende allgemeine Kenntnis der Lebensbedingungen, unter denen man steht. Sondern sie bedeutet etwas anderes: das Wissen daran, oder den Glauben daran: dass man, wenn man nur wollte, es jederzeit erfahren könnte, dass es also prinzipiell keine geheimnisvollen unberechenbaren Mächte gebe, die da hineinspielen, dass man vielmehr alle Dinge—im Prinzip—durch Berechnen beherrschen könne. Das aber bedeutet: die Entzauberung der Welt."

60. Ibidem: "dieser in der okzidentalen Kultur durch Jahrtausende fortgesetzte Entzauberungsprozess und überhaupt: dieser 'Fortschritt', dem die Wissenschaft als Glied und Triebkraft mit angehört."

61. The most prominent example here is K. Thomas' *Religion and the Decline of Magic* (London: Weidenfeld & Nicolson, 1971) (reprinted since as a Penguin book).

62. This particular issue was fought out between Brian Vickers (introduction to *Occult and Scientific Mentalities*) and Patrick Curry ("Revisions of Science and Magic").

63. One example is Patrick Curry's *Prophecy and Power: Astrology in Early Modern England* (London: Polity Press, 1989). Witchcraft has also been discussed in connection with the Scientific Revolution by some authors: H.R. Trevor-Roper, *The European Witch-Craze of the 16th and 17th Centuries* (Pelican Books, 1969); C. Webster, chapter on "Demonic Magic" in *From Paracelsus to Newton;* Brian Easlea, *Witch Hunting, Magic and the New Philosophy: An Introduction to Debates of the Scientific Revolution, 1450–1750* (Sussex: Harvester Press, 1980).

64. A.R. Hall, "Magic, Metaphysics and Mysticism in the Scientific Revolution," p. 277.

65. A.R. Hall, "The Scholar and the Craftsman in the Scientific Revolution," p. 21.

66. To be discussed briefly in section 4.4.4 of the present book.

67. P. Rossi, "Hermeticism, Rationality and the Scientific Revolution," p. 272.

68. Ibidem.

69. Ibidem, pp. 271 and 270.

70. See, for one representative example, J.L. Heilbron, *Electricity in the 17th and 18th Centuries: A Study of Early Modern Physics* (Berkeley: University of California Press, 1979), p. 30, note 51.

71. F.A. Yates, *Giordano Bruno,* p. 452.

72. F.A. Yates, *Rosicrucian Enlightenment,* pp. 277–278.

73. Yates did not work out possible consequences of such lack of insight for man's capacity truly to understand nature. The view that man should not aim for an understanding of nature until he has understood himself was expressed by Pascal in the *Pensées* (fragment 72 in Brunschvicg's numbering): "je souhaite, avant que d'entrer dans de plus grandes recherches de la nature, qu'il [l'homme] la considère une fois sérieusement et à loisir, qu'il se regarde aussi soi-même, et connaissant quelle proportion il a . . ." Also (same fragment): "Manque d'avoir contemplé ces infinis, les hommes se sont portés témérairement à la recherche de la nature, comme s'ils avaient quelque proportion avec elle." See also R. Hooykaas, "Pascal," in particular p. 131.

74. F.A. Yates, *Giordano Bruno,* pp. 454–455.

75. See for one exception three notes further down.

76. G. Sorel, *Les préoccupations métaphysiques des physiciens modernes,* esp. pp. 58–59: "Le but de la science expérimentale est donc de construire une *nature artificielle* (si on peut employer ce terme) à la place de la *nature naturelle*." (My attention was called to Sorel by J. de Kadt's study *Georges Sorel: Het einde van een mythe* [Amsterdam: Contact, 1938].)

77. In the first pages of his "Galileo and Plato," Koyré refers to P. Duhem, *Système du monde* I, pp. 194 sqq., and to P. Tannery, *Mémoires scientifiques* VI, p. 399.

78. R. Hooykaas, "Das Verhältnis von Physik und Mechanik in historischer Hinsicht" and "La Nature et l'Art," both in his *Selected Studies in History of Science.* A convenient summing-up of Pagel's and Debus' line of approach is Allen G. Debus, "The Chemical Philosophy and the Scientific Revolution."

79. R. Hooykaas, "The Rise of Modern Science: When and Why?" p. 470.

80. The topic has recently been pursued by P. Dear, "Miracles, Experiments, and the Ordinary Course of Nature."

81. S. Shapin and S. Schaffer, *Leviathan and the Air-Pump,* p. 131.

82. Ibidem, pp. 3–5 in particular; also p. 341.

83. In studies and passages devoted to Pascal and to Boyle, respectively, R. Hooykaas has repeatedly stressed that the former, unlike the latter, was not at all averse to invoking causes. For this, if for no other reason, it seems unwarranted to equate experimental science with acausal science.

84. S. Shapin and S. Schaffer, *Leviathan and the Air-Pump,* pp. 112, 163–164, 186, with references to philosophers' debates over the possibility of crucial experiments. See for an example of a historian's treatment of the issue, Dijksterhuis, *Mechanization,* IV:275–276, on Pascal's and Périer's allegedly decisive experiment with the void on the Puy de Dôme.

85. S. Shapin and S. Schaffer, *Leviathan and the Air-Pump,* pp. 77–79, 114, 128.

86. On Pascal: R. Hooykaas, "Pascal." On Huygens: C. Burch's unpublished doc-

toral dissertation "Christiaan Huygens: The Development of a Scientific Research Program in the Foundation of Mechanics," 2 vols. (University of Pittsburgh, 1981), pp. 170–181, on Huygens' idealization of phenomena in impact as compared to Descartes'; esp. Huygens' texts printed in the corresponding notes. On the issue in general: A. Van Helden, "The Birth of the Modern Scientific Instrument, 1550–1700," in particular, p. 62.

87. I take this example from my book *Quantifying Music*, p. 102, with reference to M. Mersenne, *Harmonie universelle*, livre quatriesme des instrumens, prop. 9, coroll. 1.

88. M. Daumas, *Les instruments scientifiques aux XVIIe et XVIIIE siècles* (translated as *Scientific Instruments of the Seventeenth and Eighteenth Centuries*), pp. 40–41 (27–28): "L'ingéniosité et l'habileté des artisans n'avaient d'autre objet que de perfectionner des modèles d'instruments dont le principe était, pour certains, connu depuis une dizaine de siècles." "... un épanouissement basé non plus sur l'évolution, mais sur l'invention." "Le milieu technique avait atteint un niveau de qualité suffisant pour que les problèmes nouveaux de fabrication, posés par les récentes découvertes, puissent recevoir une solution; mais encore fallait-il qu'une demande se manifestât pour que les ateliers cherchent à résoudre ces problèmes."

89. Ibidem, p. 2.

90. A. Van Helden, "The Birth of the Modern Scientific Instrument," p. 57. For microscopy this picture has meanwhile been confirmed in a study by M. Fournier, *The Fabric of Life: The Rise and Decline of Seventeenth-Century Microscopy*.

91. A pioneer in this area is P. de Clerq, who contributed various papers on this topic (chiefly but not exclusively on the 18th-century instrument trade) to *Annals of Science, Tractrix,* and a number of congress reports.

92. Here a pathbreaking study is J.H. Leopold, "Christiaan Huygens and His Instrument Makers."

93. "et ainsi nous rendre comme maîtres et possesseurs de la nature."

94. One early study in which Bacon's background in Renaissance naturalism is stressed is P. Rossi, *Francis Bacon: From Magic to Science* (on which more in section 4.4.4).

95. The idea that an applied science was practiced among the Puritans during the decades before the Restoration has been defended by Charles Webster, most notably in his book *The Great Instauration: Science, Medicine and Reform, 1626–1660* (New York: Holmes & Meier, 1976). I have consulted a succinct outline of Webster's views prepared by Harold J. Cook on pp. 265–300 of I.B. Cohen (ed.), *Puritanism and the Rise of Modern Science: The Merton Thesis*.

96. A. Koyré, *Newtonian Studies*, pp. 5–6 (footnote).

97. R.K. Merton, *Science, Technology and Society in Seventeenth-Century England*, 1970 preface, pp. xx-xxi.

98. D.S.L. Cardwell, *Turning Points in Western Technology*, chs. 2 and 3.

99. This was the topic of Hall's doctoral dissertation. See for a convenient summary, A.R. Hall, "Gunnery, Science, and the Royal Society."

100. J.D. Bernal, *Science in History*, p. 491.

101. Ibidem. The exception Bernal made for navigation is explained in the present book, section 3.6.2.

102. R.S. Westfall, "Robert Hooke, Mechanical Technology, and Scientific Investigation," p. 107.

103. A useful overview of 17th-century technology is in C.A. Davids, "Technological Change in Early Modern Europe (1500–1780)."

104. D.S. Landes, *Unbound Prometheus,* pp. 108–114.

105. C. Merchant, *The Death of Nature: Women, Ecology, and the Scientific Revolution,* p. xvii.

106. Ibidem, p. 5.

107. These passages are quoted ibidem, pp. 170 and 168, respectively. They refer back to Francis Bacon, *The Masculine Birth of Time,* translated in B. Farrington, *The Philosophy of Francis Bacon* (Liverpool: University of Liverpool Press, 1964), p. 62; and *De dignitate et augmentis scientiarum,* in *Works* (ed. Spedding, Ellis, and Heath), vol. 4, p. 296.

108. C. Merchant, *The Death of Nature,* p. 168.

109. Ibidem, pp. 171, 172. (There is a tendency throughout Merchant's book to associate conceptions of nature like those upheld by American Indians with a nonexploitative posture. An interesting repudiation of this 'myth'—widespread in the ecological movement—is to be found in an article by Kent H. Redford, "The Ecologically Noble Savage," *Cultural Survival Quarterly* 15, 1, 1991, pp. 46–48.)

110. C. Merchant, *The Death of Nature,* pp. 288–289, also the Epilogue. Evelyn Fox Keller, in chs. 2 and 3 of her book *Reflections on Gender and Science,* constructs roughly the same connection between the birth of early modern science and a changing image of feminine Nature. Rather than the consequent 'death of nature', she emphasizes the contingency of the triumph of 'masculine', power-directed science in the 17th century, discussing alchemy, with its image of the harmonious collaboration between masculine and feminine principles, as pointing toward a possible, alternative route for science which lost the battle because of contemporary shifts in man/woman relationships and in the wider economy. Londa Schiebinger, in *The Mind Has No Sex? Women in the Origins of Modern Science,* derives most of her material, by itself very informative, from the 18th century. For other reasons, too, I agree with Anita Guerrini's judgment, at the end of a review in *Isis* 82, 1, 311, 1991, pp. 133–134, that "the story of women in the origins of modern science remains to be written."

111. In my opinion, a great deal of confusion was brought to this debate by the parallel debate among philosophers of science aiming to do precisely what I affirm in the text cannot be done: to settle the issue in the abstract, as if science were a timeless entity.

112. Unlike Merton's doctoral dissertation, which is at bottom a historical inquiry (although guided by questions, and carried out with tools, common to the sociologist), the later studies are usually timeless in the sense that they claim to lay bare certain institutional features held by him to be characteristic of science as such. The historical element comes in principally as fortifying the main thesis of a particular study with a great number of examples taken from all over the past. Virtually without exception, these historical examples go no farther back than the year 1600.

113. When Merton collects his articles, he often changes their titles while preserving their content. Thus the latter article was originally entitled "Science and Technology

in a Democratic Order," and also spent part of its life under the title "Science and Democratic Social Structure." I have used the versions to be found in R.K. Merton, *The Sociology of Science: Theoretical and Empirical Investigations.*

114. R.K. Merton, *The Sociology of Science,* p. 268.

115. Ibidem, pp. 268–269.

116. Ibidem, p. 270.

117. Ibidem, p. 273.

118. Ibidem.

119. Ibidem, p. 275.

120. Ibidem, p. 276.

121. Ibidem, pp. 278–279.

122. Ibidem, p. 265.

123. S. Shapin and S. Schaffer, *Leviathan and the Air-Pump,* chs. 2 and 4, in particular. The authors often acknowledge the stimulus they have drawn from Merton's approach to the history of science.

124. This paragraph on the *practice* of open communication of scientific findings is not meant to criticize but rather to supplement the argument William Eamon makes in "From the Secrets of Nature to Public Knowledge" on what he calls (p. 333) "the ideology of openness in science," attributed by him, in conformity to Merton's point about the value of 'communism', to the onset of the Scientific Revolution.

125. E.L. Eisenstein, *The Printing Press as an Agent of Change,* pp. 634, 644–645. (See section 5.2.9 for an extended discussion of this book.)

126. M. Ornstein, *The Rôle of Scientific Societies in the Seventeenth Century.* (The author's name is given in the book as Martha Ornstein [Bronfenbrenner], the latter being the name of her husband, whom she married in 1914.) The quotation is on p. 261.

127. Ibidem, p. 39.

128. Ibidem, p. 21.

129. Ibidem, p. 54.

130. Ibidem, pp. 67–68.

131. Ibidem, p. 68.

132. Ibidem, p. 259.

133. K. van Berkel, "Universiteit en natuurwetenschap in de 17e eeuw, in het bijzonder in de Republiek."

134. J.L. Heilbron, *Electricity in the 17th and 18th Centuries,* esp. pp. 98–106.

135. I have made this case at greater length in my "Comment on 'The Universities and the Scientific Revolution: The Case of England' by Mordechai Feingold."

136. The expectation expressed in the text seems to be borne out by John Gascoigne's "A Reappraisal of the Role of the Universities in the Scientific Revolution," which appeared after the present section had been written. Although he takes pains to cover universities all across Europe rather than Oxford and Cambridge alone, the typical outcome 'greater flexibility than hitherto acknowledged' is almost all I can find in the evidence he adduces.

137. Westfall announced some provisional results of his research-in-progress in a

paper "Science and Technology during the Scientific Revolution: An Empirical Approach," contributed to a conference "The Scientific Revolution" held in Oxford, 17–20 July 1990.

138. The first of these was R.S. Westfall, "Science and Patronage: Galileo and the Telescope"; another, his "Patronage and the Publication of Galileo's *Dialogue*."

139. R. Hahn, "Changing Patterns for the Support of Scientists from Louis XIV to Napoleon," p. 407. (Two recent books pertinent to the subject of patronage of science could not be considered in the present section. One is H. Dorn, *The Geography of Science;* the other, which I have not seen at all, is B. Moran [ed.], *Patronage and Institutions: Science, Technology, and Medicine at the European Court, 1500–1750* [Rochester, N.Y.: Boydell, 1991]. The whole subject has over the past few years become rather a trend in the historiography of early modern science.)

140. These remarks sum up the point made on pp. 7–11 of T.K. Rabb, *The Struggle for Stability in Early Modern Europe.*

141. This somewhat surprising point is also made by Rabb: ibidem, pp. 13–14.

142. Ibidem, p. 36.

143. Ibidem, p. 37.

144. Only music is lacking altogether; a shortcoming which Alexander Silbiger has attempted to rectify in "Music and the Crisis of Seventeenth-Century Europe," in V. Coelho (ed.), *Music and Science in the Age of Galileo* (Dordrecht; Kluwer Academic Publishers, 1992), pp. 35–44.

145. T.K. Rabb, *The Struggle for Stability in Early Modern Europe,* pp. 116–117.

146. Ibidem, pp. 39, 49–52.

147. For the remainder of the argument on the role of science, see ibidem, pp. 107–115.

148. Ibidem, p. 111.

149. Ibidem, p. 112.

150. Ibidem, p. 114.

151. Ibidem.

152. The issue of why the late 17th-century Dutch patriciate, which counted several excellent scientists among its members (Huygens, Hudde, van Heuraet, de Witt), gave no support whatsoever to science is discussed by K. van Berkel, *In het voetspoor van Stevin,* and taken up in an essay review I devoted to this book under the title " 'Open and Wide, yet without Height or Depth.' "

153. A related attempt to connect English science of the second half of the 17th century with contemporary politics has been undertaken by James R. Jacob and Margaret C. Jacob. Both together, in "The Anglican Origins of Modern Science: The Metaphysical Foundations of the Whig Constitution," and the latter author alone, in *The Cultural Meaning of the Scientific Revolution,* argued that crucial ideas in Boyle's and Newton's work were designed in a straightforward manner to comply with contemporary political trends. Since, unlike Shapin and Schaffer, the authors have hardly made an effort to examine relevant portions of the content of the scientific works in question, I have not discussed their thesis in the present book. Also, among a variety of theses on the 'origins of early modern science' theirs is the one most narrowly confined to por-

tions of late 17th-century science in England (see also section 3.7 on recent tendencies to confine the Scientific Revolution to its English contribution).

154. S. Shapin and S. Schaffer, *Leviathan and the Air-Pump,* ch. 2, with discussions of Boyle's 'three technologies'; summed up on pp. 77–79.

155. Ibidem, pp. 107, 306. (The quotation from Sprat is referred to as from *The History of the Royal-Society,* p. 427.)

156. Ibidem, p. 152.

157. Ibidem, p. 141.

158. Ibidem, p. 108.

159. Ibidem, p. 341: "We have not taken as one of our questions, 'Why did Boyle win?'" However, the authors go on at once to answer that very question, and I do not see what aim is served by their ch. 7 if not to provide material for their answer.

160. Ibidem, p. 304.

161. The high point of these efforts is to be found ibidem, p. 293.

162. Ibidem, p. 283. Note how completely Hobbes' peculiar outlook on the causes of civil war is being adopted here.

163. Once, on p. 284, Shapin and Schaffer use the same term: "The events of 1660 began a long drawn-out search for stability by the restored régime."

164. Ibidem, p. 343.

165. P. Hazard, *La crise de la conscience Européenne (1680–1715),* vol. 2, p. 289: "Mais cette pensée critique, qui l'a nourrie? où a-t-elle pris sa force et ses audaces? et d'où vient-elle enfin?" (my translation in text).

166. The book to be discussed in the present section is J.D. Bernal, *Science in History* (London: Watts, 1954). I have used the illustrated edition in 4 vols. from 1969, with text as established in the 3rd edition of 1965. On Bernal's life and works I have consulted Gary Werskey, *The Visible College.* See also, in the present book, section 5.2.2, on the Hessen thesis, for a discussion of the impact Soviet Marxism had upon a group of Cambridge scientists, to which Bernal belonged, in 1931 (this group is the subject of Werskey's—incidentally, very apologetic—monograph).

167. Much in what follows derives from my critical reading, during the middle sixties, of such Marxist classics as *Das Kapital,* the *Manifest der kommunistischen Partei,* the entire Marx–Engels correspondence, *Herrn Eugen Dührings Umwälzung der Wissenschaft,* and *Der Achtzehnte Brumaire des Louis Bonaparte* (which, under the stimulus of J. de Kadt, I came to consider as Marx's best work and, in all likelihood, the best product of 'historical materialism' ever written). Among scores of books on Marx and Marxism, I still regard Joseph Schumpeter's discussion in Part I, "The Marxian Doctrine," of his *Capitalism, Socialism and Democracy* (New York: Harper, 1942; many editions since), as the one that, in a spirit of critical sympathy, makes the most of what is worth preserving in Marx's thought.

168. Henry Guerlac offers a curious example on p. 36 of his *Essays and Papers:* On one occasion when he related "the rise of chemistry in eighteenth-century France to some aspects of French industrial progress," a colleague of his found this interesting "mais un peu Marxiste." (Incidentally, I wonder whether the colleague in question may have been Alexandre Koyré.)

169. There was a time when thousands and thousands of intellectuals were ready to buy this kind of stuff. Note how such persuasive words as 'because' and 'therefore' suggest an argument where none whatever comes forward.

170. I owe this insight to Leopold Schwarzschild's most revealing biography of Karl Marx, *The Red Prussian* (originally *Der rote Preusse*) (New York: Scribner's, 1947); in the edition by Grosset & Dunlap which I possess the passage is on pp. 130–131 (in ch. 7, "Our Theory").

171. They have rarely received answers that do more than reassert the original statement.

172. I owe this last point, in particular, to a wonderful study of Marxist-Leninist doctrine by K. van het Reve, *Het geloof der kameraden* (The Comrades' Creed) (Amsterdam: van Oorschot, c. 1972).

173. J.D. Bernal, *Science in History,* p. 30.

174. See, e.g., Bernal's polemics against other historians of science: ibidem, pp. 45–46, 55, 376.

175. Ibidem, p. 5.

176. Ibidem, p. 47.

177. Ibidem, p. 309.

178. Ibidem, p. 373.

179. Ibidem, p. 377.

180. Ibidem, p. 389.

181. Ibidem, p. 410.

182. Ibidem, p. 435; I have more on this topic in section 5.2.4 on Zilsel.

183. Ibidem, p. 418. Another example is the treatment of Holland, characterized throughout by Bernal as economically (together with England) the farthest advanced of the world, whereas on p. 448 he baldly and quite rightly reports that Holland created no scientific center of its own.

184. Ibidem, p. 373.

185. *Science in History* may be taken as the artfully integrated aggregate of at least four distinct driving forces in Bernal's thought: (1) original Marxism (e.g., the explanation of social change in history on p. 48, which is a close paraphrase of Marx's "Preface"); (2) its Stalinist perversion (marked, *inter alia,* by a shift to voluntarism amounting to a conspiratorial conception of things which, in *Science in History,* appears in particular in vol. 4); (3) Bernal's largely original discovery of science as a contemporary force of production; and (4) the conviction of the working scientist that science is the embodiment *par excellence* of objective truth. Many inconsistencies in Bernal's book can be explained by his simultaneous adherence to these four, often contradictory, persuasions.

186. Ibidem, p. 490.

187. One example is Benjamin Farrington, who heralded Francis Bacon as the originator of 'industrial science'.

188. In particular, J.D. Bernal, *Science in History,* pp. 3, 42, 455, 491–493.

189. Ibidem, pp. 1, 27, 55.

190. Ibidem, p. 2 (taken in conjunction with a similar passage on p. 6).

191. Ibidem, p. 321.

192. Ibidem, p. 335.

193. Weber's scholarly work is in the nature partly of a torso, partly of a jigsaw puzzle. Marianne Weber's biography of her late husband provides at once a moving story, a useful guide to his work, and an almost complete bibliography. My grasp of Weber's work as a whole is not yet as solid as I wish it were, and my brief treatment in the text derives largely from my impression that, in the voluminous literature devoted to Weber by social scientists, there is nothing at all about what little Weber had to say on the natural sciences. The one exception I am aware of is the historian of science L. Pyenson, "What Is the Good of History of Science?" in which Weber appears as one of the three protagonists.

The history of translations into English of Weber's works is a morass (to which Reinhard Bendix, *Max Weber: An Intellectual Portrait* [Garden City, N.Y.: Doubleday, 1960], pp. xi–xiv, provides a guideline). The process of translation by bits and pieces has contributed greatly to dispersing still further the already hard-to-discern unity that, at bottom, underlies Weber's work. (Some recent studies in German have helped to make that unity more clearly visible.) I have consulted the German originals of Weber's work and made my own translations of the passages quoted in the text.

194. M. Weber, *Gesammelte Aufsätze zur Religionssoziologie,* vol. 1, "Vorbemerkung," p. 1: "Universalgeschichtliche Probleme wird der Sohn der modernen europäischen Kulturwelt unvermeidlicher- und berechtigterweise unter der Fragestellung behandeln: welche Verkettung von Umständen hat dazu geführt, dass gerade auf dem Boden des Okzidents, und nur hier, Kulturerscheinungen auftraten, welche doch—wie wenigstens wir uns gern vorstellen—in einer Entwicklungsrichtung von *universeller* Bedeutung und Gültigkeit lagen?"

195. Ibidem, pp. 1–2: "Eine rationale Chemie fehlt allen Kulturgebieten ausser dem Okzident."

196. Ibidem, p. 10: "In einer Universalgeschichte der Kultur ist also, für uns, rein wirtschaftlich, das zentrale Problem letztlich *nicht* die überall nur in der Form wechselnde Entfaltung kapitalistischer Betätigung als solcher: des Abenteurertypus oder des händlerischen oder des an Krieg, Politik, Verwaltung und ihren Gewinnchancen orientierten Kapitalismus. Sondern vielmehr die Entstehung des *bürgerlichen Betriebs-*kapitalismus mit seiner rationalen Organisation der *freien Arbeit.*"

197. Ibidem, pp. 10–11: "Der spezifisch moderne okzidentale Kapitalismus nun ist zunächst offenkundig in starkem Masse durch Entwicklungen von *technischen* Möglichkeiten mitbestimmt. Seine Rationalität ist heute wesenhaft bedingt durch *Berechenbarkeit* der technisch entscheidenden Faktoren: der Unterlagen exakter Kalkulation. Das heisst aber in Wahrheit: durch die Eigenart der abendländischen Wissenschaft, insbesondere der mathematisch und experimentell exakt und rational fundamentierten Naturwissenschaften. Die Entwicklung dieser Wissenschaften und der auf ihnen beruhenden Technik erhielt und erhält nun andererseits ihrerseits Impulse von den kapitalistischen Chancen, die sich an ihre wirtschaftliche Verwertbarkeit als Prämien knüpfen. Zwar nicht die Entstehung der abendländischen Wissenschaft ist durch solche Chancen bestimmt worden. Gerechnet, mit Stellenzahlen gerechnet, Algebra getrieben haben auch die Inder, die Erfinder des Positionszahlensystems, welches erst in den

Dienst des sich entwickelnden Kapitalismus im Abendland trat, in Indien aber keine moderne Kalkulation und Bilanzierung schuf. Auch die Entstehung der Mathematik und Mechanik war nicht durch kapitalistische Interessen bedingt. Wohl aber wurde die *technische* Verwendung wissenschaftlicher Erkenntnisse: dies für die Lebensordung unsrer Massen Entscheidende, durch ökonomische Prämien bedingt, welche im Okzident gerade darauf gesetzt waren. Diese Prämien aber flossen aus der Eigenart der *Sozialordung* des Okzidents. Es wird also gefragt werden müssen: aus *welchen* Bestandteilen dieser Eigenart, da zweifellos nicht alle gleich wichtig gewesen sein werden."

198. Ibidem, p. 11: "Denn es handelt sich ja in all den angeführten Fällen von Eigenart offenbar um einen spezifisch gearteten 'Rationalismus' der okzidentalen Kultur."

199. A passing hint is given in M. Weber, *Wirtschaftsgeschichte: Abriss der universalen Sozial- und Wirtschaftsgeschichte* (Munich: Duncker & Humblot, 1924), pp. 308–309: "Die Magie zu brechen und Rationalisierung der Lebensführung durchzusetzen, hat es zu allen Zeiten nur ein Mittel gegeben: grosse *rationale Prophetien*. Nicht jede Prophetie allerdings zerstört ihre Macht; aber es ist möglich, dass ein Prophet, der sich durch Wunder und andere Mittel legitimiert, die überkommenen heiligen Ordnungen durchbricht. Prophetien haben die *Entzauberung der Welt* herbeigeführt und damit auch die Grundlage für unsere moderne Wissenschaft, die Technik und den Kapitalismus geschaffen."

200. The respective places are the following. On Copernicus (with some more material that is covered at other places): *Wirtschaftsgeschichte,* in the 3rd edition (1958) the remark in question is on the penultimate page, 314. On Swammerdam and on the origin of experiment in art (also the one place where Galileo's name is mentioned): *Gesammelte Aufsätze zur Wissenschaftslehre* (Tübingen: Mohr, 1922), pp. 538–539. On Baconian utilitarianism and the rise of early modern science: ibidem, p. 399. On science and religion in the 17th century: *Gesammelte Aufsätze zur Religionssoziologie,* vol. 1, p. 188 (footnote 2). (This is the place from which Merton's doctoral thesis [see section 5.1.2] took its origin.) However, incidental remarks about science are scattered over all of Weber's work.

201. D.S. Landes, in ch. 1 of *Unbound Prometheus,* has come closer to such a treatment in a more or less Weberian vein than any other author known to me, even though his topic is the issue of why no *Industrial* Revolution occurred in other civilizations than the West.

202. M. Weber, "'Energetische' Kulturtheorien," in *Gesammelte Aufsätze zur Wissenschaftslehre,* pp. 376–402.

203. I.B. Cohen, *Revolution in Science,* pp. 390–391, 398–399, has some striking observations on the prose of this chapter in Butterfield's book. He also calls attention, on pp. 395–396, to two 'general' historians to precede Butterfield by inserting discussions of the emergence of early modern science in their general histories: Preserved Smith, and J.H. Robinson.

204. H. Butterfield, *The Origins of Modern Science,* p. 186.

205. Ibidem, p. 179.

206. Ibidem, p. 175.

207. The theme of the 16th/17th-century shift of the center of European trade and

politics has lately been much discussed in the historical literature, e.g., by Braudel and by Wallerstein. Some allusions to the shift in connection with the geographical movements of the Scientific Revolution are to be found in remarks made by Hill and by Rabb in C. Webster (ed.), *The Intellectual Revolution of the Seventeenth Century,* pp. 243–244 and 276, respectively; also by Hall in *Dictionary of the History of Science,* s.v. "Scientific Revolution." (NB The topic of the geography of science has at long last been taken up in systematic, pioneering fashion by H. Dorn in his book of that name [1991].)

208. H. Butterfield, *The Origins of Modern Science,* p. 190.

209. S. Shapin and S. Schaffer, *Leviathan and the Air-Pump,* pp. 13, 16–17.

210. Quoted ibidem, p. 323. (The quotation is referred to as from *Leviathan,* p. 31 in vol. 3 of the Molesworth ed.)

211. R.S. Westfall once made this point in conversation with me.

212. I thank Rob Wentholt for his ongoing exhortation never to lose sight of these elementary matters.

213. In his recent chapter "The Scientific Revolution," J. Schuster has repeatedly stressed (by way of a programmatic point rather than through demonstration) how important it is not to take aspects of the onward course of the Scientific Revolution as foreordained.

214. Woody Allen, *Side Effects* (New York: Random House, 1975), p. 13.

215. To avoid possible misunderstanding: It is far from me to advocate a narrow parochialism for the historian's craft. Not by chance I have so far spent my entire professional life as a historian in surroundings peopled almost exclusively by nonhistorians (scientists, philosophers, sociologists), and I have learned more than a little from their respective approaches. Perhaps by that very experience, however, I have come to believe more and more that, whereas the study of history can profit a great deal from approaches adopted (and adapted) from other domains of study, there is also an integrity of historical thought which should not be subverted thereby. It is this integrity which I wish to uphold in the face of social-constructivist doctrines that, at least for historians, are called in to render services that they, by the nature of their craft, are perfectly—even better—equipped to perform themselves.

216. T.S. Kuhn, *Essential Tension,* p. 224.

217. See note 153.

218. I. Hacking, essay review of S. Shapin and S. Schaffer, *Leviathan and the Air-Pump,* in *British Journal of the History of Science* 24, 2, 1991, pp. 235–241; quotation on p. 239. (Many thanks to Joella Yoder for calling my attention to the phrase.)

219. The letter is dated 23 April 1953; it is quoted in D. de Solla Price, *Science since Babylon* (New Haven: Yale University Press, 1963), p. 15, note 10. (Joseph Needham also made use of the letter, while in quoting it he improved Einstein's English somewhat.)

220. R. Wentholt, "The Conquest of Bias in the Human Sciences" (unpublished typescript, 1976), pp. 44–45.

221. Ibidem, p. 63.

FOUR

1. This sweeping statement about the philosophy of history derives from my reading of a number of recent Dutch works in the field by such authors as F.R. Ankersmit, W.J. van der Dussen, and C. Lorenz. Their books provide extensive surveys of current international debates in this domain, which I find quite as interesting in their own right as they are disappointing to a practicing historian in search of theoretical guidance.

2. Thus I leave out of my account altogether a particular approach of which the few fruits—though interesting and promising by themselves—seem to me to be in too premature a state as yet. This is the approach which seeks to elucidate events in the history of science by means of established features of human cognition. Two lines of attack can be distinguished:

—R. Wentholt's unpublished study of human motivation in general, even though focusing mostly on the social sciences, contains many germs of insight into how the human ability to cultivate science in a broadly objective manner derives from certain features built into the human mental outfit (with other features standing in the way of that ability; this side of the matter being stressed in particular in our section 3.7).

—Jean Piaget's stage-by-stage account of how children learn to think has been connected to the explanation of the Scientific Revolution in two different ways. Piaget himself, together with Rolando Garcia, analyzed the transition from Aristotle's ideas about motion to two successive notions of impetus and, from there, to the Newtonian conception of motion. He aimed to show how this succession ran parallel—not only where the overall mechanism of thought is concerned but also *qua* content—to how children of successive ages, when interrogated on the subject, appear to think about motion (J. Piaget and R. Garcia, *Psychogenèse et histoire des sciences,* chs. 1 and 2, in particular). His key concepts here are "the transition from pseudo-necessities and pseudo-impossibilities to logical and causal necessity," "passing from attributes to relations," and the transition from a search for last causes (*"causes dernières"*) to a conception of dynamics in terms of functional dependence and of systems of transformation (pp. 73–77).

Six years earlier, Wolfgang Krohn, in an essay "Die 'Neue Wissenschaft' der Renaissance" (The 'New Science' of the Renaissance), had employed two other concepts from Piaget's genetic epistemology—'decentration' and 'reflexive abstraction'—to explain the Scientific Revolution. Taking the historiographical debate over the significance of 14th-century theorizing for the emergence of early modern science as his point of departure, Krohn identified Renaissance thought as the period of crucial transition. To this period he attributed the occurrence of decentration and of reflexive abstraction as the required stage in the advance of the human mind toward maturity.

Given the manner in which Piaget himself was soon to apply his own conceptual apparatus to history, it is doubtful whether he can have approved of Krohn's mode of applying it. Independently of such a hypothetical verdict, however, it seems to me that Krohn's attempt suffers fatally from an overdose of schematic arrangement of secondary data. Unlike Krohn, Piaget himself (assisted by a physicist, Garcia) displays a firsthand familiarity with the ideas on motion that provide his evidence. However, I find his essay too fragmentary in its handling of the Scientific Revolution, with the historical evidence concerning impetus theorizing being squeezed too much into a preestablished, psychogenetic pattern. Hence, I believe that neither attempt is altogether satisfactory even by way of a pioneering effort.

The common feature that underlay Piaget's search for parallels between the history of science and the mental development of the child as conceived by him is the overall pattern of growth to be found in both: "le progrès du savoir ne consistant pas en simples additions mais réorganisations conditionnant les créations" (J. Piaget and R. Garcia, *Psychogenèse et histoire des sciences,* p. 93; similarly on many other pages). This surely important piece of common ground between history of science and psychogenesis (the enunciation of which was a major novelty half a century ago but is nowadays rather a commonplace) seems by itself insufficient to turn what are so far no more than remarkable parallels into more closely knit causal connections. I believe nonetheless that cognitive science is in a good position to contribute important insights into the question of why the Scientific Revolution manifested itself when it did. The first survey of the state of the art in 'cognitive history of science' is N.J. Nersessian, "Opening the Black Box: Cognitive Science and History of Science," CSL Report 53 of the Cognitive Science Laboratory of Princeton University, January 1993 (a significantly abbreviated version of this paper is forthcoming in *Osiris 10: Critical Problems and Research Frontiers in History of Science,* ed. A. Thackray, 1994). (I thank Marius Engelbrecht for an illuminating discussion of the Piaget approach to the history of science.)

3. W. Schadewaldt, *Hellas und Hesperien: Gesammelte Schriften zur antiken und zur neueren Literatur,* 2nd ed. (Zurich: Artemis, 1970), p. 601: "wieso die Griechen einerseits in der Eroberung und Erweiterung ihres Wissens von der Welt und von dem Menschen mit einem grossartigen Forscherdrang vorgestossen seien, und andererseits dann wieder vor manchen, auf ihrem eigenen Wege liegenden Erkenntnissen haltgemacht hätten, so als ob sie an irgendeine unsichtbare Glaswand stiessen, als ob irgendein geheimnisvolles Tabu ihnen den weiteren Schritt verwehrte." (I owe finding this passage to J.H.J. van der Pot, *Die Bewertung des technischen Fortschritts* [Assen: Van Gorcum, 1985], p. 44.)

4. Unfortunately D.C. Lindberg's *The Beginnings of Western Science* (Chicago: University of Chicago Press, 1992) came out too late to be considered here.

5. W. Whewell, *History* I, p. 28.

6. Ibidem, p. 20.

7. E.J. Dijksterhuis, *Mechanization,* section I:92.

8. Ibidem, I:61, 64, 95. (Dijksterhuis went much more deeply into the question of the limitations of Greek *mathematics* in an article "De grenzen der Grieksche wiskunde" (1934; reprinted in K. van Berkel [ed.], *Clio's stiefkind* [Amsterdam: Bakker, 1990].)

9. E.J. Dijksterhuis, *Mechanization,* I:60, III:1–31. See also idem, "Ad quanta intelligenda condita," pp. 118–120, where one finds the almost palpable tension in Dijksterhuis' mind which I mentioned at the end of section 2.3.2. Here he lists a number of causes for the emergence of early modern science, going on at once to tell the reader that Copernicus and Stevin and Galileo took off at exactly the point where the Greeks had stopped.

10. This view of Hooykaas' is to be encountered in all publications where he set forth his various explanations of the Scientific Revolution. See for bibliographical data, notes to sections 4.3, 5.1.1, and 5.2.8.

11. S. Sambursky, *The Physical World of the Greeks*. The original, English edition of 1956 was reprinted in 1987 without alteration except for the addition of a new introduction.

12. Ibidem, p. 235.

13. Ibidem, pp. 241, 242.

14. Ibidem, p. 243.

15. A. Koyré, "Monde de l'à-peu-près,'" p. 342: "En soi rien ne s'oppose à ce que Copernic et Galilée leur aient directement succédé."

16. The two pertinent books by Farrington are *Science in Antiquity* (Oxford: Oxford University Press, 1969) (which is the 2nd ed. of a book first published in 1936) and *Greek Science* (Harmondsworth: Penguin, 1953) (originally in two parts, of 1944 and 1949, respectively; I have used the revised reprint of 1961).

17. B. Farrington, *Greek Science*, p. 153.

18. This quotation has been compounded from two connected passages: ibidem, pp. 301 and 302.

19. Ibidem, p. 307, quoting Lynn White, Jr, "Technology and Invention in the Middle Ages," *Speculum* 15, 1940, pp. 141–159 (reprinted later in his *Medieval Religion and Technology*, pp. 1–22, with the quoted passage on p. 22).

20. Quoted in B. Farrington, *Science in Antiquity*, p. 143, with a reference to Bacon's *Collected Works* I, p. 395; the original reads: "Etenim in artificialibus natura jugum recepit ab imperio hominis; nunquam enim illa facta fuissent absque homine. At per operam et ministerium hominis conspicitur prorsus nova corporum facies et veluti rerum universitas altera sive theatrum alterum."

21. B. Farrington, *Greek Science*, p. 308.

22. M. Clagett, *Greek Science in Antiquity*, p. 182. (An interesting retrospective was written by J.T. Vallance, "Marshall Clagett's *Greek Science in Antiquity:* Thirty-five Years Later".)

23. M. Clagett, *Greek Science in Antiquity*, p. 31.

24. Ibidem.

25. Ibidem, p. 120.

26. H.F. Cohen, *Quantifying Music*, pp. 75–78.

27. G.E.R. Lloyd, *Greek Science after Aristotle*, p. 165 (see also p. 166).

28. Ibidem, p. 166.

29. Ibidem, p. 171.

30. Ibidem, p. 174.

31. Ibidem, p. 176.

32. Ibidem, p. 178.

33. *The Scientist's Role in Society* appeared in 1971. A reprint (with a new introduction on pp. xi–xxvi, but otherwise unaltered) was published in 1984. Basic data about Ben-David's life and work are in the "Eloge" contributed by Gad Freudenthal and J.L. Heilbron to *Isis* 80, 304, December 1989, pp. 659–663. This informative piece derives largely from an extensive, insightful "General Introduction: Joseph Ben-David—An Outline of His Life and Work," which Freudenthal wrote for a volume he

edited of reprinted papers by Ben-David (entitled *Scientific Growth: Essays on the Social Organization and Ethos of Science*). Here one finds not only a biographical outline intertwined with an overview of Ben-David's scholarly pursuits and an idea of their inner coherence but also an interesting explanation of the relative neglect of his work, both among historians and (even more glaringly) among sociologists of science.

34. J. Ben-David, *The Scientist's Role in Society,* p. xvi. The two possibilities are not taken to exclude one another. Examples stressing the former would be Egypt and Babylonia; the latter, China.

35. Ibidem, p. xix.

36. Ibidem, p. 38.

37. Ibidem, p. 39.

38. Ibidem, p. 40.

39. Ibidem, p. 41.

40. Ibidem.

41. G.E.R. Lloyd, *Greek Science after Aristotle,* p. 170, in particular. (H. Dorn, *The Geography of Science,* has some interesting points about patronage of science in Alexandria.)

42. J. Ben-David, *The Scientist's Role in Society,* p. 41.

43. Cited as "ein berühmtes Wort Leibniz' " in A. Maier, *Ausgehendes Mittelalter* I, p. 413; no source indicated.

44. W. Whewell, *History* I, Book IV, pp. 181–251, discusses the four features, whereas *Philosophy* II, ch. 7, is about the "Innovators of the Middle Ages."

45. When Duhem died in 1916, he had completed ten volumes, with vol. V–X appearing posthumously (the five final tomes were not published until 1954–1959). A one-volume selection in English translation appeared in 1985 under the title *Medieval Cosmology: Theories of Infinity, Place, Time, Void, and the Plurality of Worlds* (ed. and trans. R. Ariew).

46. The general thesis is contained in an "Avant-propos" entitled "Le péripatétisme, les religions et la science d'observation" (vol. IV, pp. 309–320). Curiously, this piece was not included in the translated selection cited in the previous note.

47. P. Duhem, *Système* IV, pp. 309–310: "ils ont pu croire que l'esprit humain avait atteint l'intarissable source où il lui serait désormais permis d'étancher sa soif; ils ont pu penser que l'intelligence possédait enfin la théorie, si ardemment souhaité, où tout ce qui est trouve sa place et découvre sa raison d'être."

48. Ibidem, p. 313: "Deux puissances les déterminèrent à briser le joug du Péripatétisme."

49. Ibidem, p. 314: "En face de cet enseignement [namely, Aristotle's], la Religion juive, la Religion chrétienne, la Religion musulmane, s'accordaient à déclarer aux hommes que Dieu a librement créé le Monde; qu'il le gouverne par une toute puissante providence; qu'il a fait l'homme libre, donc capable de mérite ou de démérite; qu'il lui a conféré une âme personnellement immortelle, et qu'il récompensera ou punira, durant la vie future, les actes que cette âme accomplit en la vie présente."

50. Ibidem, p. 315: "Lorsqu'elle condamnait les affirmations hérétiques du système péripatéticien, la Théologie des trois religions monothéistes ouvrait des brèches

dans la solide muraille de ce système; en ces brèches, la Science expérimentale trouvait un passage qu'elle élargissait au point qu'il permit sa libre expansion; c'est pourquoi l'historien comprendrait imparfaitement l'essor que la Science, libérée de l'Aristotélianisme, a pris au Moyen Age, s'il ne rappelait les coups de bélier dont la Théologie a secoué les murs de la prison."

51. On pp. 20 sqq. of *Système* VI Duhem gave an extensive account of the event, followed by an effort to identify common denominators in the statements declared heretical.

52. Ibidem, pp. 80–81.

53. These statements are translated quotations of Duhem's slightly inaccurate French translations, to be found ibidem, p. 66. He quoted the original Latin in *Études* II, p. 412: "Quod prima causa non posset plures mundos facere." "Quod Deus non possit movere Caelum motu recto. Et ratio est quia tunc relinqueret vacuum."

54. P. Duhem, *Système* VI, p. 66: "Étienne Tempier et son conseil, en frappant ces propositions d'anathème, déclaraient que pour être soumis à l'enseignement de l'Église, pour ne pas imposer d'entraves à la toute puissance de Dieu, il fallait rejeter la Physique péripatéticienne. Par là, ils réclamaient implicitement la création d'une Physique nouvelle que la raison des chrétiens pût accepter. Cette Physique nouvelle, nous verrons que l'Université de Paris, au XIVe siècle, s'est efforcée de la construire et qu'en cette tentative, elle a posé les fondements de la Science moderne; celle-ci naquit, peut-on dire, le 7 mars 1277, du décret porté par Monseigneur Étienne, Évêque de Paris; l'un des principaux objets du présent ouvrage sera de justifier cette assertion."

55. This book, section 2.2.4.

56. P. Duhem, *Système* VII, p. 3: "une longue suite de transformations partielles, dont chacune prétendait seulement retoucher ou agrandir quelque pièce de l'édifice sans rien changer à l'ensemble. Mais lorsque toutes ces modifications de détail eurent été faites, l'esprit humain, embrassant d'un regard le résultat de ce long travail, reconnut avec surprise qu'il ne restait rien de l'ancien palais et qu'un palais neuf se dressait à sa place."

57. Ibidem, ch. 3; see in particular p. 302, where Duhem sums up the whole issue in a nutshell.

58. This is further associated by Hooykaas with 'biblical theology' (more on the topic in section 5.1.1). The outline in the text is based upon R. Hooykaas, "Science and Theology in the Middle Ages," in particular the following passages: pp. 77–82, 88–91, 93–95, 106–108, 131–137, 162. In a much later article, "The Rise of Modern Science: When and Why?" the author briefly restated his argument on pp. 456–458.

59. A. Maier, *Ausgehendes Mittelalter* I, p. 434: "Ihre 'Ergebnisse' ... bestehen weniger in der Umwandlung des überkommenen Naturbildes nach seiner inhaltlichen Seite als vielmehr in der neuen Art und Weise, in der man die Natur zu erfassen und zu ergreifen sucht; kurz gesagt: was sich ändert, ist die *Methode* der Naturerkenntnis. Zum erstenmal wird der Versuch gemacht, Prinzipien herauszustellen, mit denen ein unmittelbares, von aller Autorität unabhängiges, individuell-empirisches Erkennen und Verstehen der Natur möglich ist." (There is an English translation of this essay; see notes 89 and 219 of ch. 2.)

60. E. Grant, *Physical Science in the Middle Ages,* p. 83.

61. Ibidem, pp. 83–84.

62. Ibidem, p. 86, where Grant attributes to the Mertonians and the Parisians a "sophisticated positivistic attitude."

63. Ibidem, p. 89.

64. This point was made by R. Hooykaas, who explores the dilemmas occupying us here in his article "The Rise of Modern Science: When and Why?", pp. 463–467; also note 24 on p. 472.

65. E.J. Dijksterhuis, *Mechanization,* IV:2. The concise treatment of Copernicus given in IV:2–19 is truly masterful. It may in many respects be read as an anticipated outline of Kuhn's *The Copernican Revolution,* which appeared seven years after the Dutch original of the *Mechanization* came out. Still, the discussion is distorted somewhat by Dijksterhuis' now familiar inconsistency over continuity and discontinuity in the birth of early modern science.

66. O. Neugebauer, *The Exact Sciences in Antiquity,* 2nd ed. (Providence, R.I.: Brown University Press, 1957), pp. 205–206. (I found this passage through G.E.R. Lloyd, *Greek Science after Aristotle,* pp. 129–130.)

67. T.S. Kuhn, *The Copernican Revolution,* p. 229 (pagination of the 2nd ed.).

68. Ibidem, p. 264.

69. Some of the complexities hinted at here at sketched out in note 71 below.

70. Copernicus' debts to humanism are discussed in detail in ch. 5, "Copernicus in Italy," of P.L. Rose, *The Italian Renaissance of Mathematics.*

71. It is not possible to discuss science in the Renaissance and its significance for the Scientific Revolution without cutting through several knots. Thus, I have ignored the very broad usage of the term 'humanism' that marks the engaging historiographical tradition going back to Jakob Burckhardt, for whom 'humanism', 'Renaissance thought', 'the discovery of the world and of man', and the birth of modern consciousness *tout court* amounted to more or less the same thing. In his superb paper "Renaissance en natuurwetenschap" E.J. Dijksterhuis showed into what intractable difficulties historians of science get entwined if they adopt such a broad conception of the term. I also learned much from a discussion in D. Weinstein's essay review "In Whose Image and Likeness? Interpretations of Renaissance Humanism." Here Burckhardt's approach is contrasted with a much narrower conception of humanism, a precise definition of which is due above all to the historian of Renaissance thought P.O. Kristeller. As Weinstein observes on p. 166, the latter expert in Renaissance thought does not conceive of humanism as a "concerted philosophical movement" but rather as "a cultural and educational enterprise whose chief concerns were rhetoric, scholarship, and literature." It is this sense of the term that I adopt throughout in the main text.

Another knot through which I cut in the present cluster of sections concerns the matter of periodization. As we know from ch. 2, there are conceptions of the Scientific Revolution which extend backward to the 15th and 14th centuries; equally, there are conceptions of the Renaissance which extend forward to the 17th century (or backward to the 12th century; also, many historians have come to deny the validity of the concept of the 'Renaissance' altogether). In my treatment I stick to the periodization that has become customary for historians of science, which comes down to labeling late 15th- and 16th-century science 'Renaissance', and late 16th- and 17th-century science 'Scientific Revolution', with the demarcation and/or overlap between the two being located at some point or other in the 16th or early 17th century. The precise determination of that point forms one bone of contention that runs across almost all literature on the Scientific Revolution, as we have equally noted throughout ch. 2.

72. There is a long tradition of discussing the topic of 'humanism and science' in each of its manifold dimensions. Dijksterhuis' discussion in *Mechanization,* Part III, ch. IA, "Humanism" (sections III:1–4) never ceases to strike me as particularly balanced and well informed.

Recently, Anthony Grafton in his book *Defenders of the Text: The Traditions of Scholarship in an Age of Science, 1450–1800* (Cambridge, Mass.: Harvard University Press, 1991) and also in an article written jointly with A. Blair, "Reassessing Humanism and Science," *Journal of the History of Ideas* 53, 4, 1992, pp. 535–540, has sought to establish two specific points about the historical relation between humanism and science: (1) in the course of their campaign for the new science men like Bacon and Descartes drew a distorted, condescending picture of humanism that still colors our perception of the cultural significance of textual criticism and suchlike concerns of humanists from the 15th century onward; (2) the advent of the new science certainly did not bring such humanist efforts to an end.

73. E.J. Dijksterhuis, *Mechanization,* III:2.

74. Ibidem, III:3 (the expression occurs only in the Dutch original).

75. Ibidem, III:2.

76. By "modestly put forward" I mean not only Rose's unassuming tone but also the fact that he did not organize his book around his thesis. Rather, his chief concern apparently was to describe the humanist movement in mathematics in as much detail as possible by gleaning his data from a very wide array of sources and glueing these scattered pieces together into one coherent pattern. Rose's more analytical points, which appear as it were separated out in my treatment in the main text, are in his own treatment almost drowned in the flood of his imposing erudition.

77. P.L. Rose, *The Italian Renaissance of Mathematics,* p. 2.

78. Thus, as usual, he virtually omits the science of music.

79. Ibidem, p. 78. In the case of *On Floating Bodies* the Italian humanists used the Moerbeke translation because the Greek original from which he had worked had meanwhile been lost, so that no more authoritative text than Moerbeke's was available (or has been until early in our own century, when Heiberg, a historian of Greek mathematics, found another Greek manuscript of the text).

80. M. Clagett, *Archimedes in the Middle Ages,* 5 vols. (Madison: University of Wisconsin Press, 1964–1984).

81. P.L. Rose, *The Italian Renaissance of Mathematics,* pp. 80–81. Among the mathematicians meant in the final sentence, Rose mentions in particular those who, like Campanus de Novara, produced those simplified versions of Euclid, Ptolemy, and the like that gained so much popularity throughout the Middle Ages.

82. Ibidem, p. 76.

83. Ibidem, p. 84 (sentence construction slightly adapted).

84. Ibidem, p. 97.

85. Ibidem, p. 214.

86. Ibidem, p. 280.

87. Ibidem, p. 294.

88. Ibidem, p. 293.

89. Ibidem, p. 294 (sentence construction slightly adapted).

90. A recent collection of articles in S. Gaukroger (ed.), *The Uses of Antiquity: The Scientific Revolution and the Classical Tradition* (Dordrecht: Kluwer, 1991).

91. M.J. Osler (ed.), *Atoms, Pneuma, and Tranquillity: Epicurean and Stoic Themes in European Thought* (Cambridge: Cambridge University Press, 1991), besides offering a range of new studies on atomism has many references to the older literature (to which 19th-century German philosophers contributed a great deal).

92. Particularly pertinent examples are papers by P. Barker, B.J.T. Dobbs, and J.C. Kassler in the two collections cited in the two notes above.

93. Randall's article "The Development of Scientific Method in the School of Padua" appeared originally in 1940 in the *Journal of the History of Ideas* and was reprinted with some revisions in 1961 in a little book with the more significant title *The School of Padua and the Emergence of Modern Science*. I have consulted this later version. The passage quoted in the text is on p. 63.

94. J.H. Randall, *The School of Padua and the Emergence of Modern Science*, pp. 55–56.

95. Ibidem, p. 25. (The word left out of the quotation in the text is 'anti-clerical'. It refers to Randall's insistence that the Aristotelian tradition at Padua was Averroist and far removed from theology—a fine point he employed further to distinguish his own thesis from Duhem's. I leave this particular subclaim and its subsequent fate in the literature out of account here. W.A. Wallace, in his "Randall *Redivivus*" [see note 98 below], has more to say about it.)

96. Ibidem, p. 63.

97. C.B. Schmitt, *Aristotle and the Renaissance,* passim (in particular, pp. 7 and 10–11). In presenting an overall survey of the domain, this booklet differs from most of Schmitt's other work in that his chosen method of expression was rather the meticulous scholarship of the research paper devoted to one or another specialized issue. Many of these were collected in three volumes issued by Variorum Reprints in 1981, 1984, and 1989, respectively.

98. See in particular W.A. Wallace, "Randall *Redivivus:* Galileo and the Padua Aristotelians," *Journal of the History of Ideas* 49, 1988, pp. 133–149 (reprinted in the collection cited in note 100 below).

99. Besides A. Carugo and A.C. Crombie, that is (see this book, section 2.4.2).

100. The articles were collected in 1991 in W.A. Wallace, *Galileo, the Jesuits and the Medieval Aristotle.* An earlier book on closely related issues is his *Prelude to Galileo* (Dordrecht: Reidel, 1981).

101. For example, E.J. Dijksterhuis in *Mechanization,* III:14–18.

102. J.H. Randall, *The School of Padua and the Emergence of Modern Science,* pp. 23, 27. Randall explicitly denied any contribution of neo-Platonism to the methodological preparation of a mathematized science in Italy (as opposed to the German-speaking countries).

103. Ibidem, p. 66.

104. Ibidem, p. 25.

105. Ibidem, p. 21. Randall's apparent unwillingness to face squarely the problem of what disquisitions on method actually could have meant for the creation of a new physics, and of whether the similarity between Galileo and Zabarella was more than a matter of a shared vocabulary, manifests itself most revealingly in his quick and most unsatisfactory dismissal (p. 34, note 8) of Koyré's fundamental objection, already made in the *Études Galiléennes,* to every case for continuity that rests for its evidence, not on actual scientific performance, but rather on methodological issues. (See this book, sections 2.4.1 and 3.1.1.)

106. W.A. Wallace, *Galileo and His Sources,* p. 347.

107. Ibidem.

108. Ibidem, p. 339.

109. A revealing passage is in W.A. Wallace, *Causality and Scientific Explanation,* vol. 1, pp. 154–155. Here, in the context of medieval disputes, Wallace grants the importance of the distinction: "Little progress, it is true, had been made in laying bare the secrets of nature"; followed at once by "but this could be attributed more to lack of skill in application than to a defect in the methods available."

110. In his "Aristotelian Influences on Galileo's Thought" (1983; reprinted in *Galileo, the Jesuits and the Medieval Aristotle*) Wallace made an effort to interpret Galileo's work at Padua as 'progressive Aristotelianism'. Again, in my opinion this can only be done by taking an almost exclusively formal view of Galileo's novel insights, while also ignoring their Archimedean roots as analyzed by Rose (this book, previous section).

111. A. Koyré, *Études Galiléennes,* p. 213.

112. Yates' pertinent works are listed in note 49 to ch. 3, with the principal collections containing historians' debates over the Yates thesis and/or surveys of those debates being listed in note 55 to ch. 3. An important preliminary consideration is advanced by B.P. Copenhaver, "Natural Magic, Hermetism, and Occultism in Early Modern Science," cautioning that historians should employ the terms 'Hermetic' and 'Hermeticist' less loosely than, on Yates' own example, has often been done in the literature called forth by her thesis.

113. For example, in his *Études Galiléennes* Koyré still interpreted Bruno in such an old-fashioned manner. In his *From the Closed World to the Infinite Universe* he appears meanwhile to have become aware that this is not the proper context for Bruno.

114. F.A. Yates, *Giordano Bruno,* p. 447.

115. Ibidem, p. 432.

116. Ibidem, pp. 435, 441, and 447, respectively.

117. Ibidem, p. 447.

118. Ibidem, pp. 448–449.

119. Ibidem, p. 449.

120. Ibidem, p. 147.

121. F.A. Yates, *Rosicrucian Enlightenment,* introduction to the Paladin edition, p. 17.

122. A.R. Hall, *Revolution in Science,* p. 34.

123. F.A. Yates, *Rosicrucian Enlightenment*, p. 230.

124. F.A. Yates, *Giordano Bruno*, p. 146.

125. The rejection of numerology as a valid mode of scientific discourse is one significant prerequisite of early modern science that seems to await its student. Some pointers toward Kepler's, Mersenne's, and Stevin's (by themselves, quite diverging) attitudes toward number as a clue to nature are indicated in such studies as E.J. Dijksterhuis, *Simon Stevin;* J.V. Field, *Kepler's Geometrical Cosmology* (Chicago: University of Chicago Press, 1988); and H.F. Cohen, *Quantifying Music.*

126. F.A. Yates, *Giordano Bruno*, p. 151.

127. The pioneer student of these matters was D.P. Walker (one of Yates' colleagues at the Warburg Institute). A number of his pertinent writings (e.g., on "Ficino's *Spiritus* and Music") appeared in a collection edited by P. Gouk and entitled *Music, Spirit and Language in the Renaissance* (London: Variorum, 1985). The state of the art in historians' thinking on matters of harmony and musical science is surveyed in the introduction to P. Gozza's anthology *La Musica nella Rivoluzione Scientifica del Seicento* (Bologna: Il Mulino, 1989).

128. F.A. Yates, *Giordano Bruno*, p. 451.

129. Ibidem, p. 155.

130. F.A. Yates, "Hermetic Tradition," p. 228.

131. F.A. Yates, *Giordano Bruno*, pp. 144–145.

132. P. Rossi, *Francis Bacon*, p. 21.

133. A.R. Hall, *Revolution in Science*, p. 33.

134. T.S. Kuhn, *Essential Tension*, p. 54.

135. F.A. Yates, "Hermetic Tradition," pp. 241, 243.

136. F.A. Yates, *Giordano Bruno*, p. 453.

137. Ibidem, p. 452.

138. F.A. Yates, "Hermetic Tradition," p. 228.

139. In 1960, when the book came out with van Gorcum, Assen, it was entitled *The History of Scepticism from Erasmus to Descartes*. Revised editions appeared in 1964 with Humanities Press and in 1968 as a Harper Torchbook. Later it was revised again and expanded into its present-day guise as *The History of Scepticism from Erasmus to Spinoza* (Berkeley: University of California Press, 1979). I have consulted the 1979 version.

140. R.H. Popkin, *The History of Scepticism from Erasmus to Spinoza*, p. 85.

141. Ibidem, p. 54.

142. Ibidem, pp. 145–146.

143. Ibidem, p. 147; similarly on p. 176.

144. Ibidem, pp. 148–149.

145. Ibidem, pp. 147–148: "For some, the age ... had set off a quest for certainty, a search for an absolutely certain foundation for human knowledge. For others the quest was only for stability, for a way of living once the quest for infallible grounds for knowledge had been abandoned." Rabb himself, on pp. 39–40 of *The Struggle for Stability in Early Modern Europe*, cited Popkin's book as a welcome confirmation of his own thesis

(even though he finds it mistaken to seek the sole origin of the 'sceptical crisis' in theological disputes).

146. I owe the point about Lucian's sceptical jesting as a source of inspiration for such works as Rabelais' *Gargantua* and *Pantagruel,* Erasmus' *Praise of Folly,* and More's *Utopia* to an essay by Frederik Bokshoorn, "Het sceptisch tekort," *Tirade* 26, 1982, 280/1, pp. 562–579. His discussion goes back in its turn to M.A. Schreech, *Rabelais* (Ithaca, N.Y.: Cornell University Press, 1979); also C. Robinson, *Lucian and His Influence in Europe* (London: Duckworth, 1979).

147. R. Hooykaas, "The Rise of Modern Science: When and Why?" p. 456.

148. P.L. Rose, *The Italian Renaissance of Mathematics,* p. 293.

149. I exclude here from discussion Hooykaas' explanation, because it draws its principal strength from its conjunction with two other partial explanations put forward by him, which come up for treatment in ch. 5.

FIVE

1. A comprehensive bibliographical essay (covering large portions of the English-language literature on the historical relationship between science and belief) is in John Hedley Brooke's recent survey of the domain, *Science and Religion: Some Historical Perspectives.* Two other widely cast, fairly recent books by historians of science are C.A. Russell, *Cross-currents: Interactions between Science and Faith* (Leicester: Inter-Varsity Press, 1985), and D.C. Lindberg and R.L. Numbers (eds.), *God and Nature: Historical Essays on the Encounter between Christianity and Science* (Berkeley: University of California Press, 1986).

2. A. Funkenstein, *Theology and the Scientific Imagination from the Middle Ages to the Seventeenth Century,* p. 3: "A new and unique approach to matters divine, a secular theology of sorts, emerged in the sixteenth and seventeenth centuries to a short career. It was *secular* in that it was conceived by laymen for laymen. Galileo and Descartes, Leibniz and Newton, Hobbes and Vico were either not clergymen at all or did not acquire an advanced degree in divinity. They were not professional theologians, and yet they treated theological issues at length. Their theology was secular also in the sense that it was oriented toward the world, *ad seculum.* The new sciences and scholarship, they believed, made the traditional modes of theologizing obsolete; a good many professional theologians agreed with them about that. Never before or after were science, philosophy, and theology seen as almost one and the same occupation." It is unfortunate that, although Funkenstein persuasively shows how the genre could be born out of the primacy allotted to professional theologians at the medieval universities, he has not chosen (see his p. 9) to turn this first-rate eye-opener of his into a full-fledged, comparative study of features common to the new genre.

3. R.S. Westfall, *Science and Religion in Seventeenth-Century England,* p. ix, preface to the 2nd ed.

4. Ibidem, pp. 219–220.

5. R. Hooykaas, *Religion and the Rise of Modern Science,* p. 108.

6. Ibidem, p. 162.

7. The booklet, so I gather from the author, is in fact a summary of a much larger, Dutch-language manuscript that has never been published.

8. Ibidem, p. 12–13.

9. Ibidem, p. 109.

10. Quoted ibidem, p. 105, from Kepler's letter of 16 March 1598 to Herwart von Hohenburg (Kepler, *Gesammelte Werke* 13, p. 193: "Ego vero sic censeo, cum Astronomi, sacerdotes dei altissimi ex parte libri Naturae simus. . .").

11. R. Hooykaas, *Religion and the Rise of Modern Science,* p. 36.

12. One example here is Pascal, to whose pertinent ideas Hooykaas devoted an inspiring study in 1939 (translated by me in 1990 as "Pascal: His Science and His Religion"). Hooykaas' brief remarks on Kepler equally go back to an early article: "Het hypothesebegrip van Kepler" (Kepler's Conception of Hypothesis; also published in 1939; never translated).

13. In particular, an anti-Copernican 'quotation' from Calvin, which gained notoriety through Andrew White but which Hooykaas showed in 1955 not to occur at the indicated place in Calvin's works, nor at any other. The matter is summed up in R. Hooykaas, *Religion and the Rise of Modern Science,* pp. 117–122, 154.

14. A sample of de Candolle's pertinent argument has been translated and printed on pp. 145–150 of I.B. Cohen (ed.), *Puritanism and the Rise of Modern Science: The Merton Thesis.*

15. A somewhat reworked version of the doctoral thesis appeared in 1938 as a paper in *Osiris.* It was reproduced in book form in 1970, with a new preface added by the author. I.B. Cohen (Merton's successor as an assistant to George Sarton, who supervised the dissertation) published an exhaustive account of the origins and fate of Merton's thesis in his introduction to an anthology he edited entitled *Puritanism and the Rise of Modern Science: The Merton Thesis,* pp. 1–111. In the notes that follow I refer to Merton's edition of 1970.

16. R.K. Merton, *Science, Technology and Society in Seventeenth-Century England,* pp. 94–95.

17. Ibidem, p. 110.

18. Ibidem, p. 136.

19. Ibidem.

20. The principal anthologies in which large portions of the debate have been assembled are C. Webster (ed.), *The Intellectual Revolution of the Seventeenth Century* (1974), and I.B. Cohen (ed.), *Puritanism and the Rise of Modern Science: The Merton Thesis* (1990). Certain other contributions to the debate were collected in C.A. Russell (ed.), *Science and Religious Belief: A Selection of Recent Historical Studies* (1973). Virtually all English-language contributions to the debate prior to 1990 are listed in the anthology by I.B. Cohen on pp. 89–111. Merton's own views on the debate, his recollections, and his ideas on the origins in his thesis of themes in the sociology of science elaborated by him in later years (e.g., the four values of science discussed in the present book in section 3.5.1) are in his "Preface: 1970" to the reprint of *Science, Technology and Society in Seventeenth-Century England,* in an article "George Sarton: Episodic Recollections by an Unruly Apprentice," *Isis* 76, 284, 1985, pp. 470–486; and

in his contribution to I.B. Cohen's anthology: "*STS:* Foreshadowings of an Evolving Research Program in the Sociology of Science."

Meanwhile the debate over the Merton thesis goes on unabated. One especially noteworthy contribution is an article by Steven Shapin, "Understanding the Merton Thesis." In 1988 a symposium was held in Jerusalem, with the papers delivered there being published by R. Feldhay and Y. Elkana as " 'After Merton': Protestant and Catholic Science in Seventeenth-Century Europe." While most papers address issues of Roman Catholicism and science, four or five are about the Merton thesis itself. H. Zuckerman, distinguishing between two Merton theses as Shapin does in the article just mentioned and as I do in the present chapter, focuses on "The Other Merton Thesis." E. Mendelsohn and D. Struik seek to put Merton's early work in context, and Merton himself publishes Sorokin's critical comments on his dissertation and his own defense at the time.

While lately the topic of Catholicism and 17th-century science has received attention, J. Morgan went deeply into the original issue of Puritanism and 17th-century science both in his article "Puritanism and Science: A Reinterpretation," *Historical Journal* 22, 1979, pp. 535–560, and in his book *Godly Learning: Puritan Attitudes towards Reason, Learning and Education, 1560–1640* (Cambridge: Cambridge University Press, 1986).

21. R.S. Westfall, *Science and Religion in Seventeenth-Century England,* p. x, preface to the 2nd ed.

22. Too often this relatively modest objective was taken to stand for the *a priori* social reduction of autonomous ideas, and as such the dispute perhaps testified above all to the misunderstandings which are apt to arise between the sociologist and the historian of ideas. (A sociologist's formulation of the view that historians have tended to overlook the differential functions of values and motives in guiding human conduct is in G.A. Abraham, "Misunderstanding the Merton thesis.") This particular variety of misunderstanding is actually a spillover from related debates over the explanation of the Scientific Revolution through needs arising from early capitalism (see sections 5.2.1–5).

23. R.K. Merton, *Science, Technology and Society in Seventeenth-Century England,* p. xxxi (which is the first page of the original preface).

24. A.R. Hall, "Merton Revisited, *or* Science and Society in the Seventeenth Century," p. 55. It is true that the passage continues "even though the provision of such an explanation was not Merton's chief or explicit concern"; yet it is equally true that this cautionary remark has since been cast to the winds both by Hall himself and by virtually all other commentators, who have gone on to take the Merton thesis for an attempt at explanation of the Scientific Revolution (*vide infra*).

25. T.K. Rabb, "Religion and the Rise of Modern Science," pp. 269–270.

26. J.L. Heilbron, "Science in the Church," p. 11.

27. T.K. Rabb, "Religion and the Rise of Modern Science," p. 279.

28. T.S. Kuhn, *Essential Tension,* pp. 58–59; see also pp. 115–118; quotation from p. 59.

29. I took these data from an obituary in *Physis* 4, 1962, pp. 159–160. The three volumes of *Geschichte der neusprachlichen wissenschaftlichen Literatur* are the fol-

lowing: *Erster Band: Die Literatur der Technik und der angewandten Wissenschaften vom Mittelalter bis zur Renaissance* (Heidelberg: Winter, 1919); *Zweiter Band: Bildung und Wissenschaft im Zeitalter der Renaissance in Italien* (Leipzig and Florence: Olschki, 1922); *Dritter Band: Galilei und seine Zeit* (Halle: Niemeyer, 1927). The whole was reprinted by Kraus Reprint (Vaduz) in 1965.

30. L. Olschki, *Geschichte der neusprachlichen wissenschaftlichen Literatur* II, p. vii.

31. Ibidem, I, p. 5: "die kulturellen Vorbedingungen der wissenschaftlichen Entwicklung aufzudecken."

32. Ibidem, I, pp. 5–6: "Sie entstand, als die Verweltlichung der Lebensformen und -anschauungen die Menschen zwang, die weltfremd gewordenen Wissenschaften für die praktische und geistige Betätigung heranzuziehen . . . . Deswegen beginnt die neusprachliche wissenschaftliche Literatur mit den angewandten und den Erfahrungswissenschaften, um, jenseits der Grenzen praktischer Notwendigkeiten angelangt, selbständig den Weg zu den reinen wissenschaftlichen Abstraktionen zu finden. Das Ende dieser Entwicklung, welcher diese Entstehungs- und Bildungsgeschichte der neueren wissenschaftlichen Prosa gewidmet is, zeigt sich im Werke Galileis und Descartes', deren Schaffen und Entdecken keine Emanation antiker und mittelalterlicher Forschungsmethoden, sondern die Fortentwicklung und der Triumph einer Idee sind."

33. Ibidem, III, pp. 156–157: "Die Probleme der Kraftersparnis und der Leistungsfähigkeit von Maschinen, der Treffsicherheit von Geschossen, des Widerstandes von Festungsbauten sind die gleichen, die schon durch zwei Jahrhunderte hindurch in der Literatur der Technik ihre Erörterung gefunden hatten. Galilei hat aber die Ueberlieferung der Werkstätten, die er durch seinen Lehrer erst kennen gelernt hatte, hauptsächlich als Gebiet der Erfahrung und Beobachtung betrachtet, um in erster Linie die theoretischen Grundlagen der mechanischen Künste festzulegen. Deshalb ist die Formulierung jener Fragen doch eine grundsätzlich verschiedene, ihre Lösung von jeder unmittelbaren Ueberlieferung der Werkstätten und der Theoretiker durchaus unabhängig, wiewohl seine Aufmerksamkeit stets wieder auf die Möglichkeiten praktischer Anwendungen der spekulativ und experimentell gewonnenen Lehren hingelenkt wird."

34. I learned from personal communication with I.B. Cohen that some adumbrations of Olschki's ideas (which escaped me) are to be found in Mach's *Mechanik*.

35. To some extent it was, as appears from I.B. Cohen's most enjoyable recollection of Olschki's protestations, against Koyré, that Galileo had not failed to carry out his experiments as described. See I.B. Cohen, "Alexandre Koyré in America: Some Personal Reminiscences," p. 58.

36. L. Olschki, *Geschichte der neusprachlichen wissenschaftlichen Literatur* III, p. 164: "die durchaus platonische Richtung seines Geistes"; p. 165: "eine platonische Methode der Naturbetrachtung"; p. 174: "so erkennt man, wie sein Verhältnis zu ihm weniger ein wissenschaftliches und gelehrtes, als ein stimmungsmässiges war, das eher vom Gefühle einer geistigen Verwandschaft, als vom Bewusstsein einer Gleichartigkeit der Meinungen gefestigt wurde. Der allgemeinen Geistesrichtung nach ist Galilei Platoniker, aber Platos literarische Eigenarten fesselten ihn viel mehr als seine eigentlichen Lehren."

37. A. Koyré, *Études Galiléennes,* p. 13; see also "Galileo and Plato," p. 17, note 3.

38. A. Koyré, "Origines," p. 12 of the English version of this article (see note 164 to ch. 2).

39. A. Koyré, "Monde de l''à-peu-près,'" p. 344: "la pensée technique du sens commun."

40. The papers in which Koyré expressed the views that follow are "Commentary on H. Guerlac's 'Some Historical Assumptions of the History of Science,'" and "Les philosophes et la machine" (the latter article can be found in his *Études d'histoire de la pensée philosophique*). The same position on Koyré as a less-than-purely internalist historian is reached on rather different grounds in Y. Elkana, "Alexandre Koyré: Between the History of Ideas and Sociology of Disembodied Knowledge."

41. A. Koyré, "Les philosophes et la machine," pp. 323–324: "Syracuse n'explique pas Archimède, pas plus que Padoue ou Florence n'expliquent Galilée."

42. Ibidem, p. 323: "même s'il est impossible, comme je le crois, de donner une explication sociologique à la naissance de la pensée scientifique."

43. Ibidem, p. 316: "Car toutes les explications, si plausibles qu'elles soient, finalement tournent en rond. Ce qui, après tout, n'est pas un scandale pour l'esprit. Il est assez normal qu'il y ait dans l'histoire—même dans l'histoire de l'esprit—des événements inexplicables, des faits irréductibles, des commencements absolus." Ibidem, p. 339: "Il est impossible, en histoire, d'évacuer le fait, et de tout expliquer."

44. See also A. Koyré, *Newtonian Studies,* p. 6: "the content of the scientific revolution of the seventeenth century, a content which we have to understand before we can attempt an explanation (whatever this may be) of its historical occurrence."

45. The two exceptions known to me are Lynn White and I.B. Cohen. See L. White, *Medieval Religion and Technology,* pp. 121–123 (with a helpful synopsis of the fate of Olschki's work on Galileo), and I.B. Cohen, "A Harvard Education," p. 17, where he remarks that Olschki is "shamefully neglected at the present time."

46. The paper appeared in *Science at the Cross Roads* (see further down in the main text for the history of this remarkable book). I have consulted a photomechanical reprint issued in 1971 with a foreword by Joseph Needham and an introduction by P.G. Werskey.

47. Cf. P.G. Werskey, "On the Reception of *Science at the Cross Roads* in England," p. xxi (where the concept of the Scientific Revolution is being invoked somewhat anachronistically).

48. B. Hessen, "The Social and Economic Roots of Newton's 'Principia,'" p. 155.

49. Ibidem, pp. 158, 161, and 164, respectively.

50. Ibidem, p. 158.

51. Ibidem, pp. 158 and 161, respectively.

52. Ibidem, p. 166.

53. Ibidem, p. 167.

54. Ibidem, p. 171.

55. Ibidem, p. 174.

56. Ibidem, pp. 163 and 164, respectively.

57. In "The Socio-political roots of Boris Hessen," Graham describes the troubles Hessen found himself in at the time of the London Congress. He apparently sought to extricate himself by means of the two following messages implicit in his paper: (1) Look how orthodox my Marxism is, and how faithfully I concur in current campaigns against the 'wreckers'; (2) if Newton, whose physics we all accept, was rooted in bourgeois interests, as I show here, then we can accept Einstein's physics, too, even though its bourgeois roots are no less apparent. (Incidentally, whereas Graham gives 1893 as Hessen's year of birth, in an article of 1978 shown to me by my one-time teacher of elementary Russian, M. Broekmeyer, the author, K.H. Delokarov, mentions 1883.) Graham also has all sorts of interesting details about power relations within the delegation, where Bukharin was much less in command than outsiders thought at the time. Also, Bukharin's programmatic paper was, on return, found heretical by his Stalinist adversaries.

58. Fascinating data (with a very slanted interpretation put upon them) on the impact of the conference upon Bernal and Needham *cum suis* are to be found in G. Werskey, *The Visible College,* more or less *passim.*

59. G.N. Clark, *Science and Social Welfare in the Age of Newton,* p. 70.

60. Ibidem, p. 72.

61. Ibidem, pp. 78–79.

62. Ibidem, p. 79.

63. Ibidem, p. 82.

64. Ibidem, p. 86.

65. Ibidem, p. 89.

66. R.K. Merton, *Science, Technology and Society in Seventeenth-Century England,* p. 137.

67. Ibidem, pp. 148, 187, and 275.

68. Ibidem, p. 180 (referring to Book II, Section VII, Scholium to Proposition 34).

69. I cannot refer this common mistake to printed evidence, but I have heard it sometimes made in conversation. A.R. Hall, however ("Merton Revisited"; see section 5.2.5), did not fall into this little trap.

70. R.K. Merton, *Science, Technology and Society in Seventeenth-Century England,* p. 199.

71. Ibidem, pp. 200–202.

72. Ibidem, p. 176.

73. Ibidem, pp. 206–207.

74. In his article "Understanding the Merton Thesis," S. Shapin has argued persuasively that the mechanism of social dynamics at the back of Merton's mind at the time was Pareto's system of sociology. An illuminating outline of Pareto's mechanism of residues and derivations is in S.E. Finer's introduction to a volume edited by him: *Vilfredo Pareto: Sociological Writings* (London: Pall Mall Press, 1966).

75. R.K. Merton, *Science, Technology and Society in Seventeenth-Century England,* pp. 231–232; see also p. 205.

76. In his foreword to the edition of 1970 (pp. xviii–xix) Merton put far more emphasis on this point than is warranted by his original text of 1938.

77. Zilsel's pertinent articles are spread over diverse journals and books. In 1976 Wolfgang Krohn translated all eight of them, assembling them in one book, *Die sozialen Ursprünge der neuzeitlichen Wissenschaft* (The Social Origins of Early Modern Science). On pp. 44–46 there is a brief note on Zilsel's life and work by Jörn Behrmann. In the notes that follow I cite first the English version, adding in brackets the corresponding page number in Krohn's collection.

78. E. Zilsel, "The Sociological Roots of Science," pp. 544–545 [pp. 49–50].

79. Ibidem, p. 545 [p. 50].

80. Ibidem, p. 546 [p. 52].

81. A characteristic passage from an author loosely connected to that circle is J.A. Schumpeter, *Capitalism, Socialism and Democracy* (ch. 11; p. 124 in the 1962 ed.).: "The rugged individualism of Galileo was the individualism of the rising capitalist class.... The spirit behind such exploits [was] the spirit of rationalist individualism, the spirit generated by rising capitalism." Some other references are given in G.N. Clark, *Science and Social Welfare in the Age of Newton;* in E.J. Dijksterhuis, "Renaissance en natuurwetenschap"; and in D. Struik, "Further Thoughts on Merton in Context," *Science in Context* 3, 1, 1989, pp. 227–238.

Two authors not treated in the present book who also came forward in the thirties with theses linking the rise of early modern science to capitalism are Franz Borkenau (who later did much better work on the Spanish civil war and on the history of European communism) and Henryk Grossmann (both briefly criticized by Dijksterhuis in *Mechanization,* III:26).

82. E.J. Dijksterhuis, "Renaissance en natuurwetenschap," pp. 4–13 (the point is summed up briefly in his *Mechanization,* III:26).

83. E. Zilsel, "The Sociological Roots of Science," p. 548 [p. 53].

84. Ibidem.

85. Ibidem, p. 550 [p. 56], p. 551 [pp. 56–57].

86. E. Zilsel, "The Origins of Gilbert's Scientific Method," p. 245 [p. 121].

87. Ibidem, p. 248 [p. 124].

88. E. Zilsel, "The Sociological Roots of Science," p. 553 [p. 59].

89. Ibidem, p. 554 [p. 59].

90. Ibidem, pp. 554–555 [pp. 59–60].

91. Olschki, however, is only mentioned by Zilsel in two footnotes to other articles of his.

92. E. Zilsel, "The Genesis of the Concept of Scientific Progress."

93. A comparable remark was made by A.R. Hall in *Revolution in Science,* p. 15: "At least in the broader perspective of history it is probably useless to try to be too definite on this question of the influence of technical progress—which is itself beyond doubt—on the modification of 'scientific' ideals and methods."

94. E. Zilsel, "The Sociological Roots of Science," p. 553 [p. 58].

95. An argument along similar lines was made by A.C. Crombie in M. Clagett (ed.), *Critical Problems in the History of Science,* pp. 68–69.

96. E. Zilsel, "The Sociological Roots of Science," p. 560 [p. 65].

97. Ibidem. (The final clause is rather crass if one considers that Weber's collected

essays on the economic ethic of the world's religious had been directed toward exploring this very problem.)

98. Ibidem.

99. H. Holorenshaw [pseud. of Joseph Needham], "The Making of an Honorary Taoist," pp. 12–13.

100. One may naturally suppose that this return to the background had something to do with the Cold War, which in the United States and most of Western Europe did much to make Marxism intellectually suspect. In "The Scientific Revolution: A Spoke in the Wheel?" Roy Porter has followed this line of thought, arguing that the concept of the Scientific Revolution fitted in much better with the political climate of the fifties and early sixties.

101. Merton commented on "The Scholar and the Craftsman in the Scientific Revolution" at the Wisconsin conference of 1957.

102. A.R. Hall, "The Scholar and the Craftsman in the Scientific Revolution," p. 16.

103. Ibidem, p. 17.

104. Ibidem, p. 20.

105. A.R. Hall, "Merton Revisited," p. 62.

106. Ibidem, pp. 63–64.

107. Ibidem, p. 72.

108. See section 6.5.2 of the present book for Joseph Needham's growing recognition that no solid connection has ever been made.

109. One more prerequisite almost entirely neglected by historians of science is to define the involved concept of 'capitalism' very carefully, and also to date its emergence with as much precision as the historical data allow.

110. Reprinted in White's book, *Medieval Religion and Technology*. There is in this book an argument (spread over "Natural Science and Naturalistic Art in the Middle Ages," pp. 23–41, and "The Medieval Roots of Modern Technology and Science," pp. 75–91) that amounts to a kind of thesis on the origins of early modern science in medieval technology. Although the argument is thought-provoking (as always with White), I have in the end decided to exclude it from treatment here, chiefly because its overreliance on the Duhem thesis makes it just as untenable in my opinion as it has always remained uninfluential.

111. Ibidem, p. 132.

112. H.F. Cohen, "Beats and the Origins of Early Modern Science."

113. Arnolt Schlick, *Spiegel der Orgelmacher und Organisten* (Mainz, 1511; facsimile edition with appended translation, Buren: Knuf, 1980), pp. 78–79: "Item fach an in ffaut im manual sein quint ascendendo cffaut / die mach dar zü nitt hoch genug / oder gantz gerade in. sonder etwas in die niedere schwebend. so vil das gehör leyden mag."

114. The 17th-century transformation of the science of music from a specimen of applied arithmetic into a matter of physical and physiological theory is the principal subject of my *Quantifying Music*.

115. K. van Berkel, *Isaac Beeckman (1588–1637) en de mechanisering van het wereldbeeld,* p. 230; the general conclusion is drawn on pp. 230–231. Further on in his

argument, to be sure, the author maintains that Beeckman did find inspiration in his craft background for the elaboration of his corpuscularian picture of the world.

116. J.D. Bernal, *Science in History,* p. 42. (The quotation from Thomson is referred to p. 199 of Lord Rayleigh's *The Life of Sir J.J. Thomson* [Cambridge: Cambridge University Press, 1942].)

117. T.S. Kuhn, *Essential Tension,* pp. 56–58.

118. Hooykaas summed up his pertinent views in *Religion and the Rise of Modern Science,* ch. 4. Earlier writings of his on the topic include *Humanisme, science et réforme: Pierre de la Ramée* (Leiden: Brill, 1958), and "Das Verhältnis von Physik und Mechanik in historischer Hinsicht," in idem, *Selected Studies in History of Science.*

119. There is a huge expert literature in historical horology, naturally including the issue of when, where, and how the mechanical clock was invented (on which more in section 6.4). Large portions of the literature were digested by D.S. Landes and expounded in a manner accessible to the layperson in his book *Revolution in Time* (on which much more in the text that follows). He discusses the invention and its dating on pp. 53–58. One later, very interesting contribution to the possible prehistory of the mechanical clock is in L. Okken, "Die technische Umwelt der frühen Räderuhr."

120. L. Olschki, *Geschichte der neusprachlichen wissenschaftlichen Literatur* III, p. 274; also the chapter on "die Briefe über geographische Ortsbestimmung" on pp. 437–447. A recent treatment of the topic is in a booklet by Silvio A. Bedini, *The Pulse of Time: Galileo Galilei, the Determination of Longitude, and the Pendulum Clock.*

121. B. Hessen, "The Social and Economic Roots of Newton's 'Principia,'" pp. 159, 175.

122. R.K. Merton, *Science, Technology and Society in Seventeenth-Century England,* p. 171.

123. E. Zilsel, "The Origins of Gilbert's Scientific Method," p. 235 [p. 112].

124. L. White, *Medieval Religion and Technology,* pp. 87, 130–131.

125. Just one example among many is S. Bedini, "Die mechanische Uhr und die wissenschaftliche Revolution," in K. Maurice and O. Mayr (eds.), *Die Welt als Uhr: Deutsche Uhren und Automaten, 1550–1650* (Munich: Deutscher Kunstverlag, 1980).

126. A. Koyré, "Monde de l'à-peu-près,'" pp. 353–362.

127. D.S. Landes, *Revolution in Time,* p. 88.

128. Treated extensively in J.G. Yoder, *Unrolling Time.*

129. This statement in its turn deserves some qualification in view of the argument made by J.H. Leopold, "Christiaan Huygens and His Instrument Makers."

130. D.S. Landes, *Revolution in Time,* p. 113.

131. A. Koyré, "Monde de l' 'à-peu-près,'" p. 353.

132. L. Mumford, *Technics and Civilization* (London: Routledge & Kegan Paul, 1934), p. 15.

133. Quoted in R. Hooykaas, "Portuguese Discoveries and the Rise of Modern Science," in idem, *Selected Studies in History of Science,* p. 598.

134. Quoted in R. Hooykaas, *Science in Manueline Style,* p. 11, note 22 (also, with modernized transcription, in "Portuguese Discoveries," p. 587).

135. See. L. de Albuquerque, "Professor R. Hooykaas and the History of Sciences

in Portugal," for the facts about Hooykaas' visits to Coimbra and for a lively expression of the novelty of his findings. Hooykaas recorded these in the following studies, among others: "Portuguese Discoveries" (originally published in 1966; the principal paper to be used here), "Humanism and the Voyages of Discovery in 16th Century Portuguese Science and Letters" (1979), and *Science in Manueline Style* (1980). An outline of the argument is in his "The Rise of Modern Science: When and Why?" (1987).

136. R. Hooykaas, "Portuguese Discoveries," p. 587.

137. These cases are discussed most extensively in *Science in Manueline Style.*

138. Quoted ibidem, p. 588 (translated by me into English with the aid of a Dutch version of these lines Hooykaas published elsewhere).

139. Ibidem, p. 590.

140. Discussed extensively by Hooykaas in his *Science in Manueline Style* (which is part of vol. 4 of *Obras completas de D. João de Castro*).

141. R. Hooykaas, "Portuguese Discoveries," p. 580.

142. Quoted by Hooykaas on p. 470 of "The Rise of Modern Science: When and Why?" (Francis Bacon, *Novum organum* I, aphorism 84 [*Works* I, p. 191]: "Neque pro nihilo aestimandum, quod per longinquas nàvigationes et peregrinationes (quae saeculis nostris increbuerunt) plurima in natura patuerunt et reperta sint, quae novam philosophiae lucem immittere possint. Quin et turpe hominibus foret, si globi materialis tractus, terrarum videlicet, marium, astrorum, nostris temporibus immensum aperti et illustrati sint; globi autem intellectualis fines inter veterum inventa et angustias cohibeantur."

143. R. Hooykaas, "The Rise of Modern Science: When and Why?" p. 471.

144. Bacon, in making the two events 'coevals', may well have had Francis Drake in mind rather than Vasco da Gama.

145. R. Hooykaas, "Portuguese Discoveries," pp. 594–596.

146. R. Hooykaas, "The Rise of Modern Science: When and Why?" p. 473.

147. Ibidem.

148. Leonard Huxley, *Life and Letters of Thomas Henry Huxley* (London: Macmillan, 1900; New York: Appleton, 1901), vol. 1, p. 219: T.H. Huxley to C. Kingsley, 23 September 1860.

149. See also R. Hooykaas, "The Rise of Modern Science: When and Why?" p. 456: "any historical events creating a favourable climate for science should be taken into account.... Some of these factors had no immediate and direct effect in science, but they created an atmosphere favourable to the reception of new ideas and methods."

150. Ibidem, pp. 472, 473.

151. E.L. Eisenstein, *The Printing Press as an Agent of Change,* p. 704.

152. Ibidem, p. 511.

153. Ibidem, p. 629.

154. Ibidem, p. 602.

155. Ibidem, p. 593.

156. Ibidem, p. 504.

157. Ibidem, p. 499.

158. Ibidem, p. 23; see on Zilsel's thesis pp. 521–522, 558, 561.

159. Ibidem, p. 559, note 115.

160. Ibidem, pp. 586–587.

161. Ibidem, pp. 517–519.

162. Ibidem, p. 520.

163. Ibidem, p. 533.

164. Ibidem, p. 685 (cf. p. 592).

165. Ibidem, p. 685.

166. Ibidem, p. 703.

167. Ibidem, pp. 702–703.

168. This is the opening passage in a survey by al-Bîrûnî of Hindu astronomy. I found it quoted in D. Chattopadhyaya (ed.), *Studies in the History of Science in India,* vol. 2, p. 511. (The translation is taken by the editor from E.C. Sachau, *Albiruni's India,* part i, ch. xiv, p. 152. I have not been able to check the quotation there.) Another acute sociologist of science to prefigure key ideas of Ben-David's was Francis Bacon: *Novum organum,* aphorisms 78–80.

169. J. Ben-David, *The Scientist's Role in Society,* p. 31. (For general information about Ben-David see note 33 of ch. 4).

170. Ibidem, pp. xxii–xxiii (introduction to the 1984 ed.), 1–2 (with a list of topics suitable for treatment by the sociologist of science), 12–14, 170, 185. See also J. Ben-David, "Puritanism and Modern Science," p. 261.

171. J. Ben-David, *The Scientist's Role in Society,* p. 45.

172. The argument occupies ch. 4 and the first half of ch. 5 of *The Scientist's Role in Society,* giving way gradually to the author's second principal concern: Having settled how the scientist's role could come into being, he goes on to analyze in the remainder of the book institutional arrangements in the successive geographical centers where science flourished most.

173. Ibidem, p. 63.

174. Ibidem, p. 65.

175. Ibidem, pp. 65–66.

176. Ibidem, pp. 68–69.

177. Ibidem, p. 66.

178. Note that the term 'scientism' Ben-David uses in this connection denotes the sense of science being taken as a symbol for wider aspirations, whereas a somewhat more customary meaning of the term is the formation of a world-picture inspired by science (see note 47 to ch. 3 of the present book; a list of meanings found in the literature has been compiled by T. Vanheste, Department of History, University of Twente).

179. Ibidem, pp. 73, 79.

180. Ibidem, p. 73.

181. J. Ben-David, "Puritanism and Modern Science," p. 260 (in this article of 1985 Ben-David expressed himself somewhat more clearly on the 'Puritan' aspect of the matter than in his—on this score—somewhat nebulous treatment in ch. 4 of *The Scientist's Role in Society*).

182. Ibidem, p. 70.

183. Ibidem, p. 81.

184. Ibidem, p. 82.

185. Somehow the book makes very hard reading. I still recall that, when I had read it for the first time, I knew that I had gone through something very special, without being able to explain why I found the book so important. After two more readings, only the preparation of a dutiful, schoolboy-like extract made me aware of what the author was seeking to accomplish and how he did it. I offer this feature of the book as one other item on the list of reasons for neglect or misunderstanding composed by Gad Freudenthal in "General Introduction: Joseph Ben-David—An Outline of His Life and Work."

186. Scientists' attitudes in Muslim civilization seem to require some qualification to such a statement, however (see sections 6.2.2–5).

187. J. Ben-David, *The Scientist's Role in Society*, p. 71.

188. Unless one breaks out of Ben-David's explanation and refers the phenomenon to a posture of rather narrow-minded utilitarianism serving as a more or less perennial impediment (obstructing, not blocking, to be sure) to the scientific enterprise in the Netherlands. See note 152 of ch. 3.

189. In his article on the Merton thesis, "Puritanism and Modern Science," Ben-David distinguished between the social-psychological aspect of the Merton thesis, which he finds weak and unproven, and its historical-institutional aspect, which he sought to elaborate himself (see pp. 255–261, in particular). It seems to me that even here Ben-David failed to bring out with sufficient cogency how much the introduction of what I call his three intermediate concepts did to clarify the whole issue.

190. This was done in particular in Kuhn's essay review "Scientific Growth: Reflections on Ben-David's 'Scientific Role,'" pp. 172–173. In my opinion, throughout this review Kuhn failed to see what Ben-David was up to in chs. 1–4 of his book, largely because he conflated Ben-David's concept of 'the social role of the scientist' with his own idea of 'paradigm' (p. 169).

191. Some remarks on the topic are made by H. Dorn, *The Geography of Science* (focusing more than seems warranted to me on royal interest in Hermetic patterns of thought).

192. This viewpoint has not been adopted in such current surveys of Galileo's work as S. Drake, *Galileo: Pioneer Scientist* (Toronto: University of Toronto Press, 1990), or M. Clavelin, *La philosophie naturelle de Galilée* (Paris: Colin, 1968). At the beginning of his ch. 3 Clavelin identifies "Greek mathematics, contemporary works on statics, and Benedetti's synthesis" as the principal roots of Galileo's thought and of the transformation he wrought (p. 119 of the English version, *The Natural Philosophy of Galileo* [Cambridge, Mass.: MIT Press, 1974]).

SIX

1. J. Needham, *The Grand Titration*, pp. 41, 54.

2. N. Sivin, in his preface to *Chinese Science*, makes some clear-cut distinctions on this point.

3. S. Nakayama, "Japanese Scientific Thought," in *Dictionary of Scientific Biography,* vol. 15, 1st supplement, pp. 728–758, and F.G. Lounsbury, "Maya Numeration, Computation, and Calendrical Astronomy," ibidem, pp. 759–818.

4. H. Dorn, *The Geography of Science,* pp. 97–109, has some interesting remarks on science in Byzantium.

5. L. White, Jr., contribution to a review symposium on *Science and Civilisation in China,* in *Isis,* 75, 276, 1984, passim. Needham adopted the expression in his "Foreword for Debiprasad Chattopadhyaya's *History of Science and Technology in Ancient India,*" p. v (1986).

6. A. Rahman, "Sixteenth- and Seventeenth-Century Science in India and Some Problems of Comparative Studies" ; D. Chattopadhyaya, introduction to his anthology *Studies in the History of Science in India.* Needham, in his "Foreword for Debiprasad Chattopadhyaya's *History of Science and Technology in Ancient India,*" indicates that he, too, now finds this an answerable question. See also Satpal Sangwan, "Why Did the Scientific Revolution Not Take Place in India?" in Deepak Kumar (ed.), *Science and Empire: Essays in Indian Context (1700–1947)* (Delhi: Anamika Prakashan, 1991), pp. 31–40.

7. A perusal of Chattopadhyaya's book *History of Science and Technology in Ancient India* (Calcutta: Firma KLM Private, 1986) does little to alter the picture, nor does the author claim in this book to have taken up the question of why no Scientific Revolution occurred in India.

8. The last suggestion is hotly rejected by the vehemently antireligious Marxist Chattopadhyaya in his *Studies in the History of Science in India,* pp. xiii-xvi. Compare J. Needham, *Clerks and Craftsmen in China and the West,* p. 20.

9. J. Needham, *The Grand Titration,* p. 176.

10. J. Needham, *Clerks and Craftsmen in China and the West,* p. 402. But in idem et al., *Science and Civilisation in China* 3, pp. xlii–xlvi, there is a much more moderate discussion of the problem of translation, in the course of which he states (p. xliii): "A probable conclusion, if not certainty, usually emerges." I discuss the general background to this divergence in sections 6.3 and 6.5.3.

11. I.A. Richards, "Towards a Theory of Translating," in A.F. Wright (ed.), *Studies in Chinese Thought* (Chicago: University of Chicago Press, 1953), p. 250. (I found this passage through George Steiner's *After Babel,* where he quotes it on p. 48.)

12. Burton Watson, *The Complete Works of Chuang Tzu* (New York: Columbia University Press, 1968); A.C. Graham, *Chuang-Tzu: The Inner Chapters* (London: Allen & Unwin, 1981); J. Needham et al., *Science and Civilisation in China* 2, section 10 (see pp. 68–70 for the translation of *wu wei;* also *The Grand Titration,* p. 210).

13. A.I. Sabra, "Science, Islamic," p. 81.

14. Ibidem, p. 82.

15. Ibidem, p. 83.

16. For more on this see the introductory chapter of A. Sayili, *The Observatory in Islam and Its Place in the General History of the Observatory.* The claim about the hospital is a disputed one: F.E. Peters, *Allah's Commonwealth,* p. 375, assigns priority to the East Roman Empire. The same author restricts the education given in the *ma-*

drasa to Islamic law (p. 680). (My attention was drawn to Peters' book by an anonymous reader consulted by the publisher.)

17. A.I. Sabra, "Science, Islamic," p. 86.

18. Ibidem, p. 87.

19. Ibidem, p. 83.

20. Ibidem, p. 84.

21. Ibidem, p. 81.

22. E. Savage-Smith, "Gleanings from an Arabist's Workshop: Current Trends in the Study of Medieval Islamic Science and Medicine," p. 248.

23. Ibidem.

24. A.I. Sabra, "Science, Islamic," p. 87.

25. Ibidem, p. 88.

26. From a conversation with Sabra (October 1989) I gather that he equally thinks that a responsible historian must shun 'why not' questions because these transcend the domain of the empirical.

27. A.I. Sabra, "Science, Islamic," p. 88.

28. Ibidem.

29. In a letter to me, dated 15 August 1988, for which I am truly grateful.

30. J.J. Saunders, "The Problem of Islamic Decadence," pp. 701–704.

31. G.E. von Grunebaum, *Islam,* p. 25.

32. Ibidem, p. 111.

33. Ibidem, p. 112.

34. Ibidem. (A few remarks on this passage are made by T.F. Glick, "George Sarton and the Spanish Arabists," *Isis* 76, 284, 1985, pp. 496–497.)

35. G.E. von Grunebaum, *Islam,* p. 114.

36. Ibidem, p. 118.

37. Ibidem, p. 115.

38. Ibidem, p. 120.

39. Ibidem, p. 121.

40. Sayili is briefly mentioned in I.B. Cohen, "A Harvard Education," pp. 16–18. E. Ihsanoglu told me about his present occupation. The *i*'s, I understand, do not render his name correctly. The letter should be a dotless *i,* which is pronounced almost without voice: more or less, 'Sáyele'.

41. Quoted in A. Sayili, "The Causes of the Decline of Scientific Work in Islam," p. 408 (I have not been able to check the original text).

42. Ibidem.

43. Ibidem, pp. 408–409.

44. Ibidem, p. 409.

45. Ibidem.

46. Ibidem, p. 410.

47. Ibidem.

48. Ibidem, p. 411.

49. Ibidem.

50. Ibidem, pp. 411–412.

51. Ibidem, p. 412.

52. Ibidem, p. 410.

53. Ibidem, p. 415.

54. Ibidem.

55. Ibidem, p. 416.

56. Sayili discusses all this on pp. 13–24 of *The Observatory in Islam.*

57. A. Sayili, "The Causes of the Decline of Scientific Work in Islam," p. 417.

58. Ibidem.

59. Ibidem, p. 420.

60. Ibidem, p. 418.

61. Ibidem, p. 420.

62. Ibidem, p. 421.

63. Ibidem, p. 422.

64. Ibidem, p. 423.

65. Ibidem.

66. Ibidem, p. 424.

67. I omit discussion of one further point having to do with the fate of the Arabic language, as I have been unable to comprehend it.

68. Ibidem, p. 428.

69. Ibidem, p. 429.

70. Shared by Sabra as well (see note 26).

71. J.J. Saunders, "The Problem of Islamic Decadence," p. 706.

72. Ibidem, p. 707.

73. Ibidem.

74. Ibidem, p. 708.

75. Ibidem, p. 711.

76. Ibidem, p. 712.

77. Ibidem, pp. 712–713.

78. Ibidem, p. 715.

79. Ibidem.

80. Ibidem, p. 716.

81. In al-Ghazzali's fierce attack against science in the name of the Islamic faith is seen the fundamental cause of the decline of Islamic science by Seyyed Hossein Nasr in his fascinating book *Science and Civilization in Islam.* Only, for Nasr, a Sufi philosopher, the inward turn that ensued does not signify a matter of decline at all (more on this thesis in section 6.6). Nasr's book taught me a great deal about the historical course of Islamic science.

82. J.J. Saunders, "The Problem of Islamic Decadence," p. 718.

83. Ibidem, p. 719.

84. Ibidem, p. 720.

85. D.S. Landes, *The Unbound Prometheus,* pp. 28–30.

86. One little hint I found is a passage by al-Bîrûnî (quoted by S.H. Nasr, *Science and Civilization in Islam,* p. 98) where some geographical errors committed by Ptolemy are listed without a further conclusion being drawn.

87. Guided by what I have read, I have assumed here all along that the Muslim world appropriated more or less the whole of the Greek heritage in science, but I am not sure whether this assumption is justified.

88. Simon Leys, *Ombres chinoises* (Paris; Union Générale d'Éditions, 1974), p. 266: "Quelles que soient les vicissitudes de la politique chinoise actuelle, il n'en reste pas moins qu'en termes de civilisation, nous devons nous mettre à l'école du monde chinois: sans l'assimilation de cette grande leçon, nous ne saurions prétendre à une humanité complète et véritable."

89. J. Needham, *Science in Traditional China,* p. 3; here can be found other data on the story in addition to those mentioned in *Science and Civilisation in China* 1, pp. 10–13.

90. J. Needham, *The Grand Titration,* p. 148.

91. The original plan of *Science and Civilisation in China* is in vol. 1, pp. xvii–xxxviii; a number of changes are announced in vol. 4, part 1, p. xxviii, with further comments on how the series is expanding in vol. 5, part 4, p. xxxi.

92. J. Needham, "The Historian of Science as Ecumenical Man," p. 3, lists the names of earlier students of aspects of Chinese science (which are treated in much more detail in surveys of the literature by means of which Needham most often introduces fresh topics in *Science and Civilisation in China*).

93. His collaborators are listed in J. Needham, *Science in Traditional China,* pp. viii, 4–6; and in idem, *Clerks and Craftsmen,* pp. xviii-xix. Every tome of *Science and Civilisation in China* has a foreword in which all acts of collaboration are faithfully acknowledged.

94. L. White, Jr., *Medieval Religion and Technology,* p. xviii.

95. J. Needham et al., *Science and Civilisation in China,* vol. 5, part 4, p. xxxviii, note b.

96. L. White, Jr., contribution to a review symposium in *Isis* on *Science and Civilisation in China,* pp. 173, 175; J. Needham, *Science in Traditional China,* p. 131; and idem, *The Grand Titration,* p. 298. (There are just a few minor changes in wording.)

97. As a result of this procedure, most quotations in the text that follows are drawn from the essay collections. In the notes I append references to corresponding passages in *Science and Civilisation in China* according to this format: [*SCC* vol. no., part no., page no.].

98. See note 10.

99. H. Holorenshaw [pseud. of J. Needham], "The Making of an Honorary Taoist," pp. 2, 20.

100. Ibidem, pp. 3–4.

101. General features of the history of technology are discussed in *SCC* 4, 1, pp. xxix–xxxi.

102. 'H. Holorenshaw', "The Making of an Honorary Taoist," p. 4.

103. J. Needham, "The Historian of Science as Ecumenical Man," p. 5.

104. J. Needham *The Grand Titration*, p. 124. A discussion of the topic is in S. Nakayama, "Joseph Needham, Organic Philosopher," in S. Nakayama and N. Sivin (eds.), *Chinese Science*, pp. 23–43.

105. J. Needham, *The Grand Titration*, pp. 126–127.

106. 'H. Holorenshaw', "The Making of an Honorary Taoist," p. 9.

107. J. Needham, *The Grand Titration*, p. 139.

108. J. Needham, *Science in Traditional China*, pp. 8–9.

109. J. Needham, *The Grand Titration*, pp. 127–135.

110. D.J. de Solla Price, "Joseph Needham and the Science of China," p. 14; also 'H. Holorenshaw', "The Making of an Honorary Taoist," p. 15.

111. J. Needham, foreword to the 2nd ed. of *Science at the Cross Roads*, p. ix.

112. The last sentence explains, I hope, why I find it necessary to relate all this political unpleasantness here, even after the demise of communist regimes almost everywhere (though not yet, at this writing, in China).

113. J. Needham, *The Grand Titration*, p. 217.

114. J. Needham, *Science in Traditional China*, p. 7.

115. J. Needham, "The Historian of Science as Ecumenical Man," p. 5.

116. L. White, Jr., contribution to a review symposium in *Isis* on *Science and Civilisation in China*, p. 173.

117. Already in 1920 Max Weber relegated this simple-minded idea of capitalism to the nursery ("Vorbemerkung," in *Gesammelte Aufsätze zur Religionssoziologie* 1, p. 4: "Es gehört in die kulturgeschichtliche Kinderstube, dass man diese naive Begriffsbestimmung ein für allemal aufgibt.").

118. J. Needham, "The Historian of Science as Ecumenical Man," p. 4.

119. J. Needham, *The Grand Titration*, p. 53.

120. *SCC* 5, 4 p. xxxvi.

121. E.g., in J. Needham, *Clerks and Craftsmen*, p. 397; idem, *The Grand Titration*, p. 16.

122. J. Needham, *The Grand Titration*, pp. 114–115; also pp. 86–113; idem, *Clerks and Craftsmen*, pp. 61, 70.

123. J. Needham, *Clerks and Craftsmen*, p. 82.

124. Ibidem, ch. 4.

125. E.g., J. Needham, *The Grand Titration*, p. 16 [*SCC* 1, pp. 223, 238–239].

126. J. Needham, "The Historian of Science as Ecumenical Man," p. 3.

127. L. White, Jr., contribution to a review symposium in *Isis* on *Science and Civilisation in China*, p. 178.

128. J. Needham, *Clerks and Craftsmen*, p. 72.

129. L. White, Jr., contribution to a review symposium in *Isis* on *Science and Civilisation in China*, pp. 178–179.

130. In the essays, in particular, the picture that arises for the transmission of technology is often allowed to color the picture for science, too, whereas in section 7, "Conditions of Travel of Scientific Ideas and Techniques between China and Europe" (*SCC* 1, pp. 150–248), the two are kept neatly distinct. This section, as far as I know, is still what comes closest to a full survey of the transmission of scientific findings across the premodern world. Still, much additional material is to be found in subsequent volumes of *SCC*, e.g., vol. 5, part 4, section 33 i, "Comparative Macrobiotics," which also has large portions on alchemy in the Islamic world.

131. A. Sayili, *The Observatory in Islam*, p. 391.

132. A. Sayili, "Islam and the Rise of the Seventeenth Century Science," p. 368.

133. Additional support for this overall conclusion may be derived from N.M. Swerdlow and O. Neugebauer, *Mathematical Astronomy in Copernicus's* De Revolutionibus, 2 vols. (New York: Springer, 1984), part 1, pp. 41–48 ("Arabic Astronomy and the Maragha School").

134. J. Needham, *Clerks and Craftsmen*, p. 15.

135. Ibidem, p. 17 [*SCC* 1, pp. 220–223].

136. Ibidem, p. 19 [*SCC* 1, pp. 151–157, 212–214].

137. There is also a speculative claim on the transmission of the usage of rectangular grids in cartography (*SCC* 3, pp. 587–590), but since Needham does not extend this claim to the making of the Scientific Revolution (and, hence, leaves it unmentioned in his pertinent essays), I omit it from treatment here.

138. *SCC* 4, 1, p. 236: "On the whole it may be said that between ancient and medieval knowledge of attraction in Europe and China there was nothing to choose."

139. J. Needham, *The Grand Titration*, p. 73 [*SCC* 4, 1, pp. xxiii–xxiv; further discussed on pp. 239–312, 330–334].

140. J. Needham, *Clerks and Craftsmen*, pp. 246–248. In *SCC* 4, 1, pp. 330–331, essentially the same conclusion is reached in a much more nuanced and cautious manner.

141. J. Needham, *The Grand Titration*, pp. 73, 74 [*SCC* 4, 1, p. 334].

142. J. Needham, "The Historian of Science as Ecumenical Man," p. 4 [*SCC* 5, 4, pp. 502–507].

143. *SCC* 5, 4, p. xxxviii.

144. J. Needham, *Clerks and Craftsmen*, p. 406.

145. J. Needham, *The Grand Titration*, p. 81 [*SCC* 3, pp. 378–379].

146. Ibidem, p. 83.

147. Ibidem, p. 85 [*SCC* 3, pp. 219–220, discussing the system only, not possible routes of transmission].

148. J. Needham, *The Grand Titration*, p. 82.

149. Ibidem, p. 83 [*SCC* 4, 2, pp. 532–546].

150. The fullest treatment is in *SCC* 4, 2, pp. 435–546.

151. J. Needham, *The Grand Titration*, p. 245 (Needham owed much of this to Lynn White's book *Medieval Technology and Social Change*, acknowledged ibidem, pp. 83–84); the second quotation is ibidem, p. 85.

152. Ibidem, p. 83.

153. J. Needham, *Clerks and Craftsmen,* p. 70 [*SCC* 4, 1, p. xxvii, expresses the idea more cautiously]. In section 7 on transmission in *SCC* 1 the principle is not yet to be found. It is my impression that Needham introduced it later, after his discovery of Su Sung's clock put his thoughts on transmission on a novel, more radical footing (cf. *SCC* 4, 2, p. 440).

154. J. Needham, *The Grand Titration,* pp. 52–53, 59 (throughout *SCC* one finds many more).

155. Not by way of an explicit methodological rule, to be sure, but as a mode of arguing in the case at hand.

156. D. Landes, *Revolution in Time,* ch. 1, entitled "A Magnificent Dead End."

157. L. Okken in "Die technische Umwelt der frühen Räderuhr," p. 113, arrives at the same conclusion along a somewhat different route: "für den Gangregler der frühen europäischen Räderuhr gibt es in der chinesischen Uhrenbau-Tradition kein rechtes Gegenstück."

158. J. Needham, *Science in Traditional China,* p. ix.

159. C. Ronan, *The Shorter Science and Civilisation in China* 1, p. 74 [*SCC* 1, pp. 223, 239]; J. Needham, *The Grand Titration,* p. 16; idem, *Science in Traditional China,* p. 9.

160. J. Needham, *Clerks and Craftsmen,* p. 21 [*SCC* 3, p. 105].

161. Quoted in J. Needham, *Clerks and Craftsmen,* p. 39.

162. Quoted in J. Needham, *The Grand Titration,* p. 64 [*SCC* 3, p. 166].

163. Ibidem, p. 326 [*SCC* 2, p. 580].

164. Ibidem, p. 311.

165. Ibidem, p. 224 [*SCC* 2, pp. 171–184].

166. J. Needham, *Clerks and Craftsmen,* p. 398 [*SCC* 3, p. 229, after de Saussure].

167. J. Needham, *The Grand Titration,* p. 44, lists some examples [*SCC* 3, section 19].

168. J. Needham, *Clerks and Craftsmen,* p. 398.

169. A discussion in *SCC* 4, 1, pp. 59–62, comes closest to such a statement.

170. J. Needham, *The Grand Titration,* pp. 24–31 [*SCC* 4, 2, pp. 29–39; here Needham stresses the very provisional nature of the survey].

171. Ibidem, p. 26.

172. Ibidem, p. 27.

173. Ibidem, p. 31.

174. Ibidem, p. 23.

175. C. Ronan, *The Shorter Science and Civilisation in China* 1, p. 187 [*SCC* 2, pp. 335–340].

176. J. Needham, *The Grand Titration,* p. 46 [*SCC* 3, pp. 171–172].

177. Ibidem.

178. Ibidem, p. 50 (two passages fused into one in the quotation).

179. Ibidem, pp. 147, 154.

180. Ibidem, p. 154.

181. Ibidem, p. 16 [*SCC* 1, pp. 3–4].

182. J. Needham, *Clerks and Craftsmen*, p. 405.

183. J. Needham, *The Grand Titration*, pp. 14–15 [*SCC* 3, pp. 150–168].

184. As indicated in *SCC* 3, pp. 154–158, the point about Leonardo was inspired by H.T. Pledge, *Science since 1500*, 2nd ed. (London: Science Museum, 1966) (originally published 1939), pp. 14–15; the thrust of the argument is derived to a considerable extent from Koyré's *Études Galiléennes*.

185. J. Needham, *The Grand Titration*, p. 15.

186. The only exception I came across is J. Needham, "The Historian of Science as Ecumenical Man," p. 1. Here Needham suddenly talks of "the 'scientific revolution' of the fifteenth and sixteenth centuries, culminating in the seventeenth." However, in one of his latest pronouncements to date, *Science in Traditional China*, p. 108, note 2, he has returned to calling Galileo "the father of the Scientific Revolution."

187. J. Needham, *The Grand Titration*, pp. 50–51.

188. Ibidem, p. 40.

189. Ibidem, pp. 214–215.

190. Ibidem, p. 216.

191. Ibidem, p. 136.

192. Ibidem, pp. 142–143.

193. Ibidem, p. 176.

194. Contribution by Needham to a congress discussion printed in A.C. Crombie (ed.), *Scientific Change*, p. 867.

195. J. Needham, *The Grand Titration*, p. 152 [*SCC* 2, pp. 130–132].

196. Ibidem, ch. 4, "Science and Society in Ancient China."

197. Ibidem, p. 176.

198. Ibidem, p. 161.

199. *SCC* 2, pp. 33–164.

200. Ibidem, pp. 68–70.

201. Quoted in J. Needham, *The Grand Titration*, p. 163 [*SCC* 2, p. 33].

202. Ibidem, p. 311.

203. Ibidem, pp. 152–153.

204. Ibidem, ch. 5.

205. At that time Wittfogel was still a Marxist. Out of concepts employed in his early book grew eventually his formidable book *Oriental Despotism: A Comparative Study of Total Power* (New Haven: Yale University Press, 1957). Here the Marxist inspiration is turned against Marxist doctrine (although this point was labored too much by critics discussing this great book that has a great deal more to offer). Needham has always refused to repudiate Wittfogel's original concepts once the latter turned against Marxism.

206. J. Needham, *The Grand Titration*, p. 39.

207. Ibidem, p. 187.

208. Ibidem, p. 186.

209. Ibidem, p. 192.

210. Ibidem, ch. 6.

211. Ibidem, p. 212.

212. Ibidem, pp. 210–211.

213. Ibidem, p. 211.

214. Ibidem, p. 212.

215. Ibidem, p. 211; see also pp. 39–41.

216. Ibidem, p. 31 [*SCC* 4, 2, pp. 39–42].

217. Ibidem, p. 212.

218. E. Zilsel, "The Genesis of the Concept of Physical Law," passim. In *SCC* 2, p. 533, note a, Needham states he hit upon the idea independently.

219. It exists in many variants, with the ground text in *SCC* 2, pp. 518–583. I rely here upon the version in ch. 8 of *The Grand Titration*.

220. Quoted in J. Needham, *The Grand Titration*, p. 322.

221. Ibidem, p. 323.

222. Quoted ibidem, p. 324 [*SCC* 2, pp. 563–564].

223. Ibidem, p. 327.

224. Ibidem, pp. 152–153.

225. Ibidem, p. 328 [*SCC* 2, pp. 582–583].

226. In 1980 (*SCC* 5, 4, pp. xxxvii–xxxviii) the same view was expounded at somewhat greater length. In 1986 it was confidently summed up once more in Needham's "Foreword for Debiprasad Chattopadhyaya's *History of Science and Technology in Ancient India*": "The answer can only be stated in social and economic terms. Only when one knows that China was characterised by bureaucratic feudalism, while Europe had military-aristocratic feudalism, seemingly stronger but in fact much weaker, and so exposed to overthrow when the time came for the rise of the bourgeoisie; then only can one begin to see why modern science, along with capitalism and the Reformation, originated in Europe and in Europe alone. How things went in India I could not attempt to say, but I would expect that apart from wars and colonialism, some concrete social and economic factors will in the end account for the fact that, in spite of wonderful past achievements, modern science did not originate there either." On the other hand, the essay of 1973, "The Historian of Science as Ecumenical Man," p. 6, expresses the same view with more reservations.

227. J. Needham, *Clerks and Craftsmen*, p. 82.

228. See also J. Needham, "The Historian of Science as Ecumenical Man," p. 6.

229. Quoted in J. Needham, *The Grand Titration*, p. 191 [*SCC* 1, p. 196, note e].

230. Ibidem, pp. 216–217. (It is interesting to compare this view of Needham's with the final lines in Weber's "Vorbemerkung".)

231. Ibidem, p. 190.

232. Ibidem, pp. 37–39; also J. Needham, *Clerks and Craftsmen*, ch. 6.

233. J. Needham, *The Grand Titration*, ch. 7.

234. Ibidem, p. 230.

235. Ibidem, p. 298 (wording corrected in accordance with *Science in Traditional China*, p. 131).

236. J. Needham, *Clerks and Craftsmen*, ch. 4 [*SCC* 4, 3, pp. 486–535].

237. Ibidem, p. 57. A much more outspoken discussion is in *SCC* 4, 3, pp. 524–528.

238. E.g., W.H. McNeill's chapter, "The Era of Chinese Predominance, 1000–1500," pp. 46–47.

239. For mathematics: *SCC* 3, pp. 50, 153–154; for astronomy: ibidem, pp. 173, 442; for navigation: *SCC* 4, 3, pp. 524–525; for science in general: *SCC* 2, p. 496; *SCC* 3, p. 457; *SCC* 4, 3, p. 526 (with botany excepted here).

240. J. Needham, *The Grand Titration*, p. 118.

241. *SCC* 3, p. 173, in particular.

242. D.S. Landes, *Revolution in Time*, p. 35.

243. Ibidem, p. 34.

244. J. Needham, *Clerks and Craftsmen*, p. 242 (spelling adapted). (Instead of B.C. and A.D. Needham most often uses − and +.)

245. J. Needham, *The Grand Titration*, pp. 119–120.

246. Ibidem, p. 120.

247. Ibidem, pp. 121–122 [*SCC* 4, 3, pp. 517–521, goes more deeply into the matter of the perennial European deficit in its trade with the East].

248. Ibidem, p. 121.

249. Ibidem, p. 21 [*SCC,* more or less passim].

250. Ibidem, p. 37.

251. C. Ronan, *The Shorter Science and Civilisation in China* 1, p. 292.

252. J. Needham, *The Grand Titration*, p. 328 [*SCC* 2, p. 458].

253. Ibidem, p. 328 [*SCC* 2, p. 583].

254. Especially so in Ronan's shortened version: *The Shorter Science and Civilisation in China* 1, pp. 230–249 [*SCC* 2, pp. 457–495].

255. *SCC* 2, pp. 216–345.

256. C. Ronan, *The Shorter Science and Civilisation in China* 1, p. 165.

257. Ibidem.

258. Ibidem, pp. 167–168.

259. E.g., *SCC* 2, p. 285. The China-born physicist Wen-yuan Qian argues in *The Great Inertia* (pp. 131 sqq., in particular) that Needham's philosophy of science is based upon a misunderstanding of modern physics.

260. E.g., *SCC* 4, 1, p. 1.

261. *SCC* 5, 4, pp. xxxvi–xxxviii, xliv.

262. N. Sivin, "Why the Scientific Revolution Did Not Take Place in China—Or Didn't It?" pp. 546–549.

263. Ibidem, p. 547.

264. J. Needham, *The Grand Titration*, p. 308.

265. Quoted ibidem.

266. J. Needham, *Clerks and Craftsmen,* p. xiv, ch. 19 (bits and pieces of the argument already make their appearance here and there in *SCC*).

267. Ibidem, p. 415.

268. Ibidem, p. 398.

269. J.D. Spence, contribution to a review symposium in *Isis* on *Science and Civilisation in China,* p. 181.

270. The episode of the Jesuits falls outside the compass of the work, which is concerned with Chinese science in the premodern world, yet it comes up occasionally at the end of the treatment of some given branch of science. By far the most extensive discussion deals with astronomy (*SCC* 3, pp. 437–461); there are also a few pages on geography (ibidem, pp. 583–586) and horology (*SCC* 4, 2, pp. 436–439, 504–506, 513–516, 524).

271. N. Sivin, "Why the Scientific Revolution Did Not Take Place in China—Or Didn't It?" pp. 546–548.

272. D.S. Landes, *Revolution in Time,* ch. 2, "Why Are the Memorials Late?"

273. Mentioned in passing in *SCC* 3, p. 52. I had occasion to read a preliminary report on Engelfriet's ongoing research.

274. I get this distinct impression from Nakayama's article "Joseph Needham, Organic Philosopher," in S. Nakayama and N. Sivin (eds.), *Chinese Science,* and from all Sivin's articles utilized in the present section. (It is striking that Colin Mackerras, who in his book *Western Images of China* [Oxford: Oxford University Press, 1989] is busy reducing any given author's idea of China to projections reflecting political positions, ignores entirely that this might apply to Needham as well.)

275. N. Sivin, preface to S. Nakayama and N. Sivin (eds.), *Chinese Science,* p. xviii.

276. N. Sivin, "Why the Scientific Revolution Did Not Take Place in China—Or Didn't It?" p. 549. (In *SCC* 5, 4, pp. xxxi–xxxvii and p. xli, Needham has succinctly rendered these viewpoints of Sivin's, which he goes on to combat as too Spengler-like.)

277. J. Needham, *The Grand Titration,* p. 12.

278. Ibidem; see also idem, *Clerks and Craftsmen,* pp. 239–240.

279. L. White, Jr., contribution to a review symposium in *Isis* on *Science and Civilisation in China,* p. 175.

280. Needham has urged this point in *SCC* 2, p. 579; also *SCC* 5, 4, pp. xxxvii–xli.

281. J. Needham, *Clerks and Craftsmen,* p. 24; also ibidem, p. 79, and idem, *The Grand Titration,* p. 65, note 1. (The invention dates have since been shifted backward in *SCC* 5, part 1, which appeared in 1985 and presents the full story of paper and printing.)

282. J. Needham, *The Grand Titration,* p. 216.

283. L. White, Jr., contribution to a review symposium in *Isis* on *Science and Civilisation in China,* p. 173.

284. J. Needham, *The Grand Titration,* p. 216.

285. A powerful statement of this proposition is in *SCC* 3, pp. 448–449.

286. A.C. Graham, "China, Europe, and the Origins of Modern Science," pp. 62–63.

287. See note 110 of ch. 5. (Maybe Landes should also be mentioned in this connection, but I keep thinking of him as an economic historian rather than as a historian of science; perhaps wrongly so.)

288. For the discussion that follows use has been made of the following literature: M. Weber, "Konfuzianismus und Taoismus"; B. Nelson, "Sciences and Civilisations, 'East' and 'West': Joseph Needham and Max Weber"; N. Sivin, "Chinesische Wissenschaft: Ein Vergleich der Ansätze von Max Weber und Joseph Needham" (also other essays in Wolfgang Schluchter [ed.], *Max Webers Studie über Konfuzianismus und Taoismus: Interpretation und Kritik* [Frankfurt a.M.: Suhrkamp, 1983]).

289. Weber was not led to this rash assertion solely by defective secondary literature, as Sivin thinks; another reason resided in his conviction that the world of Chinese thought must be regarded as an "enchanted garden" (*Zaubergarten*).

290. With good reason Weber recommended in his Einleitung to "Wirtschaftsethik der Weltreligionen" (*Gesammelte Aufsätze zur Religionssoziologie*, 1, p. 237, note 1) that one should read the study keeping the conceptual groundwork laid down in the chapter "Religionssoziologie" of *Wirtschaft und Gesellschaft* always at the back of one's mind.

291. Wen-yuan Qian, *The Great Inertia,* passim, but pp. 81–85 and 124, in particular. (In the text Chinese names have been rendered according to the Wade-Giles system of romanization, of which Needham uses a slightly modified version, not to the pinyin system increasingly in use nowadays. In the earlier system the author's name is rendered, I suspect, as Chi'en Wen-Yüan.)

292. H. Dorn, *The Geography of Science,* chs. 1–4, in particular.

293. J. Needham, *The Grand Titration,* pp. 34–35.

294. J. Needham, *Clerks and Craftsmen,* pp. 28–29.

295. In the very preface to Nasr's book, Giorgio de Santillana courteously yet unhesitatingly pointed out the unlikelihood of such a proposition.

296. S.H. Nasr, *Man and Nature,* pp. 97–98.

297. *SCC* 5, 4, p. xl.

298. *SCC* 2, p. 431.

299. This paragraph derives from a conversation held years ago.

300. Lauxtermann has no quarrel with this proposition; I take this opportunity to thank him for a decade of enlightening exchanges on the topic of what made Europe stand apart.

SEVEN

1. E.A. Burtt, *In Search of Philosophic Understanding,* p. 23.

2. In the final chapter of his "The Conquest of Bias in the Human Sciences," R. Wentholt discusses at length the mental background to this mode of operation in the social and human sciences.

3. The sociologist of technology in question is Wiebe Bijker.

4. A critical survey of those explanations that have indeed become 'standard' is in A.R. Hall, *Revolution in Science,* ch. 1, "The Problem of Cause."

5. Held at Keble College (Oxford), 17–20 July 1990 (by itself a most enjoyable and instructive meeting, to be sure).

6. Introduction to two sessions on "The Scientific Revolution in Its Social Context" at the Joint Conference of the British Society for the History of Science and the History of Science Society held in Manchester, 11–19 July 1988. A similar point is raised at the end of a very interesting article by Robert A. Hatch, "The Scientific Revolution: Paradigm Lost?" (published in *OAH Magazine of History*). While suggesting that "the Scientific Revolution may refuse to be reduced to an 'idea', an 'event', or an 'episode', or, for that matter, to yield to conceptual or social analysis," Hatch insists, too, "that a profound and enduring transformation occurred between Copernicus and Newton."

7. R.S. Westman and D.C. Lindberg, introduction to D.C. Lindberg and R.S. Westman (eds.), *Reappraisals of the Scientific Revolution*. My comments in the text are not meant to disparage a very fine collection, many of the contributions to which have been utilized in preceding chapters.

8. The general pattern of the historical debate that led to such dilution as befell the concept of the Renaissance was masterfully sketched by E.J. Dijksterhuis, "Renaissance en natuurwetenschap," pp. 14–15.

9. I thank Klaas van Berkel for his ongoing effort to make me face viewpoints such as these, thus challenging me to find out to what extent I am prepared to share them.

10. L. White, Jr., *Medieval Religion and Technology*, p. 123.

EIGHT

1. L. White, Jr., *Medieval Religion and Technology*, pp. 217, 333.

2. One example is J. Huizinga, *De wetenschap der geschiedenis* (The 'Science' of History) (Haarlem: Tjeenk Willink, 1937), pp. 47–48: "History must be called the outstanding, nonexact 'science'.... In offering an interpretation which operates all the time with hundreds of thousands of unknowns, [the historian] turns big, unresolved complexes into a synthesis, and he does so, not by means of experiment and calculation, but owing to his personal wisdom, insightfulness, and judgment of human nature." And on p. 54, when taking on the idea that the historian may operate with laws as the scientist does: "The recognition of historical causality is rarely more than an understanding of coherence—than becoming aware of a certain, never wholly fathomable conditioning."

3. A point repeatedly stressed by S. Drake, who began to make it in his "Renaissance Music and Experimental Science," *Journal of the History of Ideas* 31, 1970, pp. 483–500.

4. Compare what three very different authors had to say on the invention of the microscope/telescope. J.D. Bernal, *Science in History*, p. 425: "The means of making it [i.e., the telescope] had in fact been available for some 300 years. It seems, however, to have required the mere quantitative concentration of optical manufacture that went with the greater wealth of the sixteenth century to bring about its discovery by chance." L. White, Jr., *Medieval Religion and Technology*, p. 88: "Why didn't the fourteenth

century have the telescope? I do not know. All that can be said at present [in 1963, that is, but reprinted unaltered in 1978] is that when at last a contemporary of Galileo did achieve it, he was resting his invention on the work of late-thirteenth-century glass-makers and gem-cutters." A. Koyré, "Monde de l' 'à-peu-prés,'" pp. 350–351: "Alors, comment se fait-il que pendant quatre siècles—le télescope est du début du XVIIe— personne, ni parmi ceux qui les faisaient, ni parmi ceux qui en usaient, ne se soit jamais avisé d'essayer de tailler, ou de faire tailler, une lentille un peu plus épaise, d'une cour-bure de surface un peu plus prononcée—et d'en arriver ainsi au microscope simple qui n'apparaît que vers le début du XVIIe siècle, ou à la fin du XVIe? On ne peut pas, me semble-t-il, invoquer l'état de la verrerie. . . . Encore une fois, ce n'est pas l'insuffisance technique, c'est l'absence de l'idée qui nous fournit l'explication." Notice that, of these three authors, Koyré was the only one to offer a rationale for the delay, namely, his idea that such an invention depended on the world of the more-or-less having meanwhile been left behind.

Books and articles only mentioned in passing in one or two notes have not been included in this list.

Abraham, G.A. "Misunderstanding the Merton Thesis: A Boundary Dispute between History and Sociology." *Isis* 74, 1983, pp. 368–387.

Albuquerque, L. de. "Professor R. Hooykaas and the History of Sciences in Portugal." In *Symposium "Hooykaas and the History of Science," Held on 3 and 4 March 1977 at the Utrecht State University,* pp. 1–13. (This book also appeared as a separate issue of the journal *Janus* 64, 1977, pp. 1–129.)

Alembert, Jean Le Rond d'. s.v. "Expérimental." *Encyclopédie,* vol. 6, 1756, pp. 298–301.

Ashworth, W.B., Jr. "Natural History and the Emblematic World View." In D.C. Lindberg and R.S. Westman (eds.), *Reappraisals of the Scientific Revolution,* pp. 303–332.

Basalla, G. (ed.). *The Rise of Modern Science: External or Internal Factors?* Lexington, Mass.: Heath, 1968.

Beaujouan, G. "Alexandre Koyré, l'évêque Tempier et les censures de 1277." In P. Redondi (ed.), *History and Technology* 4, 1–4, 1987, pp. 425–429.

Bedini, S.A. *The Pulse of Time: Galileo Galilei, the Determination of Longitude, and the Pendulum Clock.* Florence: Olschki, 1991.

Ben-David, J. *The Scientist's Role in Society: A Comparative Study.* Englewood Cliffs, N.J.: Prentice Hall, 1971. (A reprint was published by the University of Chicago Press, 1984, with a new introduction.)

————. "Puritanism and Modern Science: A Study in the Continuity and Coherence of Sociological Research." In I.B. Cohen (ed.), *Puritanism and the Rise of Modern Science,* pp. 246–261 (appeared originally in 1985).

Berkel, K. van. "Universiteit en natuurwetenschap in de 17e eeuw, in het bijzonder in de Republiek" (University of Science in the 17th century; in the Republic, in Particular). In H.A.M. Snelders and K. van Berkel (eds.), *Natuurwetenschappen van Renaissance tot Darwin: Thema's uit de wetenschapsgeschiedenis,* pp. 107–130. Den Haag: Nijhoff, 1981.

————. *Isaac Beeckman (1588–1637) en de mechanisering van het wereldbeeld* (Isaac Beeckman and the Mechanization of the World Picture). Amsterdam: Rodopi, 1983.

————. *In het voetspoor van Stevin: Geschiedenis van de natuurwetenschap in Nederland, 1580–1940* (In Stevin's Footsteps: A History of Science in the Netherlands, 1580–1940). Meppel: Boom, 1985.

Bernal, J.D. *Science in History.* London: Watts, 1954 (edition used: 1969; 4 vols.).

Boas, M. *The Scientific Renaissance, 1450–1630.* London: Collins, 1962.

Briggs, R. *The Scientific Revolution of the Seventeenth Century.* London: Longman, 1969.

Brooke, J.H. *Science and Religion: Some Historical Perspectives.* Cambridge: Cambridge University Press, 1991.

Bullough, V.L. (ed.). *The Scientific Revolution.* Huntington, N.Y.: Krieger, 1970.

Burke, J.G. (ed.). *The Uses of Science in the Age of Newton.* Berkeley: University of California Press, 1983.

Burtt, E.A. *The Metaphysical Foundations of Modern Physical Science: A Historical and Critical Essay.* London: Routledge & Kegan Paul, 1972 (reprint of the 2nd ed. of 1932; 1st ed., 1924).

———. *In Search of Philosophic Understanding.* 2nd ed. Indianapolis: Hackett, 1980 (1st ed., New American Library, 1965).

Butterfield, H. *The Origins of Modern Science, 1300–1800.* Rev. ed. London: Bell, 1957. (1st ed., 1949).

Cantor, G.N. "Between Rationalism and Romanticism: Whewell's Historiography of the Inductive Sciences." In M. Fisch and S. Schaffer (eds.), *William Whewell: A Composite Portrait,* pp. 67–86. Oxford: Clarendon, 1991.

Cardwell, D.S.L. *Turning Points in Western Technology: A Study of Technology, Science and History.* New York: Watson, 1972.

Chattopadhyaya, D. (ed.). *Studies in the History of Science in India.* 2 vols. New Delhi, 1982.

Christie, J.R.R. "The Development of the Historiography of Science." In R.C. Olby *et al., Companion to the History of Modern Science,* pp. 5–22. London: Routledge, 1990.

Clagett, M. *Greek Science in Antiquity.* New York: Abelard-Schuman, 1955.

———. *The Science of Mechanics in the Middle Ages.* Madison: University of Wisconsin Press, 1959.

——— (ed.). *Critical Problems in the History of Science: Proceedings of the Institute for the History of Science at the University of Wisconsin, September 1–11, 1957.* Madison: University of Wisconsin Press, 1959.

Clark, G.N. *The Seventeenth Century.* Oxford: Oxford University Press, 1972 (reprint of the 2nd ed. of 1947; 1st ed., 1929).

———. *Science and Social Welfare in the Age of Newton.* Oxford: Oxford University Press, 1970 (reprint of the 2nd ed. of 1949; 1st ed., 1937).

Clavelin, M. "Le débat Koyré–Duhem, hier et aujourd'hui." In P. Redondi (ed.), *History and Technology* 4, 1–4, 1987, pp. 13–35.

Cohen, H.F. "Over aard en oorzaken van de 17e eeuwse wetenschapsrevolutie: Eerste ontwerp voor een historiografisch onderzoek." (On the Nature and Causes of the 17th-Century Scientific Revolution: Preliminary Draft of a Historiographical Research Project). Amsterdam: van Oorschot, 1983. (Inaugural address at Twente University of Technology, 9 June 1983.)

————. *Quantifying Music: The Science of Music at the First Stage of the Scientific Revolution, 1580–1650.* Dordrecht: Reidel, 1984.

————. "Music as a Test-Case." *Studies in History and Philosophy of Science* 16, 4, 1985, pp. 351–378.

————. "Comment on 'The Universities and the Scientific Revolution: The Case of England' by Mordechai Feingold." In *New Trends in the History of Science: Proceedings of a Conference Held at the University of Utrecht,* pp. 49–52. Amsterdam: Rodopi, 1989.

————. "'Open and Wide, yet without Height or Depth.'" *Tractrix* 2, 1990, pp. 159–165.

————. "Beats and the Origins of Early Modern Science." In V. Coelho (ed.), *Music and Science in the Age of Galileo,* pp. 17–34. Dordrecht: Kluwer Academic Publishers, 1992.

Cohen, I.B. *The Birth of a New Physics.* New York: Norton, 1985 (revised and updated ed. of the 1960 original).

————. *The Newtonian Revolution: With Illustrations of the Transformation of Scientific Ideas.* Cambridge: Cambridge University Press, 1980.

————. "A Harvard Education." *Isis* 75, 276, March 1984, pp. 13–21.

————. *Revolution in Science.* Cambridge, Mass.: Harvard University Press, 1985.

————. "Alexandre Koyré in America: Some Personal Reminiscences." In P. Redondi (ed.), *History and Technology* 4, 1–4, 1987, pp. 55–70.

———— (ed.). *Puritanism and the Rise of Modern Science: The Merton Thesis.* New Brunswick: Rutgers University Press, 1990.

Comte, A. *Philosophie première: Cours de philosophie positive, Leçons 1 à 45.* Ed. M. Serres, F. Dagognet, and A. Sinaceur. Paris: Hermann, 1975.

Copenhaver, B.P. "Natural Magic, Hermetism, and Occultism in Early Modern Science." In D.C. Lindberg and R.S. Westman, *Reappraisals of the Scientific Revolution,* pp. 261–301.

Crombie, A.C. *Augustine to Galileo: The History of Science, AD 400–1650.* London: Heinemann, 1952.

————. *Robert Grosseteste and the Origins of Experimental Science, 1100–1700.* Oxford: Clarendon Press, 1953.

————. "The Significance of Medieval Discussions of Scientific Method for the Scientific Revolution." In M. Clagett (ed.), *Critical Problems in the History of Science,* pp. 79–101.

————. "Historians and the Scientific Revolution." *Endeavour* 19, 73, January 1960, pp. 9–13.

———— (ed.). *Scientific Change: Historical Studies in the Intellectual, Social and Technical Conditions for Scientific Discovery and Technical Invention, from Antiquity to the Present. Symposium on the History of Science, University of Oxford, 9–15 July 1961.* London: Heinemann, 1963.

Curry, P. "Revisions of Science and Magic." *History of Science* 23, 1985, pp. 299–325.

Daston, L. "History of Science in an Elegiac Mode: E.A. Burtt's *Metaphysical Founda-tions of Modern Physical Science* Revisited." *Isis* 82, 313, 1991, pp. 522–531.

Daumas, M. *Les instruments scientifiques aux XVIIe et XVIIIe siècles.* Paris: Presses Universitaires de France, 1953. (Translated as *Scientific Instruments of the Seven-teenth and Eighteenth Centuries.* New York and Washington: Praeger, 1972.)

Davids, C.A. "Technological Change in Early Modern Europe (1500–1780)." *Journal of the Japan–Netherlands Institute* 3, 1991, pp. 32–44.

Dear, P. "Totius in Verba: Rhetoric and Authority in the Early Royal Society." *Isis* 76, 282, 1985, pp. 145–161.

———. "Miracles, Experiments, and the Ordinary Course of Nature." *Isis* 81, 309, 1990, pp. 663–683.

Debus, A.G. *Man and Nature in the Renaissance.* Cambridge: Cambridge University Press, 1978.

———. "The Chemical Philosophy and the Scientific Revolution." In W.R. Shea (ed.), *Revolutions in Science: Their Meaning and Relevance,* pp. 27–48. Canton, Mass.: Science History Publications, 1988.

Dictionary of the History of Science. Ed. W.F. Bynum, E.J. Browne, and R. Porter. London: Macmillan, 1981.

Dictionary of Scientific Biography. 16 vols. Ed. Ch. C. Gillispie. New York: Scrib-ner's, 1970–1980.

Dijksterhuis, E.J. *Val en worp: Een bijdrage tot de geschiedenis der mechanica van Aristoteles tot Newton.* Groningen: Noordhoff, 1924.

———. Review of A. Maier, *Die Vorläufer Galileis im 14. Jahrhundert. Isis* 41, 124, 1950, pp. 207–210.

———. *The Mechanization of the World Picture.* Oxford: Oxford University Press, 1961. (Translation of *De mechanisering van het wereldbeeld.* Amsterdam: Meulen-hoff, 1950.)

———. "Doel en methode van de geschiedenis der exacte wetenschappen" (Aim and Methods of the History of the Exact Sciences). Amsterdam: Meulenhoff, 1953 (In-augural address at Utrecht University.)

———. "Ad quanta intelligenda condita (Designed for Grasping Quantities)." *Tractrix* 2, 1990, pp. 111–125. (Translation of "Ad quanta intelligenda condita." Amsterdam: Meulenhoff, 1955; inaugural address at Leyden University.)

———. "Renaissance en natuurwetenschap" (Science and the Renaissance). *Medede-lingen der Koninklijke Nederlandse Akademie van Wetenschappen, Afdeling Let-terkunde,* Nieuwe Reeks Deel 19, no. 5, 1956, pp. 171–200.

———. "The Origins of Classical Mechanics from Aristotle to Newton." In M. Clagett (ed.), *Critical Problems in the History of Science,* pp. 163–184.

———. *Simon Stevin: Science in the Netherlands around 1600.* Ed. R. Hooykaas and G.J. Minnaert. Den Haag: Mouton, 1970 (condensed version of the Dutch original, published in 1943).

Dorn, H. *The Geography of Science.* Baltimore: Johns Hopkins University Press, 1991.

Duhem, P. *Études sur Léonard de Vinci: Ceux qu'il a lus et ceux qui l'ont lu.* 3 vols. Paris: Hermann, 1906, 1909, and 1913 (2nd impression, Paris: De Nobele, 1955).

———. *Le système du monde: Histoire des doctrines cosmologiques de Platon à Copernic.* 10 vols. Paris: Hermann, 1913–1959.

———. *Medieval Cosmology: Theories of Infinity, Place, Time, Void, and the Plurality of Worlds.* Ed. and trans. R. Ariew. Chicago: University of Chicago Press, 1985.

Eamon, W. "From the Secrets of Nature to Public Knowledge." In D.C. Lindberg and R.S. Westman (eds.), *Reappraisals of the Scientific Revolution,* pp. 333–365.

Eisenstein, E.L. *The Printing Press as an Agent of Change: Communications and Cultural Transformations in Early-Modern Europe.* 2 vols. Cambridge: Cambridge University Press, 1979.

———. *The Printing Revolution in Early Modern Europe.* Cambridge: Cambridge University Press, 1983.

Elkana, Y. (ed.). *William Whewell: Selected Writings on the History of Science.* Chicago: University of Chicago Press, 1984.

———. "William Whewell, Historian." *Rivista di Storia della Scienza* 1, 2, 1984, pp. 149–197.

———. "Alexandre Koyré: Between the History of Ideas and Sociology of Disembodied Knowledge." In P. Redondi (ed.), *History and Technology* 4, 1–4, 1987, pp. 115–148.

Farrington, B. *Greek Science.* Harmondsworth: Penguin, 1953 (originally in two parts, 1944 and 1949; edition used: the revised reprint of 1961).

———. *Science in Antiquity.* 2nd ed. Oxford: Oxford University Press, 1969 (1st ed., 1936).

Feingold, M. *The Mathematicians' Apprenticeship: Science, Universities and Society in England, 1560–1640.* Cambridge: Cambridge University Press, 1984.

Feldhay, R., and Y. Elkana (eds.). "'After Merton': Protestant and Catholic Science in Seventeenth-Century Europe." *Science in Context* 3, 1, 1989, pp. 3–302.

Finocchiaro, M.A. *Galileo and the Art of Reasoning: Rhetorical Foundations of Logic and Scientific Method.* Dordrecht: Reidel, 1980.

Fisch, M. *William Whewell, Philosopher of Science.* Oxford: Clarendon Press, 1991.

Fisch, M., and S. Schaffer (eds.). *William Whewell: A Composite Portrait.* Oxford: Clarendon Press, 1991.

Foster, M.B. "The Christian Doctrine of Creation and the Rise of Modern Natural Science." *Mind* 43, 1934, pp. 446–468 (reprint used: C.A. Russell [ed.], *Science and Religious Belief,* pp. 295–315).

Fournier, M. *The Fabric of Life: The Rise and Decline of Seventeenth-Century Microscopy.* Enschede, 1991. (Johns Hopkins University Press, forthcoming.)

Fox Keller, E. *Reflections on Gender and Science.* New Haven: Yale University Press, 1985.

Freudenthal, Gad. "General Introduction: Joseph Ben-David—An Outline of His Life and Work." In J. Ben-David, *Scientific Growth: Essays on the Social Organization and Ethos of Science,* ed. G. Freudenthal, pp. 1–25. Berkeley and Los Angeles: University of California Press, 1991.

Funkenstein, A. *Theology and the Scientific Imagination from the Middle Ages to the Seventeenth Century.* Princeton: Princeton University Press, 1986.

Gascoigne, J. "A Reappraisal of the Role of the Universities in the Scientific Revolution." In D.C. Lindberg and R.S. Westman (eds.), *Reappraisals of the Scientific Revolution,* pp. 207–260.

Gillispie, C.C. *The Edge of Objectivity: An Essay in the History of Scientific Ideas.* Princeton: Princeton University Press, 1960. (Reprint with new preface: 1990.)

Gjertsen, D. *Science and Philosophy: Past and Present.* Harmondsworth: Penguin, 1989.

Graham, A.C. "China, Europe, and the Origins of Modern Science: Needham's *The Grand Titration.*" In S. Nakayama and N. Sivin (eds.), *Chinese Science,* pp. 45–69.

Graham, L.R. "The Socio-political Roots of Boris Hessen: Soviet Marxism and the History of Science." *Social Studies of Science* 15, 1985, pp. 705–722.

Grant, E. *Physical Science in the Middle Ages.* Cambridge: Cambridge University Press, 1977 (first impression: Wiley & Sons, 1971).

Grunebaum, G.E. von. *Islam: Essays in the Nature and Growth of a Cultural Tradition.* London: Routledge, 1969 (reprint of the 2nd ed.; 1st ed., 1955).

Guerlac, H. *Essays and Papers in the History of Modern Science.* Baltimore: Johns Hopkins University Press, 1977.

Hahn, R. "Changing Patterns for the Support of Scientists from Louis XIV to Napoleon," In P. Redondi (ed.), *History and Technology* 4, 1–4, 1987, pp. 401–411.

Hakfoort, C. *Optics in the Age of Euler.* Cambridge: Cambridge University Press, forthcoming.

Hall, A.R. *The Scientific Revolution, 1500–1800: The Formation of the Modern Scientific Attitude.* London: Longmans, 1954.

———. "The Scholar and the Craftsman in the Scientific Revolution." In M. Clagett (ed.), *Critical Problems in the History of Science,* pp. 3–23.

———. "Merton Revisited, *or* Science and Society in the Seventeenth Century." *History of Science* 2, 1963, pp. 1–16 (reprint used: C.A. Russell [ed.], *Science and Religious Belief,* pp. 55–73).

———. *From Galileo to Newton, 1630–1720.* London: Collins, 1963.

———. "Magic, Metaphysics and Mysticism in the Scientific Revolution." In M.L. Righini Bonelli and W.R. Shea (eds.), *Reason, Experiment, and Mysticism in the Scientific Revolution,* pp. 275–282.

———. *The Revolution in Science, 1500–1750.* London: Longman, 1983.

———. "Gunnery, Science, and the Royal Society." In J.G. Burke (ed.), *The Uses of Science in the Age of Newton,* p. 111–141.

———. "Alexandre Koyré and the Scientific Revolution." In P. Redondi (ed.), *History and Technology* 4, 1–4, 1987, pp. 485–496.

Hankins, T.L. *Science and the Enlightenment.* Cambridge: Cambridge University Press, 1985.

Hatfield, G. "Metaphysics and the New Science." In D.C. Lindberg and R.S. Westman (eds.), *Reappraisals of the Scientific Revolution*, pp. 93–166.

Hazard, P. *La crise de la conscience Européenne (1680–1715)*. 3 vols. Paris: Boivin, 1935.

Heilbron, J.L. *Elements of Early Modern Physics*. Berkeley: University of California Press, 1982.

———. "Science in the Church." In R. Feldhay and Y. Elkana (eds.), " 'After Merton,' " pp. 9–28.

Helden, A. Van. "The Birth of the Modern Scientific Instrument, 1550–1700." In J.G. Burke (ed.), *The Uses of Science in the Age of Newton*, pp. 49–84.

Hesse, M. "Reasons and Evaluation in the History of Science." In M. Teich & R. Young (eds.), *Changing Perspectives in the History of Science*, pp. 127–147.

Hessen, B. [M.] "The Social and Economic Roots of Newton's 'Principia.' " In *Science at the Cross Roads: Papers Presented to the International Congress of the History of Science and Technology Held in London from June 29th to July 3rd, 1931, by the Delegates of the USSR*, pp. 149–212. London, 1931; (2nd ed., London: Cass, 1971).

Hill, C. "Puritanism, Capitalism and the Scientific Revolution." In C. Webster (ed.), *The Intellectual Revolution of the Seventeenth Century*, pp. 243–253.

Hooykaas, R. "Pascal: His Science and His Religion." *Tractrix* 1, 1989, pp. 115–139. (Translation of "Pascal: Zijn wetenschap en zijn religie"; first published in 1939.)

———. "Science and Theology in the Middle Ages." *Free University Quarterly* 3, 1954, pp. 77–163.

———. *Religion and the Rise of Modern Science*. Edinburgh: Scottish Academic Press, 1973 (1st ed., 1972).

———. "Humanism and the Voyages of Discovery in 16th Century Portuguese Science and Letters." *Mededelingen der Koninklijke Nederlandse Akademie van Weten-schappen, Afdeling Letterkunde*, Nieuwe Reeks Deel 42, no. 4, 1979.

———. *Science in Manueline Style: The Historical Context of D. João de Castro's Works*. Coimbra: Academia Internacional da Cultura Portuguesa, 1980. (This is a separate edition of pp. 231–426 in the 4th volume of *Obras completas de D. João de Castro*, ed. A. Cortesão and L. de Albuquerque. Coimbra, 1968–1980.)

———. *Selected Studies in History of Science*. Coimbra: Por ordem da Universidade, 1983.

———. *G.J. Rheticus' Treatise on Holy Scripture and the Motion of the Earth*. Amsterdam: North-Holland, 1984.

———. "The Rise of Modern Science: When and Why?" *British Journal for History of Science* 20, 4, 1987, pp. 453–473.

Jacob, J.R., and M.C. Jacob. "The Anglican Origins of Modern Science: The Metaphysical Foundations of the Whig Constitution." *Isis* 71, 257, 1980, pp. 251–267.

Jacob, M.C. *The Cultural Meaning of the Scientific Revolution*. New York: Knopf, 1988.

Jaki, S.L. *The Origin of Science and the Science of Its Origins*. South Bend Ind.: Regnery/Gateway, 1978.

―――. *Uneasy Genius: The Life and Work of Pierre Duhem.* The Hague: Nijhoff, 1984.

Jones, R.F. *Ancients and Moderns: A Study of the Rise of the Scientific Movement in Seventeenth Century England.* Berkeley: University of California Press, 1965 (reprint of the 2nd ed. of 1961; 1st ed., 1936).

Kant, I. *Kritik der reinen Vernunft.* 1781 (edition used, Hamburg: Meiner, 1976).

―――. *Die drei Kritiken in ihrem Zusammenhang mit dem Gesamtwerk.* Ed. R. Schmidt. Stuttgart: Kröner, 1975.

Kearney, H.F. *Science and Change, 1500–1700.* New York: McGraw-Hill, 1971.

―――. "Puritanism, Capitalism and the Scientific Revolution." In C. Webster (ed.), *The Intellectual Revolution of the Seventeenth Century,* pp. 218–242.

―――. "Puritanism and Science: Problems of Definition." In C. Webster (ed.), *The Intellectual Revolution of the Seventeenth Century,* pp. 254–261.

Kemsley, D.S. "Religious Influences in the Rise of Modern Science: A Review and Criticism, Particularly of the 'Protestant-Puritan Ethic' Theory." *Annals of Science* 24, 3, 1968, pp. 199–226 (reprint used: C.A. Russell (ed.), *Science and Religious Belief,* pp. 74–102.

Klaaren, E.M. *Religious Origins of Modern Science: Belief in Creation in Seventeenth-Century Thought.* Grand Rapids, Mich.: Eerdmans, 1977.

Koyré, A. *Études Galiléennes.* Paris: Hermann, 1939–1940 (2nd ed. used, 1966).

―――. Review of A. Maier, *Die Vorläufer Galileis im 14. Jahrhundert. Archives Internationales d'Histoire des Sciences,* 1951, pp. 769–783.

―――. *From the Closed World to the Infinite Universe.* New York: Harper, 1958 (1st ed., Baltimore: Johns Hopkins University Press, 1957).

―――. *La révolution astronomique: Copernic—Kepler—Borelli.* Paris: Hermann, 1961.

―――. *Études d'histoire de la pensée philosophique.* Paris: Colin, 1961 (2nd ed. used, Paris: Gallimard, 1971).

―――. "Commentary on H. Guerlac's 'Some Historical Assumptions of the History of Science.'" In A.C. Crombie (ed.), *Scientific Change,* pp. 847–857.

―――. *Newtonian Studies.* London: Chapman & Hall, 1965.

―――. *Études d'histoire de la pensée scientifique.* Ed. R. Taton. Paris: Presses Universitaires de France, 1966.

―――. *Metaphysics and Measurement: Essays in Scientific Revolution.* Ed. M. Hoskin. Cambridge, Mass. Harvard University Press, 1968.

―――.*De la mystique à la science: Cours, conférences et documents, 1922–1962.* Ed. P. Redondi. Paris: Éditions de l'École des Hautes Études en Sciences Sociales, 1986.

Kragh, H. *An Introduction to the Historiography of Science.* Cambridge: Cambridge University Press, 1987.

Krohn, W. "Zur soziologischen Interpretation der neuzeitlichen Wissenschaft." In E. Zilsel, Die sozialen Ursprünge der neuzeitlichen Wissenschaft, ed. and trans. W. Krohn, pp. 7–43. Frankfurt a.M.: Suhrkamp, 1976.

———. "Die 'Neue Wissenschaft' der Renaissance." In G. Böhme, W. van den Daele, and W. Krohn, *Experimentelle Philosophie: Ursprünge autonomer Wissenschafts-entwicklung,* pp. 14–128. Frankfurt a.M.: Suhrkamp, 1977.

Kuhn, T.S. *The Copernican Revolution: Planetary Astronomy in the Development of Western Thought.* Cambridge, Mass.: Harvard University Press, 1957. Reprint. New York: Random House, Vintage Books, 1959.

———. *The Structure of Scientific Revolutions.* 2nd enl. ed. Chicago: University of Chicago Press, 1970 (1st ed., 1962).

———. "Scientific Growth: Reflections on Ben-David's 'Scientific Role'" (essay review of J. Ben-David, *The Scientist's Role in Society). Minerva* 10, 1972, pp. 166–178.

———. *The Essential Tension: Selected Studies in Scientific Tradition and Change.* Chicago: University of Chicago Press, 1977.

Landes, D.S. *The Unbound Prometheus: Technological Change and Industrial Development in Western Europe from 1750 to the Present.* Cambridge: Cambridge University Press, 1969.

———. *Revolution in Time: Clocks and the Making of the Modern World.* Cambridge, Mass.: Harvard University Press, 1983.

Leopold, J.H. "Christiaan Huygens and His Instrument Makers." In H.J.M. Bos *et al.* (eds.), *Studies on Christiaan Huygens,* pp. 221–233. Lisse: Swets & Zeitlinger, 1980.

Lindberg, D.C. "The Transmission of Greek and Arabic Learning to the West." In idem (ed.), *Science in the Middle Ages,* pp. 52–90. Chicago: University of Chicago Press, 1978.

———. "Conceptions of the Scientific Revolution from Bacon to Butterfield: A Preliminary Sketch." In D.C. Lindberg and R.S. Westman (eds.), *Reappraisals of the Scientific Revolution,* pp. 1–26.

Lindberg, D.C., and R.S. Westman (eds.), *Reappraisals of the Scientific Revolution.* Cambridge: Cambridge University Press, 1990.

Lloyd, G.E.R. *Greek Science after Aristotle.* New York: Norton, 1973.

McGuire, J.E., and P.M. Rattansi. "Newton and the Pipes of Pan." *Notes and Records of the Royal Society of London,* 1966, pp. 108–143.

Mach, E. *Die Mechanik in ihrer Entwickelung historisch-kritisch dargestellt.* Leipzig: Brockhaus, 1883. (Translated by T.J. McCormack as *The Science of Mechanics: A Critical and Historical Account of Its Development.* Chicago: Open Court, 1893.)

McMullin, E. "Empiricism and the Scientific Revvolution." In C.S. Singleton (ed.), *Art, Science, and History in the Renaissance,* pp. 331–369. Baltimore: Johns Hopkins University Press, 1967.

———. "Conceptions of Science in the Scientific Revolution." In D.C. Lindberg & R.S. Westman (eds.), *Reappraisals of the Scientific Revolution,* pp. 27–92.

McNeill, W.H. "The Era of Chinese Predominance, 1000–1500." In idem, *The Pursuit of Power,* ch. 2. Oxford: Basil Blackwell, 1983.

Maier, A. *Die Vorläufer Galileis im 14. Jahrhundert.* Rome: Edizioni di "Storia e Letteratura," 1949.

———. *Zwei Grundprobleme der scholastischen Naturphilosophie.* Rome: Edizioni di "Storia e Letteratura," 1951.

———. *An der Grenze von Scholastik und Naturwissenschaft.* Rome: Edizioni di "Storia e Letteratura," 1952.

———. *Metaphysische Hintergründe der spätscholastischen Naturphilosophie.* Rome: Edizioni di "Storia e Letteratura," 1955.

———. *Zwischen Philosophie und Mechanik.* Rome: Edizioni di "Storia e Letteratura," 1958.

———. *Ausgehendes Mittelalter: Gesammelte Aufsätze zur Geistesgeschichte des 14. Jahrhunderts.* 3 vols. Rome: Edizioni di "Storia e Letteratura," 1964, 1967, and 1977 (the final volume was edited by A. Paravicini Bagliani).

———. *On the Threshold of Exact Science: Selected Writings of Anneliese Maier on Late Medieval Natural Philosophy.* Ed. and trans. S.D. Sargent. Philadelphia: University of Pennsylvania Press, 1982.

Martin, R.N.D. "The Genesis of a Mediaeval Historian: Pierre Duhem and the Origins of Statics." *Annals of Science* 33, 1976, pp. 119–129.

Mason, S.F. "Science and Religion in Seventeenth-Century England." In C. Webster (ed.), *The Intellectual Revolution of the Seventeenth Century,* pp. 197–217.

Merchant, C. *The Death of Nature: Women, Ecology, and the Scientific Revolution.* San Francisco: Harper & Row, 1980.

Merton, R.K. *Science, Technology and Society in Seventeenth-Century England.* New York: Harper & Row, 1970 (published originally in *Osiris* 4, 2, 1938, pp. 360–632).

———. "Commentary on the Paper of Rupert Hall." In M. Clagett (ed.), *Critical Problems in the History of Science,* pp. 24–29.

———. *The Sociology of Science: Theoretical and Empirical Investigations.* Ed. N.W. Storer. Chicago: University of Chicago Press, 1973.

Montucla, J.-F. *Histoire des mathématiques.* 2nd ed. 4 vols. Paris, 1799–1802. Reprint. Paris: Blanchard, 1968.

Mulligan, L. "Civil War Politics, Religion and the Royal Society." In C. Webster (ed.), *The Intellectual Revolution of the Seventeenth Century,* pp. 317–346.

Nakayama, S. "Joseph Needham, Organic Philosopher." In S. Nakayama and N. Sivin (eds.), *Chinese Science,* pp. 23–43.

Nakayama, S., and N. Sivin (eds.), *Chinese Science: Explorations of an Ancient Tradition.* Cambridge, Mass.: MIT Press, 1973.

Nasr, S.H. *Man and Nature: The Spiritual Crisis of Modern Man.* London: Unwin, 1968.

———. *Science and Civilization in Islam.* Cambridge, Mass.: Harvard University Press, 1968.

Needham, J. *The Grand Titration: Science and Society in East and West.* London: Allen & Unwin, 1969.

————. Foreword. In *Science at the Cross Roads*, pp. vii–x. 2nd ed. London: Cass, 1971.

————. "The Historian of Science as Ecumenical Man: A Meditation in the Shingon Temple of Kongosammai-in on Koyasan." In S. Nakayama and N. Sivin (eds.), *Chinese Science*, pp. 1–8.

————. [H. Holorenshaw, pseud.]. "The Making of an Honorary Taoist." In M. Teich and R. Young (eds.), *Changing Perspectives in the History of Science*, pp. 1–20.

————. *Science in Traditional China: A Comparative Perspective*. Hong Kong: The Chinese University Press, 1981.

————. Foreword. In Debiprasad Chattopadhyaya, *History of Science and Technology in Ancient India*, pp. v–viii. Calcutta: Firma KLM Private, 1986.

Needham, J., et al. *Science and Civilisation in China*. 15 vols. to date. Cambridge: Cambridge University Press, 1954–.

Needham, J., with the collaboration of Wang Ling, Lu Gwei-Djen, and Ho Ping-Yü. *Clerks and Craftsmen in China and the West: Lectures and Addresses on the History of Science and Technology*. Cambridge: Cambridge University Press, 1970.

Nelson, B. "Sciences and Civilizations, 'East' and 'West': Joseph Needham and Max Weber." In R.J. Seeger and R.S. Cohen (eds.), *Philosophical Foundations of Science*, pp. 445–493. Boston Studies in the Philosophy of Science, vol. 11. Dordrecht: Reidel, 1974.

Okken, L. "Die technische Umwelt der frühen Räderuhr." *Tractrix* 1, 1989, pp. 85–114.

Olschki, L. *Geschichte der neusprachlichen wissenschaftlichen Literatur*. Vol. 1, *Die Literatur der Technik und der angewandten Wissenschaften vom Mittelalter bis zur Renaissance*. Heidelberg: Winter, 1919. Vol. 2, *Bildung und Wissenschaft im Zeitalter der Renaissance in Italien*. Leipzig and Florence: Olschki, 1922. Vol. 3, *Galilei und seine Zeit*. Halle: Niemeyer, 1927. (All three volumes were reprinted together by Kraus Reprint, Vaduz, 1965.)

Ornstein [Bronfenbrenner], M. *The Rôle of Scientific Societies in the Seventeenth Century*. Chicago: University of Chicago Press, 1928 (originally 1913; facsimile reprint used, New York: Arno Press, 1975).

Peters, F.E. *Allah's Commonwealth: A History of Islam in the Near East, 600–1000 AD*. New York: Simon & Schuster, 1973.

Piaget J., and R. Garcia. *Psychogenèse et histoire des sciences*. Paris: Flammarion, 1983.

Pines, S. "What Was Original in Arabic Science?" In A.C. Crombie (ed.), *Scientific Change*, pp. 181–205.

Popkin, R.H. *The History of Scepticism from Erasmus to Spinoza*. Berkeley: University of California Press, 1979 (revised and expanded edition of *The History of Scepticism from Erasmus to Descartes*. Assen: van Gorcum, 1960).

Porter, R. "The Scientific Revolution: A Spoke in the Wheel?" In R. Porter and M. Teich, *Revolution in History*, pp. 290–316. Cambridge: Cambridge University Press, 1986.

Pyenson, L. "What Is the Good of History of Science?" *History of Science* 27, 1989, pp. 353–388.

Rabb, T.K. "Religion and the Rise of Modern Science." In C. Webster (ed.), *The Intellectual Revolution of the Seventeenth Century,* pp. 262–279.

――――. *The Struggle for Stability in Early Modern Europe.* Oxford: Oxford University Press, 1975.

Rahman, A. "Sixteenth- and Seventeenth-Century Science in India and Some Problems of Comparative Studies." In M. Teich and R. Young (eds.), *Changing Perspectives in the History of Science,* pp. 52–67.

Randall, J.H., Jr. "The Development of Scientific Method in the School of Padua." In idem, *The School of Padua and the Emergence of Modern Science,* pp. 13–68. Padua: Antenore, 1961. (chapter appeared originally in 1940).

Rattansi, P.M. "Some Evaluations of Reason in Sixteenth- and Seventeenth-Century Natural Philosophy." In M. Teich and R. Young (eds.), *Changing Perspectives in the History of Science,* pp. 148–166.

Redondi P. (ed.). *History and Technology* 4, 1–4, 1987, special issue "Science: The Renaissance of a History. Proceedings of the International Conference 'Alexandre Koyré,' Paris, Collège de France, 10–14 June 1986."

Revel, J.-F. *Histoire de la philosophie occidentale.* 2 vols. Paris: Stock, 1969–1970.

Righini Bonelli, M.L., and W.R. Shea (eds.), *Reason, Experiment, and Mysticism in the Scientific Revolution.* New York: Science History Publications, 1975.

Ronan, C. *The Shorter Science and Civilisation in China: An Abridgement of Joseph Needham's Original Text.* Vol. 1 (vols. I and II of the Major Series), vol. 2 (vol. III and a section of vol. IV, part I, of the Major Series), vol. 3 (a section of vol. IV, part 1, and a section of vol. IV, part 3, of the Major Series). Cambridge: Cambridge University Press, 1978–1986.

Rose, P.L. *The Italian Renaissance of Mathematics: Studies on Humanists and Mathematicians from Petrarch to Galileo.* Geneva: Droz, 1975.

Rossi, P. *Francis Bacon: From Magic to Science.* Chicago: University of Chicago Press, 1968 (Italian original: 1957).

――――. *Philosophy, Technology, and the Arts in the Early Modern Era.* New York: Harper, 1970 (Italian original: 1962).

――――. "Hermeticism, Rationality and the Scientific Revolution." In M.L. Righini Bonelli and W.R. Shea (eds.), *Reason, Experiment, and Mysticism in the Scientific Revolution,* pp. 247–273.

――――. *The Dark Abyss of Time: The History of the Earth and the History of Nations from Hooke to Vico.* Chicago: University of Chicago Press, 1984 (Italian original: 1979).

Russell C.A. (ed.). *Science and Religious Belief: A Selection of Recent Historical Studies.* London: University of London Press, 1973.

Sabra, A.I. "The Appropriation and Subsequent Naturalization of Greek Science in Medieval Islam: A Preliminary Statement." *History of Science* 25, September 1987, pp. 223–243.

――――. S.v. "Science, Islamic. *Dictionary of the Middle Ages,* ed. J.R. Strayer, vol. 11, pp. 81–89. New York: Scribner's, 1988.

Sambursky, S. *The Physical World of the Greeks.* London: Routledge & Kegan Paul, 1956. Reprint. Princeton: Princeton University Press, 1987 (Hebrew original: 1954).

Sarton, G. "East and West." In idem, *The History of Science and the New Humanism,* pp. 73–124. New York: Holt, 1931.

———. *Sarton on the History of Science.* Ed. D. Stimson. Cambridge, Mass: Harvard University Press, 1962.

Saunders, J.J. "The Problem of Islamic Decadence." *Journal of World History* 7, 1963, pp. 701–720.

Savage-Smith, E. "Gleanings from an Arabist's Workshop: Current Trends in the Study of Medieval Islamic Science and Medicine. *Isis* 79, 297, June 1988, pp. 246–272.

Sayili, A. "Islam and the Rise of the Seventeenth Century Science." *Belleten* 22, 85–88, 1958, pp. 353–368.

———. "The Causes of the Decline of Scientific Work in Islam." In idem, *The Observatory in Islam and Its Place in the General History of the Observatory,* appendix II. Ankara, 1960.

Schiebinger, L. *The Mind Has No Sex? Women in the Origins of Modern Science.* Cambridge, Mass: Harvard University Press, 1989.

Schipper, F. "William Whewell's Conception of Scientific Revolutions." *Studies in History and Philosophy of Science* 19, 1, 1988, pp. 43–53.

Schmitt, C.B. *Aristotle and the Renaissance.* Cambridge, Mass.: Harvard University Press, 1983.

Schuster, J.A. "Descartes and the Scientific Revolution, 1618–1634: An Interpretation." Doctoral dissertation, Princeton University, 1977 (available through University Microfilms International, Ann Arbor, Mich.).

———. "The Scientific Revolution." In R.C. Olby, G.N. Cantor, J.R.R. Christie, and M.J.S. Hodge (eds.), *Companion to the History of Modern Science,* pp. 217–242. London: Routledge, 1990.

Settle, T.B. "An Experiment in the History of Science." *Science* 133, 1961, pp. 19–23.

Shapere, D. *Galileo: A Philosophical Study.* Chicago: University of Chicago Press, 1974.

Shapin, S. "Understanding the Merton Thesis." *Isis* 79, 299, 1988, pp. 594–605.

Shapin, S., and S. Schaffer. *Leviathan and the Air-Pump: Hobbes, Boyle, and the Experimental Life.* Princeton: Princeton University Press, 1985.

Shapiro, B. "Latitudinarianism and Science in Seventeenth-Century England." In C. Webster (ed.), *The Intellectual Revolution of the Seventeenth Century,* pp. 286–316.

Sivin, N. Preface. In S. Nakayama and N. Sivin (eds.), *Chinese Science,* pp. xi–xxxvi.

———. "An Introductory Bibliography of Traditional Chinese Science: Books and Articles in Western Languages." S. Nakayama and N. Sivin (eds.), *Chinese Science,* pp. 279–314.

———. "Chinesische Wissenschaft: Ein Vergleich der Ansätze von Max Weber und Joseph Needham." In Wolfgang Schluchter (ed.), *Max Webers Studie über Konfuzianismus und Taoismus: Interpretation und Kritik,* pp. 342–362. Frankfurt a.M.: Suhrkamp, 1983.

————. "Why the Scientific Revolution Did Not Take Place in China—Or Didn't It?" In E. Mendelsohn (ed.), *Transformation and Tradition in the Sciences: Essays in Honor of I. Bernard Cohen*, pp. 531–554. Cambridge: Cambridge University Press, 1984.

Solla Price, D.J. de. "Joseph Needham and the Science of China." In S. Nakayama and N. Sivin (eds.), *Chinese Science*, pp. 9–21.

Sorel, G. *Les préoccupations métaphysiques des physiciens modernes.* Cahiers de la Quinzaine, série VIII, no. 16. Paris, 1905 (appeared originally as an article in *Revue de Métaphysique et de Morale*).

Spence, J.D. Contribution to a review symposium on *Science and Civilisation in China. Isis* 75, 276, 1984, pp. 180–189.

Strong, E.W. *Procedures and Metaphysics: A Study in the Philosophy of Mathematical-Physical Science in the Sixteenth and Seventeenth Centuries.* Berkeley: University of California Press, 1936.

Teeter Dobbs, B.J. *The Foundations of Newton's Alchemy, or "The Hunting of the Greene Lyon."* Cambridge: Cambridge University Press, 1975.

Teich M., and R. Young (eds.). *Changing Perspectives in the History of Science: Essays in Honour of Joseph Needham.* London: Heinemann, 1973.

Vallance, J.T. "Marshall Clagett's *Greek Science in Antiquity:* Thirty-five Years Later." *Isis* 81, 309, 1990, pp. 713–721.

Vickers, B. (ed.). *Occult and Scientific Mentalities in the Renaissance.* Cambridge: Cambridge University Press, 1984.

Wallace, W.A. *Causality and Scientific Explanation.* Vol. 1, *Medieval and Early Classical Science.* Ann Arbor: University of Michigan Press, 1972.

————. *Galileo and His Sources: The Heritage of the Collegio Romano in Galileo's Science.* Princeton: Princeton University Press, 1984.

————. *Galileo, the Jesuits and the Medieval Aristotle.* Hampshire: Variorum, 1991.

Weber, M. *Gesammelte Aufsätze zur Religionssoziologie* Vol. 1. Tübingen: Mohr, 1920. (The volume contains "Vorbemerkung," pp. 1–16; "Die protestantische Ethik und der Geist des Kapitalismus," pp. 17–206; "Die protestantischen Sekten und der Geist des Kapitalismus," pp. 207–236; "Einleitung" [to "Die Wirtschaftsethik der Weltreligionen], pp. 237–275; "Konfuzianismus und Taoismus," pp. 276–536; "Zwischenbetrachtung: Theorie der Stufen und Richtungen religiöser Weltablehnung," pp. 536–573.)

————. "Wissenschaft als Beruf." In idem, *Gesammelte Aufsätze zur Wissenschaftslehre*, pp. 524–555. Tübingen: Mohr, 1922 (originally published, 1919; originally given as lecture, 1917).

Webster, C. *From Paracelsus to Newton: Magic and the Making of Modern Science.* Cambridge: Cambridge University Press, 1982.

————, (ed.). *The Intellectual Revolution of the Seventeenth Century.* London: Routledge & Kegan Paul, 1974.

Weinstein, D. "In Whose Image and Likeness? Interpretations of Renaissance Humanism." *Journal of the History of Ideas* 33, 1972, pp. 165–176.

Wentholt, R. "The Conquest of Bias in the Human Sciences." Unpublished typescript. 1976.

Qian, Wen-yuan. *The Great Inertia: Scientific Stagnation in Traditional China.* London: Croom Helm, 1985.

Werskey, P.G. "On the Reception of *Science at the Cross Roads* in England." In *Science at the Cross Roads,* pp. xi–xxiv. 2nd ed. London: Cass, 1971.

———. *The Visible College.* London: Allen Lane, 1978.

Westfall, R.S. *Science and Religion in Seventeenth-Century England.* New Haven: Yale University Press, 1958 (2nd ed. used, Ann Arbor: University of Michigan Press, 1973).

———. *Force in Newton's Physics: The Science of Dynamics in the Seventeenth Century.* London: MacDonald, 1971.

———. *The Construction of Modern Science: Mechanisms and Mechanics.* 2nd ed. Cambridge: Cambridge University Press, 1977 (1st ed., New York: Wiley, 1971).

———. *Never at Rest: A Biography of Isaac Newton.* Cambridge: Cambridge University Press, 1983 (1st ed.: 1980).

———. "Robert Hooke, Mechanical Technology, and Scientific Investigation." In J.G. Burke (ed.), *The Uses of Science in the Age of Newton,* pp. 85–110.

———. "Science and Patronage: Galileo and the Telescope." *Isis* 76, 1985, pp. 11–30.

———. "Patronage and the Publication of Galileo's *Dialogue.*" In P. Redondi (ed.), *History and Technology* 4, 1–4, 1987, pp. 385–399.

———. "The Scientific Revolution." In S. Goldberg (ed.), *Teaching in the History of Science: Resources and Strategies,* pp. 7–12. Philadelphia: History of Science Society Publication, 1989 (appeared originally in *History of Science Society Newsletter* 15, 3, July 1986).

———. "Making a World of Precision: Newton and the Construction of a Quantitative Physics." In F. Durham & R.D. Puddington (eds.), *Some Truer Method: Reflections on the Heritage of Newton,* pp. 59–87. New York: Columbia University Press, 1990.

Westman, R.S. "Magical Reform and Astronomical Reform: The Yates Thesis Reconsidered." In R.S. Westman and J.E. McGuire, *Hermeticism and the Scientific Revolution.* Los Angeles: University of California Press, 1977.

Whewell, W. *History of the Inductive Sciences, from the Earliest to the Present Time.* 3rd ed. 3 vols. London: Parker, 1857. (1st ed., 1837.)

———. *The Philosophy of the Inductive Sciences, Founded upon Their History.* 2nd ed. 2 vols. London: Parker, 1847. (1st ed., 1840.)

White, L., Jr. *Medieval Technology and Social Change.* Oxford: Oxford University Press, 1962.

———. *Medieval Religion and Technology: Collected Essays.* Berkeley: University of California Press, 1978.

———. Contribution to a review symposium on *Science and Civilisation in China. Isis* 75, 276, 1984, pp. 172–179.

Whitrow, G.J. *Time in History: Views of Time from Prehistory to the Present Day.* Oxford: Oxford University Press, 1988.

Yates, F.A. *Giordano Bruno and the Hermetic Tradition*. Chicago: University of Chicago Press, 1964.

———. "The Hermetic Tradition in Renaissance Science." In *Ideas and Ideals in the North European Renaissance: Collected Essays*, Vol. 3, pp. 227–246. London: Routledge & Kegan Paul, 1984 (appeared originally in C.S. Singleton [ed.], *Art, Science, and History in the Renaissance*, pp. 255–274. Baltimore: Johns Hopkins University Press, 1967).

———. *The Rosicrucian Enlightenment*. London: Routledge & Kegan Paul, 1972 (edition used: Paladin, 1975).

Yoder, J.G. *Unrolling Time: Christiaan Huygens and the Mathematization of Nature*. Cambridge: Cambridge University Press, 1988.

Zilsel, E. "The Sociological Roots of Science." *American Journal of Sociology* 47, 1941/42, pp. 544–562.

———. "The Origins of Gilbert's Scientific Method." In P.P. Wiener and A. Noland (eds.), *Roots of Scientific Thought*, pp. 219–250. New York: Basic Books, 1957 (appeared originally in the *Journal of the History of Ideas* 2, 1941, p. 1–32).

———. "The Genesis of the Concept of Physical Law." *Philosophical Review* 51, 1942, pp. 245–279.

———. "The Genesis of the Concept of Scientific Progress." In P.P. Wiener and A. Noland (eds.), *Roots of Scientific Thought*, pp. 251–275. New York: Basic Books, 1957 (appeared originally in the *Journal of the History of Ideas* 6, 1945, pp. 325–349).

———. *Die sozialen Ursprünge der neuzeitlichen Wissenschaft*. Ed., intr., and trans. W. Krohn. Frankfurt a.M.: Suhrkamp, 1976.

✧ INDEX ✧

This index is designed primarily to provide ready access to historians of the Scientific Revolution and to their ideas as discussed in the text. The names of these historians are capitalized. The entries for the historians most extensively treated conclude with specific subentries listing their works (chronologically), their comments on other historians, and comparisons made with other historians (the latter two alphabetically). Page numbers in italics denote substantive discussions.

The index includes detailed entries for scientists; for branches, periods, and aspects of science; for aspects of the Scientific Revolution (e.g., its dating); for historiographical features; and for many collateral subjects. Entries for key scientists and for basic attributes of science conclude with subentries listing the historians in whose discussions they figure.

Bacon, Francis (*continued*)
and the occult, 172, 174, 180, 192,
293–4, 304, 513
on origins of early modern science, 7, 356, 427, 585n.144
as pioneer of Scientific Revolution, 1, 22–3, 25, 32–3, 36, 152–
3, 294, 572n.72
on scientific method, 22, 25, 32–3,
106, 152–4, 156, 197, 356
Bacon, Roger, 105–6, 108, 261
'Baconian Brew' (*not so called until
ch. 8*), 134, 145, 513–4, 524,
550n.333
Baconian sciences, 151, 157, 189, 304,
318, 350, 498, 504, 513, 515–6. *See
also* KUHN, on 'Baconian' and 'classical physical sciences'
BAILLY, J.-S., 23
balance, 87, 133
BALDI, Bernardino, 276–7
ballistics, 193, 307, 316, 321, 323, 325,
329, 342–3
barometer, 139, 143, 190
Barrow, Isaac, 94
al-Battani, 269
Baxter, Richard, 315, 317
BEDINI, Silvio, 352
Beeckman, Isaac, 131, 163, 204, 348–9,
362, 512
corpuscularianism of, 69, 134,
144–5, 512, 550n.336,
584n.115
Bekker, Balthasar, 555n.58
BEN-DAVID, Joseph, 15, 213, *254–9,
367–75,* 569n.33, 587n.190
explanation of birth of early modern science, 8
from Greek science, 242–3,
255–9
from rise of ideology for science, 255, 258, 320, 368–75,
414, 586nn.178, 181,
587n.189
on flourishing and decay in science, 254–5, 367, 410, 414,
459–61, 482

on sociology of science, 368, 374
on historians
MERTON, 587n.189
compared with
EISENSTEIN, 364, 375
GRUNEBAUM, 393
KOYRÉ, 327
MERTON, 320, 336, 372, 374
NEEDHAM, 459, 482
SAYILI, 396, 410
Benedetti, Giovanni Battista, 32, 44,
51–2, 109, 249, 251–2, 273–4, 276,
326, 587n.192
BERKEL, K. van, 145–6, 206–7,
348–9, 539n.130, 583–4n.115,
600n.9
BERNAL, J. D., 194, 216, *219–23,* 332,
349, 501, 562nn.183, 185, 581n.58,
600n.4
compared with
NEEDHAM, 446
WEBER, 226–7
Bessarion, Cardinal, 274
Bible, 86, 296, 309, 311–3, 394, 510
world-view in, 248–9, 259, 350
as cause of Scientific Revolution, 16, 310–4, 356–7,
374–5, 521
Bijker, Wiebe, 599n.3
biology. *See* life sciences
al-Biruni, 367, 395, 398, 416, 431,
591n.86
BOAS HALL. *See* HALL, Marie BOAS
Boethius, Anicius Manlius Severinus,
251, 271
Bohr, Niels, 17, 67, 74, 539n.123
Bonfils, 361
bookkeeping. *See* accounting
books
figures in, 359, 365
as foci of learning on nature, 360,
364–5
greatness of, 73, 286
in libraries, 274, 363, 386, 398,
415, 439
manuscript, 387, 412, 415, 476,
507

historiography of Scientific Revolution, 2, 7–8, 10–11, 16, 31, 97, 188, 288, 528n.13

    arts and crafts tradition, 324, 328–9, 342–3 (*see also under* technology)

    concluding survey of, 494–500

    Great Tradition, 15, 151, 186, 229, 247, 268, 282, 528n.18

    'monological' character of, 7–8, 240–1

    reasons for examining, 8–14

history, philosophy of, 239, 408, 500, 517, 566n.1

history of science

    contingency in, 229–32, 525 (*see also under* Scientific Revolution)

    continuity and discontinuity in, 31, 34, 47–8, 50, 54, 59, 63–5, 73–4, 97, 124, 147–50, 361, 426, 460–1, 498, 503 (*see also* Scientific Revolution, continuity and discontinuity in)

    flourishing, stagnation, and decline in, 30, 252, 254–5, 258–9, 305, 363, 367, 396, 409–10, 459–61, 509–10 (*see for specific cases under* Chinese civilization, science in; Islamic civilization, science in; science [*periods* . . .], Greek)

    and 'general' history, 15, 24, 38–40, 66, 112, 147, 178, 209–10, 214, 222–3, 227, 232, 358, 501–2, 504–5

    periodization in, 14, 76, 221, 256, 258, 409–11, 571n.71

    unity of, 494, 533n.49

Hobbes, Thomas, 94, 138, 168, 187–8, 203, 210, 212–5, 230, 281, 513, 515, 554n.46, 561n.162, 576n.2

Hook, Sidney, 527n.2

Hooke, Robert, 84, 128, 194, 290, 320, 335, 513

HOOYKAAS, R., 9, 111, 544n.201, 551n.17, 556n.83

    on art and nature, 185, 311

explanations of birth of early modern science, 8, 186, 250, 356–7, 567n.10, 576n.149

    from Greek science, 242–3, 245–6, 250, 355

    from medieval science, 246, 260, 264–6, 304, 356–7

    from Reformation, 246, 310–4, 317, 319, 356–7, 375, 577n.13

    from Voyages of Discovery, 246, 313, 328, 354–7, 415

    on manual labor, 111, 186, 311, 313, 350, 357

    on nature and on science, 245–6, 265, 310, 356–7

    on Portuguese science and letters, 354–6

    compared with DUHEM, 264

Horace, 158

hospital. *See* medicine

Hsintu Fang, 442

Hsü Kuang-ch'i, 470

Hudde, Johan, 560n.152

HUIZINGA, Johan, 13, 600n.2

Hulagu Khan, 407, 430–1

humanism, 117, 162, 205, 241, 253, 271–3, 281, 297–8, 302, 338–9, 341, 355, 369, 406, 415, 518, 571n.71, 572n.72

    'mathematical', 24, 51, 241, 247, 249, 272–9, 282, 303–5, 349, 363, 484, 508, 518, 571n.70, 572n.76

human nature, 91, 156, 179, 181–2, 292–4, 310, 367, 487, 518, 556n.73, 566n.2, 600n.2

human sciences. *See* psychology; social sciences

Hunayn ibn-Ishaq, 398

Huxley, T. H., 357

Huygens, Christiaan, 6, 130, 133, 139–40, 157, 174, 188, 190, 199–200, 204, 208, 302, 320, 548n.298, 553n.34, 560n.152

    corpuscularianism of, 69–70, 141, 512

NEEDHAM, Joseph (*continued*)
on 'correlative thinking', 464,
474–5
and critics, 424, 466, 471–4, 479,
481, 485–6
on Eastern science as authentic science, 379–80, 403, 425, 428,
565n.219
on 'feudal bureaucracy', 448–50,
451–2, 456, 483, 596n.226
on a Galileo in China, 449, 456,
458
on Greek science, 379, 439–40,
442, 458, 464
'Grand Question' of, 8–9, 16, 381,
420, 424, 447, 475, 477, 588n.5
answered via bureaucrats,
451–3, 455–6, 481, 485
answered via Confucianism vs.
Taoism, 421, 448–9, 456
answered via democracy,
447–8, 455
answered via laws of nature,
453–5, 455–6, 467
answered via merchants, 421,
449–50, 452, 455–6, 465–6,
481
answered via Western dynamism, 461–3
answers avoided/rejected by
NEEDHAM, 421, 457–61,
474–5
answers counted, weighed, and
assessed, 421, 437, 444, 449,
455–6, 461, 471, 473–4
definition of, 418, 428, 443–4,
447, 474–5
significance of, 381, 471–4, 478
on Indian science, 382, 457,
588n.6, 596n.226
intellectual development of, 332,
418, 422–6, 465, 471, 581n.58
political views, 422–4, 452,
456, 472
religious views, 422
on Jesuits in China, 427, 465, 467,
469–70

on Mediterranean world, 431, 483,
485
'law of oecumenogenesis' of,
467–71
on practical orientation of Chinese
science, 428, 439, 442
on printing, 427, 476, 598n.281
puzzle posed by work of, 419–21
on science and technology, relation between, 380, 421, 427–9,
438
on Scientific Revolution, 444–6,
595n.186
causes of, 446–7, 583n.108
place in European history of,
446–7, 461, 480–1, 485, 501
possible 'organic Scientific Revolution', 463–6, 484
on scientist in Chinese society,
441–2, 482
on social barrier in China, 441–2,
450, 452, 476
on time
conceptions of, 457–8
measurement of, 435–7, 460,
594n.153
on translation, 383–4, 421,
588n.10
on transmission of science and
technology, 427–38, 484,
594n.153
limits of, 438
onus of proof, 436–7
on Western attitudes, 425, 428,
472
on Western dynamism, 427, 435,
474, 480, 487, 519
works
*Science and Civilisation in
China,* 380–1, 418–21, 424,
428, 433–4, 438, 459, 464,
466, 469, 471, 473–4, 478,
486
ibid., specifically vol. 7: 418,
420–1, 467
Grand Titration, 420–1, 436,
441, 447–58, 461–5, 472

Noël, le Père, 159, 162
Norman, Robert, 339, 350–1
number, significance of, 101, 104, 118,
142, 273, 285, 287–90, 445,
575n.125
Nunes, Pedro, 354

objectivity, 58, 94, 201–2, 213, 217,
219, 235–6, 371, 405, 566n.2
observation. *See* astronomy, 16th/17th
century, observational; science, observational
observatory, 386–8, 407, 429–31
occultism and science, 110, 116, 118–9,
121, 169–83, 185, 215, 250, 285–96,
401–2, 416, 442, 449, 497, 503, 509,
513
Ockham, William of, 46
Ogodai (Great Khan), 525
OKKEN, L., 594n.157
Oldenburg, Henry, 206
OLSCHKI, Leonardo, 161, *322–5,*
347, 349–51, 374–5, 544n.201,
579nn.34, 35
fate of, in historiography, 8, 327–
8, 336, 339, 345–6, 369, 496,
580n.45, 582n.91
on historians: DUHEM, 324
compared with
KOYRÉ, 325–6
MACH, 324, 579n.34
ZILSEL, 340
Olschki, Leo S., 322
open society. *See* pluralism
optics, 1, 24, 72, 87, 111, 125–7, 130,
133, 139, 143, 161, 249, 361, 386,
388, 412–3, 440, 496, 507, 512, 516,
554n.48
perspective, 323, 369, 499
Oranges (Dutch stadholders), 212
Oresme, Nicole, 47–8, 53, 55, 67, 104,
148, 261, 263–4, 322, 528n.13. *See
also* Paris Terminists
organicism. *See* nature, organic conception of
organization of science, 16, 516. *See
also* institutionalization of science

ORNSTEIN [Bronfenbrenner], Martha,
204–7
OSLER, Margaret J., 542n.164
Ostwald, Wilhelm, 226, 554n.47
Ottoman Empire, 389, 402

Pacioli, Luca, 338
PAGEL, Walter, 111, 170, 185,
544n.201
PALISCA, Claude V., 131
Palissy, Bernard, 350
Pappos, 273
Paracelsus (Theophrastus Bombastus
von Hohenheim), 111, 113, 169, 174,
185, 294, 433
paradigm. *See* KUHN, on scientific
change
Pareto, Vilfredo, 581n.74
Paris Terminists, 45–53, 55–6, 62–4,
246, 260, 265, 324, 571n.62. *See
also* Buridan; DUHEM, thesis;
Galileo, historiographical position
of, and predecessors; Oresme; philosophy, nominalism; university, Paris
Pascal, Blaise, 156, 204, 308, 577n.12
conception of science, 159–60,
162, 168, 556nn.73, 83
on corpuscularianism, 144,
549n.330
scientific achievement of, 127,
133, 139, 347–8, 512
and experiment, 128, 139, 143,
186, 188, 556nn.83, 84, 86
patronage, 367, 371, 414
in China, 441–2
in Hellenist world, 253, 258, 442,
569n.41
in Islamic civilization, 363, 386–7,
414, 442
in western Europe, 152, 192, 205,
208, 233, 274, 277, 338, 363,
369–71, 504, 524, 560n.152,
587n.191
Paul V, 210
Peirce, Charles Sanders, 157
pendulum clock. *See* clock, mechanical;
Huygens, and pendulum clock

from mechanical clock (*see
under* clock, mechanical)
from medieval science (*see
under* science [*periods*],
medieval)
from nonemergence of Scientific Revolution, 240–1, 259,
341, 376–7, 424, 472 (*see also
under* East, science in; science, Greek; science [*periods*], medieval)
from printing press (*see under*
printing press)
from Protestantism (*see under*
Bible; MERTON, thesis)
from sceptical crisis (*see*
scepticism, crisis of)
from social role for scientist
(*see* scientist, role in
society of)
from technology (*see under*
technology)
from Voyages of Discovery (*see
under* Voyages of Discovery)
as change in world-view, 58, 75,
90, 92–5, 98, 101, 103–5, 153,
246, 250, 270, 284, 344, 346,
364, 522 (*see also* universe of
precision)
comparison of ideas on
cases of, 97–9, 121–2, 146–50,
232, 250–2, 283, 303–6, 319,
340, 374–5, 389, 409–17, 487,
494–500 (*see also subentries
'compared with' under entries
for individual historians*)
why applied, 9, 376, 492–3
concept of, 2, 6, 13–5, 21, 112–3,
121–2, 328, 342–3, 580n.47,
583n.100
creation of, 11, 15, 74–80, 89,
98, 122, 153, 187–8, 216, 285,
318, 321, 343–5, 493, 495,
500
dilemmas raised by, 497–9
dilution of, 496–7, 500

fifty years in life of, 488,
493–505
future handling of, 493, 500–5
ideal for, 499, 502
indispensability of, 501–2
as metaphor, 500
possible evaporation of, 13–4,
499–501
prehistory of, 2, 24, 54, 493–5,
530n.12, 531–2n.36, 532n.49
restoring coherence to, 121–47,
497–8, 502
variety of uses of, 13–4, 499
widening of, 80–4, 98, 112–21,
188, 495–7, 547n.270
conditions for, 327, 341, 346, 368,
374, 376, 417, 504
necessary, 337, 341, 472, 487,
517, 520
sufficient, 341, 487, 517, 520
contingencies in, 16, 151–2, 203,
230–4, 473, 499, 503, 514, 524,
565n.213
contingency of, 16, 137–8, 146,
234–5, 244, 259, 337, 367–8,
376, 416–7, 427, 445, 473, 487,
497–8, 503, 510, 518, 522–5,
549n.306, 558n.110
continuity and discontinuity in,
15, 34, 41–2, 44–5, 49, 54,
65–6, 68, 74, 76, 96–8, 102,
106–8, 116, 124, 147–50, 166,
232, 264, 281, 283–4, 303–4,
356, 361, 364, 498, 545n.218,
550n.1, 551n.3, 574n.105 (*see
also* history of science, continuity and discontinuity in)
critical junctures in, 285, 303,
305–6, 313, 360–1, 517, 520–1
dynamics of, 136–7, 146–7, 233,
375, 497, 503, 510–4, 523
excitement of, 99, 495–6
geographical frame of, 188, 234,
240, 283, 303–4, 306, 334, 343,
345–6, 372–3, 497–9, 565n.207
in Dutch Republic, 212, 362,

Solzhenitsyn, Alexandr, 395
SOMBART, Werner, 226, 337
Sorel, Georges, 183
Sorokin, Pitirim, 587n.20
Soto, Domingo de, 46–7
sources. *See* historiography, sources for
Soviet Union, 3, 201, 216, 331–2, 424
SPENCE, Jonathan, 469
Spencer, Herbert, 423
SPENGLER, Oswald, 426, 598n.276
Spinoza, Baruch, 296, 515
spirit (as ethereal substance), 95
    human (*see* mind, status of in scientific thought)
Sprat, Thomas, 213, 371
stability, crisis of, 210–2, 524, 561n.163. *See also* authority, crisis of
Stahl, G. E., 25, 27
Stalin, Joseph/Stalinism, 3, 216, 221, 336, 424, 562n.185, 581n.57
statics, 30, 126–7, 133, 140, 249, 276–7, 330, 386, 507, 511–2, 535n.70, 587n.192
steam engine. *See* machine
Stevin, Simon, 133, 163, 208, 248, 251, 338, 361, 512, 550n.336, 567n.9, 575n.125
stoicism, 251, 272, 279
STRONG, E. W., 92, 102
suggestions for further research, 154, 159, 163–4, 168, 179, 188–9, 191, 194–5, 203–4, 208, 216, 291, 322, 375, 413–6, 461, 481, 502–3, 575n.125. *See also* Scientific Revolution, underexploited ideas on
sun
    significance of, 288–9, 291–2, 294
    at center of universe (*see* Copernicus, heliocentric hypothesis)
superstition, liberation from, by science, 173, 176, 285. *See also* rationality, of science
Su Sung, 435, 437, 460, 475
Swammerdam, Jan, 226, 564n.200

SWERDLOW, Noel M., 10, 530n.12, 546n.226
Switzer, 234

TANNERY, Paul, 184, 533n.49
Tartaglia, Niccolò, 203, 274–6, 322
tautochrony. *See* Huygens, and pendulum clock
technology, 191, 193, 196, 218, 248–9, 259, 306, 362, 426, 446
    historiography of, 86, 342, 346, 419–1, 425, 428
    medieval, 248–9, 306, 325, 343, 346, 427, 437, 487, 519, 583n.110
    and science, 192, 220, 222, 245, 248–9, 254, 294, 333, 342–3, 349, 365, 370, 413–4, 427–9, 438, 453, 476, 582n.93
        gap between, in 17th century, 193–5, 222, 233, 333, 347–9, 351, 361, 516
    specifically empirical, 4, 86, 129, 190, 194–5, 331, 348–9, 352, 408, 452, 515–6
        as cause of Scientific Revolution, 16, 307, 321, 323–6, 338–42, 345–6, 351, 374, 435, 496, 521, 583n.110 (*see also* Galileo, background of, in engineering)
        reform of science sought in, 350, 509
        setting problems for science, 220–1, 323–4, 343, 347–8, 350, 357, 509, 511, 521
    specifically science based, 4–5, 16, 66, 87, 146, 190, 192–5, 218, 225, 516, 524, 528n.21, 564n.199
        as cause of Scientific Revolution, 329–31, 334–6, 342, 345, 351, 374
    transmission of (*see* transmission of science and technology)

superiority feelings in, 337, 403,
425, 457, 472
Westernization, 3–4, 403
WESTFALL, Richard S., 135, *136–47*,
149, 231, 543n.176, 565n.211
on nature of Scientific Revolution,
84, 111, 136–47, 160, 186, 497,
547n.270
conflict and harmony in, 137–
42, 144–7
dating of, 138–9, 549n.306
dynamics of, 136–7, 146–7, 497
meaning of 'mechanical', 139–
40, 143–5, 498
on occult in Scientific Revolution,
173, 175
on patronage in Scientific Revolution, 208
on science and religion in Scientific Revolution, 308–9
on science and technology in Scientific Revolution, 194–5
on significance of Scientific Revolution, 5
works
*Science and Religion in
Seventeenth-Century England,*
308–9, 316
Force in Newton's Physics, 136,
554n.41
*Construction of Modern Sci-
ence,* 17, 136–47, 151–2, 173,
549n.327, 550n.332
Never at Rest, 136, 143–4,
542n.171
compared with
BURTT, 143–4
DIJKSTERHUIS, 70, 143–4
HALL, 143–4
KUHN, 137, 139, 145–6, 497
WESTMAN, Robert S., 292, 500
WHEWELL, William, 21, 24, *27–39,*
54, 184, *243–4, 260–1,* 489,
531nn.23, 26, 548n.296
on Greek science, 30, 32, 242–4,
250, 252
on instruction of history, 27–8

limitations of, 37–8, 40
on medieval science, 30, 260–1
on the Revolution in Science,
32–5, 122, 494, 531–2n.36
on scientific advance, 28–31, 34,
243
on scientific revolutions, 30–1, 34,
494, 531n.23
works
*History of the Inductive Sci-
ences,* 27–8, 31–5, 37, 39,
531n.23
*Philosophy of the Inductive Sci-
ences,* 27–8, 31–2, 35, 39,
531n.23
'Transformation of Hypotheses', 34
on historians: COMTE, 35–7,
532n.47
compared with
DIJKSTERHUIS, 72
DUHEM, 45, 54
KANT, 27, 34
KUHN, 125
MACH, 40–1, 43
WHITE, Andrew D., 309, 577n.13
WHITE, Lynn, Jr., *346–7,* 580n.45
on engineering background to
early modern science, 346–7
on history of technology, 248,
425, 487, 519, 583n.110,
600–1n.4
on medieval science, 108,
583n.110, 600–1n.4
on time keeping, 352
on uniqueness of Western development, 480, 487, 519
on writing of history, vii, 478,
502, 517
on historians: NEEDHAM, 381,
419, 421, 425, 428–9, 474
Whitehead, A. N., 422, 543n.175
WHITROW, G. J., 552n.27
WIEDEMANN, Eilhard, 429
William IV (landgrave of Hesse),
352
wisdom and science, 182, 485